Discovering
Algebra
An Investigative Approach

Jerald Murdock

Ellen Kamischke

Eric Kamischke

DISCOVERING

MATHEMATICS
™

Key Curriculum Press

Innovators in Mathematics Education

Project Editor
Christian Aviles-Scott

Project Administrator
Jason Taylor

Editors
Dan Bennett, Curt Gebhard, Joan Lewis, Greer Lleuad, Crystal Mills, Kent Turner, Pam Tyson

Editorial Assistants
James A. Browne, Heather Dever, Halo Golden, Beth Masse, Shannon Miller, Susan Minarcin, Lara Wysong

Editorial Consultants
John C. Allman, Frank Aviles, Steven Chanan, Cindy Clements, William Finzer, Ladie Malek, Stacey Miceli, Leslie Nielsen

Mathematical Content Reviewers
Bill Medigovich, San Francisco, California

Professor Mary Jean Winter, Michigan State University, East Lansing, Michigan

Multicultural and Equity Reviewers
Professor Edward D. Castillo, Sonoma State University, Rohnert Park, California

Genevieve Lau, Ph.D., Skyline College, San Bruno, California

Charlene Morrow, Ph.D., Mount Holyoke College, South Hadley, Massachusetts

Arthur B. Powell, Rutgers University, Newark, New Jersey

William Yslas Velez, University of Arizona, Tucson, Arizona

Social Sciences and Humanities Reviewers
Ann Lawrence, Middletown, Connecticut

Karen Michalowicz, Ed.D., Langley School, McLean, Virginia

Scientific Content Reviewers
Andrey Aristov, M.S., Loyola High School, Los Angeles, California

Matthew Weinstein, Macalaster College, St. Paul, Minnesota

Laura Whitlock, Ph.D., Sonoma State University, Rohnert Park, California

Accuracy Checkers
Dudley Brooks, Jim Stenson

Editorial Production Manager
Deborah Cogan

Production Editor
Christine Osborne

Copy Editor
Margaret Moore

Production Director
Diana Jean Parks

Production Coordinator
Ann Rothenbuhler

Art Director and Cover Designer
Jill Kongabel

Text Designer
Marilyn Perry

Art Editor
Jason Luz

Photo Editor
Margee Robinson

Art and Design Coordinator
Caroline Ayres

Illustrators
Juan Alvarez, Sandra Kelch, Andy Levine, Nikki Middendorf, Claudia Newell, Bill Pasini, William Rieser, Sue Todd, Rose Zgodzinski

Technical Art
Precision Graphics

Compositor and Prepress
TSI Graphics

Printer
Von Hoffmann Press

Executive Editor
Casey FitzSimons

Publisher
Steven Rasmussen

Key Curriculum Press
1150 65th Street
Emeryville, CA 94608
editorial@keypress.com
http://www.keypress.com
Printed in the United States of America
10 9 8 7 6 5 4   06 05 04
ISBN 1-55953-340-4

Acknowledgements

The development and production of *Discovering Algebra* has been a collaborative effort between teaching authors and the staff at Key Curriculum Press. We are especially grateful to thousands of *Discovering Algebra* students, to summer-institute and workshop teacher participants, and to manuscript readers, all of whom provided suggestions, reviewed material, located errors, and most of all, encouraged us to continue as we field-tested the manuscript.

As authors we are grateful to the National Science Foundation for supporting our initial technology-and-writing project that led to the 1998 publication of *Advanced Algebra Through Data Exploration*. This book, *Discovering Algebra,* has been developed and shaped by what we learned during the writing of *Advanced Algebra,* its publication process, and our work with so many students, parents, and teachers searching for more meaningful algebra courses.

Over the course of our careers, many individuals and groups have been instrumental in our development as teachers and authors. The Woodrow Wilson National Fellowship Foundation provided an initial impetus for involvement in leading workshops. Publications and conferences produced by the National Council of Teachers of Mathematics and Teachers Teaching with Technology have guided the development of this curriculum. Individuals such as Frank Demana, Arne Engebretsen, Christian Hirsch, Glenda Lappan, Richard Odell, Dan Teague, Charles VonderEmbse, and Bert Waits have inspired us.

We wish to thank our families, friends, and the Interlochen Arts Academy for their encouragement and support. Our students at Interlochen Arts Academy have also played an important part in the development of this book. The support and encouragement we received from them, their parents, and Interlochen's administration have been invaluable.

We truly appreciate the confidence, cooperation, friendship, and valuable contributions offered by the Editorial and Production Departments at Key Curriculum Press. Finally, a special thanks to Key's president, Steven Rasmussen, for encouraging and publishing a technology-enhanced beginning algebra text that offers groundbreaking content and learning opportunities.

Jerald Murdock
Ellen Kamischke
Eric Kamischke

A Note from the Publisher

The mathematics we learn and teach in school has changed dramatically over the past few decades. The changes have been driven by new discoveries in mathematics and science, new research in education, changing societal needs, and the use of new technology in work and in education. The algebra you find in this book won't look exactly like the algebra you may have seen in older textbooks. It covers the same topics and includes lots of familiar equations, but the mathematics in *Discovering Algebra* also works with data from science, emphasizes techniques for data analysis, and uses technology tools such as graphing calculators to help visualize important concepts. We have included important new mathematics in this text to better prepare students for the educational and career opportunities they'll find in our fast-changing world.

For 30 years, Key Curriculum Press has developed mathematics materials for schools. By focusing on mathematics alone, our authors, our editors, and our consultants are able to keep pace with the changes in our field to produce curriculum that enables more students to succeed in school mathematics.

Over the years, in spite of the changes in mathematics and mathematics education, we have found one truth that has not changed: Students learn mathematics best when they understand the concepts behind it. With this in mind, Key Curriculum Press introduces our Discovering Mathematics series, beginning with this book, *Discovering Algebra: An Investigative Approach*. Through the investigations that are the heart of the series, students discover many important mathematical principles themselves. In the process, they come to value their own ability to succeed at mathematics; they realize that they can recreate their discoveries should they ever forget a fundamental concept; and they develop their abilities to continue discovering and learning about mathematics. And when they understand mathematics, they perform well on tests of any kind.

Many years of research, thoughtful work, and class testing have gone into the development of *Discovering Algebra*. The text has been written, piloted with students, rewritten, and piloted again. Over the course of five years, hundreds of teachers have used the trial editions of the text with many thousands of students. Thousands of pages of feedback from pilot teachers have been combined with professional, scientific, and mathematical reviews, along with results from standardized tests, to ensure that the book offers a sound conceptual framework and sufficient skill development for students.

If you are a student, as you work through this course we hope you gain knowledge for a lifetime. If you are a parent, we hope you enjoy watching your child develop mathematical power. If you are a teacher, we hope you find that *Discovering Algebra* makes a significant positive impact in your classroom. Regardless of who you are, the professional team at Key Curriculum Press wishes you success and urges you to continue your involvement and interest in mathematics and education.

Steven Rasmussen, President
Key Curriculum Press

Contents

CHAPTER

10

Quadratic Models   531

A Note to Students from the Authors

Jerald Murdock

Ellen Kamischke

Eric Kamischke

You are about to embark on an exciting mathematical journey. The goal of your trip is to reach the point at which you have gathered the skills, tools, confidence, and mathematical power to participate fully as a productive citizen in a changing world. Your life will always be full of important decision-making situations, and your ability to use mathematics and algebra can help you make informed decisions. You need skills that can evolve and adapt to new situations. You need to be able to interpret and make decisions based on numerical information, and to find ways to solve problems that arise in real life, not just in textbooks. On this journey you will make connections between algebra and the world around you.

You're going to discover and learn much useful algebra along the way. Learning algebra is more than learning facts and theories and memorizing procedures. We hope you also discover the pleasure involved in mathematics and in learning "how to do mathematics." Success in algebra is a recognized gateway to many varied career opportunities.

With your teacher as a guide, you will learn algebra by doing mathematics. You will make sense of important algebraic concepts, learn essential algebraic skills, and discover how to use algebra. This requires a far bigger commitment than just "waiting for the teacher to show you" or studying "worked-out examples."

During this journey, successful learning will come from your personal involvement, which will often come about when you work with others in small groups. Talk about algebra, share ideas, and learn from and with your fellow group members. Work and communicate with others to strengthen your understanding of the mathematical concepts presented in this book. To improve your skills as a teammate, respect differences between group members, listen carefully when others are sharing, stay focused during the process, be responsible and respectful, and share your own ideas and suggestions.

Your graphing calculator is a tool that helps you explore new ideas and answer questions that come up along the way. With the calculator, you will be able to quickly manipulate large amounts of data so that you can see the overall picture. Throughout the text you will be referred to **Calculator Notes** that will provide useful information for using this tool. In your life and future career, technology is likely to play an important role. Learning to efficiently use your graphing calculator today, and being able to interpret its output, will prepare you to successfully use more advanced technology in your future.

The text itself will be a guidebook, leading you to explore questions and giving you the opportunity to ponder. Read the book carefully, with paper, pencil, and calculator close at hand. Work through the **Examples** and answer the questions that are asked along the way. Perform the **Investigations** as you travel through the course, being careful when making measurements and collecting data. Keep your data and calculations neat and accurate so that your work will be easier and the concepts clearer in the long run. Some **Exercises** require a great deal of thought. Don't give up. Make a solid attempt at each problem that is assigned. Sometimes you will need to fill in details later, after you discuss a problem in class or with your group.

Your notebook will serve as a record of your travels. In it you will record your notes, answers to questions in the text, and solutions for your homework problems. You may also want to keep a journal of your personal impressions along the way. You can place some of your especially notable accomplishments in a portfolio, which will serve as a "photo album" of the highlights of your trip. Collect pieces of work in your portfolio as you go, and refine the contents as you make progress on your journey. Each chapter ends with a feature called **Assessing What You've Learned.** This feature gives suggestions for organizing your notebook, writing in your journal, updating your portfolio, and other ways to reflect on what you have learned.

You should expect struggles, hard work, and occasional frustration. Yet, as your algebra skills grow, you'll overcome obstacles and be rewarded with a deeper understanding of mathematics, an increased confidence in your own problem-solving abilities, and the opportunity to be creative. Features called **Project, Improving Your . . . Skills,** and **Take Another Look** will give you special opportunities to creatively apply and extend your learning. We hope that your journey through *Discovering Algebra* will be a meaningful and rewarding experience.

And now it is time to begin. You are about to discover some pretty fascinating things.

CHAPTER
0
Fractions and Fractals

You have probably seen designs like this—you may even have heard the word "fractal" used to describe them. Complex fractals are created by infinitely repeating simple processes; some are created with basic geometric shapes such as triangles or squares. With fractals, mathematicians and scientists can model the formation of clouds, the growth of trees, and human blood vessels.

OBJECTIVES

In this chapter you will

- review operations with fractions
- review operations with positive and negative numbers
- use exponents to represent repeated multiplication
- explore designs called fractals
- learn to use this book as a tool

The Same yet Smaller

A procedure that you do over and over, each time building on, is **recursive.** You'll see recursion used in many different ways throughout this book. In this lesson you'll draw a **fractal** design using a recursive procedure. After you draw the design, you'll work with fractions to examine its parts.

> Words in **bold** type are important mathematical terms. They may be new to you, so they will be explained in the text. You can also find a definition in the glossary.

Investigation
Connect the Dots

> Investigations are a very important part of this course. Often you'll discover new concepts in an investigation, so be sure to take an active role.

You will need

• a ruler
• the worksheet Connect the Dots

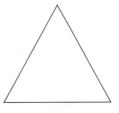

Stage 0   Stage 1

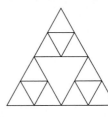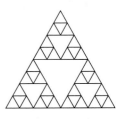

Stage 2   Stage 3

Step 1 | Examine the figures above. The starting figure is the Stage 0 figure. To create the Stage 1 figure, you join the *midpoints* of the sides of the triangle. You can locate the midpoints by counting dots to find the middle of each side. The Stage 1 figure has three small upward-pointing triangles. See if you can find all three.

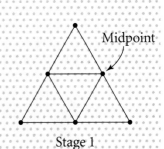

Midpoint

Stage 1

Step 2 At Stage 2, line *segments* connect the midpoints of the sides of the three upward-pointing triangles that showed up at Stage 1. What do you notice when you compare Stage 1 and Stage 2?

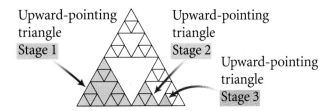

Upward-pointing triangle **Stage 1**

Upward-pointing triangle **Stage 2**

Upward-pointing triangle **Stage 3**

Step 3 How many new upward-pointing triangles are there in the Stage 3 figure?

Step 4 On your worksheet, create the Stage 4 figure. A blank triangle is provided. Connect the midpoints of the sides of the large triangle, and continue connecting the midpoints of the sides of each smaller upward-pointing triangle at every stage. How many small upward-pointing triangles are in the Stage 4 figure?

Step 5 What would happen if you continued to further stages? Describe any patterns you've noticed in drawing these figures.

> Words in *italic* are words you may have seen before or that you can probably figure out. If you need help with them, a definition is given in the glossary.

You have been using a *recursive rule*. The rule is "Connect the midpoints of the sides of each upward-pointing triangle."

If you could continue this process forever, you would create a fractal called the *Sierpiński triangle*. At each stage the small upward-pointing triangles are *congruent*—the same shape and size.

> This marker shows a convenient stopping place.

Step 6 If the Stage 0 figure has an area equal to 1, what is the area of one new upward-pointing triangle at Stage 1?

Step 7 How many different ways are there to find the combined area of the smallest upward-pointing triangles at Stage 1? For example, you could write the *addition expression* $\frac{1}{4} + \frac{1}{4} + \frac{1}{4}$. Write at least two other expressions to find this area. Use as many different operations (like addition, subtraction, multiplication, or division) as you can.

Step 8 What is the area of one of the smallest upward-pointing triangles at Stage 2? How do you know?

Step 9 How many smallest upward-pointing triangles are there at Stage 2? What is the combined area of these triangles?

Step 10 Repeat Steps 8 and 9 for Stage 3.

Step 11 If the Stage 0 figure has an area of 8, what is the combined area of

a. One smallest upward-pointing triangle at Stage 1, plus one smallest upward-pointing triangle at Stage 2?

b. Two smallest upward-pointing triangles at Stage 2, minus one smallest upward-pointing triangle at Stage 3?

c. One smallest upward-pointing triangle at Stage 1, plus three smallest upward-pointing triangles at Stage 2, plus nine smallest upward-pointing triangles at Stage 3?

Step 12 | Make up one problem like those in Step 11, and exchange it with a partner to solve.

> This marker means the investigation is done.

The Polish mathematician Waclaw Sierpiński created his triangle in 1916. But the word *fractal* wasn't used until nearly 60 years later, when Benoit Mandelbrot drew attention to recursion that occurs in nature. Trees, ferns, and even the coastlines of continents can be examined as real-life fractals.

EXAMPLE A

> Examples are important learning tools. Have your pencil in hand when you study the solution to an example. Try to do the problem before reading the solution. Work out any calculations in the solution so that you're sure you understand them.

Evan designed an herb garden. He divided each side of his garden into thirds and connected the points. He planted oregano in the labeled sections. If the whole garden has an area of 1, what is the area of one oregano section? What is the total area planted in oregano?

▶ Solution

> Often you will find questions in a solution. Try to answer these questions before you continue reading.

Because there are nine equal-size sections, each oregano section is one-ninth of the garden's area. To find the total area planted in oregano, you can either add $\frac{1}{9} + \frac{1}{9} = \frac{2}{9}$ or multiply $2 \times \frac{1}{9} = \frac{2}{9}$. So the oregano is planted in sections with a total area equal to $\frac{2}{9}$ of the garden. Can you explain how each expression represents the area?

Let's examine some features of Evan's garden in more detail. You can think of Evan's garden as a Stage 1 figure with six identical upward-pointing triangles that each have an area of $\frac{1}{9}$.

What is the area of one small upward-pointing triangle at Stage 2?

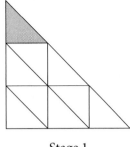

Stage 0   Stage 1   Stage 2

This feature will help you connect algebra to the people who continue to develop and use it.

History
● ● ● **CONNECTION** ● ●

Benoit Mandelbrot (b 1924) first used the word *fractal* in 1975 to describe irregular patterns in nature. You can link to a biography of Mandelbrot at **www.keymath.com/DA** .

At Stage 2, nine smaller triangles are formed in each upward-pointing triangle from Stage 1. The shaded triangle in Stage 2 has an area that is $\frac{1}{9}$ of the Stage 1 shaded triangle. This equals $\frac{1}{9}$ of $\frac{1}{9}$, which you can write as $\frac{1}{9} \times \frac{1}{9}$ and is equal to $\frac{1}{81}$.

To find combined areas, you'll be adding, subtracting, and multiplying fractions. When there are more than two operations in an expression, it can be difficult to know where to start. To avoid confusion, all mathematicians have agreed to use the **order of operations.**

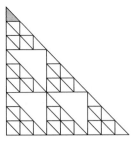

Order of Operations

1. Evaluate all expressions within parentheses.
2. Evaluate all powers.
3. Multiply and divide from left to right.
4. Add and subtract from left to right.

Go to the calculator notes whenever you see this icon. The calculator notes explain how to use your graphing calculator. You can get these notes from your teacher or log on to **www.keymath.com/DA** .

You should be able to do the calculations in this lesson with pencil and paper. Many calculators are programmed to give answers in fraction form, so use a calculator to check your answers.
[▶ 🖳 See **Calculator Note 0A.** ◀]

EXAMPLE B

If the largest triangle has an area of 1, what is the combined area of the shaded triangles?

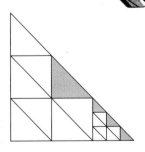

▶ Solution

The area of the larger shaded triangle is $\frac{1}{9}$.

The area of each smaller triangle is $\frac{1}{9} \times \frac{1}{9}$ or $\frac{1}{81}$.

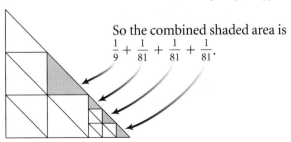

So the combined shaded area is $\frac{1}{9} + \frac{1}{81} + \frac{1}{81} + \frac{1}{81}$.

Notice that $\frac{1}{81} + \frac{1}{81} + \frac{1}{81} = \frac{3}{81}$, so the combined area is $\frac{1}{9} + \frac{3}{81}$.

To add fractions, you need *common denominators*. Since nine of the smallest triangles (with area of $\frac{1}{81}$) fit into a triangle with area of $\frac{1}{9}$, you can write $\frac{1}{9}$ as $\frac{9}{81}$. So you can rewrite the combined area as $\frac{9}{81} + \frac{3}{81}$, which equals $\frac{12}{81}$, or $\frac{4}{27}$ in *lowest terms*.

Think of another way to get the same answer. Check your method with a classmate to see if he or she agrees with you.

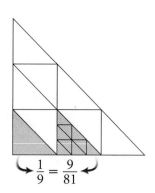

$$\frac{1}{9} = \frac{9}{81}$$

Nature
• CONNECTION •

The smallest leaves of a fern look very similar to the whole fern. This is an example of self-similarity in nature.

In the Sierpiński triangle, the design in any upward-pointing triangle looks just like any other upward-pointing triangle and just like the whole figure—they differ only in size. Objects like this are called **self-similar.** Self-similarity is an important feature of fractals, and you can find examples of self-similarity everywhere in nature.

EXERCISES

You will need your calculator for problems **1, 2,** and **4.**

Practice Your Skills

Do these calculations with paper and pencil. Check your work with a calculator.

1. Find the total shaded area in each triangle. Write two expressions for each problem, one using addition and the other using multiplication. Assume that the area of each largest triangle is 1.

   a. 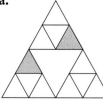   **b.** 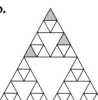   **c.** 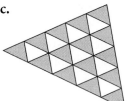   **d.**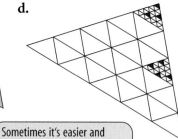

2. Write an expression and find the total shaded area in each triangle. Assume that the area of each largest triangle is 1.

   a. 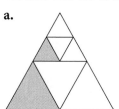   **b.** 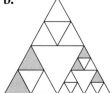   **c.** 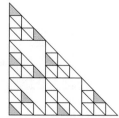   **d.**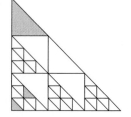

3. The first stages of a Sierpiński-like triangle are shown below.

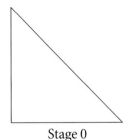 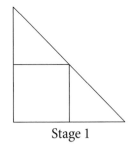 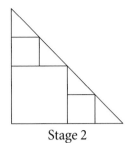 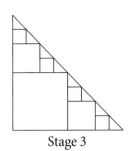

Stage 0   Stage 1   Stage 2   Stage 3

a. Draw Stage 4 of this pattern. You might find it easiest to start with a triangle that is about 8 cm or 4 in. along the bottom.

b. If the Stage 0 triangle has an area of 64, what is the area of the square at Stage 1?

c. At Stage 2, what is the area of the squares combined?

d. At Stage 3, what is the area of the squares combined?

4. Do each calculation, and check your results with a calculator. Set your calculator to give answers in fraction form.

a. $\frac{1}{3} + \frac{2}{9}$ **b.** $\frac{3}{4} + \frac{1}{2} + \frac{1}{3}$ **c.** $\frac{2}{5} \times \frac{3}{7}$ **d.** $2 - \frac{4}{9}$

► Reason and Apply

5. Suppose the fractal design at right has an area equal to 1. Copy the figure four times and shade parts to show each area.

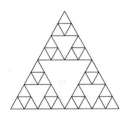
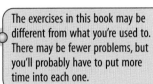

a. $\frac{1}{4}$ **b.** $\frac{3}{16}$ **c.** $\frac{5}{16}$ **d.** $1 - \frac{7}{16}$

6. You have been introduced to the Sierpiński triangle. What are some aspects of this triangle that make it a fractal?

The exercises in this book may be different from what you're used to. There may be fewer problems, but you'll probably have to put more time into each one.

7. Look at the Sierpiński-like pattern in the squares.

 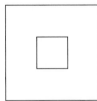 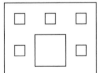

Stage 0   Stage 1   Stage 2

a. Describe in detail the recursive rule used to create this pattern.

b. Carefully draw the next stage of the pattern.

c. Suppose the Stage 0 figure represents a square carpet. The new squares drawn at each stage represent holes that have been cut out of the carpet. If the Stage 0 carpet has an area of 1, what is the total area of the holes at Stages 1 through 3?

d. What is the area of the remaining carpet at each stage?

8. Suppose the area of the original large triangle at right is 8.

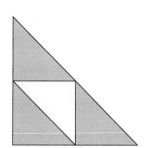

  a. Write a division expression to find the area of one of the shaded triangles. What is the area?

  b. What fraction of the total area is each shaded triangle? Use this fraction in a multiplication expression to find the area of one of the shaded triangles.

  c. What is the difference between dividing by 4 and multiplying by $\frac{1}{4}$?

  d. Write a multiplication expression using the fraction $\frac{3}{4}$ to find the combined shaded area.

9. Suppose the original large triangle below has an area of 12.

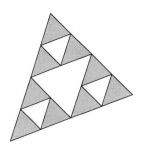

  a. What fraction of the area is shaded?

  b. Find the combined area of the shaded triangles. Write two different expressions you could use to find this area.

10. Suppose the original large triangle at right has an area of 24.

  a. What fraction of the area is the shaded triangle at the top?

  b. What fraction of the area is each smallest shaded triangle?

  c. What is the total shaded area? Can you find two ways to calculate this area?

11. Rewrite each expression below using fractions. Then draw a Sierpiński triangle and shade the area described. In each case the Stage 0 triangle has an area of 32.

  a. One-fourth of one-fourth of 32

  b. Three-fourths of one-fourth of one-fourth of 32

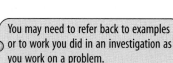
You may need to refer back to examples or to work you did in an investigation as you work on a problem.

  c. One-half of one-half of one-fourth of 32

▶ Review

12. Assume the area of your desktop equals 1. Your math book rests on $\frac{1}{4}$ of your desktop, your calculator on $\frac{1}{16}$ of your desktop, and your scrap paper on $\frac{1}{32}$ of your desktop. What area is covered by these objects? Write an addition expression and then give your answer as a single fraction in lowest terms.

13. Use the information from exercise 12 to find the area of your desk that is *not* covered by these materials. Write a subtraction expression and then give your answer as a single fraction in lowest terms.

More and More

Did you notice that at each stage of a Sierpiński design, you have more to draw than in the previous stage? The new parts get smaller, but the number of them increases quickly. Let's examine these patterns more closely.

A strong positive mental attitude will create more miracles than any wonder drug.

PATRICIA NEAL

 ## Investigation
How Many?

Explore how quickly the number of new triangles grows using multiplication repeatedly. Look for a pattern to help you *predict* the number of new triangles at each stage without counting them.

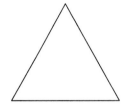   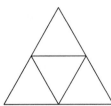   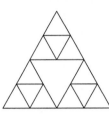   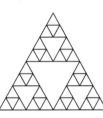   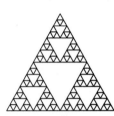

Stage 0   Stage 1   Stage 2   Stage 3   Stage 4

Step 1 | Look at the fractal designs. Count the number of new upward-pointing triangles for Stages 0 to 4. Make a table like this to record your work.

Stage	Number of new upward-pointing triangles
0	1
1	

> Throughout this course you'll record results in a table. Tables provide a useful way to keep track of your work and see patterns develop.

Step 2 | How does the number of new triangles compare to the number of new triangles at the previous stage?

Step 3 | Using your answer to Step 2, find how many new upward-pointing triangles are at Stages 5, 6, and 7.

Step 4 | Explain how you could find the number of upward-pointing triangles at Stage 15 without counting.

At each stage, three new upward-pointing triangles are drawn in each of the upward-pointing triangles from the previous stage. How is this the same as repeatedly multiplying by 3?

You can write the symbol for multiplication in different ways. For example, you can write 3×3 as $3 \cdot 3$ or $3(3)$. All three of these expressions have the same meaning. Each expression equals 9.

EXAMPLE | Describe how the number of new upward-pointing triangles is growing in this fractal.

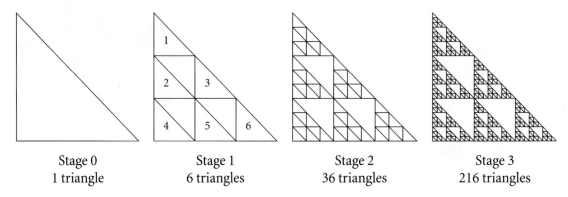

| Stage 0 | Stage 1 | Stage 2 | Stage 3 |
| 1 triangle | 6 triangles | 36 triangles | 216 triangles |

▶ **Solution** | At Stage 1, the six new upward-pointing triangles are numbered. At Stage 2, six new upward-pointing triangles are formed in each numbered Stage 1 triangle. At Stage 2, there are $6 \cdot 6$ or 36 new triangles. At Stage 3, six triangles are formed in each new upward-pointing Stage 2 triangle, so there are $36 \cdot 6$ or 216 new upward-pointing triangles.

Another way to look at the number of new upward-pointing triangles at each stage is shown in the table below.

	Number of new upward-pointing triangles		
Stage number	**Total**	**Repeated multiplication**	**Exponent form**
1	6	6	6^1
2	36	$6 \cdot 6$	6^2
3	216	$36 \cdot 6$ or $6 \cdot 6 \cdot 6$	6^3

The last number in each row of the table is a 6 followed by a small number. The small number, called an **exponent,** shows how many 6's are multiplied together. An exponent shows the number of **factors** of 6. What is the pattern between the stage number and the exponent?

Do you think the pattern applies to Stage 0? Put the number 6^0 into your calculator. [▶ 🔲 See **Calculator Note 0B** to learn how to enter exponents. ◀] Does the result fit the pattern?

How many upward-pointing triangles are there at Stage 4? According to the pattern, there should be 6^4. That's 1296 triangles! It is a lot easier to use the exponent pattern than to count all those triangles.

EXERCISES

You will need your calculator for problem **4.**

▶ Practice Your Skills

1. Write each multiplication expression in exponent form.
   a. $5 \times 5 \times 5 \times 5$
   b. $7 \times 7 \times 7 \times 7 \times 7$
   c. $3 \cdot 3 \cdot 3 \cdot 3 \cdot 3 \cdot 3 \cdot 3$
   d. $2(2)(2)$

2. Rewrite each expression as a repeated multiplication three ways: using $\times$, $\cdot$, and parentheses.
   a. 3^4
   b. 5^6
   c. $\left(\dfrac{1}{2}\right)^3$

3. Write each number with an exponent other than 1. For example, $125 = 5^3$.
   a. 27
   b. 32
   c. 625
   d. 343

4. Do the calculations. Check your results with a calculator.
   a. $\dfrac{2}{3} \cdot 12$
   b. $\dfrac{1}{3} + \dfrac{3}{5}$
   c. $\dfrac{3}{4} - \dfrac{1}{8}$
   d. $5 - \dfrac{2}{7}$
   e. $\dfrac{1}{4} \cdot \dfrac{1}{4} \cdot 8$
   f. $\dfrac{3}{64} + \dfrac{3}{16} + \dfrac{3}{4}$

▶ Reason and Apply

5. Another type of fractal drawing is called a "tree." Study Stages 0 to 3 of this tree:

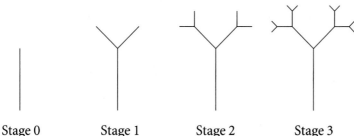

Stage 0   Stage 1   Stage 2   Stage 3

> Homework helps you reinforce what you've learned in the lesson and develops your understanding of new ideas.

   a. At Stage 1, two new branches are growing from the trunk. How many new branches are there at Stage 2? At Stage 3?

   b. How many new branches are there at Stage 5? Write your answer in exponent form.

6. Another fractal tree pattern is shown below.

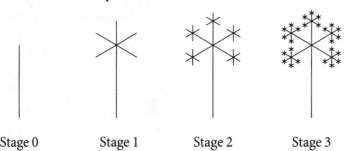

Stage 0   Stage 1   Stage 2   Stage 3

   a. At Stage 1, five new branches are growing. How many new branches are there at Stage 2?

b. How many new branches are there at Stage 3?

c. How many new branches are there at Stage 5? Write your answer in exponent form.

7. At Stage 1 of this pattern, there is one square hole. At Stage 2, there are eight new square holes.

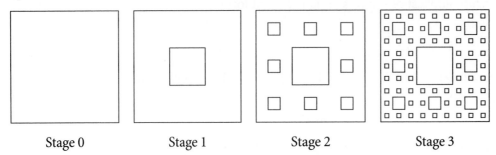

Stage 0   Stage 1   Stage 2   Stage 3

a. How many new square holes are there at Stage 3?

b. If you drew the Stage 4 figure, how many new square holes would you have to draw?

c. Write the answers to parts a and b in exponent form.

d. How many new square holes would you have to draw in the Stage 7 figure?

e. Describe the pattern between the stage number and the exponent for these figures.

f. Will the pattern you described in part e work for the Stage 1 figure? Why?

8. Study Stages 0 to 3 of this fractal "weed" pattern. At Stage 1, two new branches are created.

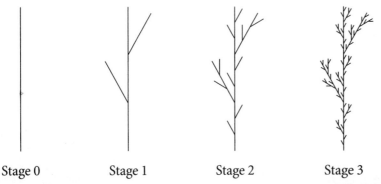

Stage 0   Stage 1   Stage 2   Stage 3

a. How many new branches are created at Stage 2?

b. How many new branches are created at Stage 3?

c. You can write the expression $2 \cdot 5^1$ to represent the number of new branches in the Stage 2 figure. Write similar expressions to represent the number of new branches in Stages 3 to 5.

d. How do the 2 and the 5 in each expression relate to the figure?

Patterns like the "weed" in problem 8 can be used to create very realistic computer-generated plants. Graphic designers can use fractal routines to create realistic-looking trees and other natural features.

9. Look again at this familiar fractal design.

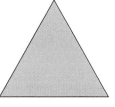 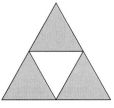 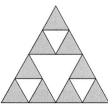 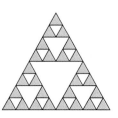

Stage 0   Stage 1   Stage 2   Stage 3

a. Make a table to calculate and record the area of one shaded triangle in each figure.

Stage number	Area of one shaded triangle	Total area of the shaded triangles
0	1	1
1		

b. Record the combined shaded area of each figure in your table.

c. Describe at least two patterns you discovered.

Review

10. Ethan deposits $2 in a bank account on the first day, $4 on the second day, and $8 on the third day. He will continue to double the deposit each day. How much will he deposit on the eighth day? Write your answer as repeated multiplication separated by dots, in exponent form, and as a single number.

11. Write a word problem that illustrates $\frac{3}{4} \cdot \frac{1}{5}$, and find the answer.

12. The large triangles below each have an area of 1. Find the total shaded area in each.

a.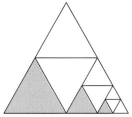

b.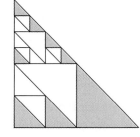

Shorter yet Longer

In fractals like the Sierpiński triangle, new enclosed shapes are formed at each stage. Not all fractals are formed this way. One example is the *Koch curve,* which is not a smooth curve, but a set of connected line segments. It was introduced in 1906 by the Swedish mathematician Niels Fabian Helge von Koch. As you explore the Koch curve, you'll continue to work with exponents.

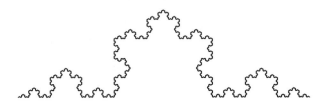

Investigation
How Long Is This Fractal?

Study how the Koch curve develops. One way to discover a fractal's recursive rule is to study what happens from Stage 0 to Stage 1. Once you know the rule, you can build, or generate, later stages of the figure.

| Stage 0 | Stage 1 | Stage 2 |

Step 1 Make and complete a table like this for Stages 0 to 2 of the Koch curve shown. How do the lengths change from stage to stage? If you don't see a pattern, try writing the total lengths in different forms.

Stage number	Number of segments	Length of each segment	Total length (Number of segments times length of segments)	
			Fraction form	Decimal form
0				

Step 2 Look at Stages 0 and 1. Describe the curve's recursive rule so that someone could recreate the curve from your description.

Step 3 Predict the total length at Stage 3.

Step 4 Find the length of each small segment at Stage 3 and the total length of the Stage 3 figure.

Stage 3

If the Koch curve were a piece of string, you could find its length by straightening the string and measuring it.

Step 5	Use exponents to rewrite your numbers in the column labeled *Total length in fraction form* for Stages 0 to 3.
Step 6	Predict the Stage 4 lengths.
Step 7	Koch was attempting to create a "curve" that was nothing but corners. Do you think he succeeded? If the curve is formed recursively for many stages, what would happen to its length?

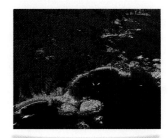

At later stages the Koch curve looks smoother and smoother. But, if you magnify a section at a later stage, it is just as jagged as at Stage 1. Mandelbrot named these figures *fractals* based on the Latin word *fractus,* meaning broken or irregular.

EXAMPLE

Look at the beginning stages of this fractal:

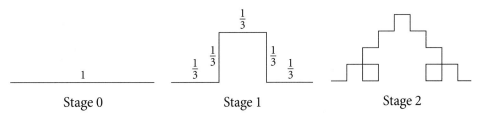

Stage 0	Stage 1	Stage 2

a. Describe the fractal's recursive rule.

b. Find its length at Stage 2.

c. Write an expression for its length at Stage 17.

▶ **Solution**

a. You compare Stage 0 and Stage 1 to get the recursive rule. To get Stage 1, you divide the Stage 0 segment into thirds. Build a square on the middle third and remove the bottom. So the recursive rule is "To get to the next stage, divide each segment from the previous stage into thirds and build a bottomless square on the middle third."

Don't forget to think through the solution and answer any questions.

b. To find the length of the fractal at Stage 2, we'll start by looking at its length at Stage 1. The Stage 1 figure has 5 segments. Each segment is $\frac{1}{3}$ long. So the total length at Stage 1 is $5 \cdot \frac{1}{3}$. You can rewrite this as $\frac{5}{3}$.

At Stage 2, you replace each of the five Stage 1 segments with five new segments. So the Stage 2 figure has $5 \cdot 5$ or 5^2 segments.

Each Stage 1 segment is $\frac{1}{3}$ long, and each Stage 2 segment is $\frac{1}{3}$ of that. So each Stage 2 segment is $\frac{1}{3} \cdot \frac{1}{3}$ or $\left(\frac{1}{3}\right)^2$ long.

So, at Stage 2, there are 5^2 segments, each $\left(\frac{1}{3}\right)^2$ long. The total length at Stage 2 is $5^2 \cdot \left(\frac{1}{3}\right)^2$. You can rewrite this as $\left(\frac{5}{3}\right)^2$.

c. Do you see the connection between the stage number and the exponent? At each stage, you replace every segment from the previous stage with five new segments. The length of each new segment is $\frac{1}{3}$ the length of a segment at the previous stage. By Stage 17, you've done this 17 times. The Stage 17 figure is $\left(\frac{5}{3}\right)^{17}$ long.

Each segment is replaced with 5 segments.

$$\left(\frac{5}{3}\right)^{17}$$

You've done this for 17 steps.

Each new segment is $\frac{1}{3}$ the length of the old segment.

To compare total lengths, express each as a decimal rounded to the hundredth's place.

Stage number	Number of segments	Length of each segment	Total length (Number of segments times length of segments)	
			Fraction form	Decimal form
0	1	1	$1 \cdot 1$	1.00
1	$1 \cdot 5 = 5^1$	$1 \cdot \frac{1}{3} = \left(\frac{1}{3}\right)^1$	$5^1 \cdot \left(\frac{1}{3}\right)^1 = \left(\frac{5}{3}\right)^1$	1.67
2	$5 \cdot 5 = 5^2$	$\frac{1}{3} \cdot \frac{1}{3} = \left(\frac{1}{3}\right)^2$	$5^2 \cdot \left(\frac{1}{3}\right)^2 = \left(\frac{5}{3}\right)^2$	2.78
⋮	⋮	⋮	⋮	⋮
17	5^{17}	$\left(\frac{1}{3}\right)^{17}$	$5^{17} \cdot \left(\frac{1}{3}\right)^{17} = \left(\frac{5}{3}\right)^{17}$	5907.84

EXERCISES

You will need your calculator for problems **1, 3, 4, 6, 7,** and **8.**

▶ Practice Your Skills

1. Evaluate each expression. Write your answer as a fraction and as a decimal, rounded to the nearest hundredth. Remember, if the third digit to the right of the decimal is 5 or higher, round up.

a. $\dfrac{5^3}{2^3}$ **b.** $\left(\dfrac{5}{3}\right)^2$ **c.** $\left(\dfrac{7}{3}\right)^4$ **d.** $\left(\dfrac{9}{4}\right)^3$

2. The fractal from the example is shown below. How much longer is the figure at Stage 2 than at Stage 1? Use the table on the previous page and find your answer as a fraction and as a decimal rounded to the nearest hundredth.

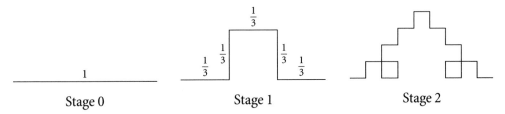

Stage 0   Stage 1   Stage 2

3. At what stage does the figure above first exceed a length of 10?

4. Evaluate each expression and check your results with a calculator.

a. $\dfrac{1}{5} + \dfrac{3}{4}$ b. $3^2 + 2^4$ c. $\dfrac{2}{3} \cdot \left(\dfrac{6}{5}\right)^2$ d. $4^3 - \dfrac{2}{5}$

▶ Reason and Apply

5. The Stage 0 figure below has a length of 1. At Stage 1, each segment has a length of $\frac{1}{4}$.

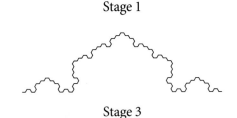

Stage 0   Stage 1

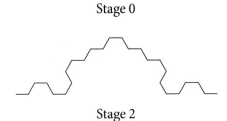

Stage 2   Stage 3

a. Complete a table like the one below by calculating the lengths of the figure at Stages 2 and 3. Give each answer as a fraction in multiplication form, as a fraction in exponent form, and as a decimal rounded to the nearest hundredth. Try to figure out the total lengths at Stages 2 and 3 without counting.

b. Which is the first stage to have a length greater than 3? A length greater than 10?

Stage number	Total length		
	Multiplication form	**Exponent form**	**Decimal form**
0	1	1^0	1
1	$5 \cdot \dfrac{1}{4} = \dfrac{5}{4}$	$\left(\dfrac{5}{4}\right)^1$	1.25
2	$5 \cdot 5 \cdot \dfrac{1}{4} \cdot \dfrac{1}{4} = \dfrac{25}{16}$		
3			

6. The Stage 0 figure below has a length of 1.

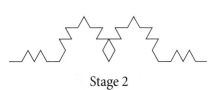

Stage 0

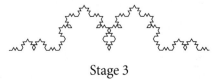

Stage 1

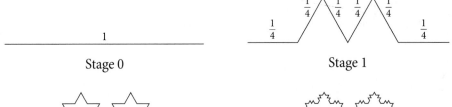

Stage 2   Stage 3

a. Complete a table like the one below by calculating the total length of the figure at each stage shown above. Give each answer as a fraction in expanded form, as a fraction in exponent form, and as a decimal rounded to the nearest hundredth.

Stage number	Total length		
	Expanded form	**Exponent form**	**Decimal form**
0	1	1^0	1
1	$6 \cdot \dfrac{1}{4} = \dfrac{6}{4} = \dfrac{3}{2}$	$6^1 \cdot \left(\dfrac{1}{4}\right)^1 = \left(\dfrac{6}{4}\right)^1 = \left(\dfrac{3}{2}\right)^1$	1.5
2			
3			

b. At what stage does the figure have a length of $\frac{243}{32}$?

c. At what stage is the length closest to 100?

7. The Stage 0 figure below has a length of 1.

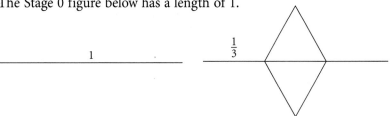

Stage 0   Stage 1   Stage 2

Stage 3   Stage 4

a. Complete a table like the one below by calculating the total length of each stage. Give each answer as a fraction in expanded form, as a fraction in exponent form, and as a decimal number rounded to the nearest hundredth. Figure out the lengths of Stages 3 and 4 without counting.

Stage number	Total length		
	Expanded form	Exponent form	Decimal form
0	1	1^0	1
1	$7 \cdot \dfrac{1}{3} = \dfrac{7}{3}$	$7^1 \cdot \left(\dfrac{1}{3}\right)^1 = \left(\dfrac{7}{3}\right)^1$	2.33
2			
3			
4			

b. At what stage does the figure have a length of $\dfrac{16{,}807}{243}$?

c. Will the figure ever have a length of 168? If so, at what stage? If not, why not?

IMPROVING YOUR **REASONING** SKILLS

As the Koch curve develops, segment length decreases as the number of segments increases.

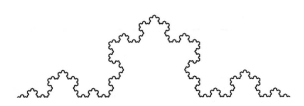

As you draw higher stages, the length of individual segments approaches, or gets closer and closer to, what number? What number does the number of line segments approach? Is it possible to draw the "finished" fractal? Why or why not? What would its total length be?

Now consider this pattern:

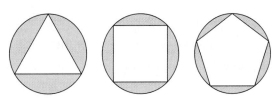

As you draw higher and higher stages, what does the length of each polygon side approach? What number does the number of sides approach? Is it possible to draw the "finished" polygon? What would it look like? What would the total perimeter of the polygon be? Is this pattern recursive? Is the result a fractal? Why or why not?

8. The figures below look a little more complicated because parts overlap. The Stage 0 figure has a length of 1.

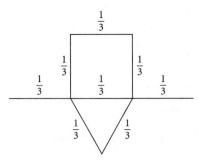

Stage 0   Stage 1

Stage 2   Stage 3

a. Complete a table like the one below by calculating the total length of the Stage 2 and Stage 3 figures shown above.

Stage number	Total length		
	Expanded form	**Exponent form**	**Decimal form**
0	1	1^0	1
1	$8 \cdot \dfrac{1}{3} = \dfrac{8}{3}$	$8^1 \cdot \left(\dfrac{1}{3}\right)^1 = \left(\dfrac{8}{3}\right)^1$	2.67
2			
3			

b. Look at how the lengths of the figures grow with each stage. Estimate how long the length will be in Stage 4. Then calculate this value.

c. At what stage does your calculator begin to use a different notation for the length?

► **Review**

Whenever possible it's a good idea to try to estimate your answer before calculating it. Estimating will help you determine whether or not your calculated answer is reasonable.

9. Write $\frac{14}{5}$ as a decimal.

10. What is $\frac{8}{3} - \frac{4}{9} \cdot \frac{3}{1}$?

11. Look at the fractal "cross" pattern below. At each stage, new line segments are drawn through the existing segments to create crosses.

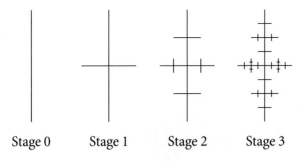

Stage 0   Stage 1   Stage 2   Stage 3

a. How many new segments are drawn in Stage 2?

b. How many new segments are drawn in Stage 3?

c. How many new segments would be drawn in Stage 4?

d. Use exponents to represent the number of new segments drawn in Stages 2 to 4.

e. In general, how is the exponent related to the stage number for Stages 2 to 4? Does this rule apply to Stage 1?

INVENT A FRACTAL!

Recursive procedures can produce surprising and even beautiful results. Consider these two fractals. (The top one was "invented" by student Andrew Riley!) Would you have expected that the Stage 1 figures would lead to the higher-stage figures?

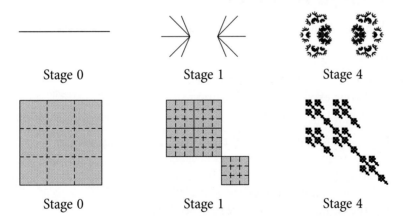

Invent your own fractal. You can start with a line segment, as in the Koch curve or Andrew's fractal. Or try a two-dimensional shape like Stage 0 of Sierpiński's triangle or the square in the "kite" above. Your project should include

▶ A drawing of your fractal at Stages 0, 1, 2, and 3 (and possibly higher).

▶ A written description of the recursive rule that generates the fractal.

▶ A table that shows how one aspect of the figure changes. Consider area, length, or the number of holes or branches at each stage.

▶ A written explanation of how to continue your table for higher stages.

THE GEOMETER'S SKETCHPAD®

The Geometer's Sketchpad® was used to create these fractals. Sketchpad™ has several tools to help you create fractals. With Sketchpad, you can quickly and easily create Sierpiński's triangle, the Koch curve, and more. Learn how to use Sketchpad and create your own fractals!

Going Somewhere?

Leslie was playing miniature golf with her friends. First she hit the ball past the hole. Then she hit it back, but it went too far and missed again. She kept hitting the ball closer, but it still missed the hole. Finally she got so close that the ball fell in.

Some number processes also get closer and closer to a final target, until the result is so close that the number rounds off to the target value or answer. You'll explore processes like these while reviewing operations with positive and negative numbers.

Investigation
A Strange Attraction

Step 1 | Each member of your group takes one of these four expressions.

$$2 \cdot \square + 1 \qquad 3 \cdot \square - 4 \qquad -2 \cdot \square + 3 \qquad -3 \cdot \square - 1$$

Step 2 | As a group, choose a starting number. Record your expression and starting number in a table like this.

Original Expression:	
Starting Number (at Stage 0):	
Stage number	Result
1	

Step 3 | Put your starting number in the box, and do the computation. This process is called **evaluating the expression,** and the result is called the **value of the expression.** When doing the computation, be sure to follow the order of operations. Check your answer with a calculator, and record it in the table as your first result. [▶ 🖳 See **Calculator Note 0C** to learn about the difference between the negative key and the subtraction key. ◀]

Step 4 | Take the result you got from Step 3, put it in the box in your expression, and evaluate your expression again. Place your new answer in the table as your second result.

Step 5 | Continue this recursive process using your result from the previous stage. Evaluate your expression. Each time, record the new result in your table. Do this ten times.

Step 6 | Draw a number line and scale it so that you can show the first five results from your table. Plot the first result from your table, and draw an arrow to the next result to show how the value of the expression changes. For example,

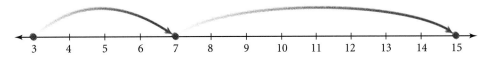

Step 7 | How do the results in your group compare?

Step 8 | Repeat Steps 1 to 6 with one of the expressions below.

$$0.5 \cdot \square - 3 \qquad 0.2 \cdot \square + 1 \qquad -0.5 \cdot \square + 3 \qquad -0.2 \cdot \square - 2$$

Step 9 | How do the results in your group compare? Do the results of these expressions differ from the results of your first expression?

In this investigation, you explored what happens when you recursively evaluate an expression. First you selected a starting number to put into your expression, then you evaluated it. Then you put your result back into the same expression and evaluated it again. Calculators, like computers, are good tools for doing these repetitive operations.

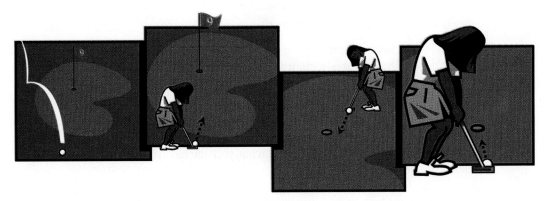

EXAMPLE A | What happens when you evaluate this expression recursively with different starting numbers?

$$0.5 \cdot \square - 2$$

▶ Solution | The starting number 1 gives

> Have your pencil and calculator in hand as you work through the solution to this example.

$$0.5 \cdot (1) - 2 = 0.5 - 2 = \boxed{-1.5}$$

$$0.5 \cdot (-1.5) - 2 = -0.75 - 2 = \boxed{-2.75}$$

$$0.5 \cdot (-2.75) - 2 = -1.375 - 2 = \boxed{-3.375}$$

$$0.5 \cdot (-3.375) - 2 = -1.6875 - 2 = \boxed{-3.6875}$$

$$0.5 \cdot (-3.6875) - 2 = -1.84375 - 2 = \boxed{-3.84375}$$

Each result of the recursion seems to get closer to a certain number. If you continue the process a few more times, you'll get -3.9219, then -3.9609, then -3.9805. What do you think will happen after even more recursions?

Using 4 as a starting number in the same expression, you get

$$0.5 \cdot (4) - 2 = 2 - 2 = \boxed{0}$$

$$0.5 \cdot (0) - 2 = 0 - 2 = \boxed{-2}$$

$$0.5 \cdot (-2) - 2 = -1 - 2 = \boxed{-3}$$

$$0.5 \cdot (-3) - 2 = -1.5 - 2 = \boxed{-3.5}$$

$$0.5 \cdot (-3.5) - 2 = -1.75 - 2 = \boxed{-3.75}$$

Again the values seem to get closer to one number, perhaps -4. If any starting number that you try eventually gets closer and closer to -4, then -4 is called an **attractor** for this expression.

To check whether -4 is an attractor value, use it as the starting number. If your result stays at -4, then it is an attractor value.

Using -4 as the starting number gives

$$0.5 \cdot (-4) - 2 = -2 - 2 = -4$$

Since you get back exactly what you started with, -4 is an attractor value for the expression $0.5 \cdot \square - 2$.

Evaluating expressions recursively does not always lead to an attractor value. Some expressions continue to grow larger when evaluated recursively, while others are difficult, if not impossible, to recognize.

EXAMPLE B

What happens when you recursively evaluate this expression for different starting numbers?

$$\square^2 - 2$$

▶ **Solution**

Choosing 1 as a starting number:

$$(1)^2 - 2 = 1 - 2 = -1$$

$$(-1)^2 - 2 = 1 - 2 = -1$$

$$(-1)^2 - 2 = 1 - 2 = -1$$

The results are all -1's, so -1 is an attractor value for this expression. On a number line the results look like this:

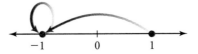

Choosing 3 as a starting number:

$$(3)^2 - 2 = 9 - 2 = 7$$
$$(7)^2 - 2 = 49 - 2 = 47$$
$$(47)^2 - 2 = 2209 - 2 = 2207$$

In this case the results get larger and larger.

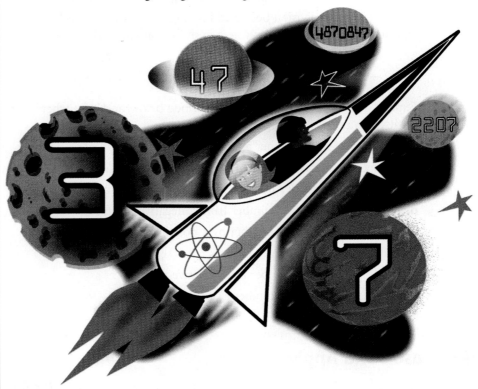

Choosing -2 as a starting number:

$$(-2)^2 - 2 = 4 - 2 = 2$$
$$(2)^2 - 2 = 4 - 2 = 2$$

So 2 is another attractor value for this expression.

Choose any other starting number, and either you'll get a series of repeating -1's or 2's, or the values will get farther apart at each stage.

With enough practice you may be able to predict the attractor values for some simple expressions without actually doing any computations. But as you try this process with more complex expressions, the results are less predictable.

EXERCISES

You will need your calculator for problems **1, 2, 3, 5, 6, 7, 8,** and **9.**

▶ Practice Your Skills

1. Do each calculation and use a calculator to check your results. Then use a number line to illustrate your answer.

   a. $-4 + 7$ **b.** $5 + (-8)$

   c. $-2 - 5$ **d.** $-6 - (-3)$

2. Do each calculation and check your results on your calculator.

   a. $-2 \cdot 5$ **b.** $6 \cdot (-4)$ **c.** $-3 \cdot (-4)$

   d. $-12 \div 3$ **e.** $36 \div (-6)$ **f.** $-50 \div (-5)$

3. Do the following calculations. Remember, if there are no parentheses you must do multiplication or division before addition or subtraction. Check your results by entering the expression exactly as it is shown on your calculator.

   a. $5 \cdot -4 - 2 \cdot -6$ **b.** $3 + -4 \cdot 7$

   c. $-2 - 5 \cdot (6 + -3)$ **d.** $(-3 - 5) \cdot -2 + 9 \cdot -3$

4. Explain how to do each operation described below, and state whether the result is a positive or a negative number.

   a. adding a negative number and a positive number

   b. adding two negative numbers

   c. subtracting a negative number from a positive number

   d. subtracting a negative number from a negative number

   e. multiplying a negative number by a positive number

   f. multiplying two negative numbers

   g. dividing a positive number by a negative number

   h. dividing two negative numbers

▶ Reason and Apply

5. Pete Repeat was recursively evaluating this expression starting with 2.

$$-0.2 \cdot \square - 4$$

   a. Check his first two stages and explain what, if anything, he did wrong.

$$-0.2 \cdot 2 - 4 = 0.4 - 4 = -3.6$$
$$-0.2 \cdot -3.6 - 4 = -0.72 - 4 = -4.72$$

   b. Redo Pete's first two stages and do two more.

   c. Now do three recursions using Pete's expression and starting with -1.

   d. Do you think this expression has an attractor value? Explain.

6. To tell whether or not an expression has an attractor value, you often have to look at the results of several different starting values.

Starting value	2	−1	10
First recursion			
Second recursion			
Third recursion			

a. Evaluate this expression for different starting values.

$$0.1 \cdot \square - 2$$

Record the results for three recursions in a table like the one shown.

b. Based on your table, do you think this expression reaches an attractor value in the long run? If so, what is it? If not, why not?

c. If you found an attractor in 6b, use your calculator to see if substituting that value in the expression gives it back to you.

7. Investigate this expression.

$$-2 \cdot \square + 1$$

Starting value	2	−1	10
First recursion			
Second recursion			
Third recursion			

a. Evaluate the expression for different starting values. Record your results in a table like the one shown.

b. Based on your table, do you think this expression reaches an attractor in the long run? If so, what is it? If not, why not?

8. Use a calculator to investigate the behavior of these expressions.
[▶🖳 See **Calculator Note 0D** for specific instructions. ◀]

a. Use recursion to evaluate each expression many times, and record the attractor value you get after many recursions.

i. $0.5 \cdot \square + 6$ **ii.** $0.5 \cdot \square - 8$ **iii.** $0.5 \cdot \square - 4$

b. Describe any connections you see between the numbers in the original expressions and their attractor values.

c. Create an expression that has an attractor value of 6.

9. Use a calculator to investigate the behavior of the expressions below.

i. $0.2 \cdot \square + 6$ **ii.** $0.2 \cdot \square - 8$ **iii.** $0.2 \cdot \square + 5$

a. Use recursion to evaluate each expression many times, and record its attractor value.

b. Describe any connections you see between the numbers in the original expressions and the attractor values.

c. Create an expression that has an attractor value of 2.25.

10. How is the recursion process like drawing the Sierpiński triangle in Lesson 0.1? How is it like creating the Koch curve in Lesson 0.3?

▶ Review

11. What is $4 - 12 \div 4 \cdot \frac{1}{2} - 5^2$?

12. Find $(-3 \cdot -4) - (-4 \cdot 2)$.

LESSON 0.5

Out of Chaos

*Nothing in nature is random....
A thing appears random only
through the incompleteness
of our knowledge.*

BARUCH SPINOZA

If you looked at the results of 100 rolls of a die, would you expect to find a pattern in the numbers? You might expect each number to appear about one-sixth of the time. But you probably wouldn't expect to see a pattern in when, for example, a 5 appears. The 5 appears **randomly,** without order. You could not create a method to predict exactly when or how often a 5 appears. As you explore seemingly random patterns, you'll review some measurement and fraction ideas.

Investigation
A Chaotic Pattern?

You will need

- a die
- a centimeter ruler
- a blank transparency and marker
- the worksheet A Chaotic Pattern?

What happens if you use a random process recursively to determine where you draw a point? Would you expect to see a pattern?

Work with a partner. One partner rolls the die. The other measures distance and marks points.

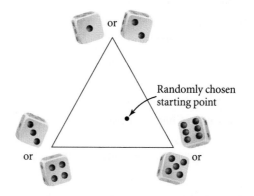

Randomly chosen
starting point

Step 1	Mark any point inside the triangle as your starting point.
Step 2	Roll the die.
Step 3	In centimeters, measure the distance from your starting point to the corner, or *vertex*, labeled with the number on the die. Take half of the distance, and place a small dot at this midpoint. This is your new starting point.
Step 4	Repeat Steps 2 and 3 until you've rolled the die 20 times. Then switch roles with your partner and repeat the process another 20 times.
Step 5	How is this process recursive?
Step 6	Describe the arrangement of dots on your paper.
Step 7	What would have happened if you had numbered the vertices of the triangle 1 and 3, 2 and 5, and 4 and 6?

For best results, measure as accurately as you can.

Step 8	Place a transparency over your worksheet. Use a transparency marker and mark the vertices of the triangle. Carefully trace your dots onto the transparency.
Step 9	When you finish, place your transparency on an overhead projector. Align the vertices of your triangle with the vertices of your classmates' triangles. This allows you to see the results of many rolls of a die. Describe what happens when you combine everyone's points. How is this like the result in other recursion processes? Is the result as random as you expected? Explain.

A *random* process can produce ordered-looking results while an orderly process can produce random-looking results. Mathematicians use the term *chaotic* to describe systematic, non-random processes that produce results that look random. Chaos helps scientists understand the turbulent flow of water, the mixing of chemicals, and the spread of an oil spill. They often use computers to do these calculations. Your calculator can repeat steps quickly, so you can use the calculator to plot thousands of points.

Step 10	Enter the Chaos program into your calculator. [▶ 🖳 See **Calculator Note 0E** for the program. To learn how to link calculators, see **Calculator Note 0F.** To learn about how to enter a program, see **Calculator Note 0G.**◀]

The program randomly "chooses" one vertex of the triangle as a starting point. It "rolls" an imaginary die and plots a new point halfway to the vertex it chose. The program rolls the die 999 more times. It does this a lot faster than you can.

Step 11	Run the program. Select an equilateral (equal-sided) triangle as your shape. When it "asks" for the fraction of the distance to move, enter $\frac{1}{2}$ or 0.5. It will take a while to plot all 1000 points, so be patient.
Step 12	What do you see on your calculator screen, and how does it compare to your class's combined transparency image?

Most people are surprised that after plotting many points, a familiar figure appears. When an orderly result appears out of a random process like this one, the figure is a *strange attractor*. No matter where you start, the points "fall" toward this shape. Many fractal designs, like the Sierpiński triangles on your calculator screen, are also strange attractors. Accurate measurements are essential to seeing a strange attractor form. In the next example, practice your measurement skills with a centimeter ruler.

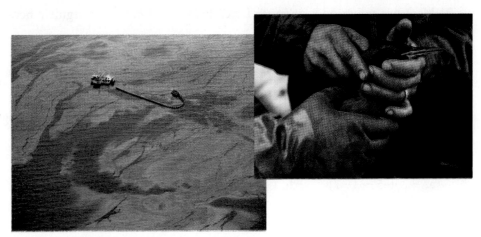

EXAMPLE

Find point *C* one-third of the way from *A* to *B*. Give the distance from *A* to *C* in centimeters.

⊢————————————————————————————————————⊣
A   *B*

▶ Solution

Have your ruler handy so that you can check the measurements. Use your calculator to check the computations.

Measuring segment *AB* shows that it is about 10.5 cm long. (Check this.) Find one-third of this length.

$$\frac{1}{3} \cdot 10.5 = \frac{10.5}{3}$$ Multiply by $\frac{1}{3}$ or divide by 3.

$$= 3.5$$ Divide.

Place a point 3.5 cm from *A*.

⊢—————————⊢————————————————————————⊣
A   3.5 cm   *B*

EXERCISES

You will need your calculator for problems **4, 5, 6, 7,** and **10.**

▶ Practice Your Skills

1. Estimate the length of each segment. Then measure and record the length to the nearest tenth of a centimeter.

a. ⊢————————————————————————⊣
   A   *B*

b. ⊢————————————————⊣
   C   *D*

c. ⊢——————————————————————⊣
   E   *F*

2. Draw a segment to fit each description.

   a. one-third of a segment 8.4 cm long

   b. three-fourths of a segment 7.6 cm long

   c. two-fifths of a segment 12.7 cm long

3. Mark two points on your paper. Label them *A* and *B*. Draw a segment between the two points.

   a. Mark a point two-thirds of the way from *A* to *B* and label it *C*.

   b. Mark a point two-thirds of the way from *C* to *B* and label it *D*.

   c. Mark a point two-thirds of the way from *D* to *A* and label it *E*.

   d. Which two points are closest together? Does it matter how long your original segment was?

4. Do these calculations. Check your results with a calculator.

   a. $-2 + 5 - (-7)$ **b.** $(-3)^2 - (-2)^3$ **c.** $\frac{3}{5} + \frac{-2}{3}$ **d.** $-0.2 \cdot 20 + 15$

Reason and Apply

[▶🖵 You'll need the program in **Calculator Note 0E** for these exercises.◀]

5. Draw a large right triangle on your paper. You can use the corner of a piece of paper or your book to help draw the right angle numbered 2. Number the vertices as shown at right.

a. Choose a point anywhere inside your triangle. This is your starting point.

b. Write the numbers 1, 2, and 3 on three small pieces of paper. Put the pieces of paper in a cup or bowl and randomly select one to choose a vertex. Measure from the starting point to the chosen vertex. Mark a new point halfway from the starting point to the vertex.

c. Let the new point be your starting point, and repeat 5b at least 20 more times.

d. Describe any pattern you see forming.

e. Run the Chaos program for problem 5 to see what happens when you plot 1000 points.

6. Draw a large square or rectangle on your paper. Number the vertices from 1 to 4 as shown at right.

a. Choose a point anywhere inside your figure. This is your starting point.

b. In order to choose a vertex to move toward, flip two different coins, such as a nickel and a penny. Use this scheme to determine the vertex:

Nickel	Penny	Vertex number
H	H	1
H	T	2
T	H	3
T	T	4

Measure the distance between your starting point and the chosen vertex. Mark a new point two-thirds of the distance to the vertex. Use this point as your new starting point. Repeat the process at least 20 more times.

c. Run the calculator simulation for problem 6 to see what happens when you plot 1000 points.

d. Describe any pattern you see forming.

e. How could you have used a die to determine which corner to move toward? What problems are there with using a die?

7. Experiment with the calculator program for each game description. For each, use the shape and fraction given. The program will start with a point inside the shape, randomly choose a vertex, and plot a point a fraction of the distance to the vertex. Describe your results and draw a sketch if possible.

a. square, $\frac{1}{2}$

b. equilateral triangle, $\frac{2}{3}$

c. square, $\frac{3}{4}$

d. right triangle, $\frac{2}{5}$

8. Suppose you are going to play a chaos game on a pentagon. Describe a process that would tell you which corner to move toward on each move.

▶ Review

9. Draw a segment that is 12 cm in length. Find and label a point that is two-thirds the distance from one of the endpoints.

10. Use a calculator to investigate the behavior of the expressions below.

i. $-0.5 \cdot \square + 3$ **ii.** $-0.5 \cdot \square + 6$ **iii.** $-0.5 \cdot \square - 9$

a. Use recursion to evaluate each expression many times and record its attractor value.

b. Describe any connections you see between the numbers in the original expressions and the attractor values.

c. Create an expression that has an attractor value of -10.

11. Look at the fractal below. The Stage 0 figure has a length of 1.

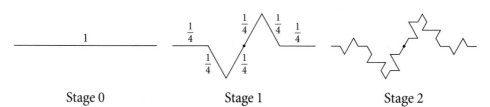

Stage 0   Stage 1   Stage 2

Complete a table like the one below for Stages 0 to 2. Use any patterns you notice to extend the table for Stages 3 and 4.

Stage number	Total length		
	Multiplication form	**Exponent form**	**Decimal form**
0	1	1^0	1
1	$6 \cdot \frac{1}{4}$	$6^1 \cdot \left(\frac{1}{4}\right)^1$	
2	$6 \cdot 6\frac{1}{4} \cdot \frac{1}{4}$		

CHAPTER 0

REVIEW

In this chapter you saw many instances of how you can start with a figure or a number, apply a mathematical rule, get a result, then apply the same rule to the result. This is called **recursion,** and it led you to find patterns in the results. When the recursive rule involved multiplication, you used an **exponent** as a shorthand way to show repeated multiplication.

Patterns in the results of recursion were often easier to see when you left them as common fractions. To add and subtract fractions, you needed a common denominator. You also needed to round decimals and measure lengths.

To **evaluate expressions** with any kind of numbers, you needed to know the **order of operations** that mathematicians use. The order is (1) evaluate what is in parentheses, (2) multiply out anything in exponent form, (3) multiply and divide as needed, and (4) add and subtract as needed. You used your knowledge of operations (add, subtract, multiply, divide, raise to a power) to write several expressions that gave the same number. Having an expression for the recursive rule helps you predict a value later in a sequence without figuring out all the values in between.

In this chapter you also got a peek at some mathematics that are new even to mathematicians, including **fractals** like the Sierpiński triangle, chaos, and strange attractors. You had to think about **random** processes and whether the long-term outcome of these processes was truly random.

EXERCISES

You will need your calculator for problems **1, 7,** and **8.**

1. Evaluate the expressions below.

a. 3^0 b. 3^1 c. 3^2 d. 3^3 e. 3^4

f. 3^8 g. 3^{12} h. $\left(\frac{1}{3}\right)^0$ i. $\left(\frac{1}{3}\right)^1$ j. $\left(\frac{1}{3}\right)^2$

k. $\left(\frac{1}{3}\right)^3$ l. $\left(\frac{1}{3}\right)^4$ m. $\left(\frac{1}{3}\right)^8$ n. $\left(\frac{1}{3}\right)^{12}$

2. Match equivalent expressions.

a. $\frac{1}{9} + \frac{1}{9} + \frac{1}{9}$ i. $\frac{35}{81}$

b. $\frac{1}{9} + \frac{1}{9} + \frac{1}{3}$ ii. $\frac{10}{27}$

c. $\frac{2}{9} + \frac{1}{9} + \frac{1}{27}$ iii. $3 \times \frac{1}{9}$

d. $\frac{4}{9} + \frac{2}{27} + \frac{3}{81}$ iv. $\frac{12}{27} + \frac{2}{27} + \frac{1}{27}$

e. $\frac{2}{81} + \frac{1}{3} + \frac{2}{27}$ v. $2 \times \left(\frac{1}{9}\right) + \frac{1}{3}$

3. Evaluate these expressions.

   a. $2 \times (24 + 12)$ **b.** $2 + 24 \times 12$ **c.** $2 - 24 + 12$

   d. $(2 + 24) \times 12$ **e.** $(2 + 24) \div 12$ **f.** $2 - (24 + 12)$

4. Write a multiplication expression equivalent to each expression below in exponent form.

   a. $\left(\dfrac{1}{3}\right)^3$ **b.** $\left(\dfrac{2}{3}\right)^4$ **c.** $(1.2)^2$ **d.** 16^5 **e.** 2^7

5. Write an addition expression that gives the combined total of shaded areas in each figure. Then evaluate the expression. The area of each figure is originally 1.

   a. 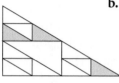   **b.** 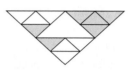   **c.** 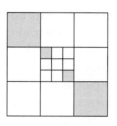   **d.**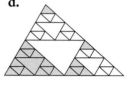

6. Draw the next stage of each fractal design below. Then describe the recursive rule for each pattern in words.

   a.

   Stage 0   Stage 1   Stage 2

   b.

   Stage 0   Stage 1   Stage 2

   c.

   Stage 0   Stage 1   Stage 2

   d.

   Stage 0   Stage 1   Stage 2

7. Look at the figures below.

 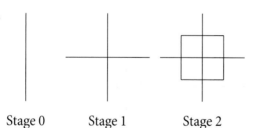 

   Stage 0   Stage 1   Stage 2

a. Complete a table like the one below.

Stage number	Total length		
	Multiplication form	Exponent form	Decimal form
0			
1			
2			

b. If you were to draw Stage 20, what expression could you write with an exponent to represent the total length? Evaluate this expression using your calculator, and round the answer to the nearest hundredth.

8. Investigate the behavior of the expression below. Use recursion to evaluate the expression several times for different starting values. You may want to record your recursions in a table. Does the expression appear to have an attractor value?

$$0.4 \cdot \square + 3$$

TAKE ANOTHER LOOK

- If a number gets larger when it is raised to a power, what kind of number is it?
- If a number gets smaller when it is raised to a power, what kind of number is it?
- What numbers stay the same when they are raised to a power?

To investigate these questions, choose positive and negative numbers, zero, and positive and negative fractions to put in the box and evaluate the expressions

$$\square^3 \text{ and } \square^4$$

You may want to use a table like the one to the right to save your results.

You know that if the denominator of a fraction increases, the value of the fraction decreases. Why is that? Are there any exceptions?

Look again at your results for the expression $\square^3$.

What would the numerator of the fraction have to be so that

$$\dfrac{\bigcirc}{\square^3} \text{ is smaller than } \square^3? \text{ Greater than } \square^3?$$

Now do the same thing with $\dfrac{\bigcirc}{\square^4}$.

Display your results in a chart.

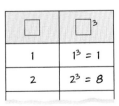

$\square$	$\square^3$
1	$1^3 = 1$
2	$2^3 = 8$

Assessing What You've Learned

KEEPING A PORTFOLIO

If you look up "assess" in a dictionary, you'll find that it means to estimate or judge the value of something. The value you've gained by the end of a chapter is not what you studied, it's what you remember and what you've gained confidence in. There are ideas you may not remember, but you will be able to reconstruct them when you meet similar situations. That's mathematical confidence.

One way to hold on to the value you've gained is to start a portfolio. Like an artist's portfolio, a mathematics portfolio shows off what you can do. It also collects work that you found interesting or rewarding (even if it isn't a masterpiece!). It also reminds you of ideas worth pursuing. The fractal designs that you figured out or invented are worth collecting and showing. Your study of patterns in fractals is a rich example of investigative mathematics that is also a good reference for how to work with fractions and exponents.

Choose one or more pieces of your work for your portfolio. Your teacher may have specific suggestions. Document each piece with a paragraph that answers these questions:

▶ What is this piece an example of?
▶ Does it represent your best work? Why else did you choose it?
▶ What mathematics did you learn or gain confidence in with this work?
▶ How would you improve the piece if you wanted to redo it?

Portfolios are an ongoing and ever-changing display of your work and growth. As you finish other chapters, don't forget to update your portfolio with new work.

Data Exploration

You are surrounded by information in many forms—in pictures, in graphs, in words, and in numbers. This information can influence what you eat, what you buy, and what you think of the world around you. This photo collage by Robert Silvers shows a lot of information.

OBJECTIVES

In this chapter you will

- interpret and compare a variety of graphs
- find summary values for a data set
- draw conclusions about a data set based on graphs and summary values
- review how to graph points on a plane
- organize and compute data with matrices

Bar Graphs and Dot Plots

This **pictograph** shows the number of CDs sold at Sheri's music store in one day. Can you tell just by looking which *type* sold the most? How many CDs of this type were sold?

This specific information, the kind Sheri may later use to make decisions, is sometimes called **data.** You use data every day when you answer questions like "Where is the cheapest place to buy a can of soda?" or "How long does it take to walk from class to the lunchroom?"

In this lesson you interpret and create graphs. Throughout the chapter, you'll learn more ways to organize and represent data.

I've always felt rock and roll was very, very wholesome music.

ARETHA FRANKLIN

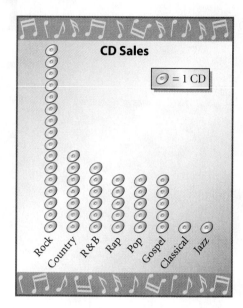

CD Sales

⊙ = 1 CD

Rock Country R&B Rap Pop Gospel Classical Jazz

EXAMPLE

Joaquin's school posts a pictograph showing how many students celebrate their birthdays that month. Here is part of this pictograph. Create a table of data and a **bar graph** from the pictograph.

Number of Birthdays in Each Month

Birth month

January

February

March

April

🚶 = 5 students

▶ **Solution**

This table lists the birthday data.

Number of Birthdays in Each Month

Jan	Feb	Mar	Apr
15	10	0	25

In the pictograph, there are three figures for January. Each figure represents five students. So 3×5 gives 15.

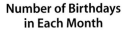

This bar graph shows the same data. The height of a bar shows the total in that **category,** in this case, a particular month. You use the *scale* on the *vertical axis* to measure the height of each bar. The vertical axis extends slightly past the greatest number of birthdays in any one month, so the data does not go beyond the scale.

Bar graphs gather data into categories and make it easy to present a lot of information in a compact form. In a bar graph you can quickly compare the quantities for each category.

In a **dot plot** each item of numerical data is shown above a number line or *horizontal axis*. Dot plots make it easy to see gaps and clusters in the data set, as well as how the data **spreads** along the axis.

In the investigation you'll gather and plot data about pulse rates. People's pulse rates vary, but a healthy person at rest usually has a pulse rate between certain values. A pulse that is too fast or too slow could tell a paramedic that a person needs immediate care.

Number of Birthdays in Each Month

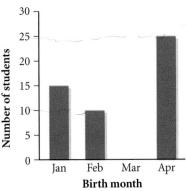

Investigation
Picturing Pulse Rates

You will need

- a watch or clock with a second hand

Use the Procedure Note to learn how to take your pulse. Practice a few times to make sure you have an accurate reading.

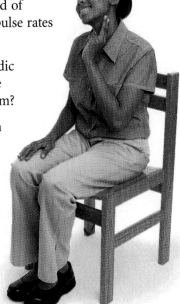

Procedure Note

How to Take Your Pulse
1. Find the pulse in your neck.
2. Count the number of beats for 15 seconds.
3. Multiply the number of beats by 4 to get the number of beats per minute (bpm). This number is your pulse rate.

Step 1 | Start with pulse-rate data for 10 to 20 students.

Step 2 | Find the **minimum** (lowest) and **maximum** (highest) values in the pulse-rate data. The minimum and maximum describe the spread of the data. For example, you could say, "The pulse rates are between 56 and 96 bpm."

Based on your data, do you think a paramedic would consider a pulse rate of 80 bpm to be "normal"? What about a pulse rate of 36 bpm?

Step 3 | Construct a number line with the minimum value near the left end. Select a scale and label equal **intervals** until you reach the maximum.

Step 4 | Put a dot above the number line for each item of your data. When a data value occurs more than once, stack the dots.

Here is an example for the data set {56, 60, 60, 68, 76, 76, 96}. Your line will probably have different minimum and maximum values.

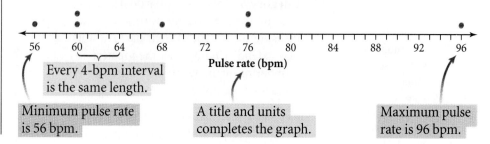

Every 4-bpm interval is the same length.

Minimum pulse rate is 56 bpm.

Pulse rate (bpm)

A title and units completes the graph.

Maximum pulse rate is 96 bpm.

The **range** of a data set is the difference between the maximum and minimum values. The data on the example graph have a range of 96–56 or 40 bpm.

Step 5 What is the range of your data? Suppose a paramedic says normal pulse rates have a range of 12. Is this range more or less than your range? What information is the paramedic not telling you when she mentions the range of 12?

Step 6 For your class data, are there data values between which a lot of points cluster? Are there any values that occur most often? What do you think these clusters would tell a paramedic? Do you think you could use your class data to say what is a "normal" pulse for all people? Why or why not?

History
● ─ **CONNECTION** ● ─

The word "statistics" was first used in the late eighteenth century to mean a branch of political science dealing with the collection of data about a state or country.

Statistics is a word used many ways. We sometimes refer to data we collect and the results we get as "statistics." For example, you could collect pulse-rate statistics from thousands of people and then determine a "normal" pulse-rate. The single value you calculate to be "normal" could also be called a statistic. You'll learn more about statistics and their usefulness in this chapter.

EXERCISES

Practice Your Skills

1. Angelica has taken her pulse 11 times in the last six hours. Her results are 69, 92, 64, 78, 80, 82, 86, 93, 75, 80, and 80 beats per minute. Find the maximum, minimum, and range of the data.

2. The table shows the percentages of the most common elements found in the human body. Make a bar graph to display the data.

Elements in the Human Body

Oxygen	Carbon	Hydrogen	Nitrogen	Calcium	Phosphorus	Other
65%	18%	10%	3%	2%	1%	1%

3. Use this bar graph to answer each question.

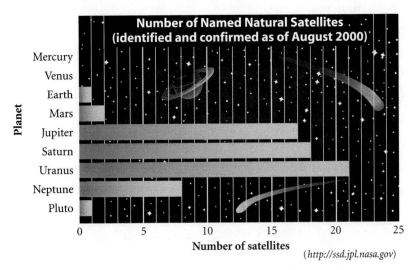

a. Which planet has the most satellites?

b. What does this graph tell you about Mercury and Venus?

c. How many more satellites does Jupiter have than Neptune?

d. How many times as many satellites does Saturn have than Mars?

4. This table shows how long it takes the students in one of Mr. Matau's math classes to get to school.

a. Construct a dot plot to display the data. Your number line should show time in minutes.

b. How many students are in this class?

c. What is the combined time for Mr. Matau's students to travel to school?

d. What is the average travel time for Mr. Matau's students to get to school?

Travel Time to School

Time (min)	Number of students
1	2
3	2
5	6
6	1
8	6
10	7
12	3
14	2
15	1

Reason and Apply

5. This graph is a dot plot of Angelica's pulse-rate data.

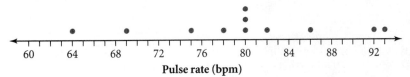

a. What pulse rate appeared most often?

b. What is the range of Angelica's data?

c. If your class followed the procedure directions for the investigation, your pulse rates should be multiples of four. Angelica's are not. How do you think she took her pulse?

d. How would your data change if everyone in your class had taken his or her pulse for a full minute? How would the dot plot be different?

e. Do you think medical professionals measure pulse rates for 1 minute or 15 seconds? Why?

6. Each graph below displays information from a recent class survey. Determine which graph best represents each description:

a. number of people living in students' homes

b. students' heights in inches

c. students' pulse rates in beats per minute

d. number of working television sets in students' homes

i.
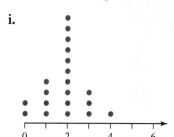

ii.
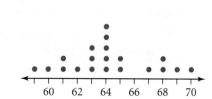

iii.
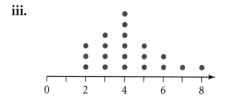

iv.
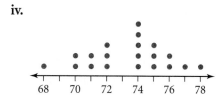

7. Reporter "Scoop" Presley of the school paper polled 20 students about their favorite type of music—classical, pop/rock, R&B, rap, or country. He delivered his story to his editor, Rose, just under the deadline. Rose discovered that Scoop, in his haste, had ripped the page with the bar graph showing his data. The vertical scale and one category were missing! Unfortunately, the only thing Scoop could remember was that three students had listed R&B as their favorite type. Reconstruct the graph so that it includes the vertical axis and the missing category with the correct count.

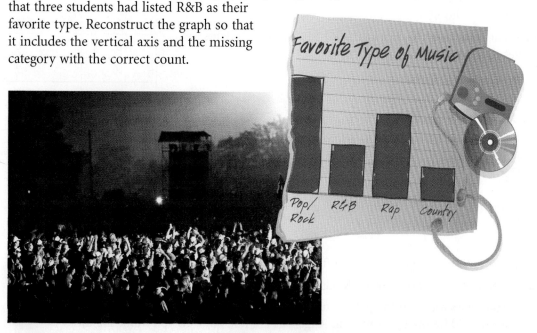

8. Suppose you collect information on how each person in your class gets to school. Would you use a bar graph or a dot plot to show the data? Explain why you think your choice would be the better graph for this information.

Review

9. Rewrite each of these multiplication expressions using exponents.

   a. $10 \times 10 \times 10 \times 10$

   b. $2 \cdot 2 \cdot 2 \cdot 5 \cdot 5 \cdot 5 \cdot 5 \cdot 5 \cdot 5$

   c. $\dfrac{3^2(3^4)}{8(8)(8)}$

10. Use the order of operations to evaluate each expression.

   a. $7 + (3 \cdot 2) - 4$ **b.** $8 + 2 - 4 \cdot 12 \div 16$ **c.** $1 - 2 \cdot 3 + 4 \div 5$

   d. $1 - (2 \cdot 3 + 4) \div 5$ **e.** $1^2 \cdot 3 + (4 \div 5)$

11. The early Egyptian *Ahmes Papyrus* (1650 B.C.) shows how to use a doubling method to divide 696 by 29. (George Joseph, *The Crest of the Peacock,* 2000, pp. 61–66)

Doubles of 29	58	116	**232**	**464**	928
Doubles of 1	2	4	**8**	**16**	32

Double the divisor (29) until you go past the dividend (696). Find doubles of 29 that sum to 696: $232 + 464 = 696$. Then sum the corresponding doubles of 1: $8 + 16 = 24$. So, 696 divided by 29 is 24.

   a. Divide 4050 by 225 with this method.

   b. Divide 57 by 6 with this method. (Hint: Use doubles and halves.)

IMPROVING YOUR **REASONING** SKILLS

Janet and JoAnn used the same data set of high and low temperatures for cities in a month in early spring. Which graph shows the low temperatures better? Is either graph better for showing the differences between high and low temperatures?

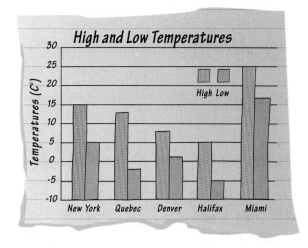

Janet's graph

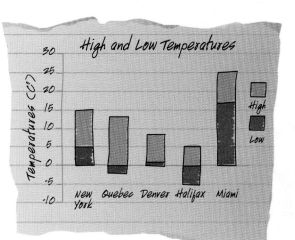

JoAnn's graph

Summarizing Data with Measures of Center

"Americans watch an average of four hours of television each day."

"Half of the participants polled had five or more people living in their home."

"When asked how many colleges they applied to, the most frequent answer given by graduating seniors was three."

Statements like these try to say what is typical. They summarize a lot of data with one number called a **measure of center.** The first statement uses the **mean,** or *average*. The second statement uses the **median,** or middle value. The third statement uses the **mode,** or most frequent value.

Investigation
Making "Cents" of the Center

You will need

- 20 to 30 pennies

In this investigation you learn to find the mean, median, and mode of a data set. You may have learned about these three measures of center in a past mathematics class.

Step 1 | Sort your pennies by mint year. Make a dot plot of the years.

Step 2 | Now put your pennies in a single line from oldest to newest. Find the median, or middle value. Does the median have to be a whole-number year? Why or why not? Would you get the same median if you arranged the pennies from newest to oldest?

Step 3 | On your dot plot, circle the dot (or dots) that represent the median. Write the value you got in Step 2 beside the circled dot or dots, and label the value "median."

Step 4 | Now stack pennies with the same mint year. The year of the tallest stack is called the mode. If there are two tall stacks, your data set is **bimodal.** If every stack had two

> ### Procedure Note
>
> **Finding the Median**
> If you have an odd number of pennies, the median is the year on the middle penny. If you have an even number of pennies, add the dates on the two pennies closest to the middle and divide by two.

pennies in it, you might say there is "no mode" because no year occurs most often. How many modes does your data set have? What are they? Does your mode have to be a whole number?

Step 5 Draw a square around the year corresponding to the mode(s) on the number line of your dot plot. Label each value "mode."

Step 6 Find the sum of the mint years of all your pennies and divide by the number of pennies. The result is called the mean. What is the mean of your data set?

Step 7 Show where the mean falls on your dot plot's number line. Draw an arrowhead under it and write the number you got in Step 6. Label it "mean."

Step 8 Now enter your data into a calculator list, and use your calculator to find the mean and the median. Are they the same as what you found using pencil and paper? [▶  See **Calculator Notes 1A and 1B.** Refer to **Calculator Note 1J** to check the settings on your calculator.◀]

Save the dot plot you created. You will use it in Lesson 1.3.

Measures of Center

Mean
The mean is the sum of the data values divided by the number of data items. The result is often called the average.

Median
For an odd number of data items, the median is the middle value when the data values are listed in order. If there is an even number of data items, then the median is the average of the two middle values.

Mode
The mode is the data value that occurs most often. Data sets can have two modes (bimodal) or more. Some data sets have no mode.

Each measure of center has its advantages. The mean and the median may be quite different, and the mode, if it exists, may or may not be useful. You will have to decide which measure is most meaningful for each situation.

EXAMPLE

This data set shows the number of people who attended a movie theater over a period of 16 days

{14, 23, 10, 21, 7, 80, 32, 30, 92, 14, 26, 21, 38, 20, 35, 21}

a. Find the measures of center.

b. The theater's management wants to compare its attendance to that of other theaters in the area. Which measure of center best represents the data?

► Solution

a. The mean is approximately 30 people.

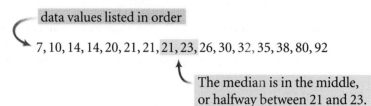

$$\frac{14+23+10+21+7+80+32+30+92+14+26+21+38+20+35+21}{16} = 30.25$$

number of data values   the mean

The median is 22 people.

data values listed in order

7, 10, 14, 14, 20, 21, 21, 21, 23, 26, 30, 32, 35, 38, 80, 92

The median is in the middle, or halfway between 21 and 23.

The most frequent value, 21 people, is the only mode.

b. To determine which measure of center best summarizes the data, look for patterns in the data and look at the shape of the graph.

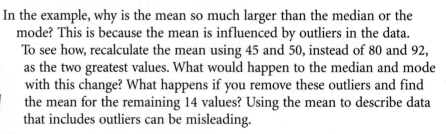

Attendance (number of people)

The dot plot clearly shows that, except for two items, the data are clustered between 7 and 38. The items with values 80 and 92 are far outside the range of most of the data and are called **outliers.**

Either the median, 22, or the mode, 21, could be used by the management to compare this theater's attendance to that of other theaters. The management could say, "Attendance was about 21 or 22 people per day over a 16-day period." The mean, 30, is too far to the right of most of the data to be the best measure of center. Yet the theater's management might prefer to use the mean of 30 in an advertisement. Why?

In the example, why is the mean so much larger than the median or the mode? This is because the mean is influenced by outliers in the data. To see how, recalculate the mean using 45 and 50, instead of 80 and 92, as the two greatest values. What would happen to the median and mode with this change? What happens if you remove these outliers and find the mean for the remaining 14 values? Using the mean to describe data that includes outliers can be misleading.

EXERCISES

▶ ## Practice Your Skills

1. Find the mean, median, and mode for each data set.

   a. {1, 5, 7, 3, 5, 9, 6, 8, 10}   b. {6, 1, 3, 9, 2, 7, 3, 4, 8, 8}

   c. {12, 6, 11, 7, 18, 5, 2, 21}   d. {10, 10, 20, 20, 20, 25}

2. Find the mean, median, and mode for each dot plot.

   a.

   b.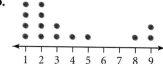

3. Students were asked how many pets they had. Their responses are shown in the dot plot below.

   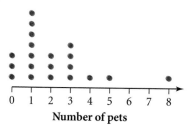

   Number of pets

   a. How many students were surveyed?

   b. What is the range of answers?

   c. What was the most common answer?

4. This graph gives the lift heights and vertical drops of the five tallest roller coasters at Cedar Point Amusement Park in Ohio.

   a. Find the mean, median, and mode for the lift heights.

   b. Find the mean, median, and mode for the vertical drops.

This is the first hill of the Mean Streak at Cedar Point.

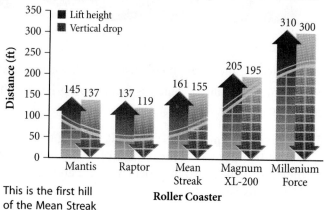

(*www.cedarpoint.com*)

5. If you purchase 16 grocery items at an average cost of $1.14, what is your grocery bill? Explain how you found the total bill.

Reason and Apply

6. Tsunamis are ocean waves caused by earthquakes. The heights of the 20 tallest tsunamis on record are listed in the table.

Tallest Tsunamis

Location of source	Year	Height (ft)	Location of source	Year	Height (ft)
Papua New Guinea	1998	45	Nankaido (Japan)	1946	20
Mindoro (Philippines)	1994	49	Kii (Japan)	1944	25
Sea of Japan	1983	49	Sanriku (Japan)	1933	93
Indonesia	1979	32	East Kamchatka (Russia)	1923	66
Celebes Sea	1976	98	South Kuril Island (Russia)	1918	39
Alaska	1964	105	Sanriku (Japan)	1896	98
Chile	1960	82	Sunda Strait (Indonesia)	1883	115
Aleutian Islands (Alaska)	1957	52	Chile	1877	75
Kamchatka (Russia)	1952	60	Chile	1868	69
Aleutian Islands (Alaska)	1946	105	Hawaii	1868	66

(*www.pmel.noaa.gov*)

(*The Hollow of the Deep-Sea Wave Off Kanagawa* by Katsushika Hokusai/Minneapolis Institute of Art Acc. No. 74.1.230)

a. Calculate the mean, median, and mode for the height data.

b. Which measure of center is most appropriate for the height data? Explain your reasoning.

7. The first three members of the stilt-walking relay team finished their laps of the race with a mean time of 53 seconds per lap. What mean time for the next two members will give an overall team mean of 50 seconds per lap?

8. Noah scored 88, 92, 85, 65, and 89 on five tests in his history class. Each test was worth 100 points. Noah's teacher usually uses the mean to calculate each student's overall score. Explain why the median is a better measure of center for Noah's work on the tests.

9. This table gives information about ten of the largest saltwater fish species in the world. The approximate mean weight of these fish is 1527.4 pounds.

a. Explain how to use the mean to find an approximate total weight for these ten fish. What is the total weight?

b. The median weight of these fish is about 1449 pounds. Assuming that no two weights are the same, what does the median tell you about the individual weights of the fish?

c. The range of weights is 1673 pounds, and the minimum weight is 991 pounds. What is the weight of the great white shark, the largest fish caught?

Largest Saltwater Fish Species

Species	Location where caught
Swordfish	Chile
Bluefin tuna	Nova Scotia
Great white shark	South Australia
Atlantic blue marlin	Brazil
Greenland shark	Norway
Black marlin	Peru
Hammerhead shark	Florida
Tiger shark	California
Pacific blue marlin	Hawaii
Mako shark	Mauritius

(*The Top 10 of Everything 2001*, p. 41)

10. Create a set of data that fits each description.

a. The mean age of a family is 19 years, and the median age is 12 years. There are five people in the family.

b. Six students in the Mathematics Club compared their family sizes. The mode was five people, and the median was four people.

c. The points scored by the varsity football team in the last seven games have a mean of 20, a median of 21, and a mode of 27 points.

11. This data set represents the ages of 20 of the highest-paid athletes in the United States as of December 31, 1998. (*www.infoplease.com*)

{35, 29, 29, 23, 46, 26, 25, 36, 69, 30, 28, 37, 37, 27, 38, 43, 29, 33, 30, 29}

a. Make a dot plot of the survey results.

b. Give the mean, median, and mode for the data.

c. Which measure of center best summarizes the data? Explain your reasoning.

▶ Review

12. Fifteen students gave their ages in months.

168   163   142   163   165   164   167   153   149   173   163   179   155   162   162

a. Would you use a bar graph, pictograph, or dot plot to display these data? Explain your reasoning.

b. Create the graph you chose in 12a.

13. Use this segment to measure or calculate in 13a–c.

a. What is the length in centimeters of the segment?

b. Draw a segment that is $\frac{2}{3}$ as long as this segment. What is the length of your new segment?

c. Draw a segment that is $\frac{1}{5}$ as long as the original segment. What is the length of this new segment?

Five-Number Summaries and Box Plots

To talk sense is to talk quantities. It is no use saying a nation is large—how large?

ALFRED NORTH WHITEHEAD

Michael Jordan of the Chicago Bulls scored the most points in the 1997–98 season in the National Basketball Association (NBA). In fact he scored almost three times as many points as the next highest scorer on his team. Does any measure of center give a complete description of how the Bulls scored as a team? A **five-number summary** could give a better picture. It uses five boundary points: the minimum and the maximum, the median (which divides the data set in half), the **first quartile** (the median of the first half), and the **third quartile** (the median of the second half).

Points Scored by Chicago Bulls Players Who Played Over 40 Games (1997–98 Season)

Chicago Bulls	Total points scored	Chicago Bulls	Total points scored
Michael Jordan	2357	Steve Kerr	376
Toni Kukoc	984	Dennis Rodman	375
Scottie Pippen	841	Randy Brown	288
Ron Harper	764	Jud Buechler	198
Luc Longley	663	Bill Wennington	167
Scott Burrell	416		

Michael Jordan

(*www.nba.com*)

EXAMPLE

Find the five-number summary for the number of points scored by the Chicago Bulls during the 1997–98 season (use the table above).

▶ **Solution**

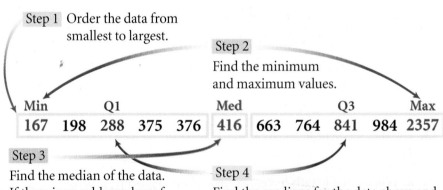

Step 1 Order the data from smallest to largest.

Step 2 Find the minimum and maximum values.

Min		Q1			Med			Q3		Max
167	198	288	375	376	416	663	764	841	984	2357

Step 3
Find the median of the data. If there is an odd number of values, consider the median value removed so you have an equal number of data on either side.

Step 4
Find the medians for the data above and below the median. The lower median is the first quartile, labeled Q1, the upper median is the third quartile, labeled Q3.

The five-number summary is 167, 288, 416, 841, 2357. This is the minimum, first quartile, median, third quartile, and maximum in order of smallest to largest. The first quartile, median, and third quartile divide the data into four equal groups. Each of the four groups has the same number of values, in this case two.

A five-number summary helps you better understand the spread of the data along the number line. It also helps you compare different sets of data. A **box plot** is a visual way to show a five-number summary. This box plot shows the data for the Bulls.

Points scored

Notice how the box plot shows the spread of data and Michael Jordan as an extreme outlier. Can you see why this type of graph is sometimes called a **box-and-whisker plot**? Can you find the five-number summary values in this box plot? In the next investigation you'll see how to use the five-number summary to construct a box plot.

Investigation
Pennies in a Box

You will need

- your dot plot from Investigation: Making "Cents" of the Center

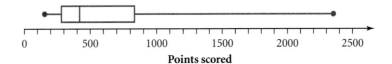

Year

The illustrations are examples only. Your box plot should look different.

Step 1   Find the five-number summary values for your penny data.

Step 2   Place a clean sheet of paper over your dot plot and trace the number line. Using the same scale will help you compare your dot plot and box plot.

Step 3   Find the median value on your number line and draw a short vertical line segment just above it. Repeat this process for the first quartile and the third quartile.

Step 4   Place dots above your number line to represent the minimum and maximum values from your dot plot.

This man is counting and bagging pennies at the U.S. Mint in Denver, Colorado.

| Step 5 | Draw a rectangle with ends at the first and third quartiles. This is the "box." Finally, draw horizontal segments that extend from each end of the box to the minimum and maximum values. These are the "whiskers." |  |

Step 6 | Compare your dot plot and box plot. On which graph is it easier to locate the five-number summary? Which graph helps you to see the spread of data better?

Remember that the first quartile, median, and third quartile divide the data items into four equal groups. Although each section has the same number of data items, your boxes and whiskers may vary in length. Some box plots will be more *symmetric* than others. When would that happen?

Step 7 | Clear any old data from your calculator and enter your penny mint years into list L₁.

Step 8 | Draw a calculator box plot. [▶ 🖵 Follow the procedure outlined in **Calculator Note 1C.**◀] Does your calculator box plot look equivalent to the plot you drew by hand?

Step 9 | Use the trace function on your calculator. What values are displayed as you trace the box plot? Are the five-number summary values the same as those you found before?

The difference between the first quartile and third quartile is the **interquartile range, or IQR.** Like the range, the interquartile range helps describe the spread in the data.

Step 10 | Complete this investigation by answering these questions:

a. What are the range and IQR of your data?

b. How many pennies fall between the first and third quartiles of the graph? What fraction of the total number of pennies is this number? Will this fraction always be the same? Explain.

c. Under what conditions will exactly $\frac{1}{4}$ of the pennies be in each whisker of the box plot?

Box plots are a good way to compare two data sets. These box plots summarize the final test scores for two of Ms. Werner's algebra classes. Use what you have learned to compare these two graphs. Which class has the greater range of scores? Which has the greater IQR? In which class did the greatest fraction of students score above 80?

Notice that neither graph shows the number of students in the class nor the individual scores. If knowing each data value is important, then a box plot is not the best choice to display your data.

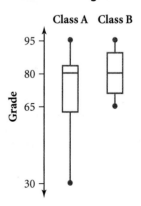

Ms. Werner's Algebra Test

EXERCISES

Practice Your Skills

1. Find the five-number summary for each data set.

   a. {5, 5, 8, 10, 14, 16, 22, 23, 32, 32, 37, 37, 44, 45, 50}

   b. {10, 15, 20, 22, 25, 30, 30, 33, 34, 36, 37, 41, 47, 50}

   c. {44, 16, 42, 20, 25, 26, 14, 37, 26, 33, 40, 26, 47}

   d. {47, 43, 35, 34, 32, 21, 17, 16, 11, 9, 5, 5}

2. Sketch each graph below on your own paper.

   i.

   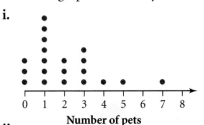

   Number of pets

   ii.

   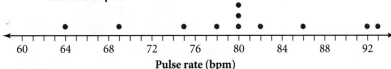

   Pulse rate (bpm)

   a. Circle the points that represent the five-number summary values. If two data points are needed to calculate the median, first quartile, or third quartile, draw a circle around both points.

   b. List the five-number summary values for each data set.

3. Give the five-number summary and create a box plot for the listed values.

   {2, 6, 4, 9, 1, 6, 4, 7, 2, 8, 5, 6, 9, 3, 6, 7, 5, 4, 8}

4. Which data set matches this box plot? (More than one answer may be correct.)

   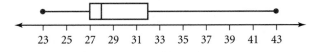

   a. {23, 25, 26, 28, 28, 28, 28, 30, 31, 33, 41, 43}

   b. {23, 23, 24, 25, 26, 27, 29, 30, 31, 33, 41, 43}

   c. {23, 27, 28, 28, 33, 43}

   d. {23, 27, 28, 28, 29, 32, 43}

5. Check your vocabulary by answering these questions.

   a. How does the term *quartile* relate to how data values are grouped when using a five-number summary?

   b. What is the name for the difference between the minimum and maximum values in a five-number summary?

   c. What is the name for the difference between the third quartile and first quartile in a five-number summary?

   d. How are outliers of a data set related to the whiskers of its box plot?

Reason and Apply

6. Stu had a mean score of 25.5 on four 30-point papers in English. He remembers three scores: 23, 29, and 27.

   a. Estimate the fourth score without actually calculating it.

   b. Check your estimate by calculating the fourth score.

   c. What is the five-number summary for this situation?

   d. Does it make sense to have a five-number summary for this data set? Explain why or why not.

7. **APPLICATION** The Toronto Raptors basketball team came in last in their division in the 1997–98 season.

**Points Scored by Toronto Raptors Players
Who Played Over 40 Games (1997–98 Season)**

Toronto Raptors	Total points scored
Kevin Willis	1305
Doug Christie	1287
John Wallace	1147
Chauncey Billups	893
Charles Oakley	711
Dee Brown	658
Gary Trent	630
Reggie Slater	625
Tracy McGrady	451
Oliver Miller	401
Alvin Williams	324
John Thomas	151

(*www.nba.com*)

   a. The five-number summary for the Chicago Bulls is 167, 288, 416, 841, 2357. (See the table and example at the beginning of this lesson.) Find the five-number summary for the Raptors.

   b. Find and compare the measures of center for the Bulls and the Raptors.

   c. Decide which measure of center best describes each team. Explain your answer.

   d. These box plots compare the points scored by the Chicago Bulls players to the points scored by the Toronto Raptors players. Compare the two teams' performance based on what you see in the graphs.

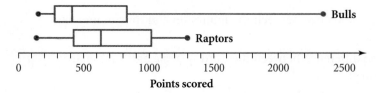

e. Remove Michael Jordan's points from the data table for the Chicago Bulls and make a new box plot. How does this new box plot compare to the original box plot for the Bulls? How does it compare to the box plot for the Raptors?

8. APPLICATION This table lists median weekly earnings of full-time workers by occupation and gender for 2000.

Median Weekly Earnings, 2000

Occupation	Men	Women
Managerial and professional specialty	$999	$697
Executive, administration, and managerial	995	684
Professional specialty	1001	708
Technical, sales, and admin. support	653	451
Technicians and related support	754	539
Sales occupations	683	379
Administrative support including clerical	552	455
Service occupations	405	313
Protective service	636	470
Precision production, craft, and repair	622	439
Mechanics and repairers	645	588
Operators, fabricators, and laborers	492	353
Machine operators, assemblers, and inspectors	498	353
Transportation and material moving	555	421
Handlers, equipment cleaners, helpers, and laborers	401	329
Farming, forestry, and fishing	342	288

(*http://stats.bls.gov*)

a. Make two box plots, one for men's salaries and one for women's salaries, above the same number line. Use them to compare the two data sets. Use the terms you have learned in this chapter.

b. What does the data tell you about women's and men's wages for the same type of work in 2000?

c. Do the box plots help you identify characteristics of the data better than the table does? Are there any aspects of the data that are better seen in the table?

d. How could you use the box plots to explain the slogan "Equal pay for equal work"?

During World War II many women took nontraditional jobs to support war industries. Some fought for and achieved equal pay for equal work.

9. These box plots display the recorded lengths of the ten longest snakes and the recorded running speeds for the ten fastest mammals.

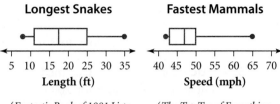

Longest Snakes

Length (ft)

(*Factastic Book of 1001 Lists*, 1999, p. 41)

Fastest Mammals

Speed (mph)

(*The Top Ten of Everything 2000*, p. 59)

a. The longest snake in the world is believed to be the reticulated python. What is the length of this snake?

b. The fifth-longest snake is the king cobra. Can you determine its length from the box plot? Explain.

c. The fastest mammal in the world is believed to be the cheetah. What is the fastest recorded speed for a cheetah?

d. Explain what each box plot tells you about the spread of the data.

e. Could these two box plots be constructed above the same number line? Explain.

f. The fifth- and sixth-fastest mammals (Grant's gazelle and Thomson's gazelle) have been recorded at the same maximum speed. About how fast can they run?

10. As a general rule, if the distance of a data point from the nearest end of the box is more than 1.5 times the length of the box (or IQR), then it qualifies as an outlier.

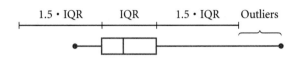

$1.5 \cdot IQR$ IQR $1.5 \cdot IQR$ Outliers

a. The five-number summary for the number of points scored by the Chicago Bulls players is 167, 288, 416, 841, 2357. What is 1.5 times the interquartile range?

b. What is the value of the first quartile less 1.5 times the interquartile range?

c. What is the value of the third quartile plus 1.5 times the interquartile range?

d. The values you found in 10b and 10c are the limits of outlier values. Identify any Chicago Bulls players who are outliers.

▶ Review

11. Create a data set for a family of five with a mean age of 22 years and a median age of 14.

12. The majority of pets in the United States are cats.

a. How many pet cats are there in the United States? Use the pictograph below.

b. How many fewer dogs are there?

c. If parakeets (11 million) were added to the pictograph below, how many pawprints would be drawn to represent parakeets?

Pets in the US

🐾 = 2 million

(*The Top 10 of Everything 2001*, p. 45)

Histograms and Stem-and-Leaf Plots

This dot plot provides information about the amount of pocket money 16 students had with them on a given day. If you collect similar data for all the students in your school, you probably wouldn't want to make a dot plot because you would have too many dots. A box plot could be used to show the spread of the data set, but it wouldn't show whether you polled 16 or 600 students.

Pocket Money

Amount of pocket money ($)

A **histogram** is related to a dot plot but more useful when you have a large data set. Histograms use columns to display how data are distributed and reveal clusters and gaps in the data. Unlike bar graphs, which use categories, the data for histograms must be numeric and ordered along the horizontal axis.

This histogram shows the same data as the dot plot.

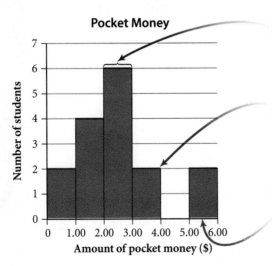

Pocket Money

The width of each column represents an interval of $1.00. These intervals are also called **bins.**

The height of each bin shows the number of students whose money falls in that interval. This is the **frequency** of each bin.

Boundary values fall in the bin to the right. That is, this bin is $5.00 to $5.99.

All **bins** in a histogram have the same width, and the columns are drawn next to each other without any space between them. A gap between columns means that there is an interval with no data items that have those values. You can't name the individual values represented in a histogram, so a histogram summarizes data. You *can* determine the total number of data items. The sum of the bar heights in the histogram above tells you the total number of students.

When you construct a histogram, you have to decide what bin width works best. These histograms show the same data set but use different bin widths. Use each of the three histograms presented to answer the question "Most students had pocket money between which values?" How do the bin widths of each histogram affect your answers?

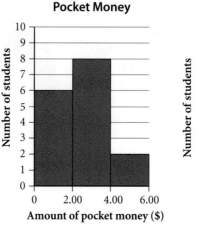

Pocket Money

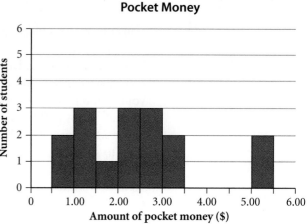

Pocket Money

Too many bins may create an information overload. Too few bins may hide some features of the data set. As a general rule, try to have five to ten bins. Of course, there are exceptions.

Investigation
Hand Spans

In this investigation you'll collect hand-span measurements and make a histogram. You'll organize the data using different bin widths and compare the results to a box plot.

Step 1	Measure your hand span in centimeters. Post your hand-span measurement in a classroom data table.
Step 2	Mark a zero point on your graph paper. Draw a horizontal axis to the right and a vertical axis up from this point.
Step 3	Scale the horizontal axis for the range of your data. Clearly divide this range into 5 to 10 equal bins. Label the boundary values of each bin.
Step 4	Count the data items that will fall into each bin. For example, in a bin from 20 cm to 22 cm, you would count all the items with values of 20.0, 20.5, 21.0, and 21.5. Items with a value of 22.0 are counted in the next bin.
Step 5	Scale the vertical axis for **frequency,** or count, of data items. Label it from zero to at least the largest bin count.
Step 6	Draw columns showing the correct frequency of the data items for each bin.

| Step 7 | Enter your hand-span measurements into list L₁ of your calculator. Create several versions of the histogram using different bin widths. [▶ 🖥 See **Calculator Note 1D** for instructions on creating a histogram.◀] |

| Step 8 | How did you select a bin width for your graph-paper histogram? Now that you have experimented with calculator bin widths, would you change the bin width of your paper graph? Write a paragraph explaining how to pick the "best" bin width. |

| Step 9 | Add a box plot of your hand-span data to both your graph-paper and calculator versions of the histogram. What information does the histogram provide that the box plot does not? Consider gaps in the data and the shape of the histogram. |

A graph that is often more useful for small data sets is a **stem plot,** or **stem-and-leaf plot.** A stem plot, like a dot plot, displays each individual item. But, like a histogram, data values are grouped into intervals or bins. You need a **key** to interpret a stem plot.

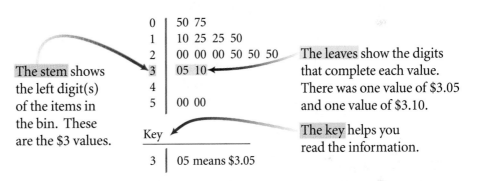

The stem shows the left digit(s) of the items in the bin. These are the $3 values.

The leaves show the digits that complete each value. There was one value of $3.05 and one value of $3.10.

The key helps you read the information.

```
0 | 50 75
1 | 10 25 25 50
2 | 00 00 00 50 50 50
3 | 05 10
4 |
5 | 00 00

Key
3 | 05 means $3.05
```

EXAMPLE

Use the stem-and-leaf plot to draw a histogram of the Canadian universities data.

▶ **Solution**

First, find the range of the data: 1995 − 1852 = 143 years. Then consider a "friendly" bin width. For a bin width of 10 years you'd need at least 15 bins, which is too many. For a bin width of 25 years you'd need 6 bins, which is more manageable.

You could start the first bin with the minimum value, 1852. However, it may be better to round down to 1850 so that each boundary will be a multiple of 25, and the centuries 1900 and 2000 will fall on boundaries.

Establishment Dates for Canadian Universities (1850–2000)

```
185 | 2  2  3  7
186 | 3  5
187 | 1  3  3  6  7  8
188 | 7  7
189 | 0  9
190 | 0  5  6  7  7  8
191 | 0  0  0  1  3  3  7  9
192 | 1  5
193 | 6
194 | 2  5  8
195 | 4  4  7  9
196 | 0  2  3  3  3  4  4  5  5  5  7  8  9  9
197 | 0  0  4  4  4  6  8  9
198 | 2
199 | 2  4  5

Key
197 | 3  means 1973
```

(*www.aucc.ca*)

Count the frequency for each bin. You could use a table like this:

Bin	1850–1874	1875–1899	1900–1924	1925–1949	1950–1974	1975–1999
Frequency	9	7	15	5	23	7

Then create your histogram.

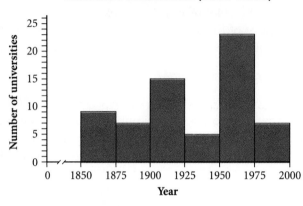

Establishment Dates for Canadian Universities (1850–2000)

This histogram may or may not use the best bin width. Use your calculator to experiment with other bin widths for the data. Which bin width do you think highlights the spread of the data? Which do you feel highlights the clustering of data? Does one bin width show an increasing trend?

EXERCISES

You will need your calculator for problem **6.**

▶ Practice Your Skills

1. Marketing consultant Maive Wishnev surveyed people attending matinee and evening ballet performances. She made the two graphs below showing the ages of attendees whom she surveyed.

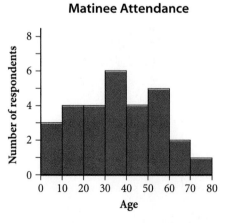

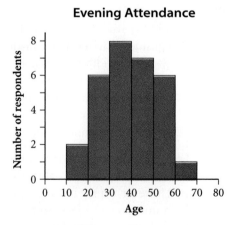

a. How many people did she survey at each performance?

b. At which performance did the ages of survey respondents vary more?

c. How many children younger than 10 responded to the survey at the evening performance?

d. What can you say about the number of 15-year-olds who were surveyed at the matinee?

2. Thirty students were asked at random to pick a number from 0 to 20. Here are the results:

{12, 7, 8, 3, 5, 7, 10, 13, 7, 10, 2, 1, 11, 12, 17, 4, 11, 7, 6, 18, 14, 17, 11, 9, 1, 12, 10, 12, 2, 15}

a. Construct two histograms for the data. Use different bin widths for each.

b. Do you notice any patterns in the data? What do the histograms tell you about the numbers that the students tend to choose?

c. Give the five-number summary for the data and construct a box plot.

d. Give the mode(s) for the data.

3. This box plot and histogram reflect the female life expectancy for several countries in 1999.

Women's Life Expectancy

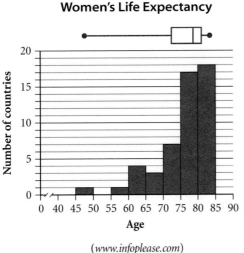

(*www.infoplease.com*)

a. How many countries are represented?

b. The right whisker of the box plot is very short. What does this mean?

c. How many countries had female life expectancies of less than 60 years?

d. How can you tell that no country had a female life expectancy of greater than 85 years?

4. Redraw the histogram for Exercise 3, changing the bin width from 5 to 10.

5. Suppose some class members measure the lengths of their ring fingers. The measurements are 6.5, 6.5, 7.0, 6.0, 7.5, 7.0, 8.5, and 7.0 centimeters.

a. Identify the minimum, maximum, and range values of the data.

b. Create a stem plot of these data values.

Reason and Apply

6. The histogram displays the passenger-car information listed in the table.

Top 20 Selling Passenger Cars in the United States in 1998

Car	Number sold	Car	Number sold
Toyota Camry	429,575	Chevrolet Lumina	177,631
Honda Accord	401,071	Ford Mustang	144,732
Ford Taurus	371,074	Nissan Altima	144,451
Honda Civic	334,562	Ford Contour	139,838
Ford Escort	291,936	Buick LeSabre	136,551
Chevrolet Cavalier	256,099	Buick Century	126,220
Toyota Corolla	250,501	Pontiac Grand Prix	122,915
Saturn	231,786	Dodge Neon	117,964
Chevrolet Malibu	223,703	Mercury Grand Marquis	114,162
Pontiac Grand Am	180,428	Nissan Maxima	113,843

(*2000 World Almanac and Book of Facts,* p. 716)

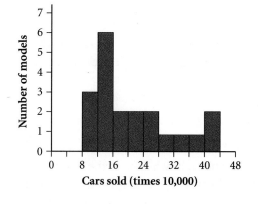

1998 Sales

a. What does 24 on the horizontal axis represent?

b. Explain the meaning of the first bin of this graph.

c. List calculator window values that would produce this graph. Give your answer in brackets like this: [Xmin, Xmax, Xscl, Ymin, Ymax, Yscl].

d. Graph this histogram on your calculator and on graph paper.

e. Use the table and your histogram to approximate the values of a five-number summary.

f. Graph a box plot on your calculator, and then sketch this plot above the histogram you drew in 6d.

7. Sketch what you think a histogram looks like for each situation below. Remember to label values and units on the axes.

a. The outcomes when rolling a die 100 times.

b. The estimates of the height of the classroom ceiling made by 100 different students.

c. The ages of the next 100 people you meet in the school hallway.

d. The 100 data values used to make the box plot below. (Use a bar width of 1.)

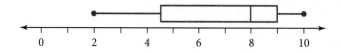

8. Create a data set with eight data values for each graph.

a.

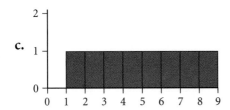

b.

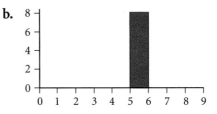

c.

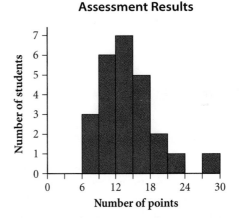

d.

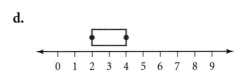

9. **APPLICATION** The histogram shows the results of an assessment on which 30 points were possible.

a. How would you assign the grades A–D?

b. Using your grading scheme, what grade would you assign the student represented by the bin farthest to the right?

c. Write a short description explaining why your grading scheme is a "fair" distribution of grades. Include comments about the measures of center, the variability, and the shape of the distribution.

Assessment Results

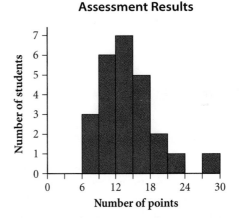

10. Chip and Dale, two algebra students, visited four different stores and recorded the prices on 1-pound bags of potato chips. They organized their data in the stem plot shown at the right.

a. What is the lowest price they found?

b. What do the entries in the third line from the top represent?

c. How many bags cost less than $2?

d. What is the most common price?

e. What is the range of prices for these chips?

Potato Chip Prices

15	0 0 9
16	9 9
17	5 9
18	5 9
19	9 9 9 9 9
20	9
21	5 9 9 9
22	5 9 9
23	9
24	
25	9

Key
—————————
25 | 9   means $2.59

▶ Review

11. Recreate on your paper this dot plot representing ages (in months) for students in an algebra class. Then add a box plot of this information to your graph.

Student Ages

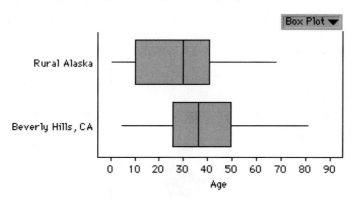

12. Create a data set of nine test scores with a five-number summary of 64, 72, 82, 82, 95 and a mean of 79.

COMPARE COMMUNITIES

The U.S. Bureau of the Census collects data on people from all over the United States. Citizens and governments use these data to make informed decisions and to develop programs that best serve diverse communities. Here are two box plots that use census data to compare the ages of 50 people from two communities.

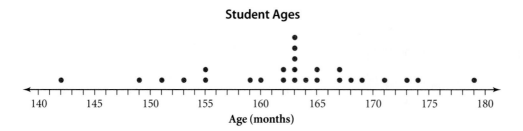

Fathom™

These box plots were made with Fathom. You can use this software to work with many more people, attributes, or communities than you could by hand. Learn how to use Fathom to make a wide range of interesting graphs.

Use data provided with Fathom™ Dynamic Statistics™ software or get worksheets from your teacher listing official census data for many people in two communities. Compare the communities using three or more attributes such as gender, age, race, ancestry, marital status, education level, or income. Your project should include

▶ A graph of each attribute, for each community. You can use any type of graph from this chapter.

▶ A summary description of the attributes for each community. Compare and contrast the two communities using values like mean, median, or range.

▶ A written explanation of how you think each community could use your graphs to make decisions.

Activity Day

Exploring a Conjecture

data
measures of center
mean
median
mode
range
outlier

five-number summary
minimum
maximum
first quartile (Q1)
third quartile (Q3)
interquartile range (IQR)

pictograph
bar graph
dot plot
box-and-whisker plot
histogram
frequency
stem-and-leaf plot

Statistics is a branch of applied mathematics dedicated to collecting and analyzing numeric data. The **data analysis** that statisticians do is used in science, government, and social services like health care. In this chapter you have learned concepts fundamental to statistics: measures of center, summary values, and types of graphs to organize and display data. Terms you have learned are in the box to the left.

Each measure or graph tells part of the story. Yet having too much information for a data set might not be helpful. Statisticians, and other people who work with data, must choose which measures and graphs give the best picture for a particular situation. Carefully chosen statistics can be informative and persuasive. Poorly chosen statistics, ones that don't show important characteristics of the data set, can be accidentally or deliberately misleading.

Activity
The Conjecture

You will need

- two books
- graph paper
- colored pencils or pens
- poster paper

A **conjecture** is a statement that might be true but has not been proven. Your group's goal is to come up with a conjecture relating two things and to collect and analyze the numeric evidence to support your conjecture or cast doubt on it.

In this activity you'll review the measures and graphs you have learned. Along the way, you will be faced with questions that statisticians face every day.

Step 1 | Your group should select two books on different subjects or with different reading levels. Flip through the books, but do not examine them in depth. State a conjecture comparing these two books. Your conjecture should deal with a quantity that you can count or measure. For example, "The history book has more words per sentence than the math book."

Step 2 | Decide how much data you'll need to convince yourself and your group that the conjecture is true or doubtful. Design a way to choose data to count or measure. For example, you might use your calculator to randomly select a page or a sentence. [▶🖥 See **Calculator Note 2A** to generate random numbers.◀]

Step 3 | Collect data from both books. Be consistent in your data collection, especially if more than one person is doing the collecting. Assign tasks to each member of your group.

Step 4 | Find the measures of center, range, five-number summary, and IQR for each of the two data sets.

Step 5 | Create a dot plot or stem-and-leaf plot for each set of data.

Step 6 | Make box plots for both data sets above the same horizontal axis.

Step 7 | Make a histogram for each data set.

Be sure that you have used descriptive units for all of your measures and clearly labeled your axes and plots before going on to the next step.

Step 8 | Choose one or two of the measures and one pair of graphs that you feel give the best evidence for or against your conjecture. Prepare a brief report or a poster. Include

a. Your conjecture.

b. Tables showing all the data you collected.

c. The measures and graphs that seem to support or disprove your conjecture.

d. Your conclusion about your conjecture.

Could you have designed your data collection process differently? Would this change have provided more persuasive results?

Suppose another group used your data but chose a different selection of measures and graphs. Could they have reached a different conclusion about the truth or falsehood of your conjecture?

Two-Variable Data

A **variable** is a trait or quantity whose value can change or vary. For example, birth dates will vary from person to person. In algebra, letters and symbols are used to represent variables. A person's birth date could be represented with the variable *b*, but the letter or symbol you choose doesn't matter. A data set that measures only one trait or quantity is called **one-variable data.** In Lessons 1.1 to 1.5, you learned to graph and summarize one-variable data.

Statisticians often collect information on *two* variables hoping to find a relationship. For example, someone's age may affect his or her pulse rate. In this lesson you explore **two-variable data** and plot points to create graphs called **scatter plots.** Plotting points and identifying the location of points on a graph are important skills for algebra.

Two-variable graphs are constructed with two axes. Each axis shows possible values for one of the two variables. On the **coordinate plane,** the horizontal axis is used for values of the variable *x*, and the vertical axis is used for values of the variable *y*. The **origin** is the point where the *x*- and *y*-axes intersect. The axes divide the coordinate plane into four **quadrants.** Each point is identified with **coordinates** (*x*, *y*) that tell its horizontal distance *x* and vertical distance *y* from the origin. The horizontal distance of value *x* is always listed first, so (*x*, *y*) is called an **ordered pair.**

The coordinates tell you to move left 1 and up 2.

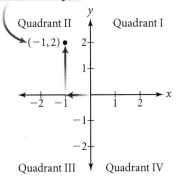

Investigation
Backed into a Corner

You will need

- 8 to 10 small round objects (coins, buttons, bottle caps)
- two centimeter rulers
- tape
- graph paper

In this investigation you'll collect two-variable data. A graph of these data points may help you to see a relationship between the two variables.

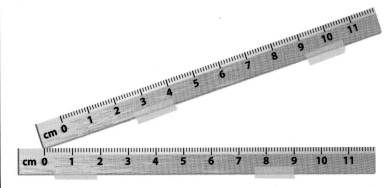

Step 1 | Tape one ruler to your desktop. Touch one vertex of the second ruler to the zero mark of the centimeter side of the first ruler and tape it in place to form an angle.

| Step 2 | Measure the diameter of one of your round objects to the nearest tenth of a centimeter. Record this measurement in a table like this one. |

Diameter (cm)	Distance from vertex (cm)

| Step 3 | Place the object inside your angle and push it toward the vertex until it fits snugly. Measure the point where the object touches the ruler that has zero at the vertex. This is the distance from the vertex. Enter this measurement, to the nearest tenth of a centimeter, in your table. |

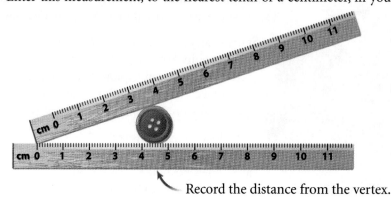

Record the distance from the vertex.

| Step 4 | Repeat Steps 2 and 3 for all of your round objects. |

| Step 5 | Create a set of axes on your graph paper. Label the axes as shown. |

| Step 6 | Scale the *x*-axis to fit all of your diameter values. For example, if the largest diameter was 5 cm, you might make each grid unit stand for 0.2 cm. Scale the *y*-axis to fit all of your distance-from-vertex values. For example, if the largest object touched the ruler at 30 cm, you might use 2 cm for each vertical grid unit. |

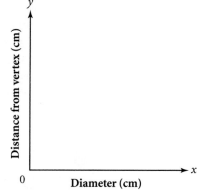

| Step 7 | Plot each piece of two-variable data from your table. Think of each row in your table as an ordered pair. Locate each point by first moving along the horizontal axis to the diameter measurement. Then move up vertically to the distance from vertex. |

| Step 8 | Describe any patterns you see in the graph. Is there a relationship between your two variables? |

| Step 9 | Enter the information from your table into two calculator lists. Make a scatter plot. The calculator display should look like the graph you drew by hand. [▶ 🖳 See **Calculator Note 1E** to learn how to display this information on your calculator screen.◀] |

This officer is using a radar device that measures the speed of oncoming cars.

There are many ways to collect data. Have you ever seen a police officer measuring the speed of an approaching car? Have you wondered how technicians measure the speed of a baseball pitch? Did you know that satellites collect information about the earth from distances over 320 miles? Each of these measurements involves the use of remote sensors that collect data. In this course you may have the opportunity to work with portable sensor equipment.

EXAMPLE

This scatter plot shows how the distance from a motion sensor to a person varies over a period of 6 seconds. Describe where the person is in relation to the sensor at each second.

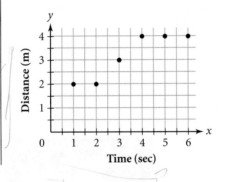

Distance From Motion Sensor

▶ **Solution**

The first point (1, 2) shows that after 1 second the person was 2 meters away from the sensor.

The next point (2, 2) indicates that after 2 seconds the person was still 2 meters away.

The point (3, 3) means that after 3 seconds the person was 3 meters away.

The point (4, 4) means that at 4 seconds the person was 4 meters away. He or she remained 4 meters away until 6 seconds had passed, as indicated by the points (5, 4) and (6, 4).

The graph in this example is a **first-quadrant graph** because all the values are positive. A lot of real-world data is described with only positive numbers, so first-quadrant graphs are very useful. However, you could graph the person's distance in front of the sensor as positive values and his or her distance *behind* the sensor as negative values. This would require more than one quadrant to show the data. If the graph showed negative values of time, how would you interpret this?

Practice Your Skills

1. Draw and label a coordinate plane so that the *x*-axis extends from −9 to 9 and the *y*-axis extends from −6 to 6. Represent each point below with a dot, and label the point using its letter name.

   $A(-5, -3.5)$ $B(2.5, -5)$ $C(5, 0)$ $D(-1.5, 4)$ $E(0, 4.5)$

   $F(2, -3)$ $G(-4, -1)$ $H(-5, 5)$ $I(4, 3)$ $J(0, 0)$

2. Sketch a coordinate plane. Label the axes and each of the four quadrants—I, II, III, and IV. Identify the axis or quadrant location of each point described.

   a. The first coordinate is positive, and the second coordinate is 0.

   b. The first coordinate is negative, and the second coordinate is positive.

   c. Both coordinates are positive.

   d. Both coordinates are negative.

   e. The coordinates are $(0, 0)$.

   f. The first coordinate is 0, and the second coordinate is negative.

3. Use your calculator to practice identifying coordinates. The program POINTS will place a point randomly on the calculator screen. You identify the point by entering its coordinates to the nearest one-half. Run the program POINTS until you can easily name points in all four quadrants. [▶ ☐ See **Calculator Note 1F.**◀]

Reason and Apply

4. This graph pictures a walker's distance from a stationary motion sensor.

Motion Sensor Readings

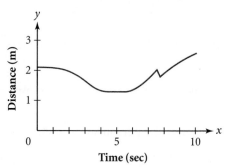

These people are running and walking in a fund-raiser to support cancer research.

   a. How far away was the walker after 2 seconds?

   b. At what time was the walker closest to the sensor?

   c. Approximately how far away was the walker after 10 seconds?

   d. When, if ever, did the walker stop?

5. Look at this scatter plot.

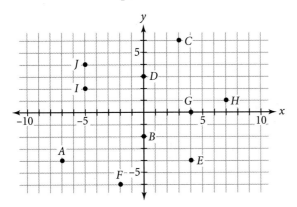

a. Name the (*x, y*) coordinates of each point pictured.

b. Enter the *x*-coordinates of the points you named in 5a into list L₁ and the corresponding *y*-coordinates into list L₂. Set your graphing window to the values suggested in the pictured graph, and graph a scatter plot of the points. We call this the scatter plot of (L₁, L₂).

c. Which points are on an axis?

d. List the points in Quadrant I, Quadrant II, Quadrant III, and Quadrant IV.

6. Write a paragraph explaining how to make a calculator scatter plot, and identify point locations in the coordinate plane.

7. APPLICATION The graph below is created by connecting the points in a scatter plot as you move left to right. [▶ 🖥 See **Calculator Note 1G** to learn how to connect a scatter plot.◀]

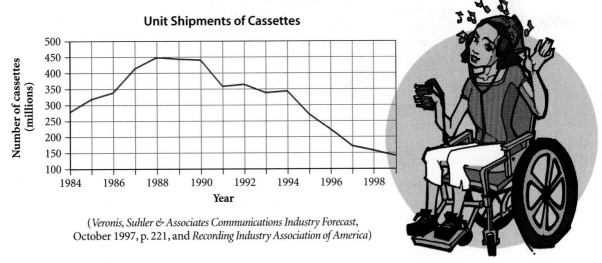

(*Veronis, Suhler & Associates Communications Industry Forecast,*
October 1997, p. 221, and *Recording Industry Association of America*)

a. Approximate the 16 data points, for which the *x*-value is the year and the *y*-value is the number of cassettes, represented on this graph.

b. Name graphing window values and create this graph on your calculator screen. If necessary, adjust your coordinates from 7a so that your graph matches the graph shown above.

c. The graph shows a pattern or trend for shipments of music cassettes. Describe any patterns you see. What do you think happened in the 1980s that would cause the patterns in this graph?

8. The data in this table show the average miles per gallon (mpg) for all U.S. automobiles during the indicated years.

a. Copy this table onto paper and calculate the years elapsed since 1960 to complete it.

b. Graph a scatter plot on your paper of points whose *x*-value is years elapsed and whose *y*-value is miles per gallon. Carefully label and scale your axes. Give your graph a title.

c. Connect each point in your scatter plot with a line segment from left to right.

d. What is the mean mpg for these data?

e. Graph a horizontal line that starts on the *y*-axis at a height equal to your answer to 8d. What does this line tell you?

f. Write a short descriptive statement about any pattern you see in these data and in your graph.

Average Miles per Gallon for All U.S. Automobiles

Year	Years elapsed	mpg
1960	0	12.4
1965		12.5
1970		12.0
1975		12.2
1980		13.3
1985		14.6
1990		16.4
1995		16.8
1996		16.9
1997		17.0
1998		17.0

(*U.S. Department of Transportation*)

9. The graph at right is a hexagon whose vertices are seven ordered pairs. Two of the points are (3, 0) and (1.5, 2.6). The hexagon is centered at the origin.

a. What are the coordinates of the other points?

b. Create this connected graph on your calculator. Add a few more points and line segments to make a piece of calculator art. Identify the points you added.

10. Xavier's dad braked suddenly to avoid hitting a squirrel as he drove Xavier to school. His speed during the trip to school is shown on the graph at right.

a. At what time did Xavier's dad apply the brakes?

b. What was his fastest speed during the trip?

c. How long did it take Xavier to get to school?

d. Find one feature of this graph that you think is unrealistic.

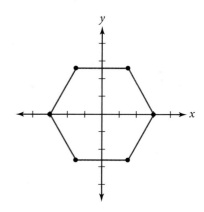

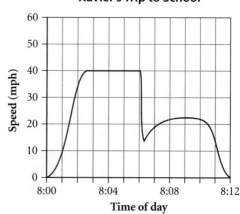

Xavier's Trip to School

Review

11. Create a data set with the specified number of items and the five-number summary values 5, 12, 15, 30, 47.

   a. 7   **b.** 10   **c.** 12

12. The table gives results for seventh-through ninth-grade students in the Third International Mathematics and Science Study.

   a. Find the five-number summary for this data set.

   b. Construct a box plot for these data.

   c. Between which five-number summary values is there the greatest spread of data? The least spread?

   d. What is the interquartile range?

   e. List the countries between the first quartile and the median. List those between the third quartile and the maximum, inclusive.

Results of the Third International Mathematics and Science Study (7th to 9th Grade)

Country	Mean score	Country	Mean score
Singapore	643	Sweden	519
Korea	607	New Zealand	508
Japan	605	England	506
Hong Kong	588	Norway	503
Belgium (Fl)	565	United States	500
Czech Republic	564	Latvia (LSS)	493
Slovak Republic	547	Iceland	487
Switzerland	545	Spain	487
France	538	Lithuania	477
Hungary	537	Cyprus	474
Russian Federation	535	Portugal	454
Canada	527	Iran, Islamic Republic	428
Ireland	527		

(IEA Third International Mathematics and Science Study [TIMSS], 1994–95)

IMPROVING YOUR **REASONING** SKILLS

A **glyph** is a symbol that presents information nonverbally. These weather glyphs show data values for several variables in one symbol. How many variables can you identify? The diagram shows data for 12 hours starting at 12:00 noon. Which characteristics would you call categorical? Which are numerical? Is it possible to show all or part of the data in one or more of the graphs you learned to use in this chapter? How would you do that? Which graphs have the greatest advantages in this situation? Why?

You can learn more about weather symbols and research your local weather conditions by using the Internet links at \boxed{\textbf{www.keymath.com/DA}} .

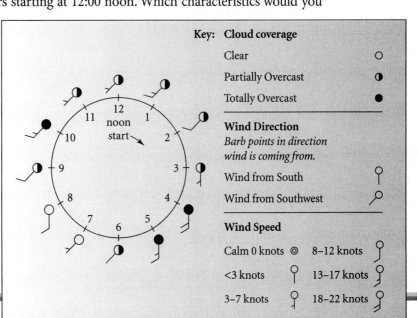

Estimating

If you read Michael Crichton's book *Jurassic Park* or if you saw the movie, you may remember that the dinosaur population grows faster than expected. The characters estimate that there will be only 238 dinosaurs. But a computer inventory shows that there are actually 292 dinosaurs!

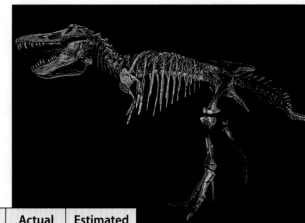

This skeleton of a Tyrannosaurus rex is at the Royal Tyrrell Museum in Alberta, Canada. Tyrannosaurs were among the largest carnivorous dinosaurs and measured up to 40 feet in length.

Dinosaurs in *Jurassic Park*

Species	Actual number	Estimated number
Tyrannosaurs	2	2
Maiasaurs	22	21
Stegosaurs	4	4
Triceratops	8	8
Procompsognathids	65	49
Othnielia	23	16
Velociraptors	37	8
Apatosaurs	17	17

Species	Actual number	Estimated number
Hadrosaurs	11	11
Dilophosaurs	7	7
Pterosaurs	6	6
Hypsilophodontids	34	33
Euoplocephalids	16	16
Styracosaurs	18	18
Microceratops	22	22
Total	292	238

(*Jurassic Park*, 1991, p. 164)

It is easy to see how the estimated numbers and the actual numbers compare by looking at this table. In this lesson you learn how to make efficient comparisons of data using a scatter plot.

Investigation
Guesstimating

You will need

- a meterstick, tape measure, or motion sensor

In this investigation, you will estimate and measure distances around your room. As a group, select a starting point for your measurements. Choose nine objects in the room that appear to be less than 5 meters away.

Step 1 | List the objects in the description column of a table like this one.

Description	x Actual distance (m)	y Estimated distance (m)
(item 1)		
(item 2)		
(item 3)		
(item 4)		

Step 2	Estimate the distances in meters or parts of a meter from your starting point to each object. If group members disagree, find the mean of your estimates. Record the estimates in your table.
Step 3	Measure the actual distances to each object and record them in the table.

Step 4	Draw coordinate axes and label actual distance on the x-axis and estimated distance on the y-axis. Use the same scale on both axes. Carefully plot your nine points. For each point, the actual distance is the x-value and the estimated distance is the y-value.
Step 5	Describe what this graph would look like if each of your estimates had been exactly the same as the actual measurement. How could you indicate this pattern on your graph?

Step 6	Make a calculator scatter plot of your data. Use your paper-and-pencil graph as a guide for setting a good graphing window.
Step 7	On your calculator, graph the line $y = x$. What does this line represent? [▶ 🖳 See **Calculator Note 1H** to graph a scatter plot and an equation simultaneously.◀]
Step 8	What do you notice about the points for distances that were underestimated? What about points for distances that were overestimated?
Step 9	How would you recognize the point for a distance that was estimated exactly the same as its actual measurement? Explain why this point would fall where it does.

Throughout this course you will create useful and informative graphs. Sometimes adding other elements to a graph as a basis for comparison can help you interpret your data. In the investigation, you added the line $y = x$ to your graph. How did this help you assess the accuracy of your estimates?

EXERCISES

You will need your calculator for problems **1, 4,** and **7.**

▶ Practice Your Skills

1. Enter the *Jurassic Park* data into your calculator. Put actual numbers into list L1 and estimated numbers into list L2 so that each (x, y) point has the form (*actual number, estimated number*).

a. Identify the two variables in this situation.

b. Graph a scatter plot of the data and record the window you used.

c. If *x* represents the actual number of dinosaurs and *y* the estimated number of dinosaurs, what equation represents the situation when the actual numbers equal the estimated numbers? Graph this equation on your calculator.

d. Are any points of the scatter plot above the line you drew in 1c? What do these points represent?

e. Are any points of the scatter plot below the line you drew in 1c? What do these points represent?

2. Lucia and Malcolm each estimated the weights of five different items from a grocery store. Each of Lucia's estimates was too low. Each of Malcolm's was too high. The scatter plot at right shows the (*actual weight, estimated weight*) data collected. The line drawn shows when an *estimate* is the same as the *actual* measurement.

a. Which points represent Lucia's estimates?

b. Which points represent Malcolm's estimates?

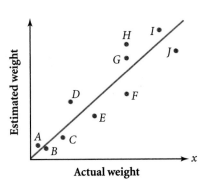

Lucia and Malcolm's Estimates

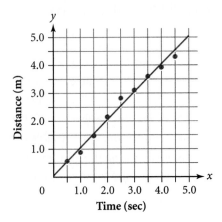

3. These points represent student estimates of temperatures in degrees Celsius for various samples of salt water. They are listed in the form (*actual temperature, estimated temperature*). Which points represent overestimates and which represent underestimates?

$A(27, 20)$	$B(-4, 2)$	$C(18, 22)$
$D(0, 3)$	$E(47, 60)$	$F(36, 28)$
$G(-2, 0)$	$H(33, 31)$	$I(-1, -2)$

4. This graph is a scatter plot of a person's distance from a motion sensor in a 5-second time period. The line is shown only as a guide.

a. Make a table of coordinates for the points pictured on this graph.

b. Describe how you would make a scatter plot of these data points on your calculator. Name the window values you would use.

c. What is the equation of the line pictured on the graph?

d. Was the distance between the person and the sensor increasing or decreasing?

Distance from Motion Sensor

Reason and Apply

5. A group of students conducted an experiment by stretching a rubber band and letting it fly. They measured the amount of stretch (cm) in each trial and recorded it as x. The distance flown (cm) was measured and recorded as y. This scatter plot shows six trials.

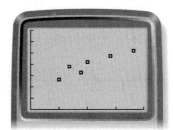

   a. Describe any relationship you see.

   b. Based on the plot, how far might the rubber band fly if they stretch it 15 cm?

   c. How far should they stretch the rubber band if the target is at 400 cm?

6. Copy this graph on your paper.

   a. Plot a point that represents someone overestimating a $12 item by $4. Label it A. What are the coordinates of A?

   b. Plot a point that represents someone underestimating an $18 item by $5. Label it B. What are the coordinates of B?

   c. Plot and label the points $C(6, 8)$, $D(20, 25)$, and $E(26, 28)$. Describe each point as an overestimate or underestimate. How far off was each estimate?

   d. Plot and label points F and G to represent two different estimates of an item priced at $16. Point F should be an underestimate of $3, and point G should be a perfect guess.

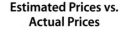
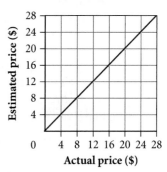

   e. Where will all the points lie that represent an estimated price of $16? Describe your answer in words and show it on the graph.

   f. Where will all the points lie that picture an actual price of $16? Describe your answer in words and show it on the graph.

   g. If x represents the actual price and y represents the estimated price, where are all the points represented by the equation $y = x$? What do these points represent?

7. Recall the *Jurassic Park* data on page 74. Enter the actual numbers and estimated numbers into two calculator lists. (You may still have the data in list L1 and list L2 from Exercise 1.) Use list L3 to calculate the estimated number minus the actual number for each dinosaur species. [▶ 🖥 See **Calculator Note 1I** for an explanation of how to use calculator lists in this way.◀]

   a. What information does list L3 give you?

   b. Use list L2 and list L3 to create a scatter plot of points in the form (*estimated number, estimated number − actual number*). Name the graphing window that provides a good display of your scatter plot.

   c. How many points are below the x-axis? What do these points represent in general?

   d. Name the coordinates of the point farthest from the x-axis. What do these coordinates tell you?

8. Draw the line for the equation $y = x$ on a coordinate grid with the *x*-axis labeled from -9 to 9 and the *y*-axis labeled from -6 to 6. Plot and label the points described.

   a. Point *A* with an *x*-coordinate of -4 and a *y*-coordinate 5 more than -4.

   b Point *B* with an *x*-coordinate of -2 and a *y*-coordinate 3 less than -2.

   c. Point *C* with an *x*-coordinate of 1 and a *y*-coordinate 4 units above the line.

   d. Now plot several points with coordinates that are opposites (*inverses*) of each other, for example, $(-5, 5)$ or $(5, -5)$. Describe the pattern of these points. Write an equation to describe the pattern.

9. **APPLICATION** This graph shows the mean SAT verbal and mathematics scores for 50 states and the District of Columbia in 1999.

   a. Which statement is true: "More states had students with higher mathematics scores than verbal scores" or "More states had students with higher verbal scores than mathematics scores"?

   b. The national average mathematics score was slightly higher than the average verbal score. This may seem contradictory to 9a. Explain.

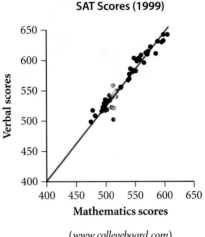

SAT Scores (1999)

(*www.collegeboard.com*)

10. Below is a scatter plot of points whose coordinates have the form (t, d). The variable *t* stands for time in seconds and *d* stands for distance in meters. The graph describes a person walking away from a motion sensor. The line $d = t$ is also graphed.

   a. Approximately how fast is the person moving? Explain how you know this.

   b. Name two different half-second intervals in which the person is moving more slowly than the rate you found in 10a.

   c. Name two different half-second intervals in which the person is moving faster than the rate you found in 10a.

   d. Name two different half-second intervals in which the person is moving at about the same rate you found in 10a.

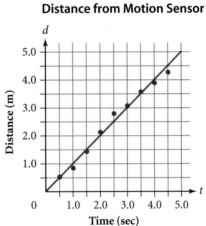

Distance from Motion Sensor

► Review

11. For each description, invent a seven-value data set so that all the values in the set are less than 10 and meet the conditions.

   a. The box plot represents data with a median that is not inside the box.

   b. The box plot represents data with an interquartile range of zero.

   c. The box plot represents data with one outlier on the left.

   d. The box plot is missing its right whisker.

12. Rocky Rhodes and his algebra classmates measured the circumference of their wrists in centimeters. Here are the data:

15.2 14.7 13.8 17.3 18.2 17.6 14.6 13.5 16.5
15.8 17.3 16.8 15.7 16.2 16.4 18.4 14.2 16.4
15.8 16.2 17.3 15.7 14.9 15.5 17.1

13	8 5
14	7 6 2 9
15	2 8 7 8 7 5
16	5 8 2 4 4 2
17	3 6 3 3 1
18	2 4

Key

10	4 means 10.4 cm

a. Rocky made the stem plot shown here. Unfortunately, he was not paying attention when these plots were discussed in class. Write a note to Rocky telling him what he did incorrectly.

b. Make a correct stem plot of this data set.

c. What is the range of this data set?

project

BASEBALL RELATIONSHIPS

Baseball statistics, or "stats," often show relationships between variables. These relationships could be used to identify the best players on a particular Major League team, or in an entire league, and may influence the strategy of managers. These scatter plots show stats of 50 National League players from the 1996 season.

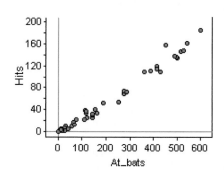
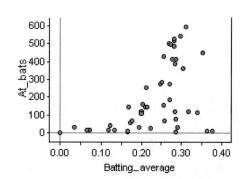

Find the stats of any one team for a particular season. Your data can come from library research, or you can use the links at **www.keymath.com/DA** . Briefly review all the stats and make at least two conjectures about relationships like "If a player appears in more games, he will have more times at bat." Use scatter plots to see if relationships truly exist between the variables. For each conjecture, your project should include

▶ A sentence stating the conjecture and a scatter plot to accompany it.

▶ An explanation of how the scatter plot either supports your conjecture or casts doubt on it. If a relationship does appear to exist, why do you think this is so?

Conclude your project by explaining how you think a baseball manager could use your graphs to strategize for a game.

Using Matrices to Organize and Combine Data

Don't agonize.

Organize.

FLORYNCE KENNEDY

Did you know that the average number of hours people work per week has been decreasing during the last 100 years? The table provides data for some countries. This table is a 6 × 2 (say "six by two"), because it has six rows of countries and two columns of years. Can you identify the entry in row 2, column 1? In which row and column is the entry 37.9?

Average Weekly Working Hours

Rows are counted top to bottom.

	Country	1900	1998
row 1	Germany	51.6	29.0
row 2	France	51.7	31.7
row 3	Japan	51.7	38.3
row 4	Netherlands	52.0	30.8
row 5	United States	52.0	37.9
row 6	Britain	52.4	35.6

This entry is in row 3, column 2.

column 1   column 2

Columns are counted left to right.

During the late 1800s and early 1900s, labor unions were formed to improve working conditions. These miners may have belonged to the American Federation of Labor, one of the first unions to include African American members.

A table is an easy way to organize data. A quicker way to organize data is to display it in a **matrix**. A matrix has rows and columns just like a table. To rewrite a table as a matrix, you simply use brackets to enclose the row and column entries. For the working-hour data, the **dimensions** of the matrix are the same as the table: 6 × 2.

In this lesson you represent different situations with matrices and explore some matrix calculations.

EXAMPLE A

a. Form a matrix from the table Average Weekly Working Hours.

b. Verify the row and column locations of the entries 31.7, 30.8, and 52.4 in matrix [A].

c. If the average person works 50 weeks per year, what were the average yearly working-hour totals for these countries in 1900 and 1998?

► **Solution**

a. Here is the 6 × 2 matrix. It is simply a table without labels.

$$[A] = \begin{bmatrix} 51.6 & 29.0 \\ 51.7 & 31.7 \\ 51.7 & 38.3 \\ 52.0 & 30.8 \\ 52.0 & 37.9 \\ 52.4 & 35.6 \end{bmatrix}$$

b. Enter the 6 × 2 matrix on your calculator. [▶ 🖳 See **Calculator Note 1L** to learn how to enter a matrix in your calculator.◀] The calculator display shows that 31.7 is located in row 2, column 2, of matrix A. By moving around in your editor, you see that the entry 30.8 is in row 4, column 2; 52.4 is in row 6, column 1.

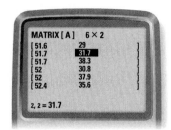

c. Use your calculator to multiply [A] by 50. [▶ 🖳 See **Calculator Note 1N** to learn how to multiply a matrix by a number.◀] This new matrix shows the result of multiplying each of the original entries by 50. The new entries are the average *yearly* working hours.

As shown in the example, it is easy to operate on all the entries in a matrix with one calculation. In the next example you will learn how you can add or subtract two matrices to help you answer questions about a situation.

EXAMPLE B

Matrix [B] provides the costs of a medium pizza, medium salad, and medium drink at two different pizzerias: the Pizza Palace and Tony's Pizzeria. Matrix [C] provides the additional charge for large items at each pizzeria.

$$[B] = \begin{array}{cc} \text{Pizza Palace} & \text{Tony's Pizzeria} \\ \downarrow & \downarrow \\ \begin{bmatrix} 8.90 & 9.10 \\ 2.35 & 2.65 \\ 1.50 & 1.60 \end{bmatrix} & \begin{array}{l} \leftarrow \text{pizza} \\ \leftarrow \text{salad} \\ \leftarrow \text{drink} \end{array} \end{array}$$

$$[C] = \begin{bmatrix} 2.50 & 2.25 \\ 1.00 & 1.25 \\ 0.65 & 0.50 \end{bmatrix} \begin{array}{l} \leftarrow \text{pizza} \\ \leftarrow \text{salad} \\ \leftarrow \text{drink} \end{array}$$

Write a matrix [D] displaying the costs at the Pizza Palace and at Tony's Pizzeria for a large pizza, large salad, and large drink.

▶ Solution

If you wanted the price of a large pizza at the Pizza Palace, you would add the medium price and the additional charge.

$$8.90 + 2.50 = 11.40$$

The totals for matrix [D] are found by adding all corresponding entries from matrix [B] and matrix [C]. [▶ See **Calculator Note 1M** to learn how to use your calculator to add and subtract matrices. ◀]

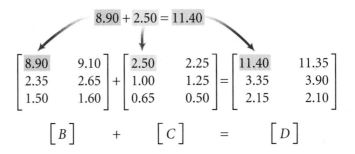

$$8.90 + 2.50 = 11.40$$

$$\begin{bmatrix} 8.90 & 9.10 \\ 2.35 & 2.65 \\ 1.50 & 1.60 \end{bmatrix} + \begin{bmatrix} 2.50 & 2.25 \\ 1.00 & 1.25 \\ 0.65 & 0.50 \end{bmatrix} = \begin{bmatrix} 11.40 & 11.35 \\ 3.35 & 3.90 \\ 2.15 & 2.10 \end{bmatrix}$$

$$\begin{bmatrix} B \end{bmatrix} \quad + \quad \begin{bmatrix} C \end{bmatrix} \quad = \quad \begin{bmatrix} D \end{bmatrix}$$

Does the order in which you add these matrices make a difference? Take a moment to calculate [B] + [C] and [C] + [B] on your calculator.

If you try adding a 6 × 2 matrix and a 3 × 2 matrix with your calculator, you'll get an error message. This is because the matrices don't have the same dimensions. There is no way to match up corresponding entries to do the operations. In order for you to add or subtract matrices, each matrix must have the same number of rows or columns.

Can you multiply two matrices? In this investigation you will discover why matrix multiplication is more complicated.

Investigation
Row-by-Column Matrix Multiplication

Recall the table and matrix for large items at the Pizza Palace and Tony's Pizzeria from Example B.

Suppose you're in charge of ordering the food for a school club party.

$$[D] = \begin{bmatrix} 11.40 & 11.35 \\ 3.35 & 3.90 \\ 2.15 & 2.10 \end{bmatrix}$$

Large-Item Prices

	Pizza Palace	Tony's Pizzeria
Large pizza	$11.40	$11.35
Large salad	$3.35	$3.90
Large drink	$2.15	$2.10

Step 1   What will be the total cost of 4 large pizzas, 5 large salads, and 10 large drinks at the Pizza Palace?

Step 2   What will be the total cost for the same order at Tony's Pizzeria?

Step 3   Describe how you calculated the total costs.

As you see, calculating the food costs for your party requires multiplication and addition. Organizing your work like this is a first step to discovering how to multiply matrices.

Step 4   Copy the calculation and replace each box with a number to find the total food cost for the same order at the Pizza Palace.

Step 5   The row matrix [A] and column matrix [B] contain all the information you need to calculate the total food cost at the Pizza Palace.

$$[A] = [4 \quad 5 \quad 10] \quad [B] = \begin{bmatrix} 11.40 \\ 3.35 \\ 2.15 \end{bmatrix}$$

Explain what matrix [A] and matrix [B] represent.

Step 6   Enter [A] and [B] in your calculator and find their *product*, [A] · [B], or

$$[4 \quad 5 \quad 10] \cdot \begin{bmatrix} 11.40 \\ 3.35 \\ 2.15 \end{bmatrix}$$

[▶ 🖳 See **Calculator Note 1P** to learn how to multiply two matrices.◀] Explain in detail what you think the calculator does to find this answer.

Step 7   Repeat Step 4 to find the total food cost at Tony's Pizzeria.

Step 8   Write the product of a row matrix and a column matrix that calculates the total food cost at Tony's Pizzeria. Use your calculator to verify that your product matches your answer to Step 7.

Step 9   Explain why the number of columns in the first matrix must be the same as the number of rows in the second matrix in order to multiply them.

Step 10   Predict the answer to the matrix multiplication problem below. Use your calculator to verify your answer. What is its meaning in the real-world context?

$$[4 \quad 5 \quad 10] \cdot \begin{bmatrix} 11.40 & 11.35 \\ 3.35 & 3.90 \\ 2.15 & 2.10 \end{bmatrix} = ?$$

Step 11   Explain how to calculate the matrix multiplication in Step 10 without using calculator matrices.

Step 12   Try this matrix multiplication:

$$\begin{bmatrix} 11.40 & 11.35 \\ 3.35 & 3.90 \\ 2.15 & 2.10 \end{bmatrix} \cdot [4 \quad 5 \quad 10]$$

In general, do you think [A] · [B] = [B] · [A]? Explain.

Step 13   Write a short paragraph explaining how to multiply two matrices.

EXAMPLE C

Is it possible to multiply each pair of matrices? If so, what is the product? If not, why not?

a. $[2 \quad 3 \quad 4] \cdot \begin{bmatrix} -1 \\ 1 \\ 5 \end{bmatrix}$

b. $[3 \quad -1] \cdot \begin{bmatrix} 2 & 4 \\ 0 & 3 \\ 5 & -6 \end{bmatrix}$

c. $[3 \quad -1] \cdot \begin{bmatrix} 2 & 0 & 5 \\ 4 & 7 & -6 \end{bmatrix}$

▶ **Solution**

a. Multiply the respective entries in the row matrix by those in the column matrix and then add the products.

$$[2 \quad 3 \quad 4] \cdot \begin{bmatrix} -1 \\ 1 \\ 5 \end{bmatrix} = [21]$$

$$(2 \cdot -1) + (3 \cdot 1) + (4 \cdot 5) = 21$$

b. This is not possible. There are not enough entries in the row matrix to match the three entries in each column.

c. Multiply the entries in the row matrix by the respective entries in each column. Each sum is a separate entry in the answer matrix.

Technology
CONNECTION

The development of 3-D video games uses matrices to process the graphics.

You multiply a 1×2 matrix by a 2×3 matrix.

$1 \times 2, \; 2 \times 3$

The inside dimensions are the same so you can multiply. The 2 row entries match up with 2 column entries.

$1 \times 2, \; 2 \times 3$

The outside dimensions tell you the dimensions of your answer.

Multiply the row by each column.

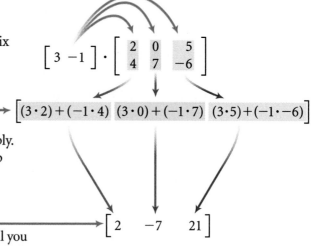

Matrix multiplication is probably too much work if all you want to do is plan a pizza party. But what if you had to manage the inventory (goods in stock) of a grocery store? Or a whole chain of grocery stores? If you needed to know the values of the inventory at each store, a matrix calculation carried out by a computer would be essential to your business.

EXERCISES

You will need your calculator for problems **4, 8,** and **12.**

▶ Practice Your Skills

Use this information for problems 1–8. Troy Aikman, Randall Cunningham, and Steve Young have been top-performing quarterbacks in the National Football League throughout their careers. The rows in matrix [A] and matrix [B] show data for Aikman, Cunningham, and Young, in that order. The columns show the number of passing attempts, pass completions, touchdown passes, and interceptions, from left to right. Matrix [A] shows stats from 1992, and matrix [B] shows stats from 1998.

$$[A] = \begin{bmatrix} 473 & 302 & 23 & 14 \\ 384 & 233 & 19 & 11 \\ 402 & 268 & 25 & 7 \end{bmatrix} \quad [B] = \begin{bmatrix} 315 & 187 & 12 & 5 \\ 425 & 259 & 34 & 10 \\ 517 & 322 & 36 & 12 \end{bmatrix}$$

(http://CNNSI.com)

1. What does the entry in row 2, column 3, of matrix [A] tell you?

2. What does the entry in row 3, column 2, of matrix [B] tell you?

3. What are the dimensions of each matrix?

4. Write a clear explanation of the procedure for entering [A] and [B] into your calculator.

5. Find [A] + [B]. What is this matrix and what information does it provide?

6. Is [A] + [B] equal to [B] + [A]? Do you think this result is always true for matrix addition? Explain.

7. Find [B] − [A]. What is this matrix and what information does it provide?

8. How can you use your calculator to find the average statistics for the three quarterbacks for these two seasons? Write a matrix expression that will give this average.

▶ Reason and Apply

9. Use matrices [A] and [B] to write the new matrices asked for in 9a–d.

$$[A] = \begin{bmatrix} 3 & -4 & 2.5 \\ -2 & 6 & 4 \end{bmatrix} \quad [B] = \begin{bmatrix} 5 & -1 & 2 \\ -4 & 3.5 & 1 \end{bmatrix}$$

   a. [A] + [B]   **b.** −[A]   **c.** 3 · [B]   **d.** the average of [A] and [B]

10. Find the matrix [B] so that this equation is valid:

$$\begin{bmatrix} -2 & 0 \\ 6 & -11.6 \\ 4.25 & 7.5 \end{bmatrix} - [B] = \begin{bmatrix} 2.8 & 2.4 \\ 2.5 & -9.4 \\ 1 & 6 \end{bmatrix}$$

11. Create a problem involving this matrix multiplication if the row matrix represents dollars and cents and the column matrix represents hours. Find the product without using your calculator's matrix menu.

$$[5.25 \quad 8.75] \cdot \begin{bmatrix} 16 \\ 30 \end{bmatrix}$$

12. APPLICATION Ms. Shurr owns three ice cream shops and wants to know how much she made on ice-cream cone sales for one day. The number of small, medium, and large cones sold at each location and the profit for each size are contained in the tables.

a. Write a quantity matrix that gives the number of each size sold at each location and a profit matrix that gives the profit for each size. What must be the same for each matrix?

b. Without using your calculator's matrix menu, find the profit from ice-cream cones for each location. Explain how you got your answer.

c. Check your answer to 12b by using your calculator to multiply the quantity matrix [A], by the profit matrix [B]. What are the dimensions of the answer matrix?

d. What do the entries in the answer matrix tell you? Convert your matrix into a table with row and column headings so that Ms. Shurr can understand the information.

e. Try to calculate [B] · [A] on your calculator. What happens? What do you think the result means?

Number of Cones Sold

	S	M	L
Atlanta	74	25	37
Decatur	32	38	16
Athens	120	52	34

Cone Profit

	Profit
S	$0.90
M	$1.25
L	$2.15

▶ **Review**

13. Create a data set that fits the information.

a. Ten students were asked the number of times they had flown in an airplane. The range of data values was 7. The minimum was 0 and the mode was 2.

b. Eight students each measured the length of their right foot. The range of data values was 8.2 cm, and the maximum value was 30.4 cm. There was no mode.

14. Mr. Chin and Mrs. Shapiro had their classes collect data on the amount of change each student had in class on a particular day. The students graphed the data on the back-to-back stem plot at right.

a. How many students are in each class?

b. Find the range of the data in each class.

c. How many students had more than $1?

d. What do the entries in the last row represent?

e. Without adding, make an educated guess which class has the most money altogether. Explain your thinking.

f. How much money does each class have?

Mrs. Shapiro's class				Mr. Chin's class					
0 0 0 0	0	0 0 0 0 5 8			
2 0	1	0 5 6			
5 5 5	2	0 0 5 5 5 7			
9 6 5 0 0	3	5 5			
5	4	0 0 0 6			
5 2 0 0	5	0 0 5 5 8			
7 3 0 0	6	0 2 5 5			
5 0	7	0 5 5 6			
2 0 0	8				
4 1	9				
4 0 0	10	0 0 5			
5 0	11	0			
6 4 1	12	1 5 5			
5 0	13				

Key

| 5 | 4 | 0 0 0 6 |

means 45¢ in Mrs. Shapiro's class; 40¢, 40¢, 40¢ and 46¢ in Mr. Chin's class

CHAPTER 1

REVIEW

In this chapter you learned how statistical measures and graphs can help you organize and make sense of **data.** You explored several different kinds of graphs—**bar graphs, pictographs, dot plots, box plots, histograms,** and **stem plots**—that can be used to represent **one-variable** data.

You analyzed the strengths and weaknesses of each kind of graph to select the most appropriate one for a given situation. A bar graph displays data that can be grouped into **categories.** Numerical data can be individually shown with a dot plot. The **spread** of data is clearly displayed with a box plot built from the **five-number summary.** A histogram uses **bins** to show the **frequency** of data and is particularly useful for large sets of data. A stem plot also groups data into intervals but maintains the identity of each data value.

You can use the **measures of center** to describe a typical data value. In addition to the **mean, median,** and **mode,** statistical measures like **range, minimum, maximum, quartiles,** and **interquartile range** help you describe the spread of a data set and identify **outliers.**

You used the **coordinate plane** to compare estimates and actual values plotted on **two-variable** plots called **scatter plots.** Here, each variable is represented on a different axis, and an **ordered pair** shows the value of each variable for a single data item. You also analyzed scatter plots for situations involving the two variables *time* and *distance.* Scatter plots allowed you to find patterns in the data; sometimes these patterns could be written as an algebraic equation.

Lastly, you learned to use a **matrix** to organize data in rows and columns, very much like a table. You discovered ways to add, subtract, and multiply matrices and learned how the **dimensions** of matrices affect these computations. Computers and calculators can use matrices to calculate with large sets of data, making it easy to answer questions about the data.

EXERCISES

1. This data set gives the number of hours of use before each of 14 batteries required recharging: {40, 36, 27, 44, 40, 34, 42, 58, 36, 46, 52, 52, 38, 36}.

   a. Find the mean, median, and mode for the data set and explain how you found each measure.

   b. Find the five-number summary for the data set and make a box plot.

2. Seven students order onion rings. The mean number of onion rings they get is 16. The five-number summary is 9, 11, 16, 21, 22. How many onion rings might each student have been served?

3. The table at right shows the mean annual wages earned by individuals with various levels of education in the United States in 1998.

a. Construct a bar graph for the data.

b. Between which two consecutive levels of education is there the greatest difference in mean annual wages? The smallest difference?

4. The table below shows the top ten scoring leaders in women's college basketball tournament history.

a. Construct a box plot for the data.

b. Are there are any outliers?

c. Which measure of center would you use to describe a typical value?

Mean Annual Wages, 1998

Level of education	Amount ($)
Did not finish high school	18,913
High school diploma only	25,257
Two-year degree (AA/AS)	33,765
Bachelor's degree (BA/BS)	45,390
Master's degree (MA/MS)	52,951
Doctorate degree	75,071

(*U.S. Census 2000*)

Leading Scorers in NCAA Tournament History (Women's Basketball, Through 1999–2000 Season)

Player, college	Total points
Chamique Holdsclaw, Tennessee	479
Bridgette Gordon, Tennessee	388
Cheryl Miller, USC	333
Janice Lawrence, Louisiana Tech	312
Penny Toler, Long Beach State	291
Dawn Staley, Virginia	274
Cindy Brown, Long Beach State	263
Venus Lacy, Louisiana Tech	263
Clarissa Davis, Texas	261
Janet Harris, Georgia	254

(*www.infoplease.com*)

5. Twenty-three students were asked how many pages they had read in a book currently assigned for class. Here are their responses: {24, 87, 158, 227, 437, 79, 93, 121, 111, 118, 12, 25, 284, 332, 181, 34, 54, 167, 300, 103, 128, 132, 345}.

a. Find the measures of center.

b. Construct histograms for two different bin widths.

c. Construct a box plot.

d. What do the histograms and the box plot tell you about this data set? Make one or two observations.

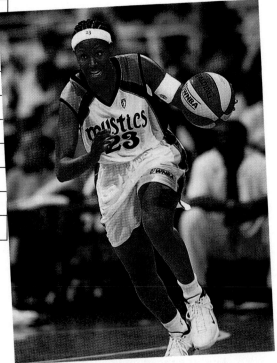

After an impressive college career, Chamique Holdsclaw went on to play professionally for the Washington Mystics.

6. Isabel made the estimates listed in the table at right for the year each item was invented.

a. Create a scatter plot of data points with coordinates having the form (*actual year, estimated year*).

b. Circle those points that picture an estimated year that is earlier than the actual year (underestimates).

c. Define your variables and write the equation of a line that would represent all estimates being correct.

7. **APPLICATION** The tables below show information for the Roxy Theater. The management is considering raising the admission prices.

Invention Dates

Item	Actual year	Estimated year
Telephone	1876	1905
Color television	1928	1960
Video disk	1972	1980
Pacemaker	1952	1945
Motion picture	1893	1915
Ballpoint pen	1888	1935
Aspirin	1899	1917
Graphing calculator	1985	1980
Compact disc	1972	1990
Car radio	1929	1940

(*2000 World Almanac*, pp. 609–610)

Current Prices

	Matinee	Evening
Adult	$5.00	$8.00
Child	$3.50	$4.75
Senior	$3.50	$4.00

Price Increases

	Matinee	Evening
Adult	$0.50	$0.75
Child	$0.50	$0.25
Senior	$0.50	$0.25

Average Attendance

Adult	Child	Senior
43	81	37

a. Convert each table to a matrix.

b. Do a matrix calculation to find the new prices after the admission increase.

c. Do a matrix calculation to find the total revenue of a matinee performance and an evening performance at the new prices.

8. The graph below shows Kayo's distance over time as she jogs straight down the street in front of her home. Point *A* is Kayo's starting point (her home).

a. During which time period was Kayo jogging the fastest?

b. Explain what the jogger might have been doing during the time interval between points *B* and *C* and between points *D* and *E*.

c. Write a brief story for this graph using all five segments.

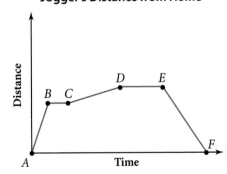

Jogger's Distance from Home

9. The table at right shows the approximate 1998 populations of the ten most populated cities in the United States.

    a. Write the approximate population of Chicago in 1998 as a decimal number.

    b. Create a bar graph of the data.

    c. Create a stem plot of the data.

    d. Create a histogram of the data.

    e. Each of the graphs you have created highlights different characteristics of the data. Briefly describe what features are unique to each graph.

The Ten Most Populated U.S. Cities, 1998

City	Population (millions)
Chicago	2.82
Dallas	1.08
Detroit	0.97
Houston	1.79
Los Angeles	3.60
New York	7.42
Philadelphia	1.44
Phoenix	1.20
San Antonio	1.11
San Diego	1.22

(www.census.gov)

10. This stem-and-leaf plot shows the number of minutes of sleep for eight students. In this plot the dots represent that the interval is divided in half at 50. Use the key to fully understand how to read this graph.

    a. What is the mean sleep amount?

    b. What is the median sleep amount?

    c. What is the mode sleep amount?

Minutes Sleeping

```
3  | 20
.  | 60 90
4  | 00
.  | 50 55 80 80
```

Key

```
3  | 20  means 320 minutes
.  | 60  means 360 minutes
```

TAKE ANOTHER LOOK

These two graphs display the same data set. Graph A is being presented by a citizen who argues that the city should use its budget surplus to buy land for a park. Graph B is being presented by another citizen who argues that a new park is not a high priority. Tell what position you would favor on this issue and what impact each graph has on your decision. Do you think either graph is deliberately misleading? What other information would you want to know?

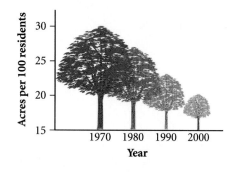

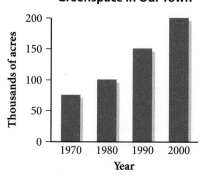

Find another graphic display in a newspaper, magazine, or voter material that seems to be "engineered" to persuade the viewer to a particular point of view. Tell how the graph could be changed for a fairer presentation.

Assessing What You've Learned

WRITE IN YOUR JOURNAL

Your course work in algebra will bring up many new ideas, and some are quite abstract. Sometimes you'll feel you have a good grip on these new ideas, other times less so.

Regular reflection on your confidence in mathematics generally, and your mastery of algebra skills in particular, will help you assess your strengths and weaknesses. Writing these reflections down will help you realize where you are having trouble, and perhaps you'll ask for help sooner. Likewise, realizing how much you know can boost your confidence. A good place to record these thoughts is in a journal—not a personal diary, but an informal collection of your feelings about what you're learning. Like a travel journal that others would find interesting to read and that will help recall the details of a trip, a mathematics journal is something your teacher will want to look at and something you'll look at again later.

Here are some questions to prompt your journal writing. Your teacher might give you other ideas, but you can write in your journal any time.

► How is what you are learning an extension of your previous mathematics courses? How is it completely new?
► What are your goals for your work in algebra? What steps can you take to achieve them?
► Can you see ways to apply what you are learning to your everyday life? To a future career?
► What ideas have you found hard to understand?

UPDATE YOUR PORTFOLIO At the end of Chapter 0, you may have started a portfolio. Now would be a good time to add one or more pieces of significant work from Chapter 1. You could choose an investigation, a homework problem, or your work on a Project or Take Another Look. It might be a good idea to include a sample of every type of graph you've learned to interpret or create.

Proportional Reasoning and Probability

Murals are just one of the many art forms around us that come to life with the help of ratios and proportions. To plan a mural, the artist draws sketches on paper then uses ratio to enlarge the image to the size of the final work.

OBJECTIVES

In this chapter you will
- use proportional reasoning to understand problem situations
- create and interpret circle graphs
- explore probability

Proportions

When you say, "I got 21 out of 24 questions correct on the last quiz," you are comparing two numbers. The **ratio** of your correct questions to the total number of questions is 21 to 24. You can write the ratio as 21 : 24 or as a fraction or decimal. The fraction bar means division, so these expressions are equivalent:

Mathematics is not a way of hanging numbers on things so that quantitative answers to ordinary questions can be obtained. It is a language that allows one to think of extraordinary questions.

JAMES BULLOCK

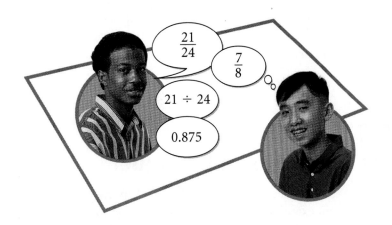

$$\frac{21}{24} \qquad \frac{7}{8}$$

$$21 \div 24$$

$$0.875$$

EXAMPLE A | Write the ratio 210 : 330 in several ways.

▶ **Solution** | $\frac{210}{330}$ or $\frac{7}{11}$ $210 \div 330$ or $7 \div 11$

To change a common fraction into a decimal fraction, divide the numerator by the denominator.

$$\begin{array}{r} .636363 \\ 330\overline{)210.000000} \\ 198\,0 \\ \hline 12\,00 \\ 9\,90 \\ \hline 2\,100 \\ 1\,980 \\ \hline 1200 \\ 990 \\ \hline \cdots \end{array}$$

When you divide 21 by 24, the decimal form of the quotient ends, or **terminates.** The ratio $\frac{21}{24}$ equals 0.875 exactly. But when you divide 210 by 330 or 7 by 11, you see a **repeating decimal** pattern 0.636363.... You can use a bar over the numerals that repeat to show a repeating decimal pattern $\frac{7}{11} = 0.\overline{63}$. [▶ 🖳 See **Calculator Note 0A** for more about converting fractions to decimals. ◀]

A **proportion** is an equation stating that two ratios are equal. For example, $\frac{2}{3} = \frac{8}{12}$ is a proportion. You can use the numbers 2, 3, 8, and 12 to write these true proportions:

$$\frac{2}{3} = \frac{8}{12} \qquad \frac{3}{2} = \frac{12}{8} \qquad \frac{3}{12} = \frac{2}{8} \qquad \frac{12}{3} = \frac{8}{2}$$

Do you agree that these are all true equations? One way to check that a proportion is true is by finding the decimal equivalent of each side. The statement $\frac{3}{8} = \frac{2}{12}$ is not true; 0.375 is not equal to $0.1\overline{6}$.

In algebra, a **variable** can stand for an unknown number, or for a set of numbers. In the proportion $\frac{2}{3} = \frac{M}{6}$, you can replace the letter M with any number, but only one number will make the proportion true. That number is unknown until the proportion is solved.

Investigation
Multiply and Conquer

You can easily guess the value of M in the proportion $\frac{2}{3} = \frac{M}{6}$. In this investigation, you examine ways to solve a proportion for an unknown number when guessing is not easy. It's hard to guess the value of M in the proportion $\frac{M}{19} = \frac{56}{133}$.

Step 1	Multiply both sides of the proportion $\frac{M}{19} = \frac{56}{133}$ by 19. Why can you do this? What does M equal?
Step 2	For each equation, choose a number to multiply both ratios by to solve the proportion for the unknown number. Then multiply and divide to find the missing value.

    a. $\frac{21}{35} = \frac{Q}{20}$ **b.** $\frac{P}{12} = \frac{132}{176}$

    c. $\frac{L}{30} = \frac{30}{200}$ **d.** $\frac{130}{78} = \frac{n}{15}$

Step 3	Check that each proportion in Step 2 is true by replacing the variable with your answer.
Step 4	In each equation of Step 2, the variables are in the numerator. Write a brief explanation of one way to solve a proportion when one of the numerators is a variable.

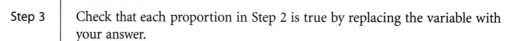

Step 5	The proportions you solved in Step 2 have been changed by switching the numerators and denominators. That is, the ratio on each side has been *inverted*. Do the solutions from Step 2 also make these new proportions true?

    a. $\frac{35}{21} = \frac{20}{Q}$ **b.** $\frac{12}{P} = \frac{176}{132}$ **c.** $\frac{30}{L} = \frac{200}{30}$ **d.** $\frac{78}{130} = \frac{15}{n}$

Step 6	How can you use what you just discovered to help you solve a proportion that has the variable in the denominator, such as $\frac{20}{135} = \frac{12}{k}$? Why does this work? Solve the equation.

Step 7 | There are many ways to solve proportions. Here are three student papers each answering the question "13 is 65% of what number?" What are the steps each student followed? What other methods can you use to solve proportions?

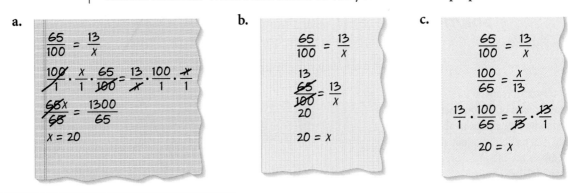

a.

$$\frac{65}{100} = \frac{13}{x}$$

$$\frac{100}{1} \cdot \frac{x}{1} \cdot \frac{65}{100} = \frac{13}{x} \cdot \frac{100}{1} \cdot \frac{x}{1}$$

$$\frac{65x}{65} = \frac{1300}{65}$$

$$x = 20$$

b.

$$\frac{65}{100} = \frac{13}{x}$$

$$\frac{13}{\frac{65}{100}} = \frac{13}{x}$$

$$20$$

$$20 = x$$

c.

$$\frac{65}{100} = \frac{13}{x}$$

$$\frac{100}{65} = \frac{x}{13}$$

$$\frac{13}{1} \cdot \frac{100}{65} = \frac{x}{13} \cdot \frac{13}{1}$$

$$20 = x$$

EXAMPLE B | Jennifer estimates that two out of every three students will attend the class party. She knows there are 750 students in her class. Set up and solve a proportion to help her estimate how many people will attend.

▶ Solution | To set up the proportion, be sure both ratios make the same comparison. Use a to represent the number of students who will attend.

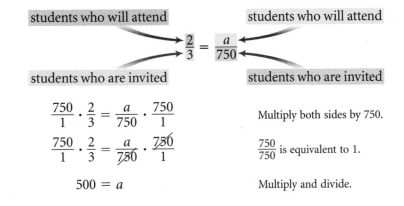

students who will attend   students who will attend

$$\frac{2}{3} = \frac{a}{750}$$

students who are invited   students who are invited

$$\frac{750}{1} \cdot \frac{2}{3} = \frac{a}{750} \cdot \frac{750}{1} \qquad \text{Multiply both sides by 750.}$$

$$\frac{750}{1} \cdot \frac{2}{3} = \frac{a}{750} \cdot \frac{750}{1} \qquad \frac{750}{750} \text{ is equivalent to 1.}$$

$$500 = a \qquad \text{Multiply and divide.}$$

Jennifer can estimate that about 500 students will attend the party.

EXAMPLE C | After the party, Jennifer found out that 70% of the class attended. How many students attended?

► **Solution**

History
● **CONNECTION** ●

The Pythagoreans, a brotherhood of philosophers begun by Pythagorus in about 520 B.C., realized that not all numbers are rational. For example, they showed that for a square one unit on a side, the diagonal $\sqrt{2}$ is *irrational.* Another irrational number is pi, the ratio of the circumference of a circle to its diameter. Pi, or π, has been used and studied for more than 4000 years.

Write and solve a proportion to answer the question "If 70 students out of 100 attended the party, how many students out of 750 attended?"

Let s represent the number of students who attended.

$$\frac{70}{100} = \frac{s}{750}$$ Write the proportion.

$$750 \cdot \frac{70}{100} = \frac{s}{750} \cdot 750$$ Multiply both sides by 750.

$$750 \cdot \frac{70}{100} = s$$ $\frac{750}{750}$ is equivalent to 1.

$$\frac{750}{10} \cdot \frac{70}{10} = s$$ 100 is 10 · 10.

$$75 \cdot 7 = s$$ You use the fact that $\frac{10}{10} = 1$ twice.

$$525 = s$$ Multiply.

Numbers that can be written as the ratio of two integers are called **rational numbers.**

EXERCISES

You will need your calculator for problem **1.**

► Practice Your Skills

1. Estimate the decimal equivalent for each fraction.

> Keep in mind these fraction and decimal equivalents:
>
> $\frac{1}{25} = 0.04 \qquad \frac{1}{20} = 0.05 \qquad \frac{1}{8} = 0.125 \qquad \frac{1}{5} = 0.2 \qquad \frac{1}{2} = 0.5 \qquad \frac{2}{1} = 2.0$

a. $\frac{7}{8}$ **b.** $\frac{13}{20}$ **c.** $\frac{13}{5}$ **d.** $\frac{52}{25}$

Now use your calculator to find the decimal number.

2. Ms. Lenz collected information about the students in her class.

Eye Color

	Brown eyes	Blue eyes	Hazel eyes
9th graders	9	3	2
8th graders	11	4	1

Write these ratios as fractions.

a. Ninth graders with brown eyes to ninth graders

b. Eighth graders with brown eyes to students with brown eyes

c. Eighth graders with blue eyes to ninth graders with blue eyes

d. All students with hazel eyes to students in both grades

3. Phrases such as miles per gallon, parts per million (ppm), and accidents per 1000 people indicate ratios. Write each ratio named below as a fraction. Use a number and a unit in both the numerator and the denominator.

a. In 2000, the McLaren was the fastest car produced. Its top speed was recorded at 240 miles per hour.

b. Pure capsaicin, a substance that makes hot peppers taste hot, is so strong that 10 ppm in water can make your tongue blister.

c. In 2000, women owned approximately 350 of every thousand firms in the United States.

d. The 2000 average income in Philadelphia, Pennsylvania, was approximately $35,500 per person.

4. Write three other true proportions using the four integers in each proportion.

a. $\dfrac{2}{5} = \dfrac{10}{25}$ **b.** $\dfrac{3}{7} = \dfrac{12}{28}$ **c.** $\dfrac{16}{20} = \dfrac{12}{15}$

5. Find the value of the unknown number in each proportion.

a. $\dfrac{24}{40} = \dfrac{T}{30}$ **b.** $\dfrac{49}{56} = \dfrac{R}{32}$ **c.** $\dfrac{52}{91} = \dfrac{42}{S}$ **d.** $\dfrac{100}{30} = \dfrac{7}{x}$

e. $\dfrac{M}{16} = \dfrac{87}{232}$ **f.** $\dfrac{6}{n} = \dfrac{62}{217}$ **g.** $\dfrac{36}{15} = \dfrac{c}{13}$ **h.** $\dfrac{220}{33} = \dfrac{60}{W}$

▶ Reason and Apply

6. APPLICATION Write a proportion for each problem, and solve for the unknown number.

a. Leaf-cutter ants that live in Central and South America weigh about 1.5 grams. One ant can carry a 4-gram piece of leaf that is about the size of a dime. If a person could carry proportionally as much as the leaf-cutter ant, how much could a 55-kilogram algebra student carry?

b. The leaf-cutter ant is about 1.27 cm long and takes strides of 0.84 cm. If a person could take proportionally equivalent strides, what size strides would a 1.65-m–tall algebra student take?

c. The 1.27-cm–long ants travel up to 0.4 km from home each day. If a person could travel a proportional distance, how far would a 1.65-m–tall person travel?

7. APPLICATION Jeremy has a job at the movie theater. His hourly wage is $7.38. Suppose 15% of his income is withheld for taxes and Social Security.

a. What percent does Jeremy get to keep?

b. What is his hourly take-home wage?

8. In a resort area during the summer months, only one out of eight people is a year-round resident. The others are there on vacation. If the year-round population of the area is 3000, how many people are there in the summer?

9. APPLICATION To make three servings of Irish porridge, you need 4 cups of water and 1 cup of oatmeal. How much of each ingredient will you need for two servings? for five servings?

10. APPLICATION When chemists write formulas for chemical compounds, they indicate how many atoms of each element combine to form a molecule of that compound. For instance, they write H_2O for water, which means there are two hydrogen atoms and one oxygen atom in each molecule of water. Acetone (or nail polish remover) has the formula C_3H_6O. The C stands for carbon.

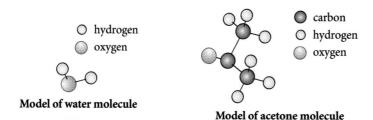

Model of water molecule

Model of acetone molecule

a. How many of each atom are there in one molecule of acetone?

b. How many atoms of carbon must combine with 470 atoms of oxygen to form acetone molecules? How many atoms of hydrogen are required?

c. How many acetone molecules can be formed from 3000 atoms of carbon, 3000 atoms of hydrogen, and 1000 atoms of oxygen?

Review

11. In the dot plot below, circle the points that represent values for the five-number summary. If a value is actually the mean of two data points, draw a circle around the two points.

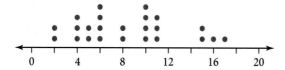

12. The *Forbes Celebrity 100* listed these ten people (and their incomes for 1999 in millions) among the celebrities that got the most media attention. Find the three measures of center for their incomes, and explain why they are so different.

George Lucas ($400)
Oprah Winfrey ($150)
Tom Clancy ($66)
Stephen King ($65)
Bruce Willis ($54)
Julia Roberts ($50)
Shania Twain ($48)
Tiger Woods ($47)
Mel Gibson ($45)
Jim Carrey ($45)

Tiger Woods

13. Use the order of operations to evaluate these expressions. Check your results on your calculator.

a. $5 \cdot -4 + 8$

b. $-12 \div (7 - 4)$

c. $-3 - 6 \cdot 25 \div 30$

d. $18(-3) \div 81$

THE GOLDEN RATIO

In this project, you'll research the amazing number mathematicians call the **golden ratio.** (There are plenty of books and web sites on the topic. Find links from www.keymath.com/DA .) Your project should include

▶ Basic information on the golden ratio, such as its exact value, why it represents a mathematically "ideal" ratio, and how to construct a golden rectangle. (Its length-to-width ratio is the golden ratio.)

▶ Some history of the golden ratio, including its role in ancient Greek architecture.

▶ At least one other interesting mathematical fact about the golden ratio, such as its relationship to the Fibonacci sequence or its own reciprocal.

▶ A report on where to find the golden ratio in the environment, architecture, or art. List items and their measurements or include prints from photographs, art, or architecture on which you have drawn the golden rectangle.

Once you've learned how to construct the golden ratio, The Geometer's Sketchpad is an ideal tool for further exploration. You can create a Custom Tool for dividing segments into the golden ratio, and then use this tool to help you construct the golden rectangle or even the golden spiral.

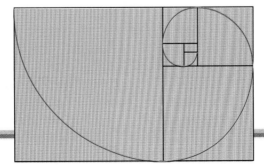

Capture-Recapture

Wildlife biologists often need to know how many deer are in a national park or the size of the perch population in a large lake. It is impossible to count each deer or fish, so biologists use a method called "capture-recapture" that uses ratios to estimate the population.

To estimate a fish population, biologists first capture some of the fish and put tags on them. Then these tagged fish are returned to the lake to mingle with all of the untagged fish. The biologist allows time for the fish to mix thoroughly, captures another **sample,** and counts how many of the sample are tagged.

The Universe is a grand book which cannot be read until one first learns to comprehend the language. . . . It is written in the language of mathematics.

GALILEO GALILEI

Investigation
Fish in the Lake

You will need

- a paper bag
- white beans
- red beans

In this investigation you'll **simulate** the capture-recapture method and examine how it works.

The bag represents a lake, the white beans are the untagged fish in the lake, and the red beans will replace white beans to represent tagged fish. Your objective is to estimate the total number of fish in the lake.

Step 1 | Reach into the lake and remove a handful of fish to tag. Count and record the number of fish you removed. Replace these fish (white beans) with an equal number of tagged fish (red beans). Return the tagged fish to the lake. Set aside the extra beans.

Step 2 | Allow the fish to mingle (seal the bag and shake it). Again remove a handful of fish, count them all, and count the number of tagged fish. In a table like this, record those counts and the ratio of tagged fish to total fish in the sample.

Tagging Simulation

Sample number	Number of tagged fish	Total number of fish	Ratio of tagged fish to total fish
1			
2			

You have taken one sample by randomly capturing some of the fish. You could use this sample to estimate the number of fish in the lake, but by taking several samples, you will get a better idea of the ratio of tagged fish to fish in the lake. Replace the fish, mix them, and repeat the sampling process four more times, filling in a row of your table each time.

Step 3	Choose one ratio to represent the five ratios. Explain how you decided this was a representative ratio.
Step 4	If you mixed the fish well, should the fraction of tagged fish in a sample be nearly the same as the fraction of tagged fish in the whole lake? Why or why not?
Step 5	Write and solve a proportion to find the number of fish in the lake. (About how many beans are in your bag?) Why is this method called capture-recapture? How accurate are predictions using this method? Why?

You can describe the results of capture-recapture situations using percents. Here are three kinds of percent problems—finding an unknown percent, finding an unknown total, and finding an unknown part. In each case, the percent equals the ratio of the part to the whole or total.

EXAMPLE A | **Finding an Unknown Percent**

In a capture-recapture process, 200 fish were tagged. From the recapture results, the game warden estimates that the lake contains 2500 fish. What percent of the fish were tagged?

largemouth bass

brook trout

yellow perch

▶ **Solution** | We know that the ratio of tagged fish (the part) to total fish in the lake (the whole) is 200 to 2500. This ratio is equivalent to the percent p of tagged fish in the sample.

$$\frac{p}{100} = \frac{200}{2500}$$ 　　　Write the proportion.

$$100 \cdot \frac{p}{100} = \frac{200}{2500} \cdot 100$$ 　　　Multiply both sides by 100.

$$p = 8$$ 　　　$\frac{100}{100} = 1$. Multiply and divide.

In the samples used for the estimate, 8% of the fish were tagged.

EXAMPLE B | **Finding an Unknown Total**

In a lake with 250 tagged fish, recapture results show that 11% of the fish are tagged. About how many fish are in the lake?

▶ Solution

We can write 11% as the ratio $\frac{11}{100}$ or 11 parts to 100 (the whole). The variable will be the denominator because the unknown quantity is the whole—the total number of fish in the lake.

$$\frac{11}{100} = \frac{250}{f}$$
Write the proportion.

$$\frac{100}{11} = \frac{f}{250}$$
Invert both ratios.

$$250 \cdot \frac{100}{11} = \frac{f}{250} \cdot 250$$
Multiply both sides by 250.

$$2273 \approx f$$
Multiply and divide.

There are approximately 2270 fish in the lake.

EXAMPLE C | **Finding an Unknown Part**

A lake is estimated to have 5000 fish after recapture experiments that showed 3% of the fish were tagged. How many fish were originally tagged?

▶ Solution

We can write 3% as the ratio $\frac{3}{100}$ or 3 parts to 100 (the whole). The variable is in the numerator because the unknown quantity is the number of tagged fish (the part).

$$\frac{t}{5000} = \frac{3}{100}$$
Write the proportion.

$$5000 \cdot \frac{t}{5000} = \frac{3}{100} \cdot 5000$$
Multiply both sides by 5000.

$$t = 150$$
Multiply and divide.

About 150 fish were tagged.

In addition to making estimates of wildlife populations, you can use proportions to estimate quantities in many other everyday situations.

EXERCISES

▶ Practice Your Skills

1. The proportion $\frac{320}{235} = \frac{g}{100}$ represents the question "320 is what percent of 235?" Write each proportion as a percent question.

a. $\frac{24}{w} = \frac{32}{100}$ **b.** $\frac{t}{450} = \frac{48}{100}$ **c.** $\frac{98}{117} = \frac{n}{100}$

2. You can write the question "What number is 15% of 120?" as the proportion $\frac{x}{120} = \frac{15}{100}$. Write each question as a proportion.

a. 125% of what number is 80?

b. What number is 0.25% of 46?

c. What percent of 470 is 72?

3. There are 1582 students attending the local high school. Seventeen percent of the students are twelfth graders. How many twelfth graders are there?

4. APPLICATION Write and solve a proportion for each situation.

    a. A biologist tagged 250 fish. Then she collected another sample of 75 fish, of which 5 were tagged. How many fish would she estimate are in the lake?

    b. A biologist estimated that there were 5500 fish in a lake in which 250 fish had been tagged. A ranger collected a sample in which there were 15 tagged fish. Approximately how many fish were in the sample the ranger collected?

▶ Reason and Apply

5. Marie and Richard played 47 games of backgammon last month. Marie's ratio of wins to losses was 28 to 19.

    a. Estimate the number of games you expect Marie to win if she and Richard play 12 more games. Explain your thinking.

    b. Write a proportion, and solve for Marie's expected number of wins if she and Richard play 12 more games.

    c. Write a proportion and solve it to determine how many games she and Richard will need to play before Richard can expect to win 30 games.

6. Jon opened a package of candies and counted them. He found 60 candies in the 1.69-ounce package.

    a. Estimate the number of candies in a 1-pound (16-ounce) bag. Explain your thinking for this estimate.

    b. Write the proportion and solve for the number of candies in a 1-pound package. Compare your approximation to your estimate in 6a.

    c. How much would 1 million candies weigh? Give your answer in pounds.

7. APPLICATION The year after a dictionary was published, two librarians notified the publisher of mistakes. One librarian found 43 errors, and the other found 62. The reprint editor noticed that 35 of the errors were mentioned by both librarians. He used the capture-recapture method to estimate the actual number of errors. How many errors are there likely to be? (Hint: The errors found by either librarian can be used as the number of tagged errors in the capture phase. The other librarian represents the recapture phase with the errors found by both being the tagged errors.)

▶ Review

8. The ratio of ninth graders to eighth graders in the class is 5 to 3. Write these ratios as fractions.

    a. Ninth graders to eighth graders

    b. Ninth graders to students in the class

    c. Eighth graders to students in the class

9. APPLICATION Sulfuric acid, a highly corrosive substance, is used in the manufacture of dyes, fertilizer, and medicine. Sulfuric acid is also used by artists for metal etching and in aqua tints. H_2SO_4 is the molecular formula for this substance. S stands for the sulfur atom. Use this information to answer each question.

a. How many atoms of sulfur, hydrogen, and oxygen are in one sulfuric acid molecule?

b. How many atoms of sulfur would it take to combine with 200 atoms of hydrogen? How many atoms of oxygen would it take to combine with 200 atoms of hydrogen?

c. If 500 atoms of sulfur, 400 atoms of hydrogen, and 400 atoms of oxygen are combined, how many sulfuric acid molecules could be formed?

This untitled drypoint and aquatint is by the American artist Mary Cassatt (1844–1926).

10. This histogram shows the ages of the first 43 presidents of the United States when they took office.

a. What is a good estimate of the median age?

b. How many presidents were younger than 50 when they took office?

c. What ages do not represent the age of any president at inauguration?

d. Redraw the histogram changing the interval width from 2 to 4.

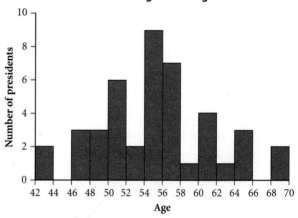

Presidents' Ages at Inauguration

11. On a group quiz, your group needs to calculate the answer to $12 - 2 \cdot 6 - 3$.
The three other group members came up with these answers:

Marta	57
Matt	−3
Miguel	30

Who, if anyone, is correct? What would you say to the other group members to convince them?

IMPROVING YOUR **REASONING** SKILLS

You have two containers of the same size; one contains juice and the other contains water. Remove one tablespoon of juice and put it into the water and stir. Then remove one tablespoon of the water and juice mixture and put it into the juice. Is there more water in the juice or more juice in the water?

2.3

Proportions and Measurement Systems

Have you ever visited another country? If so, you needed to convert your money to theirs and perhaps some of your measurement units to theirs as well. Many countries use the units of the Système Internationale, or SI, known in the United States as the metric system.

So instead of selling gasoline by the gallon, they sell it by the liter. Distance signs are in kilometers rather than in miles, and vegetables are sold by the kilo (kilogram) rather than by the pound.

Vegetables are sold at a French market.

Investigation
Converting Centimeters to Inches

You will need

- a yardstick or tape measure
- a meterstick or metric tape measure

In this investigation you will find a ratio to help you convert inches to centimeters and centimeters to inches. Then you will use this ratio in a proportion to convert some measurements from the system standard in the United States to measurements in the metric system, and vice versa.

Step 1   Measure the length or width on each of six different-sized objects, such as a pencil, a book, your desk, or your calculator. For each object, record the inch measurement and the centimeter measurement in a table like this.

Inches to Centimeters

Object	Measurement in inches	Measurement in centimeters

Step 2   Enter the measurements in inches in your calculator's list L1 and the measurements in centimeters in list L2. In list L3 enter the ratio of centimeters to inches, $\frac{L2}{L1}$, and let your calculator fill in the ratio values. [▶🖥 See **Calculator Note 1I.** ◀]

Step 3   How do the ratios of centimeters to inches compare for the different measurements? If one of the ratios is much different from the others, recheck your measurements. Choose a single representative ratio of centimeters to inches.

Step 4 | Choose a single representative ratio of centimeters to inches. Write a sentence that explains the meaning of this ratio.

Step 5 | Using your ratio, set up a proportion and convert each length.

a. 215 centimeters = x inches

b. 1 centimeter = x inches

c. 1 inch = x centimeters

d. How many centimeters high is a doorway that measures 80 inches?

In the investigation you found a common ratio or **conversion factor** between inches and centimeters. Once you've determined the conversion factor, you can convert from one system to the other by solving a proportion. If your measurements in the investigation were very accurate, the mean and median of the ratios were very close to the conversion factor, 2.54 centimeters to 1 inch.

Some conversions require several steps. The example offers a strategy called **dimensional analysis** for doing more complicated conversions.

EXAMPLE

A radio-controlled car traveled 30 feet across the classroom in 1.6 seconds. How fast was it traveling in miles per hour?

▶ **Solution**

Using the given information, you can write the speed as the ratio $\frac{30 \text{ feet}}{1.6 \text{ seconds}}$. Multiplying by 1 doesn't change the value of a number, so you can use conversion factors that you know (like $\frac{60 \text{ minutes}}{1 \text{ hour}}$) to create fractions with a value of 1. Then multiply your original ratio by those fractions to change the units.

$$\frac{30 \text{ feet}}{1.6 \text{ sec}} \cdot \frac{60 \text{ sec}}{1 \text{ min}} \cdot \frac{60 \text{ min}}{1 \text{ hour}} \cdot \frac{1 \text{ mile}}{5,280 \text{ feet}} = \frac{108,000 \text{ miles}}{8,448 \text{ hours}}$$

$$\approx \frac{12.8 \text{ miles}}{1 \text{ hour}} \text{ or } 12.8 \text{ miles per hour}$$

Each of the fractions after the first one has a value of 1 because the numerator and denominator of each fraction are equivalent: 60 sec = 1 min, 60 min = 1 hr, and 1 mi = 5280 ft. The fractions equivalent to 1 are chosen so that when units cancel, the result is miles in the numerator and hours in the denominator.

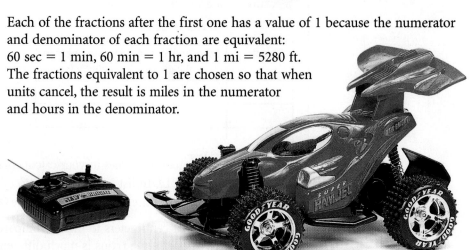

EXERCISES

▶ Practice Your Skills

1. Find the value of *x* in each proportion.

   a. $\dfrac{1 \text{ meter}}{3.25 \text{ feet}} = \dfrac{15.2 \text{ meters}}{x \text{ feet}}$
       b. $\dfrac{1.6 \text{ kilometers}}{1 \text{ mile}} = \dfrac{x \text{ kilometers}}{25 \text{ miles}}$

   c. $\dfrac{0.926 \text{ meter}}{1 \text{ yard}} = \dfrac{200 \text{ meters}}{x \text{ yards}}$
      d. $\dfrac{1 \text{ kilometer}}{0.6 \text{ mile}} = \dfrac{x \text{ kilometers}}{350 \text{ miles}}$

2. Which ratio in problem 1a is used as the conversion factor? Use it to determine how many feet you would have to run in a 50-meter dash.

3. Use dimensional analysis to change

   a. 50 meters per second to kilometers per hour.

   b. 0.025 day to seconds.

   c. 1200 ounces to tons (16 oz = 1 lb; 2000 lb = 1 ton).

4. Write a proportion and answer each question using the conversion factor 1 ounce ≈ 28.4 grams.

   a. How many grams does an 8-ounce portion of prime rib weigh?

   b. If an ice-cream cone weighs 50 grams, how many ounces does it weigh?

   c. If a typical house cat weighs 160 ounces, how many grams does it weigh?

   d. How many ounces does a 100-gram package of cheese weigh?

5. Write a proportion and answer each question using the conversion factor 1 inch = 2.54 centimeters.

   a. A teacher is 62.5 inches tall. How many centimeters tall is she?

   b. A common ceiling height is 96 inches (8 feet). About how high is this in centimeters?

   c. The diameter of a CD is 12 centimeters. What is its diameter in inches?

   d. The radius of a typical soda can is 3.25 centimeters. What is its radius in inches?

Reason and Apply

6. A group of students measured several objects around their school in both yards and meters. Use their data, shown in the table, to find a conversion factor between yards and meters.

Measurement in Yards and Meters

Yards	7	3.5	7.5	4.25	6.25	11
Meters	6.3	3.2	6.8	3.8	5.6	9.9

Use the conversion factor to answer these questions:

a. The length of a football field is 100 yards. How long is it in meters?

b. If it is 200 meters to the next freeway exit, how far away is it in yards?

c. How many yards long is a 100-meter dash?

d. How many meters of fabric should you buy if you need 15 yards?

7. a. Make a table like this showing the number of feet in lengths from 1 to 5 yards.

b. For each additional yard in your table, how many more feet are there?

c. Write a proportion that you could use to convert the measurements between y yards and f feet.

d. Use the proportion you wrote to convert each measurement.

    i. 150 yards = f feet   **ii.** 384 feet = y yards

Measurement in Yards and Feet

Yards	1	2	3	4	5
Feet					

8. A rod is a unit of measure that was used many years ago.

a. Using the table, find a common ratio that you can use to convert units between rods and feet.

b. Write a proportion using the ratio you found, and convert each measurement.

    i. 3.5 rods = x feet   **ii.** 15 feet = x rods

Measurement in Rods and Feet

Rods	1.2	2	3	4
Feet	20	33	50	66

9. APPLICATION When mixed according to the directions, a 12-ounce can of lemonade concentrate becomes 64 ounces of lemonade.

a. How many 12-ounce cans of concentrate are needed to make 120 servings if each serving is 8 ounces?

b. How many ounces of concentrate are needed to make 1 ounce of lemonade?

c. Write a proportion that you can use to find the number of ounces of concentrate based on the number of ounces of lemonade wanted.

d. Use the proportion you wrote to find the number of ounces of lemonade that can be made from a 16-ounce can of the same concentrate.

10. Recipes in many international cookbooks use metric measurements. One cookie recipe calls for 120 milliliters of sugar. How much is this in our customary unit "cups"? (There are 1000 milliliters in a liter, 1.06 quarts in a liter, and 4 cups in a quart.)

Review

11. The students in the mathematics and chess clubs worked together to raise funds for their respective groups. Together the clubs raised $480. There are 12 members in the Mathematics Club and only 8 in the Chess Club. How should the funds be divided between the two clubs? Explain your answer.

12. Draw a coordinate plane.

   a. Plot and label these points.
   $A(2, 1), B(1, -4), C(0, 4), D(-1, 0), E(-2, -1), F(-3, 1), G(4, 2)$

   b. Three of the points lie on a line. Name them and draw the line through them.

   c. Name the point whose coordinates are the median of the x-values and the median of the y-values. Plot it and label it M.

13. The box plot shows the length in centimeters of five members of the kingfisher family. Use this information to match each kingfisher to its length.

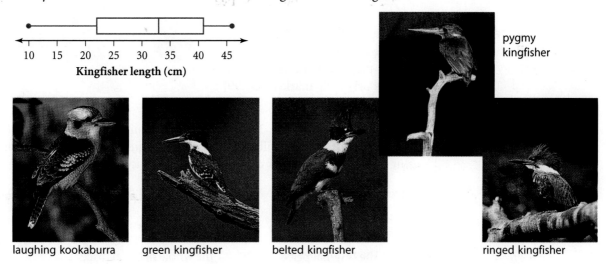

pygmy kingfisher

laughing kookaburra   green kingfisher   belted kingfisher   ringed kingfisher

- These kingfishers range in size from the tiny pygmy kingfisher to the laughing kookaburra.
- The best known kingfisher, the belted kingfisher, breeds from Alaska to Florida. It is only 2.6 centimeters longer than the mean kingfisher length.
- The ringed kingfisher, a tropical bird, is much closer to the median length than the green kingfisher.

IMPROVING YOUR **VISUAL THINKING** SKILLS

A pentomino is made up of five squares joined along complete sides. The first pentomino can be folded into an open box. The second pentomino can't.

Draw all 12 unique pentominoes, and then identify those that can be folded into open boxes.

Increasing and Decreasing

Be bold. If you're going to make a mistake, make a doozy, and don't be afraid to hit the ball.

BILLIE JEAN KING

Emily sees a sign saying "Special! Every CD is marked down 25%." She wants to know the discounted price on a $17.99 CD. Nathan works as a bagger in a grocery store, where he makes $8.25 per hour. His boss announces that all employees are getting a 2.5% raise. Nathan wants to know his new hourly rate.

These are different situations, but the pattern of reasoning to find solutions is the same for both. Example A uses a proportion to calculate a marked-down price.

EXAMPLE A

Birdbaths at the Feathered Friends store are marked down 35%. What is the cost of a birdbath that was originally marked $34.99?

▶ **Solution**

If an item is marked down 35%, then it must retain $100 - 35$ percent of its original price. That is, it will cost 65% of the original price. Set up a proportion using 65% and the ratio of cost, C, to original price.

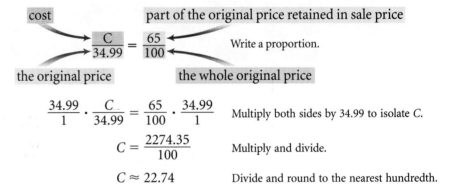

cost   part of the original price retained in sale price

$$\frac{C}{34.99} = \frac{65}{100}$$ Write a proportion.

the original price   the whole original price

$$\frac{34.99}{1} \cdot \frac{C}{34.99} = \frac{65}{100} \cdot \frac{34.99}{1}$$ Multiply both sides by 34.99 to isolate C.

$$C = \frac{2274.35}{100}$$ Multiply and divide.

$$C \approx 22.74$$ Divide and round to the nearest hundredth.

So the cost after the 35% markdown is $22.74.

Percent-increase calculations use the same reasoning as percent decreases. If there is a 6% sales tax on an item, then the cost including tax will be 106% of the price. (Remember: When a quantity increases in size, whether it is a price or a population, the new number will be greater than 100% of the original number.)

EXAMPLE B

Suppose you work at the Feathered Friends store. It's your job to make a chart showing the cost of the eight varieties of birdseed including the 6% sales tax.

All items are taxed at the same 6% rate, so the calculation will be the same for each birdseed variety. Whenever you have to repeat the same procedure over and over, it may be easier to let your calculator do the work for you in a list. You can enter the price of the feeds into list L1. In list L2 enter the correct formula for the cost including tax.

Price	Cost including 6% tax
8.25	
9.99	
10.75	
12.99	
15.99	
16.50	
17.95	
34.99	

To find the formula to enter into list L2, write a proportion. Multiply both sides by L1.

$$\frac{L_2}{L_1} = \frac{106}{100}$$

$$L_2 = L_1 \cdot \frac{106}{100}$$

The entries in list L2 will be the entries in list L1 times $\frac{106}{100}$, or 1.06. [▶  See **Calculator Note 1I.** ◀] You can set the mode on the calculator to show only two decimal places for all numbers. [▶ 🔲 See **Calculator Note 1J.** ◀]

L1	L2
8.25	8.75
9.99	10.59
10.75	11.40
12.99	13.77
15.99	16.95
16.50	17.49
17.95	19.27
34.99	37.09

Investigation
Enlarging and Reducing

You will need

- graph paper
- a ruler

Graphic artists and designers frequently make reductions and enlargements of drawings. In this investigation you'll explore how the dimensions change when a picture is enlarged or reduced.

Step 1   On graph paper, carefully draw a rectangle and one of its diagonals. Record its length, width, diagonal length, perimeter, and area in a table like this one.

Rectangle Measurements

	Original dimensions	Enlarged dimensions	Reduced dimensions
Length			
Width			
Diagonal			
Perimeter			
Area			

Step 2   Suppose you are a designer and need to make a copy that increases the length and width by 20%. Draw this new rectangle with its diagonal. Record the new enlarged length, width, diagonal, perimeter, and area.

Step 3   Now suppose you need a copy that decreases the length and width of the original rectangle by 20%. Draw this new rectangle with its diagonal. Record its new length, width, diagonal, perimeter, and area.

Step 4	Calculate the ratio of each new rectangle value to the corresponding original value. Describe how the ratios compare. Check your results with others in your group.
Step 5	Describe how other aspects of the figure, such as the diagonal, perimeter, and area, are changed when you increase or decrease the length and width of a rectangle by the same percent.
Step 6	Suppose the company that prints your school photos has a problem with its printing machines. It takes the original photo. But instead of increasing the length and width by 50% to make a large photo, it increases the length by 80% and the width by 30%. Describe what your large photo looks like.
Step 7	A tabletop can be completely covered by 100 dominoes. If the length and width are reduced by half, will it take 50 dominoes to cover it? If not, will it take more or fewer dominoes? Explain your thinking.

EXERCISES

You will need your calculator for problems **3, 5,** and **8.**

Practice Your Skills

1. If the length and width of a picture each have 150% of their value added to them, which of the statements are true and which are false?

   a. The picture is 50% wider and higher than it was before.

   b. The height and width of the picture are $2\frac{1}{2}$ times that of the original.

   c. The area of the picture is $2\frac{1}{2}$ times the area of the original.

   d. The area is more than 6 times the area of the original.

2. Write each percent change as a ratio comparing the result to the original quantity. For example, a 3% increase is $\frac{103}{100}$.

   a. 8% increase

   b. 11% decrease

   c. 12.5% growth

   d. $6\frac{1}{4}$% loss

   e. x% increase

   f. y% decrease

3. **APPLICATION** In 2000 the population of the United States was estimated to be 274,700,000. If the population grows by 1.1% per year, what would the population be in 2001? In 2002? In 2003? (Round to the nearest 1000 people.)

4. APPLICATION Justin wants to buy a computer. His mother will also use it, so she has agreed to pay 30% of the cost. The model Justin wants costs $2,649 including tax. How much will Justin need to pay?

5. This table shows the populations of the five most populous countries in the world in 2000.

Most Populous Countries in the World in 2000

Country	Population
China	1,261,832,000
India	1,014,004,000
United States	275,563,000
Indonesia	224,784,000
Brazil	172,860,000

a. The population of the world in 2000 was approximately 6,080,142,000. What percent of the world population was the total population of these five countries in 2000?

b. The population of India in 1990 was approximately 850,558,000. Approximately what percent increase does this indicate?

c. From 1990 to 2000, the population of China increased by roughly 10.8%. If the rate of population growth remains the same, what will the population be in 2010?

Reason and Apply

6. APPLICATION Raimy recently received a partial scholarship for college. Her first-year expenses, including tuition, textbooks, food, and housing, amount to $18,500. Her scholarship will cover 36% of her expenses. Raimy's parents say they can pay 35% of the balance. How much money will Raimy be responsible for?

7. APPLICATION Tamara works at a bookstore, where she earns $7.50 per hour.

a. Her employer is pleased with her work and gives her a 3.5% raise. What is her new hourly rate?

b. A few weeks later business drops off dramatically. The employer must reduce wages. He decreases Tamara's latest wage by 3.5%. What is her hourly rate now?

c. What is the final result of the two pay changes? Explain.

8. APPLICATION This table shows the revenues for six major film studios in 1992 and 1996.

a. Write out $3,115.2 million as a whole number showing all the zeros.

b. What was the increase in revenue for Walt Disney Company between 1992 and 1996? What percent is that of the 1992 profits?

c. Calculate the percent increase over the four-year period for the other studios.

d. Time Warner reported a 50.9% increase in revenue from 1992 to 1993. What was the percent increase from 1993 to 1996?

e. In 1994, Time Warner reported a 14.8% decrease in revenue from the previous year. What was its percent gain from 1994 to 1996?

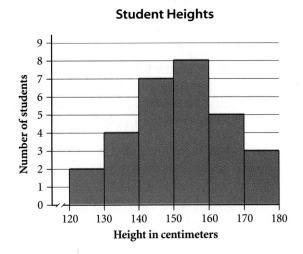

Film Studio Revenues

Film Studio	Revenue (in millions of dollars)	
	1992	1996
Walt Disney Company	3,115.2	10,505.0
20th Century Fox (News Corporation)	1,858.0	2,441.0
MCA Universal (Seagram)	4,709.4	5,413.0
Sony	2,475.3	3,003.0
Time Warner	3,945.0	6,103.0
Viacom	248.3	3,493.4

(*The Veronis, Shuler & Associates Communications Industry Report,* October 1997, p. 155)

Review

9. This histogram gives the height of the students in Mr. Moore's algebra class. Create a data set that corresponds to the graph.

Student Heights

10. APPLICATION Ryan needs to be able to easily make larger or smaller portions of the pancake recipe used in the Pancake Company, where he is a chef. Here is a matrix he created of the ingredients:

$$\begin{bmatrix} 150 \\ 2 \\ 18 \\ 3.5 \\ 200 \\ 19 \\ 125 \end{bmatrix}$$

Cups flour
Cups salt
Cups sugar
Cups baking powder
Eggs
Cups melted butter
Cups milk

He wants to be able to make a half-size recipe, a recipe that is 50% larger than the original, and a double recipe.

a. What 1×3 matrix should Ryan multiply the ingredient matrix by to calculate how much of each ingredient to use in the different-size portions?

b. Ryan can't remember if the order matters when he multiplies matrices. In what order should he multiply the original matrix and the 1×3 matrix to get a meaningful answer?

c. Calculate the resulting ingredient matrix, and label its rows and columns.

11. Draw a coordinate plane on graph paper. Label the axes and each of the four quadrants—I, II, III, and IV.

a. Mark points $W(5, 3)$ and $Q(-4, -2)$, and label them with their letter names. Sketch a rectangle that has diagonally opposite vertices W and Q. List the coordinates of the other two vertices, and identify the axis or quadrant location of each.

b. Sketch at least one other rectangle that has W and Q as vertices. List the coordinates of the other two vertices and identify their locations.

IMPROVING YOUR **REASONING** SKILLS

Three children went camping with their parents, a dog, and a tin of cookies just for the children. They agreed to share the cookies equally.

The youngest child couldn't help thinking about the cookies, so alone she divided the cookies into three piles. There was one left over. She gave it to the dog, took her share, and left the rest of the cookies in the cookie tin.

A little later the middle child took the tin where he could be alone and divided the cookies into three piles. There was one left over. He gave it to the dog and took his share.

Not too long after that, the oldest child went alone to divide the cookies. When she made three piles, there was one left over, which the dog got. She took her share and put the rest back in the tin.

After dinner, the three children "officially" divided the contents of the cookie tin into three piles. There was one left over, which they gave to the dog.

What is the smallest number of cookies that the tin might have first contained?

Circle Graphs and Relative Frequency Graphs

Circle graphs, like bar graphs, summarize data in categories. **Relative frequency graphs** also summarize data in categories, but instead of including the actual number for each category, they compare the number in that category to the total for all the categories. Relative frequency graphs show fractions or percents, not values.

Investigation
Circle Graphs and Bar Graphs

You will need

- graph paper
- a protractor
- a compass or circle template
- a ruler

The bar graph shows the approximate land area of the seven continents.

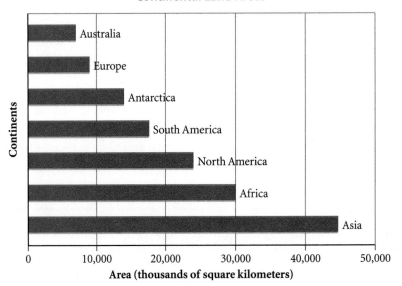

Continental Land Areas

Step 1	Determine from the bar graph the approximate area of each continent and the total land area.
Step 2	Convert the data in the bar graph to a circle graph. Use the fact that there are 360 degrees in a circle. Write proportions to find the number of degrees in each sector of the circle graph.
Step 3	Convert the data in the bar graph to a relative frequency circle graph. Instead of showing the land area, the graph will show percents of total land area.
Step 4	Convert the data in the bar graph to a relative frequency bar graph that shows percents rather than land areas.
Step 5	Compare the graphs you made with the original graph. What advantages are there to each kind of graph?

EXAMPLE

Randy has been asked to create a graphical display showing the distribution of the library's collection in six categories. His boss has asked him to create two rough drafts. Together they will decide which one to finalize for the display.

Here is the data:

Library Collection

Category	Number of Items
Children's fiction	35,994
Children's nonfiction	28,106
Adult fiction	48,129
Adult nonfiction	69,834
Media	11,830
Other	5,766
Total	199,659

▶ **Solution**

Randy puts the number of items in each category into list L1. He wants the calculator to determine in list L2 the number of degrees needed for each sector. He writes a proportion to find the number of degrees in the sector for a particular category.

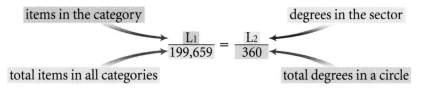

items in the category   degrees in the sector

$$\frac{L1}{199,659} = \frac{L2}{360}$$

total items in all categories   total degrees in a circle

By multiplying both sides of the proportion by 360, he finds the formula to enter into list L2.

$$L2 = L1 \cdot \frac{360}{199,659}$$

His calculator quickly determines the number of degrees for each sector of the circle graph. Using a protractor to measure the angles, Randy creates the graph.

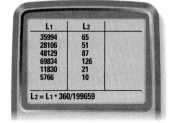

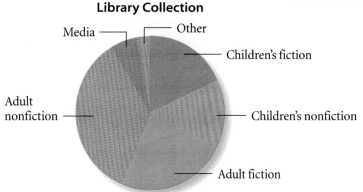

Library Collection

For a relative frequency graph, Randy finds the percent of the total represented by each category. He uses list L1 again, and the proportion

$$\frac{L1}{199{,}659} = \frac{L3}{100}$$

He solves for L3 and enters the formula that will give him the percent.

L1	L2	L3
35994	65	18
28106	51	14
48129	87	24
69834	126	35
11830	21	6
5766	10	3

L3 = L1 * 100/199659

He makes a relative frequency circle graph by putting these percents in his circle graph. He uses the same percents to create a relative frequency bar graph.

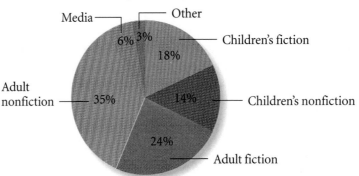

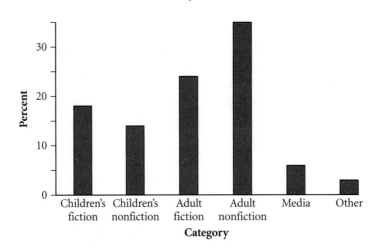

Randy and his boss decide to use the circle graph. They think it shows better how each category compares to the library's total collection.

In Chapter 1, you created box plots that gave a visual summary of the data. Like relative frequency graphs, they didn't contain the actual data values.

EXERCISES

Practice Your Skills

1. **APPLICATION** There are four basic blood types. The distribution of these types in the general population is shown in the relative frequency circle graph. In a city of 75,000 people, how many people with each blood type would you expect to find?

Blood Types

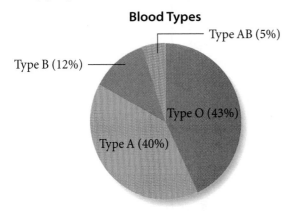

Type AB (5%)
Type B (12%)
Type O (43%)
Type A (40%)

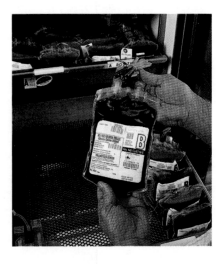

2. Which data set matches the relative frequency circle graph at right?

a. {15, 18, 22, 25, 28}   **b.** {20, 24, 30, 36, 45}

c. {12, 18, 24, 30, 36}   **d.** {9, 12, 18, 20, 24}

3. In the bar graph of the library's collection created in the example, the bar for adult fiction represents 24%. Could there be a situation where all the bars represented 24%? Explain your thinking.

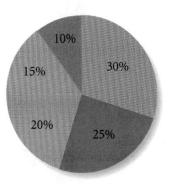

10%
15%
30%
20%
25%

Reason and Apply

4. A manufacturer states that it produces colored candies according to the percents listed in this table. Create a circle graph to show this information.

Colored Candies Manufactured

Orange	Yellow	Blue	Red	Green	Brown
10%	20%	10%	20%	10%	30%

5. Chloe bought a small package of these candies and counted the number of each color. Her count is shown in the table at right.

Chloe's Colored Candies

Orange	Yellow	Blue	Red	Green	Brown
11	10	4	12	7	14

    a. Construct a relative frequency bar graph for Chloe's package of candies.

    b. Construct a relative frequency bar graph that shows on one graph both Chloe's small package of candies and the percents stated by the candy manufacturer. Use one color for the bar representing Chloe's candies and a different color for the bar representing the manufacturer's. Include a key showing what the two bar colors mean. What conclusions can you make?

6. Match each bar graph with its corresponding relative frequency circle graph. Try to do this without calculating the actual percents.

a.
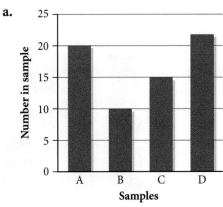

b.
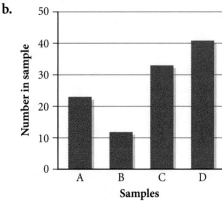

c.
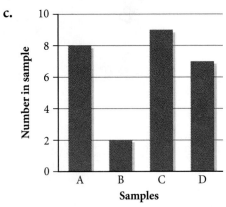

d.
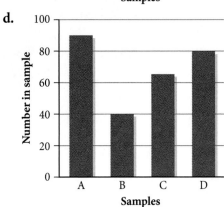

i.
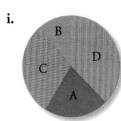

ii.
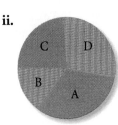

iii.
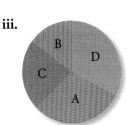

iv.
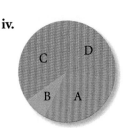

7. This table shows the number of students in each grade at a high school.

Class Size

Ninth grade	Tenth grade	Eleventh grade	Twelfth grade
185	175	166	150

a. What percent of the school is represented in each grade?

b. At semester break, the student population is counted again. The ninth grade has increased by 2%, the tenth grade has decreased by 1.5%, the eleventh grade has increased by 2.5%, and the twelfth grade has decreased by 2%. How many students are in each grade at the beginning of the second semester? By what percent has the total school population changed? What is the actual change in the number of students?

c. Make a relative frequency circle graph for the situation at the beginning of the year and another circle graph for the situation at the beginning of the second semester. How has the distribution of students changed?

Review

8. Evaluate these expressions. Check your results on your calculator.

   a. $-3 \cdot -7 \cdot (8 + -5)$
   b. $(-5 - 7) \cdot -4 + 17$
   c. $6 + -2 \cdot (7 + 9)$
   d. $14 - (25 + 8 - 32)$
   e. $1 + 6 \cdot 5 - 11$
   f. $33 + 5 \cdot -8$

9. Astrid works as an intern in a windmill park in Holland. She has learned that the anemometer gives off electrical pulses and that the pulses are counted each second. The ratio of pulses per second to wind speed in meters per second is always 4.5 to 1.

   a. If the wind speed is 40 meters per second, how many pulses per second should the anemometer be giving off?

   b. If the anemometer is giving off 84 pulses per second, what is the wind speed?

10. **APPLICATION** In 1999 there were 3142 counties in the United States. Here is data on five counties:

Fastest-Growing Counties between 1990 and 1999

County	1990 population	1999 population	Change from 1990 to 1999	Percent growth
Douglas County, CO	60,391	156,860		
Forsyth County, GA	44,083	96,686		
Elbert County, CO	9,646	19,757		
Park County, CO	7,174	14,218		
Henry County, GA	58,741	113,443		

(U.S. Bureau of the Census, 2000)

a. For each county, calculate the change in population from 1990 to 1999, and use it to calculate the percent of growth. Which county had the largest percent of growth in this time period?

b. Los Angeles County in California is the largest county in the country. Its population was 9,329,989 in 1999 and 8,863,164 in 1990. By what percent did the population of Los Angeles County grow during those nine years?

c. How does the growth of Los Angeles County compare to the population growth of the fastest-growing county? Which do you think is a better representation of the growth of a county, the percent of change or the actual number by which it grew?

San Fernando Valley, part of Los Angeles County

Probability Outcomes and Trials

Meteorologists know that out of ten days with a particular set of atmospheric conditions, it will probably rain on three of those days. When they see that set of conditions, they say, "The chance of rain is 30%."

The theory of probability is at bottom only common sense reduced to calculation.

PIERRE SIMON DE LAPLACE

The chance that something will happen is called its **probability.** The probability of an outcome is a ratio of the number of ways or times (in this case days) that outcome (rain) will occur, to the total number of ways or times (in this case days) under consideration. You can express that ratio as a percent.

The chance of rain can never be less than 0% or more than 100%, so a probability expressed as a decimal will always be between 0 and 1.

EXAMPLE

As part of Shandra's job with the forest service she tagged a total of 1470 squirrels last year. She tagged 820 black male squirrels, 100 black female squirrels, 380 gray male squirrels, and 170 gray female squirrels. If this distribution accurately reflects the squirrel population, what is the probability that the next squirrel she tags will be a gray male squirrel? A female squirrel? A red squirrel?

▶ **Solution**

The probability of the next squirrel tagged being a gray male squirrel can be expressed as the ratio

$$\frac{\textit{number of gray male squirrels tagged}}{\textit{total number of squirrels tagged}} = \frac{380}{1470} \approx 0.26$$

The probability of the next squirrel being female is

black female squirrels   gray female squirrels

$$\frac{\textit{number of female squirrels tagged}}{\textit{total number of squirrels tagged}} = \frac{100 + 170}{1470} \approx 0.18$$

During the last year Shandra hasn't tagged any red squirrels. Based on that information, the probability of a squirrel being red is 0.

$$\frac{\textit{number of red squirrels tagged}}{\textit{total number of squirrels tagged}} = \frac{0}{1470} = 0$$

In the example, each time a squirrel is tagged is a **trial.** Shandra conducted 1470 individual trials. Each possible result of a trial, is an **outcome.** Possible outcomes include a gray male squirrel, a black squirrel, or not a female squirrel.

An **observed probability** such as tagging a squirrel is based on experience or on collected data. Its usefulness is limited by the amount of data collected and the assumptions that conditions will remain the same.

The forest service can't know for sure the kinds and numbers of squirrels in the forest. If it did, it would use the known quantities to calculate the **theoretical probability** that the next squirrel to be tagged is gray. The notation P(gray squirrel) stands for the theoretical probability that the outcome of a trial will be "gray squirrel."

$$P(\text{gray squirrel}) = \frac{\textit{number of gray squirrels in the forest}}{\textit{total number of squirrels in the forest}}$$

Investigation
Candy Colors

You will need

- a packet of colored candies
- a paper bag

In the previous lesson you learned about *relative frequency.* This investigation will show you why the observed probability is sometimes also called the relative frequency. You will also work with known quantities to calculate theoretical probabilities.

Step 1 | Use a table like this one that lists the candy colors across the top to record the results of each trial.

	Experimental Outcomes						Total trials
	Red	Orange					
Tally							40
Experimental frequency							
Observed probability (relative frequency)							

Put the candies in the paper bag, then randomly select a candy by reaching into the bag without looking and removing a candy. Record the color as a tally mark, then replace the candy into the bag before the next person reaches in. Take turns removing, tallying the color, and replacing pieces of candy for a total of 40 trials. Your total for each color category is called its **experimental frequency.**

Record the experimental frequency for each outcome (color) in your table.

Step 2

From the experimental frequencies and the total number of trials (40), you can calculate the relative frequency or observed probability of each color. For instance, the observed probability of removing a red candy will be

$$\frac{number\ of\ red\ candies\ drawn}{total\ number\ of\ trials}$$

Record the observed probability in the bottom row of the table. Do you see why observed probability is also called relative frequency? How can you show these numbers as percents? What should the total be?

When all the candies are put into a bag, drawing one candy from the bag has several possible outcomes—the different colors listed on your table. Each individual candy is equally likely to be drawn, but some colors have a higher probability of being drawn than others.

Step 3

Make a second table.

	Outcomes						
	Red	Orange					Total
Number of candies counted							
Theoretical probability							

Dump out all the candies and count the number of candies of each color. Record this information in the top row.

Step 4

Use the known quantities in the first row to calculate the theoretical probability of drawing each candy color. For example,

$$P(R) = \frac{number\ of\ red\ candies\ in\ the\ bag}{total\ number\ of\ candies\ in\ the\ bag}$$

Record the results in the bottom row of the table.

Step 5

Is one color most likely to be drawn? Least likely? Explain the differences you found in the theoretical probabilities of drawing the different colors from the bag. What should their total be?

Step 6

Write a paragraph comparing your results for the theoretical probabilities you just calculated to the relative frequencies you calculated from your experiment.

In probability, an outcome is something that may or may not happen. When looking at the probability of a particular outcome, you must first ask yourself, "What outcomes are possible?" and "Are the outcomes equally likely?" If a packet of candy has exactly the same number of each color of candy, outcomes for each individual color, such as "green," are **equally likely.**

EXERCISES

You will need your calculator for problem **10**.

Practice Your Skills

1. For each trial, list the possible outcomes.

   a. tossing a coin

   b. rolling a die with faces numbered 1–6

   c. the sum when rolling 2 six-sided dice

   d. spinning the pointer on a dial divided into sections A–E

2. The table below shows the distribution by fragrance of candles in a 20-candle assortment pack.

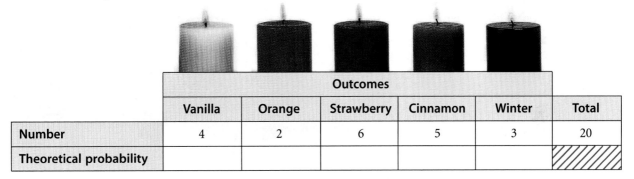

	Outcomes					
	Vanilla	**Orange**	**Strawberry**	**Cinnamon**	**Winter**	**Total**
Number	4	2	6	5	3	20
Theoretical probability						

   a. Copy the table and record in the bottom row the probability of selecting at random that type of candle.

   b. Suppose these 20 candles are put into a box. If you reach into the box without looking, what is the probability that you will pull out either a strawberry or a cinnamon candle? In other words, what is P(S or C)?

   c. What is P(W or S or V)?

   d. Suppose all 20-candle assortment packs made by this company have the same number of each type of candle listed above. If you empty ten assortment packs into a huge box, what is P(C) for the huge box? Explain why this is so.

3. One hundred tiny cubes were dropped onto a circle like the one at right, and all 100 cubes landed inside the circle. Twenty-seven cubes were completely or more than halfway inside the shaded region.

   a. Based on what happened, what is the observed probability of a cube landing in the shaded area?

   b. What is the theoretical probability in this situation? Explain your answer.

4. Igba-ita ("pitch and toss") is a favorite recreational game in Africa. In one version of Igba-ita, four cowrie shells are thrown in an effort to get a favorable outcome of all four up or all four down. Now coins are often used instead of cowrie shells and the name has changed to Igba-ego ("money toss"). Using four coins, what are the chances for an outcome in which all four land heads up or all four land tails up?

(Claudia Zaslovsky, *Africa Counts*, 1973, p. 113)

You can learn about other cultural games with the links at **www.keymath.com/DA** .

5. Draw and label a segment like the one below.

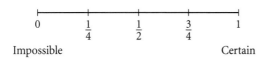

$$0 \qquad \frac{1}{4} \qquad \frac{1}{2} \qquad \frac{3}{4} \qquad 1$$

Impossible   Certain

Plot and label points on your segment to represent the probability for each situation.

a. You will eat breakfast tomorrow morning.

b. It will rain or snow sometime during the next month in your hometown.

c. You will be absent from school fewer than five days this school year.

d. You will get an A on your next mathematics test.

e. The next person to walk in the door will be under 30 years old.

f. Next Monday every teacher at your school will give 100 free points to each student.

g. Earth will rotate once on its axis in the next 24 hours.

6. Suppose that 350 beans are randomly distributed on the rectangle shown at right and that 136 beans lie either totally inside the shaded region or more than halfway inside. Use this information to approximate the area of the shaded region.

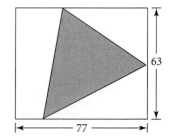

▶ Reason and Apply

7. Dr. Lynn Rogers of the North American Bear Center does research on bear cub survival. He observed 35 litters in 1996. The distribution of cubs is shown in this table.

Bear Litter Study

Number of cubs	1	2	3	4
Number of litters	2	8	22	3

(*The North Bearing News,* July 1997)

a. Describe a trial for this situation. Name one outcome.

b. Is each outcome equally likely? Explain.

c. Based on the given information, what is the probability that a litter will be three cubs?

8. Twenty randomly chosen high school students were asked to estimate the percent of students in their school who are planning to attend college. Base your answers to the questions on their responses.

Student Responses			
25	45	60	90
70	75	50	33
35	20	65	65
55	80	85	70
65	50	75	60

   a. Draw a dot plot to organize the data.

   b. What are the chances that the next student asked will give an estimate of at least 75%?

   c. If there are 4500 students in this high school, how many students do you think will give an estimate greater than or equal to 50%?

9. In the Wheel of Wealth game, contestants spin a large wheel like the one at right to see how much money each question is worth.

   a. What is the probability that a contestant will have a question worth $500?

   b. What is the probability that a contestant will have a question worth less than $500?

   c. If one contestant spins the wheel and it lands in the $400 section, what is the probability that the next contestant will spin the wheel and have a question worth more than $400?

10. The Candy Coated Carob Company produces six different-colored candies with colors distributed as shown in the circle graph.

Carob Candy colors

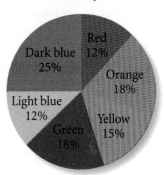

   a. You could use the numbers between 1 and 100 to represent all the candies and choose numbers in this range to represent the percent of each color. For example, because P(red) = 12%, let the numbers from 1 to 12 represent a red candy. The next interval, which will represent orange, has to have 18 numbers because P(orange) = 18%. Therefore, let this interval be the numbers from 13 to 30. Identify intervals to represent the yellow, green, light-blue, and dark-blue candies. Make a table like this one and fill in the intervals.

	Outcomes						Total
	Red	Orange	Yellow	Green	Light blue	Dark blue	
Interval	1 to 12						///
Number of candies							50
Probability							///

   b. Enter a calculator routine that will generate a list of 50 random integers from 1 to 100. [▶🖳 See **Calculator Note 2A.** ◀]

   c. Determine how many of each color you have in your collection. (You may want to sort, or order, your list first. [▶🖳 See **Calculator Note 2B.** ◀] Record the results in your table.

   d. Calculate the probability of selecting a particular color at random from your package.

▶ Review

11. Give four pairs of coordinates that would create a shape like this when connected.

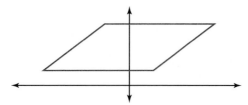

12. Michelle's hair grows $\frac{1}{2}$ inch each month.

    a. About how fast is that in meters per year? Use 2.54 cm = 1 in. and any other conversion factors you need.

    b. About how many years would it take for her hair to grow 1 meter?

13. **APPLICATION** The star hitter on the baseball team at City Community College was batting .375 before the start of a three-game series. During the three games, he came to the plate to bat eleven times. In these eleven plate appearances, he walked twice and had a sacrifice bunt. He either got a hit or made an out in his other plate appearances. If his batting average was the same at the end of the three-game series as at the beginning, how many hits did he get? (Note: Batting average is calculated by dividing hits by times at bat; sacrifice bunts and walks do not count as times at bat.)

PROBABILITY, GENES, AND CHROMOSOMES

How are we or how are we not like our parents? In this project you will use probability ratios to describe how an individual's gender, eye color, color blindness, blood type, or other characteristic can be traced to his or her parents.

As you know, the number of girls and boys is not equal in every family. However, if a couple could have hundreds of children, over the long run, half would be girls and half boys. The same simple relationship does not hold for eye color. Parents who both have brown eyes may have a child with blue eyes. Or a father with blue eyes and a mother with brown eyes may never have a child with blue eyes.

Research the difference between the way gender is determined and the way eye color, or another human trait, is determined. Write a paper or develop a presentation that describes these probabilities.

Random Outcomes

*Prediction is difficult,
especially of the future.*

NIELS BOHR

Mario is an entomologist who
studies the behavior of bees. He
videotapes bees leaving a hive for
one day and counts 247 bees flying
east and 628 bees flying west. Can he
predict what the next bee will do?
Can he use the results of his study to
predict approximately how many of
the next 100 bees will fly east? Will
the videotape counts be the same if
he repeats the study a few days later? Is
there a pattern to the bees' flying
direction, or does it appear that the bees
fly randomly either east or west?

An outcome is **random** when you can't predict what will happen on the next trial. If
three bees always fly west after each eastbound bee, then this action is not random.
When a bee leaves a hive, he is following instinct and instructions from other bees.
To the bee, his actions are not random. But unless an observer can see a pattern
and predict the direction of the next bee, the pattern is random to the observer.

EXAMPLE

In the story, the entomologist cannot predict what the next bee will do. But he
can use the results of his study to predict approximately how many of the next
100 bees will fly east.

▶ Solution

The observed probability that a bee will fly west is

$$\frac{number\ of\ bees\ that\ flew\ west}{all\ bees\ observed} = \frac{628}{628\ +\ 247} = \frac{628}{875}$$

The observed probability that a bee will fly east is

$$\frac{number\ of\ bees\ that\ flew\ east}{all\ bees\ observed} = \frac{247}{628\ +\ 247} = \frac{247}{875}$$

Mario can calculate the *probability* of what the next bee will do, but he can't
predict its *actual direction*. From Mario's perspective the outcome is random.

The probability ratio $\frac{247}{875}$ is about 0.28, or 28%.

He can expect about 28 out of 100 bees to fly east. But this is a probability, not a
fact. He should not be surprised by 26 or 30 bees flying east. But if 50 or more
of the bees fly east, he might conclude that the conditions have changed and his
observations of yesterday no longer help him determine the probabilities for
today.

When you toss a coin, you cannot predict whether it will show heads or tails
because the outcome is random. You do know, however, that there are two equally
likely outcomes—heads or tails. Therefore, you know that the theoretical
probability of getting a head is $\frac{1}{2}$. What happens when you toss a coin many times?

Investigation
Calculator Coin Toss

In this investigation you will compare a theoretical probability with an observed probability from 100 trials. You will look at how the observed probability is related to the number of trials.

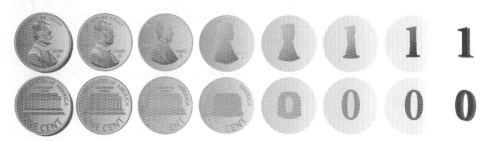

You could do this investigation by tossing coins 100 times, or you can use your calculator to simulate tossing many coins in a very short time.

Step 1 To number your tosses, enter the sequence of numbers from 1 to 100 into list L1 on your calculator. [▶ 🖳 See **Calculator Note 2B.** ◀]

Step 2 If a calculator randomly chooses 0 or 1, that's just like flipping a coin and getting tails or heads. Let 0 represent tails and 1 represent heads. Enter 100 randomly generated 0's and 1's into list L2. [▶ 🖳 See **Calculator Note 2A.** ◀]

Step 3 Display the cumulative sum of list L2 (number of heads) in list L3. [▶ 🖳 See **Calculator Note 2B.** ◀]

The table below shows an example in which the result of nine tosses was T, H, T, H, H, T, T, H, T. The numeral 1 in list L2 indicates heads. What does it mean if the eighth and ninth values in list L3 are both 4?

Step 4 To calculate the ratio of heads to total number of tosses, enter $\frac{L3}{L1}$ in list L4. What does this ratio represent?

Number of flips (L1)	Result of last flip (L2)	Total number of heads (L3)	Total heads / Total tosses (L4)
1	0	0	0
2	1	1	0.50
3	0	1	0.33
4	1	2	0.50
5	1	3	0.60
6	0	3	0.50
7	0	3	0.43
8	1	4	0.50
9	0	4	0.44
⋮	⋮	⋮	⋮

Step 5	Create a scatter plot using list L1 as the *x*-values and list L4 as the *y*-values. Name an appropriate graphing window for this plot.
Step 6	Enter the theoretical probability of tossing a head in Y1 on the Y= screen. Graph your equation on the same screen as your scatter plot from Step 5.
Step 7	Compare your plot to that of other members of your group. Describe what appears to happen after 100 trials. What would you expect to see if you continued this experiment for 150 trials? Make a sketch of your predicted graph of 150 trials. Run the calculator simulation. [▶ See **Calculator Note 2C.** ◀] Compare the results to your prediction.
Step 8	Explain what happens to the relationship between the theoretical probability and the observed probability as you do more and more trials.

If you tossed a coin many times, you would expect the ratio of the number of heads to the number of tosses to be close to $\frac{1}{2}$. The more times you toss the coin, the closer the ratio of heads to total tosses will be to $\frac{1}{2}$. With random events, patterns often emerge in the long run, but these patterns do not help predict a particular outcome.

When flipping a coin, you know what the theoretical probabilities are for heads and tails. However, in some situations you cannot calculate the theoretical probability of an outcome. After performing many trials, you can determine an observed probability based on your experimental results.

EXERCISES

You will need your calculator for problems **5** and **7**.

▶ Practice Your Skills

1. **APPLICATION** Suppose there are 180 twelfth graders in your school, and the school records show that 74 of them will be attending college outside their home state. You conduct a survey of 50 twelfth graders, and 15 tell you that they will be leaving the state to attend college. What is the theoretical probability that a random twelfth grader will be leaving the state to attend college? Based on your survey results, what is the observed probability? What could explain the difference?

2. **APPLICATION** Last month it was estimated that a lake contained 3500 rainbow trout. Over a three-day period a park ranger caught and tagged 100 fish. Then, after allowing two weeks for random mixing, she caught 100 more rainbow trout and found that 3 of them had tags.

   a. What is the probability of catching a tagged trout?

   b. What assumptions must you make to answer 2a?

   c. Based on the number of tagged fish she caught two weeks later, what is the park ranger's observed probability?

3. Suppose 250 people have applied for 15 job openings at a chain of restaurants.

   a. What fraction of the applicants will get a job?

   b. What fraction of the applicants will not get a job?

   c. Assuming all applicants are equally qualified and have the same chance of being hired, what is the probability that a randomly selected applicant will get a job?

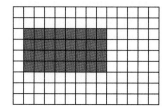

and full-time. Low stress work environment. Excellent benefits package. Call 555-7231, M-F, 9 to 5.

Restaurant Workers

Many positions open at new restaurant chain in desirable downtown location, from entry level to management. Excellent benefits. Apply in person to Dave Lee, 2100 Buena Vista Avenue. No phone calls or emails please.

Reason and Apply

4. If 25 randomly plotted points landed in the shaded region shown in the grid, about how many points do you estimate were plotted?

5. In a random walk, you move according to rules with each move being determined by a random process. The simplest type of random walk is a one-dimensional walk where each move is either one step forward or one step backward on a number line.

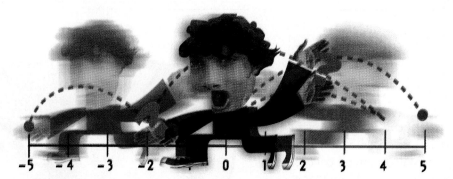

   a. Start at 0 on the number line and flip a coin to determine your move. Heads means you take one step forward to the next integer, and tails means you take one step backward to the previous integer. What sequence of six tosses will land you on the number-line locations $+1, +2, +1, +2, +3, +2$?

   b. Explore a one-dimensional walk of 100 moves using a calculator routine that randomly generates $+1$ or -1.

   In list L1, generate random numbers with 1 representing a step forward and -1 representing a step backward. Describe what you need to do with list L1 to show your number-line location after every step. [▶🖳 See **Calculator Notes 2A** and **2B.** ◀]

   c. Describe the results of your simulation. Is this what you expected?

   d. Would increasing the number of steps affect the results? Explain.

6. A thumbtack can land "point up" or "point down."

 a. When you drop a thumbtack on a hard surface do you think the outcomes will be equally likely? If not, what would you predict for P(up)?

 b. Drop a thumbtack 100 times onto a hard surface, or drop 10 thumbtacks 10 times. Record the frequency of "point up" and "point down." What are your observed probabilities for the two responses?

 c. Make a prediction for the probabilities on a softer surface like a towel. Repeat the experiment over a towel. What are your observed probabilities?

7. APPLICATION A teacher would like to use her calculator to randomly assign her 24 students to 6 groups of 4 students each. Create a calculator routine to do this.

▶ Review

8. The points listed here form a letter of the alphabet when they are connected in the proper order. What letter is it?

   $(-1, 1)$ $(-1, -1)$ $(-1, -2)$ $(1, 1)$ $(2, 0)$ $(0, 1)$ $(-1, -4)$
   $(0, -2)$ $(-1, 0)$ $(-1, -3)$ $(2, -1)$ $(1, -2)$ $(-1, -5)$ $(-1, -6)$

9. APPLICATION Zoe is an intern at Yellowstone National Park. One of her jobs is to estimate the chipmunk population in the campground areas. She starts by trapping 60 chipmunks, giving them a checkup, and banding their legs. A few weeks later, Zoe traps 84 chipmunks. Of these, 22 have bands on their legs. How many chipmunks should Zoe estimate are in the campgrounds?

10. If a number has an exponent, write it in standard form. If a number is in standard form, write it with an exponent other than 1.

 a. 4^3 b. $\left(\dfrac{1}{6}\right)^2$ c. $\left(\dfrac{3}{4}\right)^2$

 d. 27   e. $\dfrac{1}{125}$ f. $\dfrac{4}{81}$

11. Explain how to use probability to find the area of the irregular shape in the rectangle.

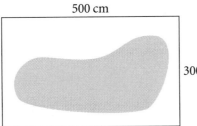

500 cm

300 cm

CHAPTER 2 REVIEW

In this chapter you explored and analyzed relationships among ratios, proportions, and percents. You used a **variable** to represent an unknown number, defined a proportion using the variable, and then solved the proportion to determine the value of the variable. You learned that a ratio of two integers is a **rational number** and that decimal representations of rational numbers either **terminate** or have a **repeating** pattern.

You can also use ratios as **conversion factors** to change from one unit of measure to another. You used **dimensional analysis** to convert units such as miles per hour to meters per second.

You constructed a relative frequency graph using your knowledge of ratios, proportions, and percents. Circle graphs and bar graphs can be **relative frequency graphs** that allow you to compare different categories in a data set proportionally.

Probability values for an **outcome** are ratios between 0 and 1 that compare the number of successful outcomes to the total number of trials. You learned that the more **samples** you select or **trials** you do in an experiment, the closer your **observed probability** will be to the **theoretical probability.** You investigated **random** outcomes. You saw that probability values can help you predict what will happen if you do many trials, but they will not help you predict the next outcome.

EXERCISES

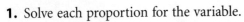

1. Solve each proportion for the variable.

   a. $\dfrac{5}{12} = \dfrac{n}{21}$

   b. $\dfrac{15}{47} = \dfrac{27}{w}$

   c. $\dfrac{2.5}{3} = \dfrac{k}{6.2}$

2. A group of 350 students were surveyed, and their eye colors are shown on the graph below right. Approximately how many students have each eye color?

3. Jeff can build 7 birdhouses in 5 hours. Write three different proportions that you could use to find out how long it would take him to build 30 birdhouses.

Survey of Eye Color

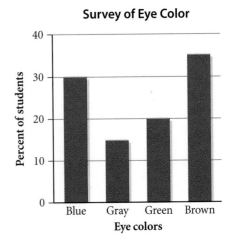

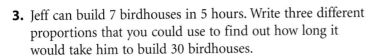

4. **APPLICATION** Tuition costs at several different colleges are listed in the table. Costs have been predicted to go up 3.7% for next year.

This year
$2,860
$3,580
$8,240
$9,460
$11,420
$22,500
$26,780

a. What will the costs be next year?

b. Explain how you can use the list feature of your calculator to quickly find the estimated cost next year for each school.

5. Plot the point (6, 3) on a graph.

a. List four other points where the *y*-coordinate is 50% of the *x*-coordinate. Plot them on the same graph.

b. Describe the pattern formed by the points.

6. In a fairy tale written by the Brothers Grimm, Rapunzel has hair that is about 20 ells in length (1 ell = 3.75 feet) by the time she is 12 years old. In the story, Rapunzel is held captive in a high tower with a locked door and only one window. From this window, she lets down her hair so that people can climb up.

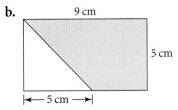

a. Approximately how long was Rapunzel's hair in feet when she was 12 years old?

b. If Rapunzel's hair grew at a constant rate from birth, approximately how many feet did her hair grow per month?

7. Find the area of each shaded region. Then determine the probability of a random point landing in the shaded region of each figure.

a.
```
   5 cm        5 cm
┌──────────┬──────────┐
│          │          │  1.5 cm
├──────────┼──────────┤
│░░░░░░░░░░│          │  2.5 cm
└──────────┴──────────┘
```

b.
```
         9 cm
┌────────────────────┐
│░░░░░░░░░░░░░░░░░░░░░│
│░░░░░░░░░░░░░░░░░░░░░│  5 cm
│░░░░░░░░░░░░░░░░░░░░░│
└────────────────────┘
  |←── 5 cm ──→|
```

8. **APPLICATION** This year the leftover candy at Sal's Candy Mart was marked down 15% after Halloween.

a. If bags of candy originally sold for $2.49, $1.89, and $3.29 before Halloween, what did they sell for after Halloween?

b. Write a proportion to calculate the sale price for any bag of leftover candy at Sal's when you know the original price. Use your proportion to verify your answers to 8a.

c. Sal's original prices were twice the wholesale cost minus 1 cent. What were the wholesale prices?

d. What percent profit did he make on the marked-down candy?

9. At a state political rally, someone said "We should raise test scores so that all students are above the state median." Analyze this statement. What is the probability of this happening?

10. Each year the Spanish club has a fund-raising raffle. First, second, and third prizes are $525, $125, and $25. The net gain is how much money you actually win if you deduct the cost of the ticket. Net gains and their respective probabilities are shown in the table.

Raffle Chances

Won	Net gain	Probability
$525	$500	1%
$125	$100	5%
$25	$0	10%
0	−$25	84%

  a. What is the cost of one raffle ticket?

  b. If you buy one ticket, what is the probability that you will win more than $50?

  c. If only 100 tickets are sold, what would be the net winnings or losses for the group of 100 buyers?

11. **APPLICATION** A thirteenth-century Chinese manuscript (*Shu-shu chiu-chang*) contains this problem: You are sold 1,534 shih of rice but find that millet is mixed with the rice. In a sample of 254 grains, you find 28 grains of millet. About how many shih of rice do you have? How much millet did you buy? (Ulrich Libbrecht, *Chinese Mathematics in the Thirteenth Century*, 1973, p. 79)

TAKE ANOTHER LOOK

You can make an interesting picture of some fractions using their decimal equivalents and a circle.

Calculate the decimal equivalent of $\frac{1}{7}$. On a circle like this one, create a pattern using the decimal equivalent of $\frac{1}{7}$. Start with your pencil at the point on the circle corresponding to the digit in the tenths place of the decimal equivalent. Use a ruler to draw a line segment from that point to the point corresponding to the digit in the hundredths place. Then draw a line from that point to the point corresponding to the digit in the thousandths place, then on to the ten-thousandths place, and so on. Continue until the pattern starts to repeat. Describe the pattern you have created.

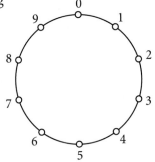

Repeat the process for the other fractions with a denominator of 7. What do you notice?

Explore and see if there are other fractions that create similar circle pictures. What do these fractions have in common?

Draw a symmetrical, repeating pattern. What fraction does it correspond to?

Assessing What You've Learned

ORGANIZE YOUR NOTEBOOK

You've been creating tables and answering questions as you do the investigations. You've been working the exercises and taking quizzes and tests. You've made notes on things that you want to remember from class discussions. Are those papers getting folded and stuffed into your book or mixed in with work from other classes? If so, it's not too late to get organized. Keeping a well-organized notebook is a habit that will improve your learning.

Your notebook should help you organize your work by lesson and chapter and give you room to summarize. Look through your work for a chapter and think about what you have learned. Write a short summary of the chapter. Include in the summary the new words you learned and things you learned about the graphing calculator. Write down questions you still have about the investigations, exercises, quizzes, or tests. Talk to classmates about your questions, or ask your teacher.

WRITE IN YOUR JOURNAL Add to your journal by expanding on a question from one of the investigations or exercises. Or use one of these prompts:

▶ It has been said that over half of the problems on the SAT Mathematics Test can be solved using proportional reasoning. Discuss why you think this might be true.

▶ Describe the progress you are making toward the goals you have set for yourself in this class. What things did you do and learn in this chapter that are helping you achieve those goals? What changes might you need to make to help keep you on track?

UPDATE YOUR PORTFOLIO Find the best work you have done in Chapter 2 to add to your portfolio. Choose at least one piece of work where you used proportions to solve a percent problem and at least one probability investigation or exercise. Choose one relative frequency graph you made. You might decide to put the graph with the graphs you selected for your Chapter 1 portfolio.

3

Variation and Graphs

Ratios and proportions describe many aspects of music composition and production. Most musicians today rely on electronic devices to generate, record, or perform music, and to amplify the sound of acoustic instruments. The individual levers and dials of a mixing board, for instance, create variations in the quality of sound.

OBJECTIVES

In this chapter you will

- learn what rates are and use them to make predictions
- study how quantities vary directly and inversely
- use equations and graphs to represent variation
- solve real-world problems using variation

Using Rates

Most people earn their income by working in a job. Yet the amount people are paid varies. Some workers have fixed yearly salaries while others are paid by the hour. How often a worker gets a paycheck varies too. One person may be paid weekly while another is paid monthly.

Measurement began our might.

WILLIAM BUTLER YEATS

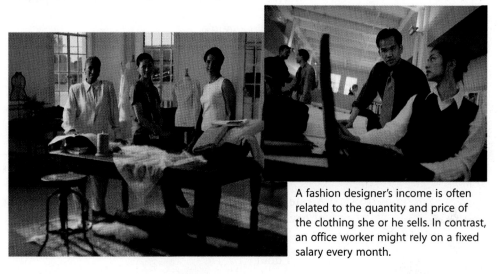

A fashion designer's income is often related to the quantity and price of the clothing she or he sells. In contrast, an office worker might rely on a fixed salary every month.

How could you compare the income of two workers based only on a paycheck? How could a worker find out how much she or he would earn for a particular time spent working? In this lesson you will learn about a type of ratio that is useful for answering questions like these.

Investigation
Off to Work We Go

You will need

• graph paper

Lacy works as a temporary employee. Last week, she earned $300 for 20 hours of work. In the first part of this investigation you will examine Lacy's pay and predict how much she will earn under different circumstances.

Step 1 | Write Lacy's pay as a ratio in the form
$$\frac{pay}{hours\ worked}.$$

Step 2 | Using your ratio from Step 1, write and solve a proportion to find out what Lacy's pay would be for a week of full-time work (40 hours). Is there a shorter method to solve your proportion? For what other number of hours of work could you use it?

Step 3	Use your ratio from Step 1 to find out how much Lacy earns for 1 hour of work.
Step 4	Write Lacy's pay for 1 hour of work as a ratio. Then write and solve a proportion to find out how much Lacy would be paid for 3 hours of work. Find another way to calculate how much Lacy will be paid for 3 hours of work.
Step 5	Complete a table like this for 1 to 5 hours of work. Let x represent time worked in hours and y represent pay in dollars, and plot the points you get from the information in your table. Do you notice any patterns?

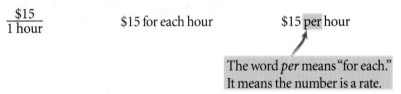

Lacy's Pay

Time worked (hours)	Pay ($)
1	15
2	

A **rate** is a ratio with 1 in the denominator. In Steps 3 and 4, you found Lacy's rate of pay. Here are three ways to write this rate.

$$\frac{\$15}{1 \text{ hour}}$$ $15 for each hour $15 per hour

The word *per* means "for each." It means the number is a rate.

Step 6	How would you use the rate to find Lacy's pay for x hours?

Joseph is another temporary employee. He was paid $513 for 38 hours of work. In this part of the investigation you will compare Joseph's pay to Lacy's.

Step 7	Find Joseph's rate of pay.
Step 8	Create a table for Joseph's pay for 1 to 5 hours. Plot the points on the same graph as Lacy's pay. How do Lacy's and Joseph's rates of pay affect what you see on the graph?
Step 9	Explain how a rate makes it convenient to compare two workers' pay.

A rate of pay is just one example of a rate. Your weekly allowance, the cost per pound of shipping a package, and the number of cookies per box are all rates. Rates make calculations easier in many real-life situations. For instance, grocery stores price fruits and vegetables by the pound. Rates make comparisons easier in other situations. For instance, a baseball player's batting average is a rate of hits per times at bat. What other rates have you seen in this book?

On average, the basketball player scores 19.1 points per game.

$$\frac{19.1 \text{ points}}{1 \text{ game}}$$

My brother drove at a speed of 65 mph.

$$\frac{65 \text{ miles}}{1 \text{ hour}}$$

These cookies have 35 grams of fat in one serving of 5 cookies.

$$\frac{35 \text{ grams}}{1 \text{ serving}} \text{ or}$$
$$\frac{35 \text{ grams}}{5 \text{ cookies}} = \frac{7 \text{ grams}}{1 \text{ cookie}}$$

EXAMPLE

Otto's car began the week with a full tank of gas. Otto drove to and from work, a total of 320 miles. He then filled his tank with 14.7 gallons of gas.

a. How many miles per gallon (mpg) does Otto's car get?

b. How far could Otto drive using only 5 gallons of gasoline?

c. When Otto's car recently broke down, he rented a newer model car. Otto drove the same distance, 320 miles, but it took only 11.5 gallons of gas to fill the tank. What was the rate of miles per gallon for the rental car? Which car can go more miles on 10 gallons of gas?

▶ **Solution**

Otto's car used 14.7 gallons to go 320 miles.

a. The ratio is $\frac{320 \text{ miles}}{14.7 \text{ gallon}}$. So you divide miles by gallons.

$$320 \div 14.7 \approx 21.8$$

Otto's car gets approximately 21.8 mpg or $\frac{21.8 \text{ miles}}{1 \text{ gallon}}$.

b. Multiply the rate by 5. You can use dimensional analysis to help verify the units in your answer.

$$\frac{21.8 \text{ miles}}{1 \text{ gallon}} \cdot \frac{5 \text{ gallons}}{1} = 109 \text{ miles}$$

c. $320 \div 11.5 \approx 27.8$

The rate of the rental car is approximately 27.8 mpg. This is greater than 21.8 mpg. The rental car can go farther than Otto's car on 1 gallon of gas, so it can also go farther than his car on 10 gallons of gas.

Consumer
● CONNECTION ●

Many factors influence the rate at which cars use gas, including size, age, and driving conditions. Advertisements for new cars often give the average mpg for city traffic (slow, congested) and highway traffic (fast, free flowing). These rates help consumers make an informed purchase.

Compare this vehicle to others in the FREE FUEL ECONOMY GUIDE available at the dealer.

CITY MPG **25** Fuel Economy Information HIGHWAY MPG **33**

Estimated Annual Fuel Cost: $723

When you are finding a rate, consider which unit would be most helpful in the denominator based on the questions to be answered. In part a of the example, miles is in the numerator and gallons is in the denominator. If you switch the quantities in the numerator and denominator, you get a different rate.

$$\frac{320 \text{ miles}}{14.7 \text{ gallons}} \approx 21.8 \text{ miles per gallon} \qquad \frac{14.7 \text{ gallons}}{320 \text{ miles}} \approx 0.046 \text{ gallon per mile}$$

How are these rates related? Which rate would you use to calculate the answer to the question "How many gallons of gas should Otto put in his car to drive 200 miles?" Why would you choose that rate? Is it possible to calculate the answer using the other rate?

EXERCISES

You will need your calculator for problem **10.**

▶ Practice Your Skills

1. Tab and Crystal both own cats.

   a. Tab buys a 3-pound bag of cat food every 30 days. At what rate does his cat eat the food?

   b. Crystal buys a 5-pound bag of cat food every 45 days. At what rate does her cat eat the food?

   c. Whose cat, Tab's or Crystal's, eats more food per day?

2. If Ray drove 240 miles in 4 hours, what was his average rate of speed?

3. Jocelyn's old car gets only 10 miles per gallon.

   a. What is the car's rate of gasoline use in gallons per mile?

   b. How many gallons of gas does the car need to go 220 miles?

   c. How far could Jocelyn drive on 15 gallons of gas?

▶ Reason and Apply

4. Find a rate for each situation. Then use the rate to answer the question.

   a. Kerstin drove 350 miles last week and used 12.5 gallons of gas. How many gallons of gas will he use if he drives 520 miles this week?

   b. Angelo drove 225 miles last week and used 10.7 gallons of gas. How far can he drive this week using 9 gallons of gas?

5. On his Man in Motion World Tour, Canadian Rick Hansen wheeled himself 24,901.55 miles to support spinal cord injury research, rehabilitation, and wheelchair sport. He covered 4 continents and 34 countries in two years, two months, and two days. Learn more about Rick's journey with the link at www.keymath.com/DA .

China was one of the many countries through which Rick Hansen traveled during the Man in Motion World Tour.

a. Find Rick's average rate of travel in miles per day. (Assume there are 365 days in a year and 30.4 days in a month.)

b. How much farther would Rick have traveled if he had continued his journey for another $1\frac{1}{2}$ years?

c. If Rick continued at this same rate, how many days would it take him to travel 60,000 miles? How many years is that?

6. **APPLICATION** Wynonna bought a prepaid calling card for $5. It allows her to make 80 minutes' worth of calls.

a. Estimate the prepaid calling card's cost per minute. Explain your thinking.

b. Calculate the actual cost per minute.

c. After purchasing her prepaid card, Wynonna made a 15-minute call to her friend Samson. How much is the card worth now?

d. Wynonna's home telephone service provides a calling card that costs 8¢ per minute. Which calling card would you recommend she use? Why?

e. Wynonna has made plans for a 7-day vacation. She expects to make 30 minutes of calls each day. How many prepaid calling cards should she take along?

7. **APPLICATION** Marie and Tracy bought boxes of granola bars for their hiking trip. They noticed that the tags on the grocery-store shelf use rates.

a. Each tag above uses two rates. Identify all four rates.

b. A box of Crunchy Granola Bars contains 6 bars. Is the price per bar correct?

c. A box of Chewy Granola Bars contains 8 bars. Use the information on the tag to find the number of ounces per bar.

d. A box of Crunchy Granola Bars weighs 10 ounces. What is the price per ounce?

e. If Marie and Tracy like Crunchy Granola Bars as much as they like Chewy Granola Bars, which should they buy? Explain your answer.

8. **APPLICATION** Portia drove her new car 425 miles on 10.8 gallons of gasoline.

a. What is the car's rate of gasoline consumption in miles per gallon?

b. If this is the typical mileage for Portia's car, how much gas will it take for a 750-mile vacation trip?

c. If gas costs $1.35 per gallon, how much will Portia spend on gas on her vacation?

d. The manufacturer advertised that the car would get 30 to 35 miles per gallon. How does Portia's mileage compare to the advertised estimates?

9. APPLICATION Chris' 85-pound black Labrador retriever, Tootsie, eats 40 pounds of dog food every 2 weeks. Each 40-pound bag costs $36.

    a. Write two ratios for this situation.

    b. How much does it cost to feed Tootsie per year?

    c. If Cathy's 60-pound golden retriever, Clara, eats the same amount per pound of body weight as Tootsie, how much does Clara eat per week?

    d. How much less does Cathy spend on dog food each year than Chris?

10. Emily bought a used scooter to use on short trips. She timed her travel on the scooter from home to her favorite places. Then she used a map to estimate the distances. She made this table:

Travel Time with Scooter

Location	Time (minutes)	Distance (miles)
Music store	15	2
Juice bar	3.75	0.5
Thrift store	22.5	3
Jeremy's house	37.5	5

    a. Enter the table values into your calculator in two lists. Use list L_1 for the times and list L_2 for the distances. Make a scatter plot of points with coordinates in the form (*time, distance*). What patterns do you notice?

    b. In list L_3, calculate $L_2 \div L_1$. What pattern do you notice? Explain in words what the values in list L_3 represent.

    c. Find Emily's average speed on the scooter in miles per hour. Your findings in 10b should be helpful.

▶ Review

11. Solve each proportion for x.

    a. $\dfrac{x}{3} = \dfrac{7}{5}$ **b.** $\dfrac{2}{x} = \dfrac{9}{11}$ **c.** $\dfrac{x}{c} = \dfrac{d}{e}$

12. In 1997, there were an estimated 25,779,000 foreign-born people living in the United States. Of those people, 6,822,000 were born in Asian countries. (*www.census.gov*)

    a. What percent of foreign-born people in the United States were born in Asia?

    b. The Philippines was the birthplace of 16.6% of the Asian-born people living in the United States. How many people who lived in the United States in 1997 were born in the Philippines?

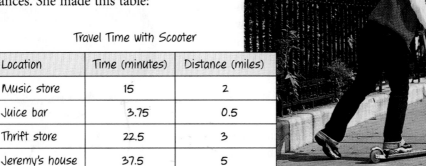

13. Hannah bought a small bag of jelly beans to share with her friends, Asha, Annelise, Patrick, and Jamal. They counted the candies and got the results shown on the blue note paper.

Small Bag
12 yellow
13 purple
11 green
14 orange
10 red

a. Draw a circle graph to show the proportion of the different colors of candies in the bag.

b. Then the friends bought a large bag of jelly beans to share. They counted the number of each color in the large bag and got the results shown on the green note paper.

Draw a relative frequency bar graph for the large bag of jelly beans.

Large Bag
78 yellow
96 purple
72 green
76 orange
104 red

c. Compare the distribution of colors in the small bag to the distribution in the large bag.

d. If the five friends share the jelly beans in the large bag, how many pieces should each one get? Will it be fair if each one gets all the jelly beans of one color?

14. This table shows information that Ms. Osborne collected about the students in her art class.

Ms. Osborne's Class

	Right-handed	Left-handed	Ambidextrous
Boys	10	4	1
Girls	9	2	2

a. Write the ratio of right-handed boys to right-handed girls.

b. Write the ratio of left-handed boys to left-handed girls.

c. Write the ratio of ambidextrous boys to boys in the class.

d. Write the ratio of ambidextrous girls to girls in the class.

e. Draw a relative frequency graph to represent this data set. Explain why you chose to draw either a circle graph or a bar graph.

IMPROVING YOUR **REASONING** SKILLS

This problem is adapted from an ancient Chinese book, *The Nine Chapters of Mathematical Art*.

A city official was monitoring water use when he saw a woman washing dishes in the river. He asked, "Why are there so many dishes here?" She replied, "There was a dinner party in the house." His next question was "How many guests attended the party?" The woman did not know but replied, "Every two guests shared one dish for rice. Every three guests used one dish

This is a detail from the 17th-century Chinese scroll painting *Landscapes of the Four Seasons* by Shen Shih-Ch'ing.

for broth. Every four guests used one dish for meat. And altogether there were sixty-five dishes used at the party." How many guests attended the party?

Direct Variation

L E S S O N

3.2

In Lesson 3.1, you worked with rates. Rates have a 1 in the denominator so they are convenient to calculate with. You also see patterns when you use a rate to make tables and graphs. In this investigation you'll use algebra to understand these patterns better.

Investigation
Ship Canals

You will need

• graph paper

In this investigation you will use data about canals to draw a graph and write an equation that states the relationship between miles and kilometers. You'll see several ways of finding the information that is missing from this table.

Longest Ship Canals

Canal	Length (miles)	Length (kilometers)
Albert (Belgium)	80	129
Alphonse XIII (Spain)	53	85
Houston (Texas)	50	81
Kiel (Germany)	62	99
Main-Danube (Germany)	106	171
Moscow-Volga (Russia)	80	129
Panama (Panama)	51	82
St. Lawrence Seaway (Canada/U.S.)	189	304
Suez (Egypt)	101	
Trollhätte (Sweden)		87

(*The Top 10 of Everything 1998*, p. 57)

Step 1 Carefully draw and scale a pair of coordinate axes for the data in the table. Let x represent the length in miles and y represent the length in kilometers. Plot points for the first eight coordinate pairs.

Step 2 What pattern or shape do you see in your graph? Connect the points to illustrate this pattern. Explain how you could use your graph to approximate the length *in kilometers* of the Suez Canal and the length *in miles* of the Trollhätte Canal.

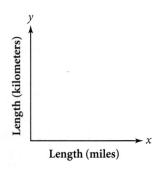

Step 3 On your calculator, make a plot of the same points and compare it to your hand-drawn plot. Use list L_1 for lengths in miles and list L_2 for lengths in kilometers. [▶☐ See **Calculator Note 1E** to review this type of plot. ◀]

146 CHAPTER 3 Variation and Graphs

| Step 4 | Use list L3 to calculate the ratio $\frac{L_2}{L_1}$. [▶️🖥️ See **Calculator Note 1I** to review using lists to calculate this way. ◀] Explain what the values in list L3 represent. If you round each value in list L3 to the nearest tenth, what do you get? |
| Step 5 | Use the rounded value you got in Step 4 to find the length in kilometers of the Suez Canal. Could you also use your result to find the length in miles of the Trollhätte Canal? |

The number of kilometers is the same in every mile, so the value you found is called a **constant.**

Step 6	How can you change x miles to y kilometers? Using variables, write an equation to show how miles and kilometers are related.
Step 7	Use the equation you wrote in Step 6 to find the length in kilometers of the Suez Canal and the length in miles of the Trollhätte Canal. How is using this equation like using a rate?
Step 8	Graph your equation on your calculator. [▶️🖥️ See **Calculator Note 1H** to review graphing equations. ◀] Compare this graph to your hand-drawn graph. Why does the graph go through the origin?
Step 9	Trace the graph of your equation. [▶️🖥️ See **Calculator Note 1H** to review tracing equations. ◀] Approximate the length in kilometers of the Suez Canal by finding when x is approximately 101 miles. Trace the graph to approximate the length in miles of the Trollhätte Canal? How do these answers compare to the ones you got from your hand-drawn graph?
Step 10	Use the calculator's table function to find the missing lengths for the Suez Canal and the Trollhätte Canal. [▶️🖥️ See **Calculator Note 3A** to learn about the table function. ◀]
Step 11	In this investigation you used several ways to find missing values— approximating with a graph, calculating with a rate, solving an equation, and searching a table. Write several sentences explaining which of these methods you prefer and why.

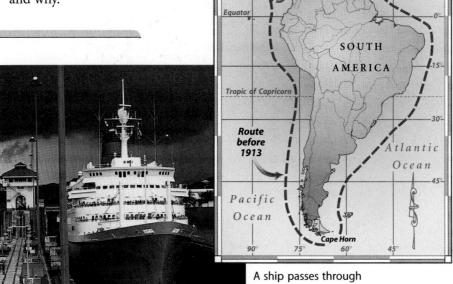

History
● **CONNECTION** ●

The Panama Canal allows ships to cross the strip of land between the Atlantic and Pacific Oceans. Before the canal was completed in 1913, ships had to sail thousands of miles around the dangerous Cape Horn, even though only 50 miles separate the two oceans.

A ship passes through the Panama Canal.

Ratios, rates, and conversion factors are closely related. In this investigation you saw how to change the ratio $\frac{129 \text{ km}}{80 \text{ mi}}$ to a rate of approximately 1.6 kilometers per mile. You can also use that rate as a conversion factor between kilometers and miles. The numbers in the ratio vary, but the resulting rate remains the same, or constant. Kilometers and miles are **directly proportional**—there will always be the same number of kilometers in every mile. When two quantities vary in this way, they have a relationship called **direct variation.**

Direct Variation

An equation in the form $y = kx$ is a **direct variation.** The quantities represented by x and y are **directly proportional,** and k is the **constant of variation.**

You can represent any ratio, rate, or conversion factor with a direct variation. Using a direct variation equation or graph is an alternative to solving proportions. A direct variation equation can also help you organize calculations with rates.

EXAMPLE

A grocery store advertises a sale on soda.

a. Write a rate for the cost per six-pack.

b. Write an equation showing the relationship between the number of six-packs purchased and the cost.

c. How much will 15 six-packs cost?

d. Sol is stocking up for his restaurant. He bought $210 worth of soda. How many six-packs did he buy?

▶ **Solution**

a. The ratio given is $\frac{\$6.00}{4 \text{ six-packs}}$. This simplifies to a rate of $1.50 per six-pack.

b. Use x for the number of six-packs and y for the cost in dollars. Write a proportion.

$$\frac{y}{x} = \frac{1.50}{1} \qquad \text{y corresponds to 1.50 and x corresponds to 1.}$$

$$x \cdot \frac{y}{x} = \frac{1.50}{1} \cdot x \qquad \text{To isolate y, multiply both sides by x.}$$

$$y = 1.50x \qquad \text{Your result is a direct variation equation.}$$

The constant, k, is the rate $1.50 per six-pack. Does every point on the graph of this equation make sense in this situation?

Would you get the same equation if you started with the proportion $\frac{y}{x} = \frac{6.00}{4}$? Do you see another way to find the equation once you know the rate?

Cost of Soda

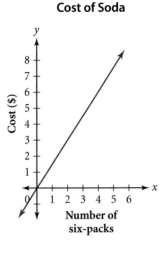

c. You can trace the graph to find the point where $x = 15$, or you can substitute 15 into the equation for x, the number of six-packs.

$$y = 1.50(15) = 22.50$$

Fifteen six-packs cost $22.50.

d. You can search the table until you close in on the value of x that gives the y-value of 210. Or you can substitute 210 into the equation for y, the cost in dollars.

$$210 = 1.50x$$ Substitute 210 for y into the original equation.

$$\frac{210}{1.50} = \frac{1.50x}{1.50}$$ To isolate x, divide both sides by 1.50.

$$140 = x$$ Simplify.

Sol purchased 140 six-packs for $210.

EXERCISES

You will need your calculator for problems **1, 2, 3, 6,** and **9.**

▶ Practice Your Skills

Let x represent distance in miles and y represent distance in kilometers. Enter the equation $y = 1.6x$ into your calculator. Use it for problems 1–3.

1. Trace the graph of $y = 1.6x$ to find each missing quantity. Adjust the window settings as you proceed.

   a. 25 miles ≈ ☐ kilometers   **b.** 120 kilometers ≈ ☐ miles

2. Use the calculator table function to find the missing quantity.

   a. 55 miles ≈ ☐ kilometers   **b.** 450 kilometers ≈ ☐ miles

3. Find the missing values in this table. Round each value to the nearest tenth.

Distance (miles)	Distance (kilometers)
	4.5
7.8	
650.0	
	1500.0

4. Solve each equation for x.

    a. $14 = 3.5x$ **b.** $x = 45(0.62)$ **c.** $\frac{x}{7} = 0.375$ **d.** $\frac{12}{x} = 0.8$

5. The equation $c = 1.25f$ shows the direct variation relationship between the length of fabric and its cost. The variable f represents the length of the fabric in yards, and c represents the cost in dollars.

Use this equation to answer these questions.

    a. How much does $2\frac{1}{2}$ yards of fabric cost?

    b. How much fabric can you buy for $5?

    c. What is the cost of each additional yard of fabric?

Christo (b 1935, Bulgaria) and Jeanne-Claude (b 1935, Morocco) are environmental sculptors who wrap large objects and buildings in fabric. This is the German Reichstag in 1995.

► Reason and Apply

6. APPLICATION Market A sells 7 ears of corn for $1.25. Market B sells a baker's dozen (13 ears) for $2.75.

    a. Copy and complete the tables below showing the cost of corn at each market.

Market A

Ears	7	14	21	28	35	42
Cost						

Market B

Ears	13	26	39	52	65	78
Cost						

    b. Let x represent the number of ears of corn and y represent cost. Find equations to describe the cost of corn at each market. Use your calculator to plot the information for each market on the same set of coordinate axes. Round the constants of variation to three decimal places.

    c. If you wanted to buy only one ear of corn, how much would each market charge you? How do these prices relate to the equations you found in 6b?

    d. How can you tell from the graphs which market is the cheaper place to buy corn?

Why is 13 called a baker's dozen? In the 13th century, bakers began to form guilds to prevent dishonesty. To avoid the penalty for selling a loaf of bread that was too small, bakers began giving 13 whenever a customer asked for a dozen.

7. Bernard Lavery, a resident of the United Kingdom, has held several world records for growing giant vegetables. The graph shows the relationship between weight in kilograms and weight in pounds.

a. Use the information in the graph to complete the table.

Bernard Lavery's Vegetables

Vegetable	Weight (kilograms)	Weight (pounds)
Cabbage	56	
Summer squash		108
Zucchini		64
Kohlrabi	28	
Celery	21	
Radish		28
Cucumber	9	
Brussels sprout		18
Carrot	5	

(*The Top 10 of Everything 1998*, p. 98)

Relationship Between Kilograms and Pounds

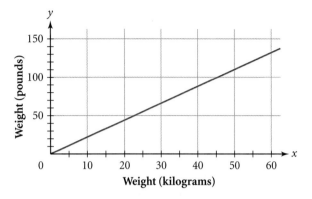

b. Calculate the rate of pounds per kilogram for each vegetable entry from the table. Use the rate you think best represents the data as the constant of variation. Write an equation to represent the relationship between pounds and kilograms.

c. Use the equation you wrote in 7b to find the weight in kilograms for a pumpkin that weighs 6.5 pounds.

d. Use the equation you wrote in 7b to find the weight in pounds for an elephant that weighs 3600 kilograms.

e. How many kilograms are in 100 pounds? How many pounds are in 100 kilograms?

8. As part of their homework assignment, Thu and Sabrina each found equations from a table of data relating miles and kilometers. One entry in the table paired 150 kilometers and 93 miles. From this pair of data values, Thu and Sabrina wrote different equations.

a. Thu wrote the equation $y = 1.61x$. How did he get it? What does 1.61 represent? What do x and y represent?

b. Sabrina wrote $y = 0.62x$ as her equation. How did she get it? What does 0.62 represent? What do x and y represent?

c. Whose equation would you use to convert miles to kilometers?

d. When would you use the other student's equation?

9. APPLICATION If you're planning to travel to another country, you will need to learn about its monetary system. This table gives some exchange rates that tell how many of each monetary unit are equivalent to one U.S. dollar.

International Monetary Units

Country	Monetary unit	Exchange rates (units per American dollar)
Brazil	real	1.790
Germany	mark	2.140
Italy	lira	2119.150
Japan	yen	108.770
Mexico	peso	9.350
Pakistan	rupee	45.400
United Kingdom	pound	0.670

(Federal Reserve Bank of New York for August 2, 2000)

a. Make a list of ten items and the price of each item in U.S. dollars. Enter these prices into list L_1 on your calculator.

b. Choose one of the countries in the table and convert the U.S. dollar amounts in your list to that country's monetary unit. Use list L_2 to calculate these new values from list L_1.

c. Using list L_3, convert the values in list L_2 back to the values in list L_1.

d. Describe how you would convert marks to liras.

e. Check the business section of a newspaper for current exchange rates for five countries of your choice.

10. If you travel at a constant speed, the distance you travel is directly proportional to your travel time. Suppose you walk 3 miles in 1.5 hours.

a. How far would you walk in 1 hour?

b. How far would you walk in 2 hours?

c. How much time would it take you to walk 6 miles?

d. Represent this situation with a graph.

e. What is the constant of variation in this situation, and what does it represent?

f. Define variables and write an equation that relates time to distance traveled.

▶ Review

11. U.S. speed limits are posted in miles per hour (mph). Germany's Autobahn has stretches where speed limits are posted at 130 kilometers per hour.

a. How many miles per hour is 130 kilometers per hour?

b. How many kilometers per hour is 25 mph?

c. If the United States used the metric system, what speed limit do you think would be posted for 65 mph?

12. APPLICATION Cecile started a business entertaining at children's birthday parties. As part of the package, Cecile comes in costume and plays games with the children. She also makes balloon animals and paints each child's face. When she started the business, she charged $3.50 per child, but she is rethinking what her charges should be so that she will make a profit.

a. The average children's party takes about 3 hours. Cecile wants to make at least $12 an hour. What is the minimum number of children she should arrange to entertain at a party at her current rate?

b. The balloons and face paint cost Cecile about 60¢ per child. What percent is that of the fee per child?

c. Cecile decided to raise her rates so that the cost of supplies for each child is only 10% of her fee. If the supplies for the party cost 60¢ per child, what should she charge per child?

13. Ms. Zany sometimes plays the "homework game" with her class as a way to check that students have done their homework. The students sit at tables numbered 1 to 8. Ms. Zany spins a spinner numbered from 1 to 10. (Each number is equally likely to be spun.) If she spins the number of a table (1 to 8), she checks the homework of the students sitting at that table. If she spins a 9, she doesn't check anyone's homework. If she spins a 10, she checks everyone's homework.

a. What is the probability that your homework will get checked if you are sitting at table 5 in Ms. Zany's class?

b. Does the probability of getting your homework checked depend on the number of your table? Why or why not?

IMPROVING YOUR **REASONING** SKILLS

Can all unit conversion problems be modeled by direct variation? Consider this table of temperatures in degrees Fahrenheit and degrees Celsius.

Calculate the ratios of degrees Fahrenheit to degrees Celsius for each city. How do they compare to the kilometers-to-miles ratios you found in the investigation? Plot degrees Celsius on one axis and degrees Fahrenheit on the other axis. How is this graph similar to the other graphs in this lesson? How is it different? Water freezes at 0°C. Is zero also the temperature at which water freezes in degrees Fahrenheit? Is the relationship between degrees Fahrenheit and degrees Celsius a direct variation?

City	°C	°F
Athens	30	86
Barcelona	18	64
Buenos Aires	11	52
Cairo	35	95
Johannesburg	15	59
Phoenix	40	104
Rio de Janeiro	25	77

LESSON
3.3

Scale Drawings and Similar Figures

Objects in mirror are closer than they appear.

MANDATED BY FEDERAL LAW
ON CONVEX MIRRORS ON
AUTOMOBILES

To design a building, an architect makes a scale drawing to show what the floor plan will look like. In the finished building the floor plan will look like the scale drawing, only larger. Maps of towns and cities are scale drawings that show how far and in which directions the roads go. Even school pictures come in large, medium, and small sizes. You can think of them as scale images of yourself. In all of these situations, a rate or **scale factor** relates the measurements of the drawing or image to the measurements of the real thing. When you create or interpret scale drawings, you use direct variation.

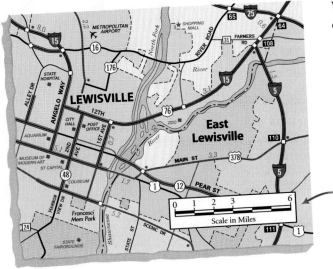

The scale of a map helps you determine actual measurements on the ground. Sometimes the scale is part of the map's legend.

Investigation
Floor Plans

You will need

• a centimeter ruler

This scale drawing shows the floor plan of an apartment. Use it to investigate how rate and direct variation apply to scale drawings.

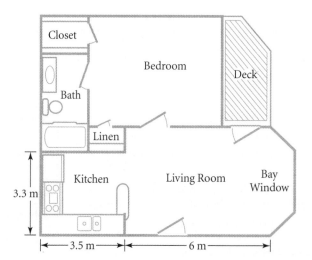

Step 1	Using a centimeter ruler, measure the three lengths given on the floor plan. Use a table to pair each actual length in meters with the scale drawing measurement in centimeters.
Step 2	Compare each pair of measurements in a ratio. $$\frac{actual\ measurement\ in\ meters}{scale\ drawing\ measurement\ in\ centimeters}$$ Use your calculator to convert each ratio to a decimal. What units should you apply to each decimal number? How do the numbers compare? Explain your findings.
Step 3	State the scale for the drawing and explain how you got it. Then write this scale as a rate.

Step 4	Measure the length and width of the bedroom in centimeters on the scale drawing. Use the scale you wrote in Step 3 to calculate the dimensions of the actual bedroom.
Step 5	Write a direct variation equation to relate the scale drawing to the actual apartment. Use this equation to calculate the actual dimensions of the other rooms.

Step 6	The manager of the apartment complex wants a small model of this apartment to show apartment hunters. She wants the 6-meter wall of the apartment to be 10 centimeters in the model. Write an equation that she can use to convert from the scale drawing to the model.
Step 7	On a clean sheet of paper, draw an accurate floor plan for the model of the apartment. Think carefully about how to find the length of each wall and the angles where the walls meet. Compare your floor plan to the scale drawing in the book. Describe how they are alike and how they are different.

Career
● CONNECTION ●

There are several ways to do three-dimensional scale drawings. Some drawings show the object from the top, bottom, and each side. Others show the object with a perspective drawing. Architects use Computer Aided Design (CAD) software to help with these drawings.

The floor plan you drew in the investigation should have the same *shape* as the scale drawing in the book—your floor plan is just bigger. It is the same shape because you made the angles the same. It is proportionally bigger because you used the same ratio to increase the length of every wall. Your drawing and the one in the book are **similar figures.** Similar polygons have sides that are proportional and angles that are *congruent.* Would the apartment's actual floor plan be similar to the scale drawing? What might happen during construction that would make the apartment *not* similar to the drawing?

EXAMPLE A

Here are two similar figures.

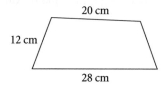

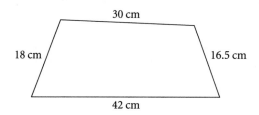

a. Write an equation to find the missing length.

b. Find the missing length on the smaller figure.

▶ Solution

a. Begin with a ratio using a pair of corresponding sides whose lengths are given, for example, 20 cm and 30 cm. Let x represent an unknown length from the smaller figure, and let y represent the corresponding unknown length from the larger figure.

$$\frac{y}{x} = \frac{30}{20}$$

Set up a proportion. y and 30 are both measurements from the large figure, and x and 20 are both measurements from the small figure.

$$x \cdot \frac{y}{x} = \frac{30}{20} \cdot x$$

To isolate y, multiply both sides by x.

$$y = 1.5x$$

Write as a direct variation.

So the equation $y = 1.5x$ can be used to find missing lengths. 1.5 is the scale factor.

b. The missing length on the smaller figure corresponds to 16.5 cm on the larger figure. Remember that in our equation, x represents a measurement on the smaller figure and y represents a measurement on the larger figure.

$$16.5 = 1.5x$$

Substitute 16.5 for y.

$$\frac{16.5}{1.5} = \frac{1.5x}{1.5}$$

To isolate x, divide both sides by 1.5.

$$x = 11$$

Reduce.

The missing length is 11 cm.

EXAMPLE B | Sean and Jon are making a map of their school. After taking many measurements, they decided that 1 inch on their map should represent 20 feet on the actual school grounds. Write an equation to help them make all the conversions. Explain how a graph of this equation could help Sean and Jon make their map.

▶ **Solution** | In a direct variation, if you substitute a value for x, you only have to multiply to find y. Since their measurements are currently in feet, Sean and Jon should let x represent actual lengths in feet and y represent scale measurements in inches.

$$\frac{y}{x} = \frac{1}{20}$$ y corresponds to 1 inch and x to 20 feet.

$$x \cdot \frac{y}{x} = \frac{1}{20} \cdot x$$ To isolate y, multiply both sides by x.

$$y = \frac{1}{20}x$$ Write as a direct variation.

With a graph of this equation, Sean and Jon can trace along the line to read each measurement. For example, the point (50, 2.5) means that a distance of 50 feet on school grounds is 2.5 inches on the map. They can also use the table function to find map distances.

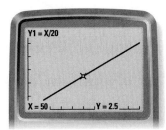

If you have to make several conversions, using equations, list calculations, graphs, and tables can save you time.

Career
● CONNECTION ●

Many artists create very large works. Often the artist will begin with a smaller model or sketch and then apply properties of similarity to make the larger object. Maya Lin (shown below) used both sketches and a model to design the Vietnam Veterans Memorial in Washington, D.C.

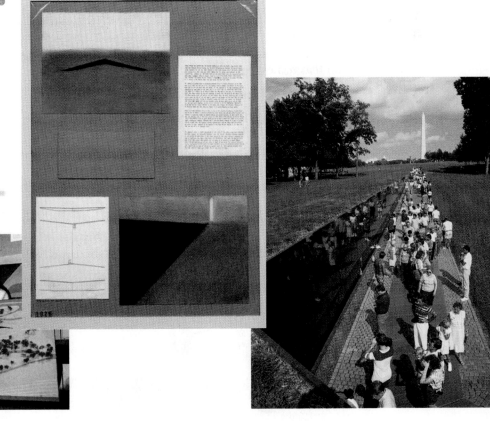

Practice Your Skills

1. Use the equation $y = \frac{1}{20}x$ to find the unknown value.

   a. Find y if $x = 15$.   **b.** Find y if $x = 40$.

   c. Find x if $y = 5$.   **d.** Find x if $y = 5.4$.

2. Write two different proportions that you can use to find the length of the missing side in the similar triangles shown. Verify that both equations produce the same result.

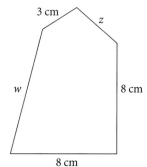

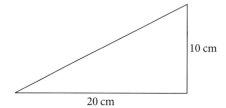

15 cm

x

20 cm

10 cm

3. These polygons are similar:

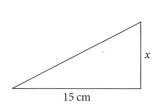

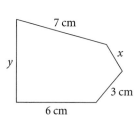

3 cm

z

w

8 cm

8 cm

7 cm

y

x

3 cm

6 cm

   a. Write an equation that relates lengths on the larger polygon to lengths on the smaller polygon.

   b. Use your equation to find the unknown side lengths.

4. **APPLICATION** Model train scales are given as $\frac{\text{length of model in feet}}{\text{length of train in feet}}$.

   An N-gauge model train has a scale of 1 : 160. An HO-gauge model train has a scale of 1 : 87.

   a. If an N-gauge caboose is 3 inches long, what is the length of the actual caboose?

   b. What is the length of an HO scale model of the caboose in part a?

5. APPLICATION The scale on a map reads "1 inch = 15 miles."

    a. How far apart are two towns that are 2.8 inches apart on the map?

    b. Suppose you know that Acme and Bates are 22 miles apart. How far apart should they be on the map?

    c. On the map, a distance on the highway is labeled as 47 miles. How long is the distance on the map?

    d. On the map, the distance across a large lake is 3.5 inches. How many miles across is the lake?

▶ Reason and Apply

6. Find two pairs of rectangles that are similar. Explain how you know they are similar.

i.
3 in.
2 in.

ii.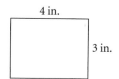
4 in.
3 in.

iii.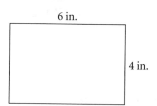
6 in.
4 in.

iv.
2 ft
4 ft

v.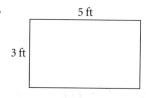
5 ft
3 ft

vi.
2 ft
1 ft

7. The polygons below are rhombuses (parallelograms with four equal sides). Find two pairs of rhombuses that are similar. Explain how you know they are similar.

i.
120° 60°
60° 120°
2.0 cm

ii.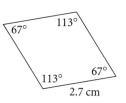
67° 113°
113° 67°
2.7 cm

iii.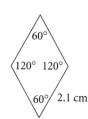
60°
120° 120°
60° 2.1 cm

iv.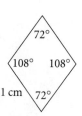
72°
108° 108°
2.1 cm 72°

v.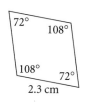
72° 108°
108° 72°
2.3 cm

vi.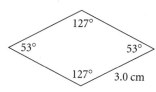
127°
53° 53°
127° 3.0 cm

8. You know that similar figures must have proportional sides *and* congruent angles.

    a. Draw two polygons that have congruent angles but are *not* similar.

    b. Draw and label two polygons that have proportional sides but are *not* similar.

9. **APPLICATION** When objects block sunlight, they cast shadows, and similar triangles are formed. On a sunny day, Sunanda and Chloe measure the shadow of the school flagpole. It is 8.5 meters long. Chloe is 1.7 meters tall and her shadow is 2.1 meters long.

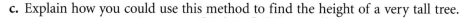
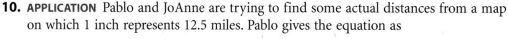
These two triangles are similar.

a. Sketch and label your own diagram to represent this situation.

b. Using the similar triangles in your diagram, write a proportion and find the height of the flagpole.

c. Explain how you could use this method to find the height of a very tall tree.

10. **APPLICATION** Pablo and JoAnne are trying to find some actual distances from a map on which 1 inch represents 12.5 miles. Pablo gives the equation as

$$y = \frac{1}{12.5} x$$

and JoAnne writes the equation as

$$y = 12.5 \cdot x$$

a. What does each variable represent in Pablo's equation?

b. What does each variable represent in JoAnne's equation?

c. Could you use either equation to convert a map distance of 4.7 inches to the actual distance in miles? How many miles does 4.7 inches represent?

11. **APPLICATION** For his social studies class, Tommy has to make scale drawings of the five longest rivers. Each drawing must fit on a piece of notebook paper, and all the drawings need the same scale so that the river lengths can be compared.

Five Longest Rivers

Name of river (location)	Length (km)
Nile (Tanzania/Uganda/Sudan/Egypt)	6670
Amazon (Peru/Brazil)	6448
Yangtze-Kiang (China)	6300
Mississippi-Missouri-Red Rock (United States)	5971
Yenisey-Angara-Selenga (Mongolia/Russia)	5540

(*The Top 10 of Everything 2001*, p. 20)

Murchison Falls, Nile River

a. Suggest a scale that Tommy could use. Be sure the longest river will fit on one piece of paper.

b. Use your scale to write a direct variation equation. Graph the equation on your calculator. Use the graph to help Tommy convert each river length for his drawing.

12. You know that similar figures have proportional sides. What about perimeter and area?

Side length	Perimeter	Area
6		
4		
3		
2		
8		

  a. On graph paper draw five squares, each with the side length given in the table. Copy and complete the table.

  b. Choose any two squares from 12a. Compare the ratio of the side lengths to the ratio of the perimeters. Do the same comparison for a different pair of squares. Do you see a relationship?

  c. Choose any two squares from 12a. Compare the ratio of the side lengths to the ratio of the areas. Do the same comparison for a different pair of squares. Do you see a relationship?

  d. If you drew two squares such that the side length of one was five times the side length of the other, how would their perimeters compare? How would their areas compare?

  e. Write at least two conjectures about the relationships among side lengths, perimeters, and areas of two different-size squares.

13. Percent problems are another type of direct variation.

  a. Thirty percent of the students in an algebra class have pets. If there are 24 students in the class, how many have pets? Set up a proportion for this problem.

  b. Rewrite the proportion from 13a using *s* for the number of students in the class and *p* for the number of students with pets.

  c. Solve the proportion in 13b for the variable *p*.

  d. Explain how to write a general direct variation equation for percent problems using these terms: total, part, percent.

14. If 2874 people in a town own bicycles and that number is 74% of the town's population, how many people are in the town? Use a direct variation equation to solve this problem.

Bicycles are the most popular form of transportation in Beijing, China.

Review

15. **APPLICATION** Two cats eat one 14-pound bag of cat food every six weeks. The bag costs $12.98.

  a. What is the cost of feeding both cats per day?

  b. How many pounds of cat food will one cat eat per year? (Assume both cats eat the same amount.)

  c. How much does it cost to feed both cats for one year, and how much will their owner spend on bags of cat food?

16. Find the five-number summary for this data set:

47	28	11	74	58	63	85
36	39	17	27	75	48	

17. A bag of fruit-flavored candies contains 8 banana, 8 orange, 6 lemon, 12 cherry, 10 grape, and 4 mango. If you select a candy at random, what is the probability that you will choose a mango-flavored candy from this package?

18. An ancient Indian Sutra describes a formula for multiplication that can be summarized as "vertically and crosswise." (Śrī Bhāratī Kṛṣṇa Tīrthajī, *Vedic Mathematics*, 1992, p. 39)

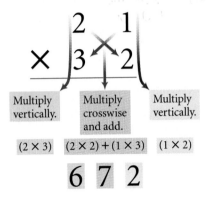

Multiply vertically.	Multiply crosswise and add.	Multiply vertically.
(2 × 3)	(2 × 2) + (1 × 3)	(1 × 2)

So, 21 · 32 = 672.

Use this method to multiply these numbers:

a. 12 · 13   **b.** 21 · 14   **c.** 42 · 31

project

MAKE YOUR OWN MAP

Maps are useful in planning trips, giving directions, and estimating distances. You can get maps at many stores, auto clubs, gas stations, and even on the Internet (see the resources at www.keymath.com/DA).

Apply the skills you've learned in this chapter to make a map of your trip to school. Whether you walk, bike, bus, or drive to school, your first step will be to measure the distance of your path to school, including how far you travel between changes in direction. Choose an appropriate scale based on the measurements you record.

Your hand-drawn map should include

▶ Your home and school.

▶ A legend to show the scale factor you used.

▶ The path you take to school including the actual measurements of each part of the route.

On your map, draw a straight line from your home to school. Use the legend to calculate this distance measured on the ground. Compare this to the distance of your current path. Can you follow this straight-line path? Is your current path to school the shortest?

Inverse Variation

*One person's constant is
another person's variable.*

SUSAN GERHART

In each relationship you have worked with in this chapter, if one quantity increased, so did the other. If one quantity decreased, so did the other. If the working hours increase, so does the pay. The shorter the trip, the less gas the car needs. These are direct relationships. Do all relationships between quantities work this way? Can two quantities be related so that increasing one causes the other to decrease?

Try opening your classroom door by pushing on it close to the hinge. Try it again farther from the hinge. Which way takes more force? As the distance from the hinge *increases*, the force needed to open the door *decreases*. This is an example of an *inverse* relationship.

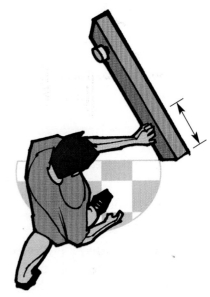

 ## Investigation
Seesaw Nickels

You will need

- a pencil
- a 12-inch ruler
- nine nickels
- tape

If a grown man and a small child sit on opposite ends of a seesaw, what happens? Would changing or moving the weight on one end of the seesaw affect the balance? You'll find out as you do the experiment in this investigation.

Step 1 | On a flat desk or table, try to balance the ruler across a pencil near the ruler's 6-inch mark.

Step 2 | Stack two nickels on the ruler so that they're centered 3 inches to the right of the pencil. You may need to tape them in place.

Step 3	Place one nickel on the left side of the ruler so that it balances the two right-side nickels. Be sure that the ruler stays centered over the pencil. How far from the pencil is this one nickel centered?
Step 4	Repeat Step 3 for two, three, four, and six nickels on the left side of the ruler. Measure to the nearest $\frac{1}{2}$ inch. Copy and complete this table.

Left side		Right side	
Number of nickels	Distance from pencil	Number of nickels	Distance from pencil
1		2	3
2		2	3
3		2	3
4		2	3
6		2	3

Step 5	As you increase the number of nickels on the left side, how does the distance from the balance point change? What relationships do you notice?

Step 6	Make a new table and repeat the investigation with three nickels stacked 3 inches to the right of center. Does the same relationship seem to hold true?

Step 7	Review the data in your tables. How does the number of nickels on the left and their distance from the pencil compare to the number of nickels on the right and their distance from the pencil? In each of your tables, do quantities remain constant? Write a sentence using the words *left nickels, right nickels, left distance,* and *right distance* to explain the relationship between the quantities in this investigation. Define variables and rewrite your sentence as an equation.
Step 8	Explain why you think this relationship between the number of nickels on the left side and the distance from the pencil is an *inverse* relationship.

In the investigation you worked with a fundamental principal of seesaws, or levers. You probably discovered this equation:

(*left nickels*) · (*left distance*) = (*right nickels*) · (*right distance*)

You could also write this relationship as a proportion.

$$\frac{left\ nickels}{right\ nickels} = \frac{right\ distance}{left\ distance} \quad or \quad \frac{left\ nickels}{right\ distance} = \frac{right\ nickels}{left\ distance}$$

Can you use what you know about solving proportion problems to show that all three of these equations are equivalent?

Look closely at the proportions above. How do they differ from the proportions you have written so far? When you wrote proportions for direct relationships, you had to make sure that the numerator and denominator of each ratio corresponded in the same way. In this inverse relationship, the proportion has ratios that correspond in the opposite (or inverse) way. The numerators and denominators seem to be flipped. These are called inverse proportions.

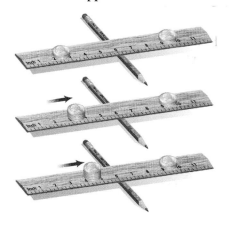

In the investigation, as the number of coins on the left side increased, the coins' distance from the pencil decreased. "Increasing" and "decreasing" show the inverse relationship.

EXAMPLE A

Tyline measured the force needed as she opened a door by pushing at various distances from the hinge. She collected the data shown in the table. Find an equation for this relationship. (A newton, abbreviated N, is the metric unit of force.)

Distance (cm)	Force (N)
40.0	20.9
45.0	18.0
50.0	16.1
55.0	14.8
60.0	13.3
65.0	12.3
70.0	11.6
75.0	10.7

▶ Solution

Enter the data into two calculator lists and graph points. The graph shows a curved pattern that is different from the graph of a direct relationship. If you study the values in the table, you can see that as distance increases, force decreases. The data pairs of this relationship might have a constant product like your data in the investigation.

The *y*-axis is being used for force (N).

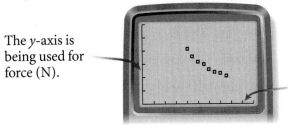

The *x*-axis is being used for distance from the hinge (cm).

Calculate the products in another list. Their mean is approximately 810, so use that to represent the product. Let x represent distance and y represent force.

Distance (cm)	Force (N)	Force · Distance (N-cm)
40	20.9	836.0
45	18.0	810.0
50	16.1	805.0
55	14.8	814.0
60	13.3	798.0
65	12.3	799.5
70	11.6	812.0
75	10.7	802.5

$$xy = 810$$ The product of distance and force is 810.

$$\frac{xy}{x} = \frac{810}{x}$$ Divide both sides by x.

$$y = \frac{810}{x}$$ Now you have the $y=$ form so you can enter this equation in your calculator.

Science
CONNECTION

Scientists use precise machines to measure the amount of force needed to pull or push. Manufacturers use these tools to test the strength of products like boxes. You can also measure force with simple tools like a spring scale. This box shows a certificate that gives the results from several force tests.

Graph the equation. Does it go through all of the points? Why do you think the graph is not a perfect fit? It is a good practice to explore small changes to your equation's constant. A slightly different value might give an even better fit.

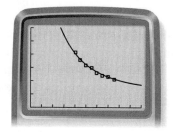

The two variables in the inverse relationships you have seen have a constant product, so you can write an **inverse variation** equation. You can represent the constant product with k just like you use k to represent the constant ratio of a direct variation. The graph of an inverse variation is always curved and will never cross the x- or y-axis. Why couldn't x or y be zero?

Inverse Variation

An equation in the form $y = \frac{k}{x}$ is an **inverse variation.** Quantities represented by x and y are **inversely proportional,** and k is the **constant of variation.**

EXAMPLE B

Ohm's law states that when the electromotive force E (in volts) is constant, the current i (in amperes) is inversely proportional to the resistance R (in ohms). A current of 24 amperes is flowing through a conductor whose resistance is 6 ohms.

Every knob or lever of this sound recording console regulates electric resistance in a current. The resistance varies directly with voltage and inversely with current.

a. What current flows in the system if the resistance increases to 8 ohms?

b. What is the resistance of the conductor if a current of 7.2 amperes is flowing?

▶ **Solution**

You can show the formula as

$$i = \frac{E}{R}$$

E is being used instead of k as the constant in the inverse variation equation.

$$i = \frac{144}{R}$$

The product of i and R is always the same, so use the product of 24 and 6 to get 144 for the value of E.

Use this equation to answer the questions.

a. $\qquad i = \frac{144}{8}$ $\qquad$ Substitute 8 for R.

$\qquad i = 18$ $\qquad$ Divide.

The current flowing through a conductor with resistance 8 ohms is 18 amperes.

b. $\qquad 7.2 = \frac{144}{R}$ $\qquad$ Substitute 7.2 for i.

$\qquad \frac{1}{7.2} = \frac{R}{144}$ $\qquad$ To get R in the numerator, invert both ratios. The proportion is still true.

$\qquad 144 \cdot \frac{1}{7.2} = \frac{R}{144} \cdot 144$ $\qquad$ To isolate R, multiply both sides by 144.

$\qquad 20 = R$ $\qquad$ Multiply and divide.

The resistance of the conductor when a current of 7.2 amperes is flowing through it is 20 ohms.

> ## Practice Your Skills

1. Rewrite each equation in *y*= form.

   a. $xy = 15$ **b.** $xy = 35$ **c.** $xy = 3$

2. Two quantities, *x* and *y*, are inversely proportional. When $x = 3$, $y = 4$. Find the missing coordinates for the points below.

   a. $(4, y)$ **b.** $(x, 2)$ **c.** $(1, y)$ **d.** $(x, 24)$

3. Find five points that satisfy the inverse variation equation $y = \frac{20}{x}$. Graph the equation and the points to make sure the coordinates of your points are correct.

4. **APPLICATION** The amount of time it takes to travel a given distance is inversely proportional to how fast you travel.

   a. How long would it take to travel 90 miles at 30 mph?

   b. How long would it take to travel 90 miles at 45 mph?

   c. How fast would you have to go to travel 90 miles in 1.5 hours?

5. Henry noticed that the more television he watched, the less time he spent doing homework. One night he spent 1.5 hours watching TV and 1.5 hours doing homework. Another night he spent 2 hours watching TV and only 1 hour doing homework. To try to catch up, the next night he spent only a half hour watching TV and 2.5 hours doing homework. Is this an inverse variation? Explain why or why not.

> ## Reason and Apply

6. For each table of *x*- and *y*-values below, decide if the values show a direct variation, an inverse variation, or neither. Explain how you made your decision. If the values represent a direct or inverse variation, write an equation.

a.

x	y
2	12
8	3
4	6
3	8
6	4

b.

x	y
2	24
6	72
0	0
12	144
8	96

c.

x	y
4.5	2.0
0	9.0
3.0	3.0
9.0	0
6.0	1.5

d.

x	y
1.3	15.0
6.5	3.0
5.2	3.75
10.4	1.875
7.8	2.5

7. **APPLICATION** In Example A, you learned that the force in newtons needed to open a door is inversely proportional to the distance in centimeters from the hinge. For a heavy freezer door, the constant of variation is 935 N-cm.

   a. Find the force needed to open the door by pushing at points 15 cm, 10 cm, and 5 cm from the hinge.

   b. Describe what happens to the force needed to open the door as you push at points closer and closer to the hinge. How does the change in force needed compare as you go from 15 cm to 10 cm and from 10 cm to 5 cm?

   c. How is your answer to 7b shown on the graph of this equation?

8. Emily and her little brother Sid are playing on a seesaw. Sid weighs 65 pounds. The seesaw balances when Sid sits on the seat 4 feet from the center and Emily sits on the board $2\frac{1}{2}$ feet from the center.

   a. About how much does Emily weigh?

   b. Sid's friend Seogwan sits with Sid at the same end of the seesaw. They weigh about the same. Can Emily balance the seesaw with both Sid and Seogwan on it? If so, where should she sit? If not, explain why not.

9. To use a double-pan balance, you put the object to be weighed on one side and then put known weights on the other side until the pans balance.

   a. Explain why it is useful to have the balance point halfway between the two pans.

   b. Suppose the balance point is off-center, 15 cm from one pan and 20 cm from the other. There is an object in the pan closest to the center. The pans balance when 7 kg is placed in the other pan. What is the weight of the unknown object?

10. **APPLICATION** The student council wants to raise $10,000 to purchase computers. All students are encouraged to participate in a fund-raiser, but it is likely that some will not be able to.

   a. Pick at least four numbers to represent how many students might participate. Make a table showing how much each student will have to raise if each participant contributes the same amount.

   b. Plot the points represented by your table on a calculator graph. Find an equation to fit the points.

   c. Suppose there are only 500 students in the school. How would this number of students affect your graph? Sketch a graph to show this limitation.

11. A tuning fork vibrates at a particular frequency to make the sound of a note in a musical scale. If you strike a tuning fork and place it over a hollow tube, the vibrating tuning fork will cause the air inside the tube to vibrate, and the sound will get louder. Skye found that if you put one end of the tube in water and raise and lower it, the loudness will vary. She used a set of tuning forks and, for each one, recorded the tube length that made the loudest sound.

Tuning Fork Experiment

Note	Frequency (hertz)	Tube length (cm)
A_4	440.0	84.6
C_5	523.3	71.1
D_5	587.3	63.4
F_5	698.5	53.3
G_5	784.0	47.5

a. Graph the data on your calculator and describe the relationship.

b. Find an equation to fit the data. Explain how you did this and what your variables and constants represent.

c. The last tuning fork in Skye's set is A_5 with a frequency of 880.0 hertz. What tube length should produce the loudest sound?

12. APPLICATION To squeeze a given amount of air into a smaller and smaller volume, you have to apply more and more pressure. Boyle's law describes the inverse variation between the volume of a gas and the pressure exerted on it. Suppose you start with a 1-liter open container of air. If you put a plunger at the top of the container without applying any additional pressure, the pressure inside the container will be the same as the pressure outside the container, or 1 atmosphere (atm).

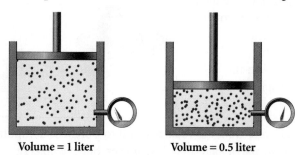

Volume = 1 liter   Volume = 0.5 liter

a. What will the pressure in atmospheres be if you push the plunger down until the volume of air is 0.5 liter?

b. What will the pressure in atmospheres be if you push the plunger down until the volume of air is 0.25 liter?

c. Suppose you exert enough pressure so that the pressure in the container is 10 atm. What will the volume of the air be?

d. What would you have to do to make the pressure inside the container less than 1 atm?

e. Graph this relationship, with pressure (in atmospheres) on the horizontal axis and volume (in liters) on the vertical axis.

▶ Review

13. APPLICATION A CD is on sale for 15% off its normal price of $13.95. What is its sale price? Write a direct variation equation to solve this problem.

14. Calcium and phosphorus play important roles in building human bones. A healthy ratio of calcium to phosphorus is 5 to 3.

  a. If Mario's body contains 2.5 pounds of calcium, how much phosphorus should his body contain?

  b. About 2% of an average woman's weight is calcium. Kyle weighs 130 pounds. How many pounds of calcium and phosphorus should her body contain?

FAMILIES OF RECTANGLES

On each set of axes below, two rectangles are drawn with a common vertex at the origin. On the left, the rectangles are similar. On the right, the rectangles have the same area. If you draw more rectangles following the same patterns and then connect their upper-right vertices, what kinds of curves will you get? Write an equation for each pattern.

THE GEOMETER'S SKETCHPAD®

With The Geometer's Sketchpad, you can construct families of rectangles and other polygons. Commands like Trace and Locus can help you dynamically create curves.

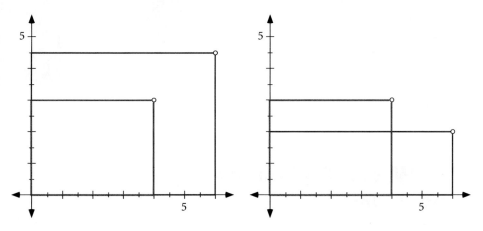

Explore at least four different families of rectangles or families of other shapes. Draw each family on its own set of axes and then describe the pattern in words. Connect corresponding vertices and see whether a curve is formed. (In mathematics, a straight line is actually considered a type of curve.) For each curve, write an equation or describe it in words. Summarize your findings in a paper or presentation.

Activity Day

Variation with a Bicycle

The Tour de France is a demanding bicycle race through Switzerland, Germany, and France. For 23 days, cyclists ride 3630 kilometers on steep mountain roads before crossing the finish line in Paris. The cyclists rely on their knowledge of gear shifting and bicycle speeds.

Many bicycles have several speeds or gears. In a low gear, it's easier to pedal uphill. In a high gear, it's harder to pedal, but you can go faster on flat surfaces and down hills. When you change gears, the chain shifts from one sprocket to another. In this activity you will discover the relationships among the bicycle's gears, the numbers of times you pedal, and the teeth on the sprockets.

Activity
The Wheels Go Round and Round

You will need

- a meterstick or metric tape measure
- a multispeed bicycle

In Steps 1–5, you'll analyze the effect of the rear sprockets.

Procedure Note

Changing Gears

Each time you change gears on the bike, you may have to turn the bike right side up and rotate the pedal and crankshaft a few times until the gear change takes effect. Then you turn the bike upside down to start observing and recording data.

Rear sprocket assemblies

Tooth

Crankshaft

Front sprocket assemblies

Step 1	Shift the bicycle into its lower gear.
Step 2	Count the number of teeth on the front and rear sprockets in use. Record your numbers in a table like this one:

Number of teeth on front sprocket	Number of teeth on rear sprocket	Number of revolutions of rear wheel for one revolution of pedals

Step 3	Line up the air valve, or a chalk mark on the tire of the rear wheel, with part of the bicycle frame. This will be the "starting point." Rotate the pedal through one complete revolution and stop the wheel immediately. Estimate the number of wheel revolutions to the nearest tenth and enter it into the table.

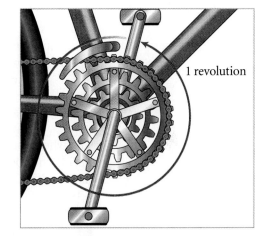

1 revolution

Step 4	Shift gears so that the chain moves onto the next rear sprocket. Do not change the front sprocket. Repeat Steps 2 and 3. Record your data in a new row of your table. Repeat this process for each rear sprocket on your bicycle.
Step 5	Describe how the number of teeth on the rear sprocket affects how the wheel turns. What kind of variation is this? Plotting the data on your calculator may help you to see this relationship. Define variables and write an equation that relates the number of wheel revolutions to the number of teeth on the rear sprocket. Explain the meaning of the constant in this equation.

In Steps 6–10, you'll analyze the effect of the front sprockets.

Step 6	Shift the bicycle into its lowest gear again.
Step 7	Count and record the number of teeth on the sprockets in use in a second table.
Step 8	As you did in Step 3, record the number of wheel revolutions for one revolution of the pedal crankshaft.
Step 9	Keep the chain on the same rear sprocket and shift gears so that the chain is placed onto the next front sprocket. Repeat Steps 7 and 8. In the second table you should have one row of data for each front sprocket.
Step 10	Describe how the number of teeth on the front sprocket affects the turning of the wheel. What kind of variation models this relationship? Plot the data on your calculator to verify your answer. Define variables and write an equation that relates the number of teeth on the front sprocket to the number of wheel revolutions. What is the meaning of the constant in this equation?

Now you'll see why gear shifting is such an important strategy in a bicycle race.

Step 11 | Find a proportion relating the number of front teeth, rear teeth, wheel revolutions, and pedal revolutions. Use it to predict the number of wheel revolutions for a gear combination you have not tried yet. Test your prediction by doing the experiment on this gear combination.

Step 12 | Explain why different gear ratios result in different numbers of rear wheel revolutions. Why is it possible to go faster in a high gear?

Step 13 | Find the circumference of the rear wheel in centimeters. How far will the bicycle travel when the wheel makes one revolution? How many revolutions will it take to travel 1 kilometer without coasting?

Step 14 | For the lowest and highest gear, how many times do you need to rotate the pedals for the bike to travel 1 kilometer? (Hint: Write a proportion or other equation involving the gear ratio and the number of revolutions of the pedals and the wheel.)

The wheels on Lance Armstrong's bicycle made roughly 1.6 million revolutions during the 2000 Tour de France. If he hadn't coasted or changed gears, he could have pedaled more than 1.5 million times.

Averaging 39 kilometers per hour, American Lance Armstrong became the 2000 Tour de France champion. He completed the race in 92 hours, 33 minutes, and 8 seconds in spite of his struggle with cancer.

CHAPTER 3

REVIEW

In this chapter you learned about variation. From your knowledge of ratios and proportions, you learned the concept of **rate.** You solved application problems with rates involving measurement units, pay, monetary conversions, and gasoline consumption.

You learned that quantities are **directly proportional** when an increase in one value leads to a proportional increase in another. These quantities form a **direct variation** when their *ratio* is constant. The constant ratio is called a **constant of variation.** When you graphed a direct relationship, you discovered a line that always passes through the origin.

For an **inverse variation,** the *product* of two quantities is constant. In this relationship, an increase in one variable causes a decrease in the other. The graph of the relationship is curved rather than straight, and the graph does not touch either axis.

You combined what you've learned about ratios, proportions, and direct variation to interpret scale drawings. You learned how to calculate the **scale factor** and how to use it in making your own drawings. You also investigated **similar figures** and learned how to find the length of missing sides using proportion and direct variation.

EXERCISES

You will need your calculator for problems **7** and **10.**

1. **APPLICATION** Nicholai's car burns 13.5 gallons of gasoline every 175 miles.
   a. What is the car's fuel consumption rate?
   b. At this rate, how far will the car go on 5 gallons of gas?
   c. How many gallons does Nicholai's car need to go 100 miles?

2. **APPLICATION** Two dozen units in an apartment complex need to be painted. It takes 3 gallons of paint to cover each apartment.
   a. How many apartments can be painted with 36 gallons?
   b. How many gallons will it take to paint all 24 apartments?

3. On many packages the weight is given in both pounds and kilograms. The table shows the weights listed on a sample of items.

Kilograms	1.5	0.7	2.25	11.3	3.2	18.1	5.4
Pounds	3.3	1.5	5	25	7	40	12

a. Use the information in the table to find an equation that relates weights in pounds and kilograms. Explain what the variables represent in your equation.

b. Use your equation to calculate the number of kilograms in 30 pounds.

c. Calculate the number of pounds in 25 kilograms.

4. Consider this graph of a sunflower's height above ground.

Plant Height

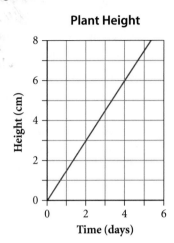

a. How tall was the sunflower after 5 days?

b. If the growth pattern continues, how many days will it take the plant to reach a height of 25 cm?

c. Write an equation to represent the height of the plant after any number of days.

5. APPLICATION The scale on a map reads "1 inch = 21 miles."

a. How far apart on the map are two towns that are actually 47 miles apart?

b. A lake on the map is $\frac{3}{4}$-inch wide. How wide is the actual lake?

6. Find the missing side lengths of the similar figures.

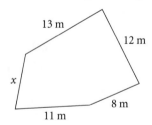
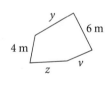

7. Use the table of values to answer each question.

a. Is the data in the table related by a direct variation or an inverse variation? Explain.

b. Find an equation to fit the data. You may use your calculator graph to see how well the equation fits the data.

c. Use your equation to predict the value of y when x is 32.

x	y
12	4
5	9
16	3
22	2
9	5
43	1

8. In the formula $d = vt$, d represents distance in miles, v represents rate in miles per hour (mph), and t represents time in hours. Use the word *directly* or *inversely* to complete each statement. Then write an equation for each.

a. If you travel at a constant rate of 50 mph, the distance you travel is _____ proportional to the time you travel.

b. The distance you travel in exactly 1 hour is _____ proportional to your rate.

c. The time it takes to travel 100 miles is _____ proportional to your rate.

9. Kris and Robbie are sitting on a 10-foot seesaw and have it perfectly balanced. At 100 pounds, Kris must sit 3.2 feet away from the center to balance Robbie, who weighs 80 pounds.

a. How far from the center is Robbie sitting?

b. Robbie's 30-pound dog jumps into his lap. Can Robbie still balance the seesaw with Kris? How?

10. APPLICATION Boyle's law describes the inverse variation between the volume of a gas and the pressure exerted on it. In the experiment shown, a balloon with a volume of 1.75 liters is sealed in a bell jar with 1 atm of pressure. As air is pumped out of the jar, the pressure decreases, and the balloon expands to a larger volume.

Pressure = 1 atm   Pressure = 0.8 atm

Air pumped out

a. Find the volume under 0.8 atm of pressure.

b. Find the pressure when the balloon's volume is 0.75 liter.

c. Write an equation that calculates the volume in liters from the pressure in atmospheres.

d. Graph this relationship on your calculator, then sketch it on your own paper. Show on your graph the solutions to 10a and b.

MIXED REVIEW

11. APPLICATION Sonja bought a pair of 210-cm cross-country skis. Will they fit in her ski bag, which is $6\frac{1}{2}$ feet long? Why or why not?

12. Fifteen students counted the number of letters in their first and last names. Here is the data set.

| 6 | 15 | 8 | 12 | 8 | 17 | 9 | 7 |
| 13 | 15 | 14 | 9 | 16 | 15 | 10 |

a. Make a histogram of the data with a bin width of 2.

b. If you selected one of the students at random, what is the probability that his or her name has 8 or 9 letters?

13. Make a circle graph of these letter grades for an algebra class.

Describe how you used the concept of direct variation in drawing the graph.

Grade	A	B	C	D	F
Number of students	8	10	12	6	4

14. Evaluate these expressions.

a. $-3 \cdot 8 - 5 \cdot 6$ **b.** $(-2 - (-4)) \cdot 8 - 11$ **c.** $7 \cdot 8 + 4 \cdot (-12)$ **d.** $11 - 3 \cdot 9 - 2$

15. California has many popular national parks. This table shows the number of visitors in thousands to national parks in 1999.

   a. Find the mean number of visitors.

   b. What is the five-number summary for the data?

   c. Create a box plot for the data.

   d. Identify any parks in California that are outliers in the numbers of visitors they had. Explain why they are outliers.

Park Attendance

National park	Visitors (thousands)
Channel Islands	607
Death Valley	1228
Joshua Tree	1316
Kings Canyon	560
Lassen Volcanic	354
Redwood	370
Sequoia	873
Yosemite	3494

(*U.S. National Park Service*)

Joshua Tree National Park, California

16. A ball is randomly selected from a bin that contains balls numbered from 1 to 99.

   a. What is the probability that the number is even?

   b. What is the probability that the number is divisible by three?

   c. What is the probability that the number contains a 2?

   d. What is the probability that the number has only one digit?

Lassen Volcanic National Park, California

17. Remember that Ohm's law states that electrical current is inversely proportional to the resistance. A current of 18 amperes is flowing through a conductor whose resistance is 4 ohms.

   a. What is the current that flows through the system if the resistance is 8 ohms?

   b. What is the resistance of the conductor if a current of 12 amperes is flowing?

18. APPLICATION Amber makes $6 an hour at a sandwich shop. She wants to know how many hours she needs to work to save $500 in her bank account. On her first paycheck, she notices that her net pay is about 75% of her gross pay.

   a. How many hours must she work to earn $500 in gross pay?

   b. How many hours must she work to earn $500 in net pay?

TAKE ANOTHER LOOK

The equation $y = kx$ is a *general equation* because it stands for a whole family of equations such as $y = 2x$, $y = \frac{1}{4}x$, even $y = \pi x$. (Does $C = \pi d$ look more familiar?)

What might k be in each line of these graphs? (If you're stumped, choose a point on the line, then divide its y-coordinate by its x-coordinate. Remember, $k = \frac{y}{x}$ is equivalent to $y = kx$.)

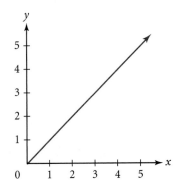 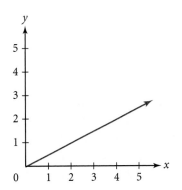 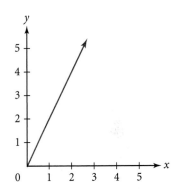

Most real-world quantities, like time and distance, are measured or counted in positive numbers. If two positive quantities vary directly, their graph $y = kx$ is in Quadrant I, where both x and y are positive. Because the quotient $\frac{y}{x}$ of two positive numbers x and y must be positive, the constant k is positive.

But the graph at right also shows a direct variation. What can you say about k for this graph?

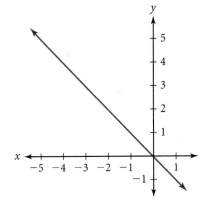

What relationship do you see between the lines in each of these situations shown below? Between the k-values?

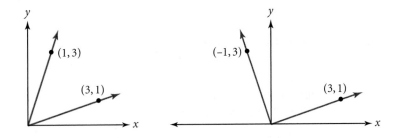

Finally, do you have a direct variation if $k = 0$? Why or why not?

Assessing What You've Learned

PERFORMANCE ASSESSMENT

This chapter has been about direct and inverse variation, so assessing what you've learned really means checking to see if you can tell how two quantities vary, what that variation looks like on a graph, and what the equation of the relationship is. If you have two direct variations using the same units, you should be able to compare them on the same graph and tell how their equations are different. Can you do one of the investigations in this chapter on your own? Can you get the equations for two variations from the word descriptions of the relationships? Showing that you can do tasks like these is sometimes called "performance assessment."

Review the Investigation "Off to Work We Go" in Lesson 3.1 about time and pay, and the Investigation "Floor Plans" in Lesson 3.3 about scale drawings. Reconstruct the graphs and drawings to see if these skills have become easier for you. Get help with any part of the process that you're not sure of. Review the Investigation "Seesaw Nickels" in Lesson 3.4 about the relationship between weight and distance from the balance point. Is it clear to you how inverse variation is different from direct variation? Graph at least one direct variation and one inverse variation for a classmate, a family member, or your teacher. Tell them how you can show the relationship in a table and in an equation as well as in a graph. Tell what the constant means in each equation.

ORGANIZE YOUR NOTEBOOK Make sure your notebook contains notes on each investigation from this chapter. Your notes should give the results of this investigation as well as some steps along the way. Be sure you have written down the definitions of new words such as *direct variation* and *inverse variation*. Are your notes complete enough that you could write a chapter summary from them?

WRITE IN YOUR JOURNAL Add to your journal using one of these prompts:
▶ Does graphing relationships help you understand how the quantities vary? Do you understand variation between quantities better when you look at a graph or when you read an equation?
▶ Tables of values, graphs, equations, and word descriptions are four ways to tell about a variation. What other mathematical ideas can you show in more than one way?

UPDATE YOUR PORTFOLIO Choose a variation that you studied in this chapter. Describe the relationship in words. Show it as a table of *x*- and *y*-values, as an equation, and as a graph. Be sure you have defined your variables and carefully labeled the graph. Add this work to your portfolio. Also, you may want to add your work from a project or the Activity Day.

Linear Equations

Weavers repeat steps when they make baskets and mats, creating patterns of repeating shapes. This process is not unlike recursion. In the top photo, a mat weaver in Myanmar executes a traditional design with palm fronds. You see bowls crafted by Native American artisans in the bottom photo.

OBJECTIVES

In this chapter you will

- review the rules for order of operations
- describe number tricks using algebraic expressions
- write recursive routines
- study rate of change
- learn to write equations for lines
- use equations and tables to graph lines
- solve linear equations

Order of Operations and the Distributive Property

Ty and Melinda both entered 4 + 6 · 3 into their calculators. Ty's simple, four-function calculator showed 30 when he pressed =. Melinda's graphing calculator showed 22 when she pressed ENTER. What did the two calculators do differently? Which calculator is correct? How do you know whether you should add first or multiply first? According to the order of operations, the correct answer is 22 because the rules tell you to multiply before adding.

Order of Operations

1. Evaluate expressions within parentheses or other grouping symbols.
2. Evaluate all powers.
3. Multiply and divide from left to right.
4. Add and subtract from left to right.

These rules ensure that everyone who correctly does a given calculation gets the same answer. A scientific or graphing calculator, which lets you finish entering an entire expression before it does the calculation, follows these rules. Understanding the rules will help you perform calculations correctly, with or without a calculator.

EXAMPLE A

Evaluate the expression $\frac{18.7 + 11.3}{5}$ without a calculator. Then enter the expression into your calculator to see if you get the same answer.

▶ **Solution**

The fraction bar is a grouping symbol, so the numbers in the numerator form a group.

$$\frac{18.7 + 11.3}{5} = \frac{30}{5} \qquad \text{Add the grouped numbers.}$$

$$= 6 \qquad \text{Now divide.}$$

In order for your calculator to recognize the grouping, you need to enter parentheses around the numerator: (18.7 + 11.3)/5 ENTER .

What's wrong with entering the expression as 18.7 + 11.3/5?

Does entering the keystrokes 18.7 + 11.3 ENTER /5 ENTER produce the correct result?

Investigation
Cross-number Puzzle

You will need

● the worksheet Cross-number Puzzle

Working the cross-number puzzle in this investigation will help you review and practice the rules for order of operations. Remember that you have to perform operations within the numerator or denominator of a fraction before you do the division indicated by the fraction bar. The square root symbol, $\sqrt{}$, can be used as a grouping symbol. Use parentheses where necessary. Part of your challenge is to figure out how to enter the entire expression into your calculator so that you get the correct answer without having to calculate part of the expression first.
[▶ 🖥 See **Calculator Note 4A** to review the instant replay command. ◀]

Write your answer in fraction form or decimal form if the clue asks for it.
[▶ 🖥 See **Calculator Note 0A** for help converting answers from decimal numbers to fractions and vice versa. ◀] Each negative sign, fraction bar, or decimal point occupies one square in the puzzle; commas are not part of the answer. For instance, an answer of 2508.5 needs six squares.

Be sure to work all the problems, even if an answer is entirely filled in by the time you get to its clue. That way you can check your work.

Across

1. $\frac{2}{3}$ of 159,327

3. $\frac{-1 + 17^2}{4 + 2^2}$

4. $4835 - 541 + 1284$

6. $\frac{3 + 140}{3 \cdot 14}$ (fraction form)

7. $8075 - 3(42)$

9. $\sqrt{6^2 + 8^2}$

11. $\frac{740}{18.4 - 2.1 \cdot 9}$

12. 57^3

Down

1. $9(-7 + 180)$

2. $\left(\frac{9}{2}\right)\left(\frac{17}{5} + \frac{25}{4}\right)$ (fraction form)

4. $3 - 3(12 - 200)$

5. $9 \cdot 10^2 - 9^2$

8. $15 + 47(922)$

10. $25.9058 \cdot 20/4 - 89$ (decimal form)

11. $1284 - \frac{877}{0.2}$

In an expression like $3(4 + 2)$, you multiply 3 by the sum inside the parentheses. You can do this by finding the sum first:

$$3(4 + 2) = 3(6) \qquad \text{Add.}$$
$$= 18 \qquad \text{Multiply.}$$

Or multiply 3 by each number in the sum and then add:

$$3(4 + 2) = 3 \cdot 4 + 3 \cdot 2 \qquad \text{Multiply through the parentheses.}$$
$$= 12 + 6 \qquad \text{Evaluate.}$$
$$= 18 \qquad \text{Add.}$$

The second method uses the **distributive property.** Think of "distributing" the number outside the parentheses to all the numbers inside. The figure at right shows a visual model of $3(4 + 2) = 18$.

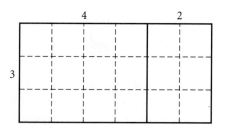

In an expression like $3(4 + 2)$, you multiply 3 by the sum inside the parentheses. You can do this two ways.

Method 1   Find the sum first:

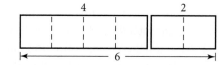

$4 + 2 = 6$

Then multiply 3 by the sum:

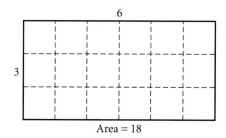

Area = 18

$3(6) = 18$

Method 2   Multiply 3 by each number in the sum:

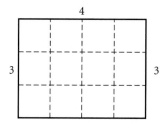

$3(4) + 3(2)$

Then add the products:

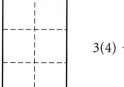

Area = 18

$12 + 6 = 18$

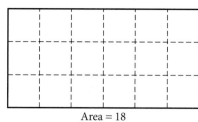

Distributive Property

For any values of a, b, and c, this equation is true:

$$a(b + c) = a(b) + a(c)$$

We say that a is "distributed through the parentheses." The values of a, b, and c can be 2, 5.6, $\frac{2}{3}$, -7, or 0, and the distributive property still holds true. Sometimes we call it "the distributive property of multiplication over addition" or "the distributive property over subtraction," depending on these values.

The distributive property is useful for doing mental math.

EXAMPLE B

Find these amounts without using your calculator.

a. four CDs at $14.95 each

b. a 15% tip on a $24 restaurant bill

▶ **Solution**

In each situation, use the distributive property.

a. Each CD is a nickel less than $15. So the cost is about $4 \cdot 15 = 60$. Subtract the four nickels (20 cents) to get $59.80.

Written out, the calculation looks like this:

$$4(15 - 0.05) = 4(15) - 4(0.05)$$ Distribute 4 over the subtraction 15 − 0.05.

$$= 60 - 0.20$$ Multiply.

$$= 59.80$$ Subtract.

b. Think of 15% as 10% + 5%. Ten percent of $24 is $2.40. Five percent is half that, or $1.20. So the total tip is $3.60.

Written out, the calculation looks like this:

$$(0.10 + 0.05)24 = 0.10(24) + 0.05(24)$$ Distribute 24 over the addition 0.10 + 0.05.

$$= 2.40 + 1.20$$ Multiply.

$$= 3.60$$ Add.

An **algebraic expression** is a symbolic representation of mathematical operations that can involve both numbers and variables. You can use the distributive property to rewrite some algebraic expressions without parentheses.

EXAMPLE C

The cost of tickets to a play varies depending on where the seats are located in the theater. The service charge for ordering tickets by phone is $3 per ticket, regardless of the ticket price.

a. Let the variable T represent the cost of one ticket. Write two equivalent expressions for the cost of four tickets ordered by phone. One expression should use parentheses and the other should not.

This is the Balet Folklórico performing at the University of Guadalajara.

b. Make up four ticket prices. Use your calculator to check that both of the expressions you wrote in part a give the same total prices for four tickets.

▶ **Solution**

a. One expression for the cost of four tickets is $4(T + 3)$. You can rewrite the expression without parentheses by distributing the 4 to get $4T + 12$.

b. Enter this list of ticket prices in list L₁:
{18, 30, 45, 65}. The calculator screen at right shows the list of prices stored in list L₁; then it shows two expressions (using L₁ in place of the variable T) evaluated. They give the same results. It costs $84 to order four $18 tickets by phone, $132 to order four $30 tickets, and so on.

EXAMPLE D

Rewrite the expression $2 - 3(4 + L_1)$ in an equivalent form without parentheses. Enter and store a short list like {1, −3, −10, 5} in list L₁ to verify your result on your calculator.

▶ **Solution**

The expression $2 - 3(4 + L_1)$ is really a family of expressions because list L₁ holds four values.

$$2 - 3(4 + L_1) = 2 + (-3)(4 + L_1)$$

Change from subtracting to adding a negative number.

$$= 2 + (-3)(4) + (-3)(L_1)$$

Distribute the –3 over the addition $4 + L_1$.

$$= 2 + (-12) + (-3L_1)$$

Multiply.

$$= -10 - 3L_1$$

Add.

Evaluating $-10 - 3L_1$ for each number stored in list L₁ gives -13, -1, 20, and -25.

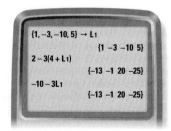

EXERCISES

You will need your calculator for problems **1, 2, 10,** and **12.**

▶ Practice Your Skills

1. Evaluate these expressions without a calculator. Use your calculator to check your answers.

   a. $\dfrac{-27 + 39}{4}$ **b.** $2 + (-3)(4)$ **c.** $\dfrac{3}{4} - \dfrac{1}{2}$ **d.** $\dfrac{12}{4 - 6}$

2. Check your understanding of the rules for order of operations.

   a. Evaluate the expression $-4(13 - 2 \cdot 3^2)$ without using your calculator.

   b. Enter the expression into your calculator, and confirm both the correct order of operations you used and your computed answer.

3. Check your understanding of the distributive property.

   a. Write an expression equivalent to $2(L + W)$.

   b. Verify that your answer is equivalent to $2(L + W)$ by substituting 15 for L and 4 for W into both the original expression and your answer to 3a.

4. Describe two different procedures to find the value of the expression $8075 - 3(37 + 5)$. Use the distributive property to show that the two ways are equivalent.

5. **APPLICATION** Jasmine makes \$7 per hour in her part-time job at the corner store. She worked 4 hours on Monday, 3 hours on Wednesday, and 5 hours on Saturday.

   a. How much did she earn for the week?

   b. Explain two ways you can answer the question in 5a. Why are the answers the same?

6. **APPLICATION** Use the distributive property to find the 15% tip amount for these food bills.

   a. \$36   **b.** \$21   **c.** \$48.50

7. **APPLICATION** Find the 15% tip amount for these food bills by writing and solving proportions. For example, $\dfrac{15}{100} = \dfrac{x}{36}$.

   a. \$36   **b.** \$21   **c.** \$48.50

▶ Reason and Apply

8. Peter and Seija evaluated the expression $37 + 8 \cdot \dfrac{6}{2}$. Peter said the answer was 135. Seija said it was 61. Who is correct? What error did the other person make?

9. In what order would you perform the operations to evaluate these expressions and get the correct answers?

   a. $9 + 16 \cdot 4.5 = 81$ **b.** $18 \div 3 + 15 = 21$ **c.** $3 - 4(-5 + 6^2) = -121$

10. You can use your calculator's instant replay command when you need to evaluate an expression for several different sets of values. This command is helpful because it recalls previous entries one at a time and displays them on the screen.

[▶ 🖥 See **Calculator Note 4A** to review the instant replay command. ◀]

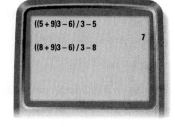

a. Enter and evaluate the expression $((5 + 9)3 − 6)/3 − 5$. What is the answer?

b. Use the instant replay function to get back to the original expression. Replace each 5 with 8 and reevaluate the expression. What's your answer this time?

c. Go back to the last expression you entered. Use the insert key and replace each 8 with 25. Reevaluate the expression. Now what's your answer?

d. Go back to the last expression you entered. Use the delete key and replace each 25 with 2. Reevaluate the expression. What answer did you get?

e. Use the instant replay function repeatedly to return to the original expression in 10a. Replace each 5 with the same number of your choice. Reevaluate the expression once more. What's your answer?

f. What did you notice about the answers to 10a–e? Why do you think this happened?

11. Insert operation signs, parentheses, or both into each string of numbers to create an expression equal to the answer given. Keep the numbers in the same order as they are written. Write an explanation of your answer, including information on the order in which you performed the operations.

a. 3 2 5 7 = 18

b. 8 5 6 7 = 13

12. Using a list in an expression is the same as writing as many expressions as there are numbers in the list. Use the distributive property to rewrite each family of expressions without parentheses. Then enter $\{−3, 4.5, 10\}$ into list L_1 to check answers.

a. $3(L_1 − 4.2)$

b. $14 + 3(L_1 − 4.2)$

c. $14 − 3(L_1 − 4.2)$

13. Write and then evaluate an expression that performs the following sequence of operations: Add 6 and 3, multiply by the square of 4, divide by the sum of 8 and 2, and then subtract 9.

14. This problem is sometimes called Einstein's problem: "Use the digits 1, 2, 3, 4, 5, 6, 7, 8, 9 and any combination of the operation signs $(+, −, \cdot, /)$ to write an expression that equals 100. Keep the numbers in consecutive order and do not use parentheses."

Here is one solution:

$$123 − 4 − 5 − 6 − 7 + 8 − 9 = 100$$

Your task is to find another one.

15. The table displays data for a bicyclist's distance from home during a four-hour bike ride.

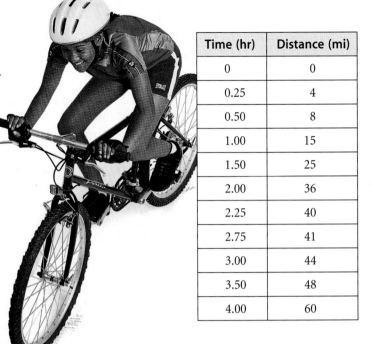

Time (hr)	Distance (mi)
0	0
0.25	4
0.50	8
1.00	15
1.50	25
2.00	36
2.25	40
2.75	41
3.00	44
3.50	48
4.00	60

   a. Make a scatter plot of the data.

   b. Find the bicyclist's average speed.

   c. Find an equation that models the data and graph it on the scatter plot.

   d. At what times might the bicyclist be riding downhill or pedaling uphill? Explain.

16. You are helping to design boxes for game balls that are 1 inch in diameter. Your supervisor wants the balls in one rectangular layer of rows and columns.

   a. Use a table to show all the ways to package 24 balls in rows and columns.

   b. Plot these points on a graph.

   c. Do the possible box dimensions represent direct or inverse variation? Explain how you know.

   d. Represent the situation with an equation. Does the equation have any limitations?

   Learn about a career in packaging science with the links at www.keymath.com/DA .

17. **APPLICATION** A restaurant menu item is $12.95. You have a coupon for 20% off the price. Sales tax is 8% of the discounted amount. Find the amounts for the discount, tax, and a 15% tip that is based on the original amount. What is the total bill?

Writing Expressions and Undoing Operations

All change is not growth; all movement is not forward.

ELLEN GLASGOW

Try this trick: Think of a number from 1 to 25. Add 9 to it. Multiply the result by 3. Subtract 6 from the current answer. Divide this answer by 3. Now subtract your original number. No matter what number you chose, your final result should be 7! Surprised? Try the trick again, starting with a different number. Did you get 7 again? Did all your classmates also get 7? How does this number trick work? The illustration at right shows the sequence this number trick generates if your original number is 11. If you understand the rules for order of operations and a little algebra, you can determine why this trick works and design your own number tricks.

Think of a number from 1 to 25. **11**

Add 9 to it. **20**

Multiply the result by 3. **60**

Subtract 6 from the current answer. **54**

Divide this answer by 3. **18**

Now subtract your original number. **7**

Investigation
The Math Behind the Number Trick

Are you convinced that the number trick you just did works regardless of the number you start with? Would it still work if you chose a decimal number, a fraction, or a negative number? You can test the trick on several numbers at once by using a calculator list with several different starting numbers.

Step 1 Enter a list of at least four different numbers on the calculator home screen and store this list in list L₁. In the example at right, the list is {20, 1.2, −5, 4}, but you should try different numbers. Perform the operations on your own starting numbers. The last operation is to subtract your original number.

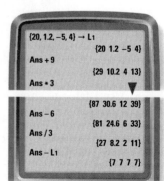

Step 2 Explain how the last operation is different from the others.

Step 3 Number tricks like this work because certain operations, such as multiplication and division, get "undone" in the course of the trick. Which step undoes Ans · 3?

You can use the scheme below to help you figure out why any number trick works. The symbol +1 represents one positive unit. You can think of n as a variable or as a container for different unknown starting numbers.

Stage		Description
1	n	Pick a number.
2	n +1 +1 +1	
3	n n +1 +1 +1 +1 +1 +1	
4	n n +1 +1	
5	n +1	
6	+1	Subtract the original number.
7	+1 +1 +1	

Step 4 | Explain what happens at each stage. The descriptions are provided for Stages 1 and 6.

Step 5 | At which stage or stages will everyone's result be the same? Explain why this happens.

Step 6 | Use your original list of starting numbers and record the results of this number trick at each stage.

Step 7 | Invent your own number trick that has at least five stages. Test it on your calculator with a list of at least four different numbers to make sure all the answers are the same. When you're convinced the number trick is working, try it on the other members of your group.

The math behind number tricks can help you understand the roles that variables and expressions play in algebra. A single expression can represent an entire number trick.

EXAMPLE A

Investigate this calculator routine to see if it is a number trick. Pick several numbers to store into list L1 and use it as the variable x. For example, use $\{1, -2, 0, 0.3\}$ to apply the operations on four numbers at the same time.

a. Describe each stage of the routine in words.

b. List the sequence of numbers this routine generates if you let x equal -5.

x	ENTER
Ans $\cdot$ (-10)	ENTER
Ans $+ 35$	ENTER
Ans$/(-10)$	ENTER
Ans $- 14.5$	ENTER
Ans $- x$	ENTER

c. Write a sequence of expressions that shows each stage of this routine. Use x to represent your starting number.

d. Pick a new starting number and store it in x. Find the number this routine produces.

▶ **Solution**

a. The first column in the table describes the routine in words.

b. The second column shows the sequence of numbers generated by a starting number of -5.

c. The third column gives the sequence of expressions for the routine for a starting number x. Notice that the final expression contains all the information about the stages of the trick in a concise, symbolic form. You can also begin to see why the number trick works. The division by -10 "undoes" the multiplication by -10, and by subtracting x in the last stage you'll get the same result no matter what x is.

a. Description	b. Sequence	c. Expression
Pick the starting number.	-5	x
Multiply by -10.	50	$-10x$
Add 35.	85	$-10x + 35$
Divide by -10.	-8.5	$\dfrac{-10x + 35}{10}$
Subtract 14.5.	-23	$\dfrac{-10x + 35}{-10} - 14.5$
Subtract your original number.	-18	$\dfrac{-10x + 35}{-10} - 14.5 - x$

d. The routine always results in -18.

EXAMPLE B

Consider the expression

$$4\left(\frac{x + 7}{4} + 5\right) - x + 13$$

a. Write in words the number trick that the expression describes.

b. Test the number trick to be sure you get the same result no matter what number you choose.

c. Which operations that undo previous operations make this number trick work?

▶ **Solution**

a. Pick a number, x.
Add 7.
Divide the answer by 4.
Add 5 to this result.
Multiply the answer by 4.
Subtract your original number.
Then add 13.

b. One way to test the trick is to enter the
expression into your calculator. Be sure to use
parentheses for grouping symbols like the
fraction bar:

$$4((x + 7)/4 + 5) - x + 13$$

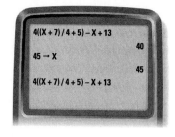

Press (ENTER) to evaluate the expression for
whatever value you now have stored in x.
Your answer should be 40. Store a different value for x, then use the instant
replay function to recall the expression. Press (ENTER) to evaluate the expression
with your new x-value. You should get 40 again. Use a list to test the trick on
several numbers at the same time.

c. The multiplication by 4 undoes the division by 4. You start with x and later
you subtract x, so it doesn't matter what value you choose for x.

An **equation** is a statement that says the value of one number or algebraic
expression is equal to the value of another number or algebraic expression. You can
represent the number trick in Example B with the equation

$$4\left(\frac{x + 7}{4} + 5\right) - x + 13 = 40$$

The 40 on the right of the equal sign is the result of evaluating the expression on
the left for some value of x. If x can be replaced by a number that makes the
equation true, then that number is a **solution** to the equation. This is a very
unusual equation, because it's true no matter what number x is. The equation has
infinitely many solutions. That's what makes it a trick!

Of course, not every algebraic expression represents a number trick. In most
algebraic expressions that you'll evaluate, your final result will depend on what your
original number is.

EXAMPLE C

Consider the expression
$$\frac{18 - 2(x + 3)}{6}$$

a. Rewrite the expression by changing the subtraction to addition.

b. Find the value of the expression if x is 30. Write an equation that sets the
expression equal to this result.

c. Solve the equation $\frac{18 + (-2)(x + 3)}{6} = 15$. That is, find the value of x that makes
the equation true.

▶ Solution

a. Subtracting is the same as adding a negative quantity, so you can rewrite the
expression as $\frac{18 + (-2)(x + 3)}{6}$.

b. One way to evaluate the expression is to
perform its operations in order, as shown on
the calculator screen.

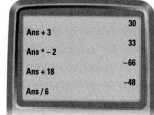

The result of the final operation is -8.
The equation is
$$\frac{18 + (-2)(x + 3)}{6} = -8$$

c. You can solve the equation by working backward, **undoing** each operation you did in part b. Determine how to undo each operation and perform them on the number 15 in reverse order.

1. List the operations on x.　　　　**2.** Start with the answer and work backward to undo each operation.

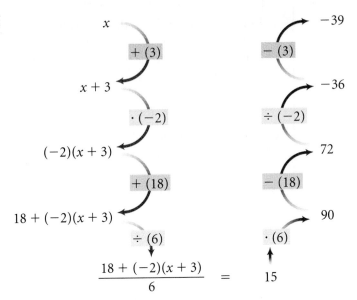

$$\frac{18 + (-2)(x + 3)}{6} = 15$$

Ans * 6	15
	90
Ans – 18	
	72
Ans / –2	
	–36
Ans – 3	

So $x = -39$. You may find it helpful to use this format when solving equations by undoing operations.

Equation: $\dfrac{18 + (-2)(x + 3)}{6} = 15$　　　　Work backwards

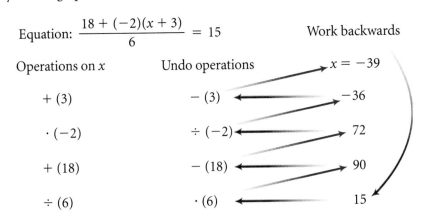

Operations on x　　　Undo operations

Operations on x	Undo operations	
$+ (3)$	$- (3)$	$x = -39$
$\cdot (-2)$	$\div (-2)$	-36
$+ (18)$	$- (18)$	72
$\div (6)$	$\cdot (6)$	90
		15

EXERCISES

You will need your calculator for problems **1, 5,** and **8.**

Practice Your Skills

1. Evaluate each expression without a calculator. Then check your result with your calculator.

   a. $-4 + (-8)$

   b. $(-4)(-8)$

   c. $-2(3 + 9)$

   d. $5 + (-6)(-5)$

   e. $(-3)(-5) + (-2)$

   f. $\dfrac{-15}{3} + 8$

   g. $\dfrac{23 - 3(4 - 9)}{-2}$

   h. $\dfrac{-4(7 + (-8))}{8} - 6.5$

   i. $\dfrac{6(2 \cdot 4 - 5) - 2}{-4}$

2. Sondro entered $75 + 81 + 71 + 78 + 83/5$ to find his average score on five tests. He got 321.6. Is this his correct average? If not, what did he do wrong, and what is his correct average?

3. Evaluate each expression if $x = 6$.

   a. $2x + 3$

   b. $2(x + 3)$

   c. $5x - 13$

   d. $\dfrac{x + 9}{3}$

4. For each equation identify the order of operations. Then work backward through the order of operations to find x.

   a. $\dfrac{x - 3}{2} = 6$

   b. $3x + 7 = 22$

   c. $\dfrac{x}{6} - 20 = -19$

5. Marteen gave Juwan these instructions: Pick a number, add 3, multiply by 4, and subtract 5.

   a. Juwan began with 2 and got 15. He then entered the expression $2 + 3 \cdot 4 - 5$ into his calculator and got 9. What's wrong?

   b. What expression should Juwan enter into his calculator so that he will get 15?

6. Justine asked her group members to do this calculation: Pick a number, multiply by 5, and subtract 2. Quentin got 33 for an answer. Explain how Justine could determine what number Quentin picked. What number did Quentin pick?

Reason and Apply

7. Daxun, Lacy, Claudia, and Al are working on a number trick. Here are the number sequences their number trick generates:

Description	Daxun's sequence	Lacy's sequence	Claudia's sequence	Al's sequence
Pick the starting number.	14	−5	−8.6	x
	19	0		
	76	0		
	64	−12		
	16	−3		
	2	2		

a. Describe the stages of this number trick in the first column.

   b. Complete Claudia's sequence.

   c. Write a sequence of expressions for Al in the last column.

8. In the scheme below, the symbol +1 represents $+1$ and the symbol -1 represents -1. The symbol n represents the original number.

Stage		Description
1	n	
2	n -1 -1 -1	
3	n n -1 -1 -1 -1 -1 -1	
4	n n -1 -1	
5	n -1	
6	-1	Subtract the original number.
7	+1 +1 +1	

   a. Explain what is happening as you move from one stage to the next. The explanation for Stage 6 is provided.

   b. At which stage will everyone's result be the same? Explain.

   c. Verify that this trick works by using a calculator list and an answer routine.

   d. Write an expression similar to the one shown in the solution to part c of Example A to represent this trick.

9. Jo asked Jack and Nina to try two other number tricks that she had invented for homework. Their number sequences are shown in the tables. Use words to describe each stage of the number tricks.

   a. Number Trick 1:

Description	Jack's sequence	Nina's sequence
Pick the starting number.	5	3
	10	6
	30	18
	36	24
	12	8
	7	5
	2	2

   b. Number Trick 2:

Description	Jack's sequence	Nina's sequence
	-10	10
	-8	12
	-24	36
	-15	45
	-30	30
	-60	60
	-10	10

10. Marcella wrote an expression for a number trick.

    a. Describe Marcella's number trick in words.

    b. Pick a number and use it to do the trick. What answer do you get? Pick another number and do the trick again. What is the "trick"?

Marcella's Trick
$\dfrac{4(x-5)+8}{2} - x + 6$

11. Write your own number trick with at least six stages.

    a. No matter what number you begin with, make the trick result in -4.

    b. Describe the process you used to create the trick.

    c. Write an expression for your trick.

12. The final answer to the number trick shown at right is 3. Starting with the final number, work backward, from the bottom to the top, undoing the operation at each step.

    a. What is the original number?

    b. How can you check that your answer to 12a is correct?

    c. What is the original number if the final result is 15?

    d. What makes this sequence of operations a number trick?

x	_____
Ans · 8	_____
Ans + 9	_____
Ans / 4	_____
Ans + 5.75	14
Ans / 2	7
Ans − 4	3

13. Write your own number trick so that, no matter what number you begin with, the result will always be that number.

14. The sequence of operations at right isn't a number trick. It will always give you a different final answer depending on the number you start with.

    a. What is the final value if you start with 18?

    b. What number did you start with if the final answer is 7.6?

    c. Describe how you got your answer to 14a.

    d. Let x represent the number you start with. Write an algebraic expression to represent this sequence of operations.

    e. Set the expression you got in 14d equal to zero. Then solve the equation for x. Check that your solution is correct by using the value you got for x as the starting number. Do you get zero again?

Ans + 10	_____
Ans · 2	_____
Ans − 12	_____
Ans / 5	_____

15. Consider the expression

$$\frac{5(x+7)}{3}$$

    a. Find the value of the expression if $x = 8$. List the order you performed the operations.

    b. Solve the equation $\dfrac{5(x+7)}{3} = -18$ by undoing the sequence of operations in 15a.

16. Consider the expression

$$\frac{2.5(x-4.2)}{5} - 4.3$$

    a. Find the value of the expression if $x = 8$. Start with 8 and use the order of operations.

    b. Solve the equation $\dfrac{2.5(x-4.2)}{5} - 4.3 = 5.4$ by undoing the sequence of operations in 16a.

Review

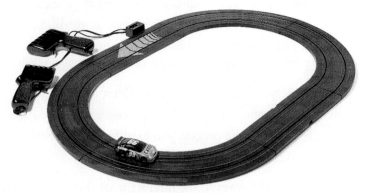

17. An HO-scale electric slot car set advertises that its cars travel in scale speeds of 200 miles per hour. If the HO scale factor is 1 : 87, find the car's actual speed in

   a. Feet per minute.

   b. Centimeters per second.

18. In the 15th century, Arab and Persian mathematicians popularized the use of an algebraic relations method of multiplication. These mathematicians multiplied two numbers, *a* and *b*, by using the relations method: $ab = 10b - (10-a)b$.

   Here's an example:

$$7 \times 8 = 10 \times 8 - (10 - 7) \times 8$$
$$= 10 \times 8 - 3 \times 8$$
$$= 80 - 24$$
$$= 56$$

   (Q. Mushtaq and A. L. Tan, *Mathematics: The Islamic Legacy,* 1990, p. 97)

   Does this method work with fractions? Decimals? Negative numbers? Pick two numbers and multiply them using this method. Discuss the advantages and disadvantages of using algebraic relations.

19. Natalie works in a shop that sells mixed nuts. Alice drops by and decides to buy a bag of mixed nuts with

$\frac{3}{4}$ cup of almonds

$\frac{2}{3}$ cup of cashews

$\frac{1}{2}$ cup of pecans

   a. How many cups of mixed nuts will there be in the bag?

   b. Almonds cost $6.98 a cup, cashews cost $7.98 a cup, and pecans cost $4.98 a cup. What is the cost of Alice's bag of nuts?

Recursive Sequences

The Empire State Building in New York City has 102 floors and is 1250 feet high. How high up are you when you reach the 80th floor? You can answer this question using a recursive sequence. In this lesson you will learn how to analyze geometric patterns, complete tables, and find missing values using numerical sequences.

A **recursive sequence** is an ordered list of numbers defined by a starting value and a rule. You generate the sequence by applying the rule to the starting value, then applying it to the resulting value, and repeating this process.

A mathematician, like a painter or a poet, is a maker of patterns. If his patterns are more permanent than theirs, it is because they are made of ideas.

G. H. HARDY

EXAMPLE A

The measurements in the table represent heights above and below ground at different floor levels in a 25-story building. Write a **recursive routine** that provides the sequence of heights −4, 9, 22, 35, . . . , 217, . . . that corresponds to the building floor numbers 0, 1, 2, ... Use this routine to find each missing value in the table.

Floor number	Basement (0)	1	2	3	4	. . .	10	. . .		. . .	25
Height (ft)	−4	9	22	35		. . .		. . .	217	. . .	

▶ Solution

The starting value is −4 because the basement is 4 feet below ground level. Each floor is 13 feet higher than the floor below it, so the rule for finding the next floor height is "add 13 to the current floor height."

The calculator screen shows how to enter this recursive routine on your calculator. Press −4 (ENTER) to start your number sequence. Press +13 (ENTER). The calculator automatically displays Ans +13 and computes the next value. Simply pressing (ENTER) again applies the rule for finding successive floor heights. [▶ ⬜ See **Calculator Note 0D.** ◀]

How high up is the 10th floor? Count the number of times you press (ENTER) until you reach 10. Which floor is at a height of 217 feet? Keep counting until you see that value on your calculator screen. What's the height of the 25th floor? Keep applying the rule by pressing (ENTER) and record the values in your table.

Investigation
Recursive Toothpick Patterns

In this investigation you will learn to create and apply recursive sequences by modeling them with puzzle pieces made from toothpicks.

Consider this pattern of triangles.

  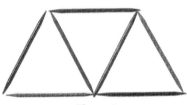

Figure 1   **Figure 2**   **Figure 3**

Step 1 Make Figures 1–3 of the pattern using as few toothpicks as possible. How many toothpicks does it take to reproduce each figure? How many toothpicks lie on the perimeter of each figure?

Step 2 Copy the table with enough rows for six figures of the pattern. Make Figures 4–6 from toothpicks by adding triangles in a row and complete the table.

	Number of toothpicks	Perimeter
Figure 1		
Figure 2		

Step 3 What is the rule for finding the number of toothpicks in each figure? What is the rule for finding the perimeter? Use your calculator to create recursive routines for these rules. Check that these routines generate the numbers in your table.

Step 4 Now make Figure 10 from toothpicks. Count the number of toothpicks and find the perimeter. Does your calculator routine give the same answers? Find the number of toothpicks and the perimeter for Figure 25.

Next, you'll see what sequences you can generate with a new pattern.

Step 5 Design a pattern using a row of squares, instead of triangles, with your toothpicks. Repeat Steps 1–4 and answer all the questions with the new design.

Step 6 Choose a unit of measurement and explain how to calculate the area of a square made from toothpicks. How does your choice of unit affect calculations for the areas of each figure?

Now you'll create your own puzzle piece from toothpicks. Add identical pieces in one direction to make the succeeding figures of your design.

Step 7　Draw Figures 1–3 on your paper. Write recursive routines to generate number sequences for the number of toothpicks, perimeter, and area of each of six figures. Record these numbers in a table. Find the values for a figure made of ten puzzle pieces.

Step 8　Write three questions about your pattern that require recursive sequences to answer. For example: What is the perimeter if the area is 20? Test your questions on your classmates.

You have been writing number sequences in table columns in this investigation. Remember that you can also display sequences as a list of numbers like this:

1, 3, 5, 7, . . .

Each number in the sequence is called a term. The three periods indicate that the numbers continue.

EXAMPLE B　Find the missing values in each sequence.

a. 7, 12, 17, __ , 27, __ , __ , 42, __ , 52

b. 5, 1, −3, __ , −11, −15, __ , __ , −27, __

c. −7, __ , −29, __ , −51, −62, __ , −84, __

d. 2, −4, 8, −16, 32, __ , 128, −256, __ , __

How many hidden numbers can you find?

▶ Solution　**a.** The starting value is 7 and you add 5 each time to get the next number. The missing numbers are shown in red.

starting value

$+5$　$+5$　$+5$　$+5$　$+5$　$+5$　$+5$　$+5$　$+5$

7,　12,　17,　22,　27,　32,　37,　42,　47,　52

b. The starting value is 5 and you subtract 4 each time to get the next number. The missing numbers are shown in red.

starting value

-4　-4　-4　-4　-4　-4　-4　-4　-4

5,　1,　−3,　−7,　−11,　−15,　−19,　−23,　−27,　−31

c. The starting value is −7. The difference between the fifth and sixth terms shows you subtract 11 each time.

starting value

-11　-11　-11　-11　-11　-11　-11　-11

−7,　−18,　−29,　−40,　−51,　−62,　−73,　−84,　−95

d. Adding or subtracting numbers does not generate this sequence. Notice that the numbers double each time. Also, they switch between positive and negative signs. So the rule is to multiply by -2. Multiply 32 by -2 to get the first missing value of -64. The last missing values are 512 and -1024.

starting value

$$\downarrow \quad \overset{\cdot(-2)}{\frown} \quad \overset{\cdot(-2)}{\frown} \quad \overset{\cdot(-2)}{\frown} \quad \overset{\cdot(-2)}{\frown} \quad \overset{\cdot(-2)}{\frown} \quad \overset{\cdot(-2)}{\frown} \quad \overset{\cdot(-2)}{\frown} \quad \overset{\cdot(-2)}{\frown} \quad \overset{\cdot(-2)}{\frown}$$

$$2, \quad -4, \quad 8, \quad -16, \quad 32, \quad -64, \quad 128, \quad -256, \quad 512, -1024$$

EXERCISES

You will need your calculator for problems **1–10, 12,** and **13.**

▶ Practice Your Skills

1. Evaluate each expression without using your calculator. Then check your result with your calculator.

   a. $-2(5 - 9) + 7$

   b. $\dfrac{(-4)(-8)}{-2}$

   c. $\dfrac{5 + (-6)(-5)}{-7}$

2. Use the distributive property to rewrite each family of expressions without using parentheses. Enter $\{2, 5, 6\}$ into list L1 and $\{-3.6, -0.5, 12\}$ into list L2 on your calculator, and use them to check your answers numerically.

   a. $3(\text{L1} + 2)$

   b. $-4(6 - \text{L2})$

   c. $-7(\text{L1} - 3)$

3. Consider the sequence of figures made from a row of pentagons.

    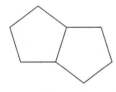 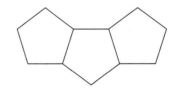

   | Figure 1 | Figure 2 | Figure 3 |

   a. Copy and complete the table for five figures.

   b. Write a recursive routine to find the perimeter of each figure. Assume each side is 1 unit long.

   c. Find the perimeter of Figure 10.

   d. Which figure has a perimeter of 47?

Figure #	Perimeter
1	5
2	8
3	

4. Find the first six values generated by the recursive routine

   -14.2 `ENTER`

   $\text{Ans} + 3.7$ `ENTER`, `ENTER`, . . .

5. Write a recursive routine to generate each sequence. Then use your routine to find the 10th term of the sequence.

   a. $3, 9, 15, 21, \ldots$

   b. $1.7, 1.2, 0.7, 0.2, \ldots$

   c. $-3, 6, -12, 24, \ldots$

   d. $384, 192, 96, 48, \ldots$

Reason and Apply

6. APPLICATION In the Empire State Building the longest elevator shaft reaches the 86th floor, 1050 feet above ground level. Another elevator takes visitors to the observation area on the 102nd floor, 1224 feet above ground level.

   a. Write a recursive routine that gives the height above ground level for each of the first 86 floors. Tell what the starting value and the rule mean in terms of the building.

   b. Write a recursive routine that gives the heights of floors 86 through 102. Tell what the starting value and the rule mean in this routine.

   c. When you are 531 feet above ground level, what floor are you on?

   d. When you are on the 90th floor, how high up are you? When you are 1137 feet above ground level, what floor are you on?

7. The diagram at right shows a sequence of gray and white squares each layered under the previous one.

   a. Explain how the sequence 1, 3, 5, 7, . . . is related to the areas of these squares.

   b. Write a recursive routine that gives the sequence 1, 3, 5, 7, ...

   c. Use your routine to predict the number of additional unit squares you would need to enlarge this diagram by one additional row and column. Explain how you found your answers.

   d. Use your routine to predict the 20th number in the sequence 1, 3, 5, 7, ...

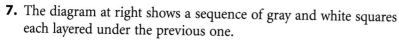

   e. The first term in the sequence is 1, and the second is 3. Which term is the number 95? Explain how you found your answer.

8. Imagine a tilted L-shaped puzzle piece made from 8 toothpicks. Its area is 3 square units. Add puzzle pieces in the corner of each "L" to form successive figures of the design. In a second figure, the two pieces "share" two toothpicks so that there are 14 toothpicks instead of 16.

 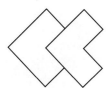 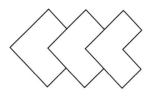

        Figure 1   **Figure 2**   **Figure 3**

   a. As you did in the investigation, make a table with enough columns and rows for the number of toothpicks, perimeter, and area of each of six figures.

   b. Write a recursive routine that will produce the number sequence in each column of the table.

   c. Find the number of toothpicks, perimeter, and area of Figure 10.

   d. Find the perimeter and area of the figure made from 152 toothpicks.

9. **APPLICATION** The table gives some floor heights in a building.

Floor	...	−1	0	1	2	...		...	25
Height (m)	...	−3	1	5	9	...	37	...	

a. How many meters are between the floors in this building?

b. Write a recursive routine that will give the sequence of floor heights if you start at the 25th floor and go to the basement (floor 0). Which term in your sequence represents the height of the 7th floor? What is the height?

c. How many terms are in the sequence in 9b?

d. Floor "−1" corresponds to the first level of the parking substructure under the building. If there are five parking levels, how far underground is level 5?

10. Consider the sequence __, −4, 8, __, 32, ...

a. Find two different recursive routines that could generate these numbers.

b. For each routine, what are the missing numbers? What are the next two numbers?

c. If you want to generate this number sequence with exactly one routine, what more do you need?

11. Positive multiples of 7 are generally listed as 7, 14, 21, 28, ...

a. If 7 is the 1st multiple of 7 and 14 is the 2nd multiple, then what is the 17th multiple?

b. How many multiples of 7 are between 100 and 200?

c. Compare the number of multiples of 7 between 100 and 200 with the number between 200 and 300. Does the answer make sense? Do all intervals of 100 have this many multiples of 7? Explain.

d. Describe two different ways to generate a list containing multiples of 7.

12. Some babies gain an average of 1.5 pounds per month during the first 6 months after birth.

a. Write a recursive routine that will generate a table of monthly weights for a baby weighing 6.8 pounds at birth.

b. Write a recursive routine that will generate a table of monthly weights for a baby weighing 7.2 pounds at birth.

c. How are the routines in 12a and 12b the same? How are they different?

d. Copy and complete the table of data for this situation.

Age (mo)	0	1	2	3	4	5	6
Weight of Baby A (lb)	6.8						
Weight of Baby B (lb)	7.2						

e. How are the table values for the two babies the same? How do they differ?

13. Write recursive routines to help you answer 13a–d.

    a. Find the 9th term of $1, 3, 9, 27, \ldots$.

    b. Find the 123rd term of $5, -5, 5, -5, \ldots$.

    c. Find the term number of the first positive term of the sequence $-16.2, -14.8, -13.4, -12, \ldots$.

    d. Which term is the first to be either greater than 100 or less than -100 in the sequence $-1, 2, -4, 8, -16, \ldots$?

Review

14. The table gives the normal monthly precipitation for three of the soggiest cities in the United States.

    a. Display the data in three box plots, one for each city, and use them to compare the precipitation for the three cities.

    b. What information do you lose by displaying the data in a box plot? What type of graph might be more helpful for displaying the data?

Precipitation for Three Cities

Month	Precipitation (in)		
	Portland	San Francisco	Seattle
January	5.4	4.1	5.4
February	3.9	3.0	4.0
March	3.6	3.1	3.8
April	2.4	1.3	2.5
May	2.1	0.3	1.8
June	1.5	0.2	1.6
July	0.7	0.0	0.9
August	1.1	0.1	1.2
September	1.8	0.3	1.9
October	2.7	1.3	3.3
November	5.3	3.2	5.7
December	6.1	3.1	6.0

(*2000 New York Times Almanac*, pp. 480–481)

It's a rainy day in Portland, Oregon.

15. **APPLICATION** Andy has a part-time job at the hardware store. He earns $7.25 an hour, but on Sundays and holidays he earns "time and a half." This means that for every hour he works on Sundays and holidays, he gets paid for 1.5 hours.

    a. Last week, Andy worked 8 hours a day Monday through Thursday, he had Friday off, and then he worked 6 hours on Sunday. What are his earnings for the week?

    b. Explain two ways that you can answer the question in 15a. Use the distributive property in one of your explanations.

16. Central High School is selling lottery tickets to raise money for a new sound system. There are 18 prizes for the lottery, and the students and teachers sold 2400 tickets.

    a. What fraction of the people who bought tickets will win a prize?

    b. What fraction of the people who bought tickets will not win a prize?

    c. What is the probability that someone who buys one ticket will win a prize?

Linear Plots

In this lesson you will learn that the starting value and the rule of a recursive sequence take on special meaning in certain real-world situations. When you add or subtract the same number each time in a recursive routine, consecutive terms in the sequence change by a constant amount. Using your calculator, you will see how these two important features of a sequence let you generate data for tables quickly. You will also plot points from these data sets and learn that the starting value and rule relate to specific characteristics of the graph.

In most sciences, one generation tears down what another has built, and what one has established, the next undoes. In mathematics alone, each generation builds a new story to the old structure.

HERMANN HANKEL

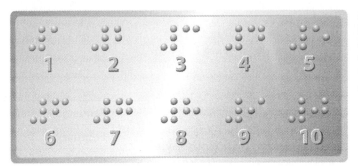

Many elevators use Braille symbols. This alphabet for the blind was developed by Louis Braille (1809–1852).

EXAMPLE

You walk into an elevator in the basement of a building. Its control panel displays "0" for the floor number. As you go up, the numbers increase one by one on the display, and the elevator rises 13 feet for each floor. The table shows the floor numbers and their heights above ground level.

Floor number	Height (ft)
0 (basement)	−4
1	9
2	22
3	35
4	48
. . .	. . .

a. Write recursive routines for the two number sequences in the table. Enter both routines into calculator lists.

b. Define variables and plot the data in the table for the first few floors of the building. Does it make sense to connect the points on the graph?

c. What is the highest floor with a height less than 200 feet? Is there a floor that is exactly 200 feet high?

▶ Solution

a. The starting value for the floor numbers is 0, and the rule is to add 1. The starting value for the height is −4, and the rule is to add 13. You can generate both number sequences on the calculator using lists.

Press {0, −4} and press **ENTER** to input both starting values at the same time. To use the rules to get the next term in the sequence, press {Ans(1) + 1, Ans(2) + 13} **ENTER**.
[▶ 🖳 See **Calculator Note 4B**. ◀]

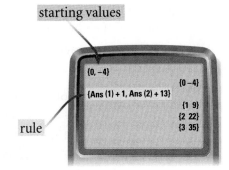

starting values

rule

{0, −4}

{Ans (1) + 1, Ans (2) + 13}

{0 −4}
{1 9}
{2 22}
{3 35}

These commands tell the calculator to add 1 to the first term in the list and to add 13 to the second number. Press **ENTER** again to compute the next floor number and its corresponding height as the elevator rises.

b. Let x represent the floor number and y represent the floor's height in feet. Mark a scale from 0 to 5 on the x-axis and -10 to 80 on the y-axis. Plot the data from the table. The graph starts at $(0, -4)$ on the y-axis. The points appear to be in a line. It does not make sense to connect the points because it is not possible to have a decimal or fractional floor number.

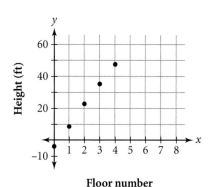

Floor number

c. The recursive routine generates the points $(0, -4)$, $(1, 9)$, $(2, 22)$, . . . , $(15, 191)$, $(16, 204)$, ... The height of the 15th floor is 191 feet. The height of the 16th floor is 204 feet. So, the 15th floor is the highest floor with a height less than 200 feet. No floor is exactly 200 feet high.

Notice that to get to the next point on the graph from any given point, move right 1 unit on the x-axis and up 13 units on the y-axis. The points you plotted in the example showed a **linear relationship** between floor numbers and their heights. In what other graphs have you seen linear relationships?

Investigation
On the Road Again

You will need

- the worksheet On the Road Again Grid

A family wants to meet for a camping trip. Mom and Dad are at the campsite when they realize they forgot to pack the tent. Their son and daughter have just left their apartments and cannot be reached by phone. Will Mom and Dad get home before their son and daughter can call from the campsite? When and where will they pass each other on the highway? In this investigation you will learn how to use recursive sequences to answer questions like these.

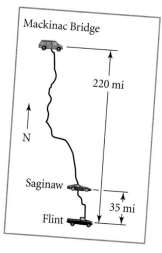

A green minivan starts at the Mackinac Bridge and heads south for Flint on Highway 75. At the same time, a red sports car leaves Saginaw and a blue pickup truck leaves Flint. The car and the pickup are heading for the bridge. The minivan travels 72 miles per hour. The pickup travels 66 mph. The sports car travels 48 mph.

Step 1	Find each vehicle's average speed in miles per minute (mi/min).
Step 2	Write recursive routines to find each vehicle's distance from Flint at each minute. What are the real-world meanings of the starting value and the rule in each routine? Use calculator lists.
Step 3	Make a table to record the highway distance from Flint for each vehicle. After you complete the first few rows of data, change your recursive routines to use 10-minute intervals for up to 4 hours.

Highway Distance from Flint

Time (min)	Minivan (mi)	Sports car (mi)	Pickup (mi)
0			
1			
2			
5			
10			
20			

Procedure Note

After you enter the recursive routine in the calculator, press ENTER five or six times. Copy the data displayed on your calculator screen onto your table. Repeat this process.

Step 4	Define variables and plot the information from the table onto a graph. Mark and label each axis in 10-minute intervals, with time on the horizontal axis. Using a different color for each vehicle, plot its (*time, distance*) coordinates.
Step 5	On the graph, do the points for each vehicle seem to fall on a line? Does it make sense to connect each vehicle's points in a line? If so, draw the line. If not, explain why not.

Use your graph and table to find the answers for Steps 6–10.

Step 6	Where does the starting value for each routine appear on the graph? How does the recursive rule for each routine affect the points plotted?
Step 7	Which line represents the minivan? How can you tell?
Step 8	Where are the vehicles when the minivan meets the first one headed north?
Step 9	How can you tell by looking at the graph whether the pickup or the sports car is traveling faster? When and where does the pickup pass the sports car?

Step 10	Which vehicle arrives at its destination first? How many minutes pass before the second and third vehicles arrive at their destinations? How can you tell by looking at the graph?
Step 11	What assumptions about the vehicles are you making when you answer the questions in the previous steps?
Step 12	Consider how to model this situation more realistically. What if the vehicles are traveling at different speeds? What if one vehicle runs out of gas or breaks down? What if one driver stops to get a bite to eat? What if the vehicles' speeds are not constant? Discuss how these questions affect the recursive routines, tables of data, and their graphs.

EXERCISES

You will need your calculator for problems **1, 4, 6, 7, 8,** and **10.**

Practice Your Skills

1. Find out if each expression is positive or negative without using your calculator. Then check your answer with your calculator.

   a. $-35(44) + 23$

   b. $(-14)(-36) - 32$

   c. $25 - \dfrac{152}{12}$

   d. $50 - 23(-12)$

   e. $\dfrac{-12 - 38}{15}$

   f. $24(15 - 76)$

2. List the terms of each number sequence of y-coordinates for the points shown on each graph. Then write a recursive routine to generate each sequence.

   a.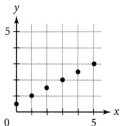

   b.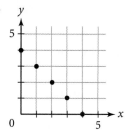

   c.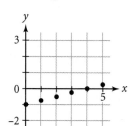

   d.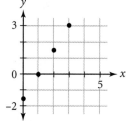

3. Make a table listing the coordinates of the points plotted in 2b and d.

4. Plot the first five points represented by each recursive routine in 4a and b on separate graphs.

   a. $\{0, 5\}$ (ENTER)

   $\{\text{Ans}(1) + 1, \text{Ans}(2) + 7\}$ (ENTER); (ENTER), . . .

   b. $\{0, -3\}$ (ENTER)

   $\{\text{Ans}(1) + 1, \text{Ans}(2) - 6\}$ (ENTER); (ENTER), . . .

   c. On which axis does each starting point lie? What is the x-coordinate of each starting point?

   d. As the x-value increases by 1, what happens to the y-coordinates of the points in each sequence in 4a and b?

5. Consider the expression

$$\frac{4 - 5(x + 3)}{6}$$

   a. Use the order of operations to find the value of the expression if $x = 9$.

   b. Set the expression equal to 12. Solve for x by undoing the sequence of operations.

6. The direct variation $y = 2.54x$ describes the relationship between two standard units of measurement where y represents centimeters and x represents inches.

   a. Write a recursive routine that would produce a table of values for any whole number of inches. Use a calculator list.

   b. Use your routine to complete the missing values in this table.

Inches	Centimeters
0	0
1	2.54
2	
	35.56
17	

▶ Reason and Apply

7. **APPLICATION** A car is moving at a speed of 68 miles per hour from Dallas toward San Antonio. Dallas is 272 miles from San Antonio.

   a. Write a recursive routine to create a table of values relating time to distance from San Antonio for 0 to 5 hours in 1-hour intervals.

   b. Graph the information in your table.

   c. What is the connection between your plot and the starting value in your recursive routine?

   d. What is the connection between the coordinates of any two consecutive points in your plot and the rule of your recursive routine?

   e. Draw a line through the points of your plot. What is the real-world meaning of this line? What does the line represent that the points alone do not?

   f. When is the car within 100 miles of San Antonio? Explain how you got your answer.

   g. How long does it take the car to reach San Antonio? Explain how you got your answer.

8. **APPLICATION** A long-distance telephone carrier charges $1.38 for international calls of 1 minute or less and $0.36 for each additional minute.

   a. Write a recursive routine using calculator lists to find the cost of a 7-minute phone call.

   b. Without graphing the sequence, give a verbal description of the graph showing the costs for calls that last whole numbers of minutes. Include in your description all the important values you need in order to draw the graph.

9. These tables show the changing depths of two submarines as they come to the surface.

USS Alabama

Time (sec)	0	5	10	15	20	25	30
Depth (ft)	−38	−31	−24	−17	−10	−3	4

USS Dallas

Time (sec)	0	5	10	15	20	25	30
Depth (ft)	−48	−40	−32	−24	−16	−8	0

a. Graph the data from both tables on the same set of coordinate axes.

b. Describe what you found by graphing the data. How are the graphs the same? How are they different?

c. Does it make sense to draw a line through each set of points? Explain what these lines mean.

d. What is the real-world meaning of the point (30, 4) for the USS *Alabama*?

10. Each geometric design is made from tiles arranged in a row.

Triangle   Rhombus   Pentagon   Hexagon

a. Make a table like the one shown. Find the number of tile edges on the perimeter of each design, and fill in ten rows of the table. Look for patterns as you add more tiles.

Tile Edges on the Perimeter

Number of tiles	Triangle	Rhombus	Pentagon	Hexagon
1	3	4	5	6
2				
3				

b. Write a recursive routine to generate the values in each table column.

c. Find the perimeter of a 50-tile design for each shape.

d. Draw four plots on the same coordinate axes using the information for designs of one to ten tiles of each shape. Use a different color for each shape. Put the number of tiles on the horizontal axis and the number of edges on the vertical axis. Label and scale each axis.

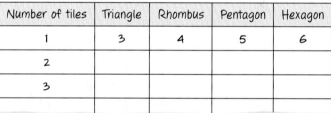

e. Compare the four scatter plots. How are they alike, and how are they different?

f. Would it make sense to draw a line through each set of points? Explain why or why not.

11. A bicyclist, 1 mile (5280 feet) away, pedals toward you at a rate of 600 feet per minute for 3 minutes. The bicyclist then pedals at the rate of 1000 feet per minute for the next 5 minutes.

   a. Describe what you think the plot of time against distance from you will look like.

   b. Graph the data using 1-minute intervals for your plot.

   c. Invent a question about the situation, and use your graph to answer the question.

▶ Review

12. Consider the expression
$$\frac{5.4 + 3.2(x - 2.8)}{1.2} - 2.3$$

   a. Use the order of operations to find the value of the expression if $x = 7.2$.

   b. Set the expression equal to 3.8. Solve for x by undoing the sequence of operations you listed in 12a.

13. Isaac has an easy time remembering how to convert from Celsius to Fahrenheit. He multiplies the temperature in Celsius by 9, divides the result by 5, and then adds 32.

   a. Write clear directions for converting Fahrenheit to Celsius.

   b. Write an expression for each conversion.

14. **APPLICATION** Karen is a U.S. exchange student to Austria. She wants to make her favorite pizza recipe for her host family, but she needs to convert the quantities to the metric system. Instead of using cups for flour and sugar, her host family measures dry ingredients in grams and liquid ingredients in liters. Karen has read that 4 cups of flour weigh 1 pound.

In her dictionary, Karen looks up conversion factors and finds that 1 ounce ≈ 28.4 grams, 1 pound ≈ 454 grams, and 1 cup ≈ 0.236 liter.

   a. Karen's recipe calls for $\frac{1}{2}$ cup water and $1\frac{1}{2}$ cups flour. Convert these quantities to metric units.

   b. Karen's recipe says to bake the pizza at 425°. Convert this temperature to degrees Celsius.

15. Draw and label a coordinate plane with each axis scaled from −10 to 10.

   a. Represent each point named with a dot, and label it using its letter name.

   $A(3, -2)$ $B(-8, 1.5)$ $C(9, 0)$ $D(-9.5, -3)$ $E(7, -4)$
   $F(1, -1)$ $G(0, -6.5)$ $H(2.5, 3)$ $I(-6, 7.5)$ $J(-5, -6)$

   b. List the points in Quadrant I, Quadrant II, Quadrant III, and Quadrant IV. Which points are on the x-axis? Which points are on the y-axis?

   c. Explain how to tell which quadrant a point will be in by looking at the coordinates. Explain how to tell if a point lies on one of the axes.

4.5

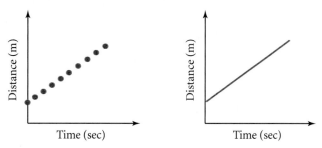

Time-Distance Relationships

Time-distance relationships are some of the most useful applications of algebra. You worked with this topic in the investigation and exercises of Lesson 4.4. Now you will explore this type of relationship in more depth by considering many walking scenarios. You'll learn the mathematics behind the starting position, speed, direction, and final position of a walker.

Activity
Walk the Line

A **graph sketch** tells what quantities the axes represent but does not have specific number scales. Both of these graph sketches show that distance increases as time passes. Which graph sketch better represents a walk? Why?

Imagine that you have a 4-meter measuring tape (or four metersticks) stretched out on the floor.

A motion sensor measures your distance from the tape's 0-mark as you walk, and it graphs the information. On the calculator graphs below, the horizontal axis shows time for 0–6 seconds and the vertical axis shows distance for 0–4 meters.

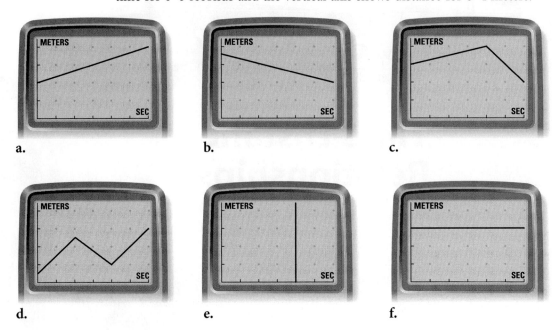

a.   b.   c.

d.   e.   f.

| Step 1 | Write a set of walking instructions, if possible, for each graph. Tell where the walk begins, how fast the person is walking, and whether the person is walking toward or away from the 0-mark. |
| Step 2 | If it is not possible to write walking instructions for a graph or graphs, tell which one(s) and why it can't be done. |

Now you'll sketch graphs based on walking instructions or data.

Step 3 | Graph a walk from each set of instructions.

a. Start at the 2.5-meter mark and stand still.

b. Start at the 3-meter mark and walk toward the 0-mark at a constant rate of 0.5 meters per second.

c. Start at the 0.5-meter mark and walk toward the 4-meter mark at 0.25 meters per second.

Table a

Time (sec)	Distance (m)
0	0.8
1	1.0
2	1.2
3	1.4
4	1.6
5	1.8
6	2.0

Table b

Time (sec)	Distance (m)
0	4.0
1	3.6
2	3.2
3	2.8
4	2.4
5	2.0
6	1.6

Step 4 | Write a set of walking instructions based on the data in Tables a and b, and sketch graphs of the walks.

Step 5 | Write recursive routines for both tables in Step 4. Explain how the starting values and rules show up in the walking instructions, graphs, and tables.

Now it's time for your group to act out each of the graphs and walking instructions. The object is not to match each graph exactly but to approximately match its shape. Your group will need a space about 4 meters long and 1.5 meters wide (13 feet by 5 feet). In this space, tape to the floor a 4-meter measuring tape or four meter sticks end-to-end.

Procedure Note

The walker follows the instructions the director gives. The timer begins timing when the director says to do so and counts the seconds out loud for the walker. The recorder records the walker's position for each second in the table. The coach guides the walker.

Step 6 | Now sketch a graph to represent these walking instructions: Start at the 0.5-meter mark and walk toward the 4-meter mark at 0.75 meters per second for 2 seconds, then walk toward the 0-meter mark at 0.5 meters per second for 2 seconds. Then walk to the 0.5-meter mark in the final 2 seconds.

Step 7 | In your group, choose a walker, a director, a timer, a recorder, and a coach. Make a table to record distances from the 0-m mark for 0–6 seconds. [▶🖳 If your group is using a motion sensor see **Calculator Note 4C.** ◄]

Step 8 | Choose three walks to act out: one from a graph, one from a table, and one from a set of instructions. After acting out each walk, discuss what you could have done to better model each situation. Then repeat the walk, rotating jobs.

project

PASCAL'S TRIANGLE

The first five rows of Pascal's triangle are shown.

```
            1
          1   1
        1   2   1
      1   3   3   1
    1   4   6   4   1
```

The triangle can be generated recursively. The sides of the triangle are 1's, and each number inside the triangle is the sum of the two diagonally above it.

Complete the next five rows of Pascal's triangle. Research its history and practical application. What is the connection between Sierpiński's triangle and Pascal's triangle? Can you find the sequence of triangular numbers in Pascal's triangle? What is its connection to the Fibonacci number sequence? Present your findings in a paper or a poster.

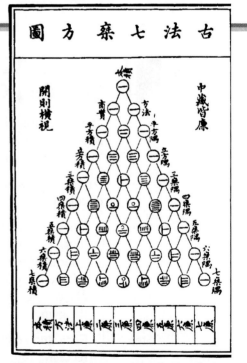

What became known as Pascal's triangle was first published in *Siyuan yujian xicao* by Zhu Shijie in 1303. This ancient version actually has one error. Can you find it?

Linear Equations and the Intercept Form

So far in this chapter you have used recursive routines, graphs, and tables to model linear relationships. In this lesson you will learn how to write linear equations from recursive routines. You'll begin to see some common characteristics of linear equations and their graphs, starting with the relationship between exercise and calorie consumption.

Different physical activities cause people to burn calories at different rates depending on many factors such as body type, height, age, and metabolism. Coaches and trainers consider these factors when suggesting workouts for their athletes.

Investigation
Working Out with Equations

Manisha starts her exercise routine by jogging to the gym. Her trainer says this activity burns 215 calories. Her workout at the gym is to pedal a stationary bike. This activity burns 3.8 calories per minute.

First you'll model this scenario with your calculator.

Step 1 | Use calculator lists to write a recursive routine to find the total number of calories Manisha has burned after each minute she pedals the bike. Include the 215 calories she burned on her jog to the gym.

Step 2 | Copy and complete the table using your recursive routine.

Step 3 | After 20 minutes of pedaling, how many calories has Manisha burned? How long did it take her to burn 443 total calories?

Manisha's Workout

Pedaling time (min)	Total calories burned
x	y
0	215
1	
2	
20	
30	
45	
60	

Next you'll learn to write an equation that gives the same values as the calculator routines.

Step 4	Write an expression to find the total calories Manisha has burned after 20 minutes of pedaling. Check that your expression equals the value in the table.
Step 5	Write and evaluate an expression to find the total calories Manisha has burned after pedaling 38 minutes. What are the advantages of this expression over a recursive routine?
Step 6	Let x represent the pedaling time in minutes, and let y represent the total number of calories Manisha burns. Write an equation relating time to total calories burned.
Step 7	Check that your equation produces the corresponding values in the table.

Now you'll explore the connections between the linear equation and its graph.

Step 8	Plot the points from your table on your calculator. Then enter your equation into the Y= menu. Graph your equation to check that it passes through the points. Give two reasons why drawing a line through the points realistically models this situation. [▶ 🖩 See **Calculator Note 1H** to review how to plot points and graph an equation. ◀]
Step 9	Substitute 538 for y in your equation to find the elapsed time required for Manisha to burn a total of 538 calories. Explain your solution process. Check your result.
Step 10	How do the starting value and the rule of your recursive routine show up in your equation? How do the starting value and the rule of your recursive routine show up in your graph? When is the starting value of the recursive routine also the value where the graph crosses the y-axis?

The equation for Manisha's workout shows a linear relationship between the total calories burned and the number of minutes pedaling on the bike. You probably wrote this linear equation as

$$y = 215 + 3.8x \qquad \text{or} \qquad y = 3.8x + 215$$

The form $y = a + bx$ is the **intercept form.** The value of a is the **y-intercept,** which is the value of y when x is zero. The intercept gives the location where the graph crosses the y-axis. The number multiplied by x is b, which is called the **coefficient** of x.

In the equation $y = 215 + 3.8x$, 215 is the value of a. It represents the 215 calories Manisha burned while jogging before her workout. The value of b is 3.8. It represents the rate her body burned calories while she was pedaling. What would happen if Manisha chose a different physical activity before pedaling on the stationary bike?

You can also think of direct variations in the form $y = kx$ as equations in intercept form. For instance, Sam's trainer tells him that swimming will burn 7.8 calories per minute. When the time spent swimming is 0, the number of calories burned is 0, so a is 0 and drops out of the equation. The number of calories burned is proportional to the time spent swimming, so you can write the equation $y = 7.8x$.

The constant of variation k is 7.8, the rate at which Sam's body burns calories while he is swimming. It plays the same role as b in $y = a + bx$.

EXAMPLE A

Suppose Sam has already burned 325 calories before he begins to swim for his workout. His swim will burn 7.8 calories per minute.

a. Create a table of values for the calories Sam will burn by swimming 60 minutes and the total calories he will burn after each minute of swimming.

b. Define variables and write an equation in intercept form to describe this relationship.

c. On the same set of axes, graph the equation for total calories burned and the direct variation equation for calories burned by swimming.

d. How are the graphs similar? How are they different?

▶ Solution

a. The total numbers of calories burned appear in the third column of the table. Each entry is 325 plus the corresponding entry in the middle column.

b. Let y represent the number of total calories burned, and let x represent the number of minutes Sam spends swimming.

$$y = 325 + 7.8x$$

Sam's Swim

Swimming time (min)	Calories burned by swimming	Total calories burned
0	0	325
1	7.8	332.8
2	15.6	340.6
20	156	481
30	234	559
45	351	676
60	468	793

c. The direct variation equation is $y = 7.8x$. Enter it into Y1 on your calculator. Enter the equation $y = 325 + 7.8x$ into Y2. Check to see that these equations give the same values as the table by looking at the calculator table.

d. The lower line shows the calories burned by swimming and is a direct variation. The upper line shows the total calories burned. It is 325 units above the first line because, at any particular time, Sam has burned 325 more calories. Both graphs have the same value of b, which is 7.8 calories per minute. The graphs are similar because both are lines with the same steepness. They are different because they have different y-intercepts.

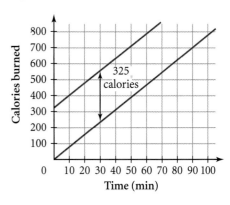

Can you tell what different values of a in the equation $y = a + bx$ will do to the graph?

EXAMPLE B

There is a linear relationship between the air temperature outside an airplane and the plane's altitude in meters. The temperature at sea level is 14.7°C. The temperature drops 7 degrees for every 1000-meter increase in altitude.

a. Define variables and write an equation in intercept form for this relationship.

b. Use your equation to evaluate expressions to complete a table of values for altitudes of 0, 1000, 2000, . . . , 6000 meters.

c. Explain the real-world meaning of the values of a and b in your equation.

d. Graph the relationship. Show how to find the temperature from the graph when the altitude is 4200 meters.

e. Use the graph to find a coordinate where the temperature is approximately 0°C. What is the real-world meaning of this coordinate?

f. Use the graph to find the altitude that gives a temperature of −10°C.

▶ **Solution**

a. If x represents altitude in meters and y represents the outside temperature in °C, the equation for the relationship is $y = 14.7 - 0.007x$.

b. Substitute the altitude values for x in the equation to get the y-values, which are the corresponding temperatures.

c. The real-world meaning of a is the temperature at altitude 0 meters, or sea level. The value of b, $\frac{-7}{1000}$ or -0.007, is the rate at which the temperature (in °C) changes for each meter increase in altitude.

Altitude (m)	Temperature (°C)
x	y
0	14.7
1000	7.7
2000	0.7
3000	−6.3
4000	−13.3
5000	−20.3
6000	−27.3

d. Trace on the graph to find the x-value 4200. The corresponding y-value is -14.7. At an altitude of 4200 meters the temperature is $-14.7°C$. The window shown is $[0, 7000, 1000, -25, 20, 5]$. [▶ 🔲 See **Calculator Note 1H**. ◀]

e. Move the cursor until the y-value is 0. You will see that the x-value is 2100. (This point is where the graph crosses the x-axis.) It means that the plane's altitude is 2100 meters when the temperature is $0°C$ outside.

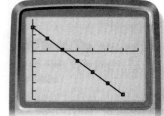

f. Move the cursor to a point on the graph where the y-value is approximately -10, and find that the corresponding x-value is approximately 3529. So the plane's altitude is about 3529 meters when the temperature is $-10°C$ outside.

In linear equations it is sometimes helpful to say which variable is the input variable and which is the output variable. The horizontal axis represents the input variable, and the vertical axis represents the output variable. In Example B, the input variable x represents altitude so the x-axis is labeled altitude, and the output variable y represents temperature so the y-axis is labeled temperature. What are the input and output variables in the investigation and in Example A?

EXERCISES

You will need your calculator for problems **2, 3, 5, 6, 9**, and **10**.

▶ Practice Your Skills

1. Match the recursive routine in the first column with the equation in the second column.

a. 3 (ENTER)
    Ans + 4 (ENTER); (ENTER), . . .

i. $y = 4 - 3x$

b. 4 (ENTER)
    Ans + 3 (ENTER); (ENTER), . . .

ii. $y = 3 + 4x$

c. -3 (ENTER)
    Ans − 4 (ENTER); (ENTER), . . .

iii. $y = -3 - 4x$

d. 4 (ENTER)
    Ans − 3 (ENTER); (ENTER), . . .

iv. $y = 4 + 3x$

2. You can use the equation $d = 24 - 45t$ to model the distance from a destination for someone driving down the highway, where distance d is measured in miles and time t is measured in hours. Graph the equation and use the trace function to find the approximate time for each distance given in 2a and b.

a. $d = 16$ mi   **b.** $d = 3$ mi

c. What is the real-world meaning of 24?

d. What is the real-world meaning of 45?

Some rental cars have in-dash navigation systems.
© 2000 Hertz System, Inc. Hertz is a registered service mark and trademark of Hertz System, Inc.

3. You can use the equation $d = 4.7 + 2.8t$ to model a walk in which the distance from a motion sensor d is measured in feet and the time t is measured in seconds. Graph the equation and use the trace function to find the approximate distance from a motion sensor for each time value given in 3a and b.

    a. $t = 12$ sec   **b.** $t = 7.4$ sec

    c. What is the real-world meaning of 4.7?   **d.** What is the real-world meaning of 2.8?

4. Undo the order of operations to find the x-value in each equation.

    a. $3(x - 5.2) + 7.8 = 14$ **b.** $3.5\left(\dfrac{x - 8}{4}\right) = 2.8$

5. Use the distributive property to rewrite each expression without parentheses. Use list $L_1 = \{-3.5, 2.5, 11\}$ and list $L_2 = \{-2.8, 4.2, 21\}$ to verify your answer.

    a. $-2(L_1 - 5)$ **b.** $-4(-L_1 + L_2)$

▶ Reason and Apply

6. **APPLICATION** Louis is beginning a new exercise workout. His trainer shows him the calculator table with x-values showing his workout time in minutes. The Y_1-values are the total calories Louis burned while running, and the Y_2-values are the number of calories he wants to burn.

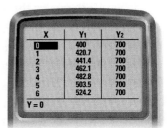

    a. Find how many calories Louis has burned before beginning to run, how many he burns per minute running, and the total calories he wants to burn.

    b. Write a recursive routine that will generate the values listed in Y_1.

    c. Use your recursive routine to write a linear equation in intercept form. Check that your equation generates the table values listed in Y_1.

    d. Write a recursive routine that will generate the values listed in Y_2.

    e. Write an equation that generates the table values listed in Y_2.

    f. Graph the equations in Y_1 and Y_2 on your calculator. Your window should show a time of up to 30 minutes. What is the real-world meaning of the y-intercept in Y_1?

    g. Use the trace function to find the approximate coordinates of the point where the lines meet. What is the real-world meaning of this point?

7. Jo mows lawns after school. She finds that she can use the equation $P = -300 + 15N$ to calculate her profit.

    a. Give some possible real-world meanings for the numbers -300 and 15 and the variable, N.

    b. Invent two questions related to this situation and then answer them.

    c. Explain why the equation $P = 15N - 300$ provides the same values as the equation $P = -300 + 15N$.

    d. Identify the variables.

8. As part of a physics experiment, June threw an object off a cliff and measured how fast it was traveling downward. When the object left June's hand, it was traveling 5 meters per second, and it sped up as it fell. The table shows a partial list of the data she collected as the object fell.

Time (sec)	Speed (m/sec)
0	5.0
0.5	9.9
1.0	14.8
1.5	19.7

a. Write an equation to represent the speed of the object.

b. What was the object's speed after 3 seconds?

c. If it were possible for the object to fall long enough, how many seconds would pass before it reached a speed of 83.4 meters per second?

d. What limitations do you think this equation has in modeling this situation?

9. APPLICATION Manny has a part-time job as a waiter. He makes $45 per day plus tips. He has calculated that his average tip is 12% of the total amount his customers spend on food and beverages.

a. Define variables and write an equation in intercept form to represent Manny's daily income in terms of the amount his customers spend on food and beverages.

b. Graph this relationship for food and beverage amounts between $0 and $900.

c. Write and evaluate an expression to find how much Manny makes in one day if his customers spend $312 for food and beverages.

d. What amounts spent on food and beverages will give him a daily income between $105 and $120?

10. APPLICATION Paula is cross-training for a triathlon in which she cycles, swims, and runs. Before designing an exercise program for Paula, her coach consults a table listing rates for calories burned in various activities.

Cross-training activity	Calories burned (per min)
Walking	3.2
Bicycling	3.8
Swimming	6.9
Jogging	7.3
Running	11.3

Paula's Exercise Program

Workout 1   Bike for 30 minutes
Workout 2   Swim for 15 minutes
Workout 3   Jog for 20 minutes

a. Write and graph an equation in intercept form to find the number of calories burned for each minute of cycling. How many calories has Paula burned by completing Workout 1?

b. After finishing Workout 1, Paula begins the swimming workout. Write and graph an equation to find the total number of calories she has burned so far during each minute of swimming. How many calories has she burned by completing Workout 2? How many calories has she burned by completing both workouts?

c. Write and graph an equation to find the total number of calories Paula has burned during each minute of jogging. How many calories has she burned by completing Workout 3? How many calories has she burned after completing all three workouts?

d. If Paula walks for half an hour before she begins her exercise program, how does that affect the equations and graphs in 10a–c?

Review

11. At a family picnic, your cousin tells you that he always has a hard time remembering how to compute percents. Write him a note explaining what percent means. Use these problems as examples of how to solve the different types of percent problems, with answers for each.

    a. 8 is 15% of what number?

    b. 15% of 18.95 is what number?

    c. What percent of 64 is 326?

    d. 10% of what number is 40?

12. **APPLICATION** Carl has been keeping a record of his gas purchases for his new car. Each time he buys gas, he fills the tank completely. Then he records the number of gallons he bought and the miles since the last fill-up. Here is his record:

Carl's Purchases

Miles traveled	Gallons	$\frac{\text{miles}}{\text{gallon}}$
363	16.2	
342	15.1	
285	12.9	

    a. Copy and complete the table by calculating the ratio of miles per gallon for each purchase.

    b. What is the average rate of miles per gallon for Carl's new car so far?

    c. The car's tank holds 17.1 gallons. To the nearest mile, how far should Carl be able to go without running out of gas?

    d. Carl is planning a trip across the United States. He estimates that the trip will be 4230 miles. How many gallons of gas can Carl expect to buy?

13. Match each recursive routine to a graph. Explain how you made your decision and tell what assumptions you made.

    a. 2.5 (ENTER)
       Ans + 0.5 (ENTER); (ENTER), . . .

    b. 1.0 (ENTER)
       Ans + 1.0 (ENTER); (ENTER), . . .

    c. 2.0 (ENTER)
       Ans + 1.0 (ENTER); (ENTER), . . .

    d. 2.5 (ENTER)
       Ans − 0.5 (ENTER); (ENTER), . . .

i.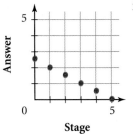
ii.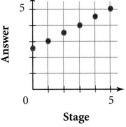
iii.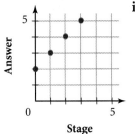
iv.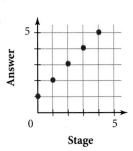

14. Bjarne is training for a bicycle race by riding on stationary bicycles with time-distance readouts. He is riding at a constant speed. The graph shows his accumulated distance and time as he rides.

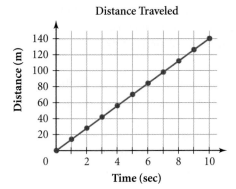

Distance Traveled

Time (sec)	Distance (m)
1	
2	
3	
4	
5	
6	
7	
8	
9	
10	

a. How fast is Bjarne bicycling?

b. Copy the table and show the distance for each time.

c. Write a recursive routine for Bjarne's ride.

d. Looking at the graph, how do you know that Bjarne is neither slowing down nor speeding up during his ride?

e. If Bjarne keeps up the same pace, how far will he ride in one hour?

Bicyclists race through San Luis Obispo, California.

15. Consider the expression $\frac{4(y - 8)}{3}$.

a. Find the value of the expression if $y = 5$. Make a table to show the order of operations.

b. Solve the equation $\frac{4(y - 8)}{3} = 8$, by undoing the sequence of operations.

Linear Equations and Rate of Change

In this lesson you will continue to develop your skills with equations, graphs, and tables of data by exploring the role that the value of b plays in the equation

$$y = a + bx$$

How can it be that mathematics, being after all a product of human thought independent of experience, is so admirably adopted to the objects of reality?

ALBERT EINSTEIN

You have already studied the intercept form of a linear equation in several real-world situations. You have used them to relate calories to minutes spent exercising, floor heights to floor numbers, and distances to time. So, defining variables is an important part of writing equations. Depending on the context of an equation, its numbers take on different real-world meanings. Can you recall how these equations modeled each scenario?

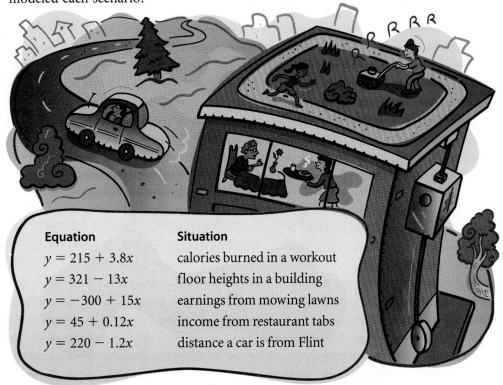

Equation	Situation
$y = 215 + 3.8x$	calories burned in a workout
$y = 321 - 13x$	floor heights in a building
$y = -300 + 15x$	earnings from mowing lawns
$y = 45 + 0.12x$	income from restaurant tabs
$y = 220 - 1.2x$	distance a car is from Flint

Winds of 40 mph blow on North Michigan Ave. in 1955 Chicago.

In most linear equations, there are different output values for different input values. This happens when the coefficient of x is not zero. You'll explore how this coefficient relates input and output values in the examples and the investigation.

In addition to giving the actual temperature, weather reports often indicate the temperature you *feel* as a result of the wind chill factor. The wind makes it feel colder than it actually is. In the next example you will use recursive routines to answer some questions about wind chill.

EXAMPLE A

The table relates the approximate wind chills for different actual temperatures when the wind speed is 20 miles per hour. Assume the wind chill is a linear relationship for temperatures between −5° and 35°.

Temperature (°F)	−5	0	5	10	15	20	25	30	35
Wind chill (°F)	−45	−38	−31			−10		4	11

a. What are the input and output variables?

b. Use calculator lists to write a recursive routine that generates the table values. What are the missing entries?

c. What is the change in temperature from one table entry to the next? What is the corresponding change in the wind chill?

▶ **Solution**

a. The input variable is the actual air temperature in °F. The output variable is the temperature you feel as a result of the wind chill factor.

b. The recursive routine to complete the missing table values is $\{-5, -45\}$ (ENTER) and $\{\text{Ans}(1) + 5, \text{Ans}(2) + 7\}$ (ENTER). The calculator screen displays the missing entries.

```
{5     −31}
{10    −24}
{15    −17}
{20    −10}
{25     −3}
{30      4}
{35     11}
```

c. For every 5° increase in temperature, the wind chill increases 7°.

In Example A, the rate at which the wind chill drops can be calculated from the ratio $\frac{7}{5}$. In other words, it feels 1.4° colder for every 1° drop in air temperature. This number is the rate of change for a wind speed of 20 mph. The **rate of change** is equal to the ratio of the change in output values divided by the corresponding change in input values.

How does the rate of change differ with various wind speeds?

Investigation
Wind Chill

In this investigation you'll use the relationship between temperature and wind chill to explore the concept of rate of change and its connections to tables, scatter plots, recursive routines, equations, and graphs.

The data in the table represent the approximate wind chill temperatures in degrees Fahrenheit for a wind speed of 15 miles per hour. Use this data set to complete each task.

Temperature (°F)	Wind chill (°F)
−5	−38
0	−31.25
1	−29.9
2	−28.55
5	−24.5
15	−11
35	16

Step 1 | Define the input and output variables for this relationship.

| Step 2 | Plot the points and describe the viewing window you used. |

| Step 3 | Write a recursive routine that gives the pairs of values listed in the table. |

| Step 4 | Copy the table. Complete the third and fourth columns of the table by recording the changes between consecutive input and output values. Then find the rate of change. |

Input	Output	Change in input values	Change in output values	Rate of change
−5	−38	//////	//////	//////
0	−31.25	5	6.75	$\frac{+6.75}{+5} =$
1	−29.9	1	1.35	
2	−28.55		1.35	$\frac{+1.35}{+1} =$
5	−24.5	3		
15	−11		13.5	$\frac{+13.5}{+10} =$
35	16			

High wind speeds in Saskatchewan, Canada, drop temperatures below freezing.

| Step 5 | Use your routine to write a linear equation in intercept form that relates wind chill to temperature. Note that the starting value, −38, is not the y-intercept. How does the rule of the routine appear in your equation? |

| Step 6 | Graph the equation on the same set of axes as your scatter plot. Use the calculator table to check that your equation is correct. Does it make sense to draw a line through the points? Where does the y-intercept show up in your equation? |

| Step 7 | What do you notice about the values for rate of change listed in your table? How does the rate of change show up in your equation? In your graph? |

| Step 8 | Explain how to use the rate of change to find the actual temperature if the weather report indicates a wind chill of 10.6° with 15-mph winds. |

The rate of change in the wind chill factor for 15-mph winds is 1.35. The rate for 20-mph winds is 1.4. What is the rate of change for a wind chill factor of 35 mph?

EXAMPLE B

The wind chill data for a wind speed of 35 miles per hour is shown in the table.

a. Add three columns to the table to record the change in input values, change in output values, and corresponding rate of change.

b. Use the table and a recursive routine to write a linear equation in intercept form.

c. What are the real-world meanings of the values for a and b in your equation?

Input	Output
Temperature (°F) (L₁)	Temperature (°F) (L₂)
−25	−89
−20	−81.25
10	−34.75
15	−27
25	−11.5
35	4

▶ **Solution**

a. You can create a table to find that the wind chill increases 7.75° for every 5° increase in temperature. The rate of change, or value of b, is 1.55.

Input	Output	Change in input values	Change in output values	Rate of change
Temperature (L₁)	Wind Chill (L₂)			
−25	−89			
−20	−81.25	5	7.75	$\frac{+7.75}{+5} = 1.55$
10	−34.75	30	46.5	$\frac{+46}{+30} = 1.55$
15	−27	5	7.75	$\frac{+7.75}{+5} = 1.55$
25	−11.5	10	15.5	$\frac{+15.5}{+10} = 1.55$
35	4	10	15.5	$\frac{+15.5}{+10} = 1.55$

The wind chill factor drops temperatures well below zero in Antarctica.

b. The recursive routine

$\{-25, -89\}$ **ENTER**,

$\{\text{Ans}(1) + 5, \text{Ans}(2) + 7.75\}$ **ENTER**, **ENTER**, **ENTER**, **ENTER**, **ENTER**

gives the result $\{0, -50.25\}$. So the y-intercept, or value of a, is -50.25. Then the equation in intercept form is

$$y = -50.25 + 1.55x$$

where x represents the air temperature in °F, and y represents the wind chill temperature you feel.

Note that the starting value of the recursive routine is not the same as the value of the y-intercept in the equation.

c. The value of a is the wind chill you feel when the temperature is 0°F. The value of b states that it feels 1.55° colder for every 1° drop in air temperature.

Because the values for rate of change increase with wind speed, you feel colder when the wind blows faster. But actually, wind speeds greater than 45 mph result in very little additional cooling effect. Also, wind chill factors are usually reported only for temperatures less than 35°F. Think about what limitations a linear equation has when you are modeling these situations.

EXERCISES

You will need your calculator for problems **1, 4, 5, 10,** and **12.**

Practice Your Skills

1. Copy and complete the table of output values for each equation.

a. $y = 50 + 2.5x$

Input x	Output y
20	
−30	
16	
15	
−12.5	

b. $L_2 = -5.2 - 10 \cdot L_1$

L_1 x	L_2 y
0	
−8	
24	
−35	
−5.2	

2. Use the equation $w = -52 + 1.6t$ to approximate the wind chill temperatures for a wind speed of 40 miles per hour. Find the wind chill temperature w for each actual temperature t given in 2a and b.

a. $t = 32°$

b. $t = 12°$

c. What is the real-world meaning of 1.6?

d. What is the real-world meaning of −52?

3. Describe what the rate of change looks like in each graph.

a. the graph of a person walking at a steady rate toward a motion sensor

b. the graph of a person standing still

c. the graph of a person walking at a steady rate away from a motion sensor

d. the graph of one person walking at a steady rate faster than another person

4. Use the "Easy" setting of the INOUT game on your calculator to produce four data tables. Copy each table and write the equation you used to match the data values in the table. [▶ ▣ See **Calculator Note 4D** to learn how to run the program. ◀]

▶ Reason and Apply

5. Each table below shows a different input-output relationship.

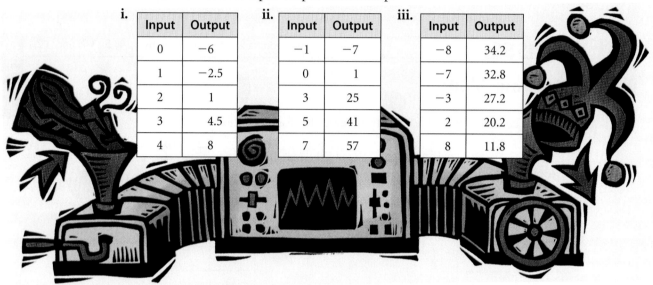

i.

Input	Output
0	−6
1	−2.5
2	1
3	4.5
4	8

ii.

Input	Output
−1	−7
0	1
3	25
5	41
7	57

iii.

Input	Output
−8	34.2
−7	32.8
−3	27.2
2	20.2
8	11.8

  a. Find the rate of change in each table. Explain how you found this value.

  b. For each table, find the output value that corresponds to an input value of 0. What is this value called?

  c. Use your results from 5a and b to write an equation in intercept form for each table.

  d. Use a calculator list of input values to check that each equation actually produces the output values shown in the table.

6. The wind chill temperatures for a wind speed of 30 mph are given in the table.

Temperature (°F)	−5	0	5	10	15	20	25	30	35
Wind chill (°F)	−56	−48	−40	−32	−24	−16	−8	0	8

  a. Define input and output variables and graph the data.

  b. Find the rate of change. Explain how you got your answer.

  c. Write an equation in intercept form.

  d. Plot the points and graph the equation on the same set of axes. How are the graphs for the points and the equation similar? How are they different?

7. The equation $y = 35 + 0.8x$ gives the distance a sports car is from Flint after x minutes.

  a. How far is the sports car from Flint after 25 minutes?

  b. How long will it take until the sports car is 75 miles from Flint? Show how to find the solution using two different methods.

8. You can use the equation $7.3x = 200$ to describe a rectangle with an area of 200 square units like the one shown. What are the real-world meanings of the numbers and the variable in the equation? Solve the equation for x and explain the meaning of your solution.

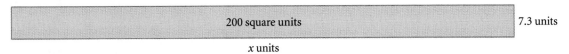

200 square units   7.3 units

x units

9. The total area of the figure at right is 1584 square units. You can use the equation $1584 - 33x = 594$ to represent an area of 1584 square units less the area of $33x$ square units. The area remaining is 594 square units.

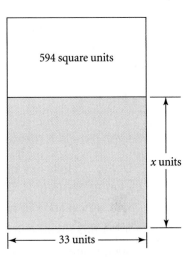

594 square units

x units

33 units

   a. What is the area of the shaded rectangle?

   b. Write the equation you would use to find the height of the shaded rectangle.

   c. Solve the equation you wrote in 9b to find the height of the shaded rectangle.

10. Use the "Medium" setting of the INOUT game on your calculator to produce four data tables. Copy each table and write the equation you used to match the data values in the table. [▶ ▣ See **Calculator Note 4D.** ◀]

▶ Review

11. Show how you can solve these equations by using an undoing process. Check your results by substituting the solutions into the original equations.

   a. $-15 = -52 + 1.6x$ **b.** $7 - 3x = 52$

12. APPLICATION To plan a trip downtown, you compare the costs of three different parking lots. ABC Parking charges $5 for the first hour and $2 for each additional hour. Cozy Car charges $3 an hour, and The Corner Lot charges a $15 flat rate for a whole day.

   a. Make a table similar to the one shown. Write recursive routines to calculate the cost of parking up to 10 hours at each of the three lots.

Hours Parked	ABC Parking	Cozy Car	The Corner Lot
1			
2			
3			

   b. Make three different scatter plots on the same pair of axes showing the parking rates at the three different lots. Use a different color for each parking lot. Put the hours on the horizontal axis and the cost on the vertical axis.

   c. Compare the three scatter plots. Under what conditions is each parking lot the best deal for your trip? Use the graph to explain.

   d. Would it make sense to draw a line through each set of points? Explain why or why not.

13. Today while Don was swimming, he started wondering how many lengths he would have to swim in order to swim different distances. At one end of the pool, he stopped, gasping for breath, and asked the lifeguard. She told him that 1 length of the pool is 25 yards and that 72 lengths is 1 mile. As he continued swimming, he wondered:

a. Is 72 lengths really a mile? Exactly how many lengths would it take to swim a mile?

b. If it took him a total of 40 minutes to swim a mile, what was his average speed in feet per second?

c. How many lengths would it take to swim a kilometer?

d. Last summer Don got to swim in a pool that was 25 meters long. How many lengths would it take to swim a kilometer there? How many for a mile?

14. APPLICATION Holly has joined a video rental club. After paying $6 a year to join, she then has to pay only $1.25 for each new release she rents.

a. Write an equation in intercept form to represent Holly's cost for movie rentals.

b. Graph this situation for up to 60 movie rentals.

c. Video Unlimited charges $60 for a year of unlimited movie rentals. How many movies would Holly have to rent for this to be a better deal?

LEGAL LIMITS

To make a highway accessible to more vehicles, engineers reduce its steepness, also called its **gradient,** or simply grade. This highway was designed with switchbacks so the gradient would be small.

A gradient is the inclination of a roadway to the horizontal surface. Research the federal, state, and local standards for the allowable gradients of highways, streets, and railway routes.

Find out how gradients are expressed in engineering terms. Give the standards for roadway types designed for vehicles of various weights, speeds, and engine power in terms of rate of change. Describe the alternatives available to engineers to reduce the gradient of a route in hilly or mountainous terrain. What safety measures do they incorporate to minimize risk on steep grades? Bring pictures to illustrate a presentation about your research, showing how engineers have applied standards to roads and routes in your home area.

Solving Equations Using the Balancing Method

Thinking in words slows you down and actually decreases comprehension in much the same way as walking a tightrope too slowly makes one lose one's balance.

LENORE FLEISCHER

In the previous two lessons, you learned about rate of change and the intercept form of a linear equation. In this lesson you'll learn symbolic methods to solve these equations. You've already seen the calculator methods of tracing on a graph and zooming in on a table. These methods usually give approximate solutions. Working backward to undo operations is a symbolic method that gives exact solutions. Another symbolic method that you can apply to solve all kinds of equations is the **balancing method.** In this lesson you'll investigate how to use the balancing method to solve linear equations. You'll discover that it's closely related to the method of working backward.

Investigation
Balancing Pennies

You will need

- pennies
- three paper cups

Here is a visual model of the equation $2x + 3 = 7$. A cup represents the variable x and pennies represent numbers. Assume that each cup has the same number of pennies in it and that the containers themselves are weightless.

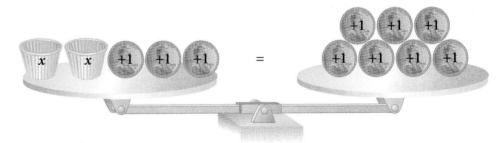

Step 1 | How many pennies must be in each cup if the left side of the scale balances with the right side? Explain how you got your answer.

Your answer to Step 1 is the solution to the equation $2x + 3 = 7$. It's the number that can replace x to make the statement true. In Steps 2 and 3, you'll use pictures and equations to show stages that lead to the solution.

Step 2 | Redraw the picture above, but with three pennies removed from each side of the scale. Write the equation that your picture represents.

| Step 3 | Redraw the picture, this one showing half of what was on each side of the scale in Step 2. There should be just one cup on the left side of the scale and the correct number of pennies on the right side needed to balance it. Write the equation that this picture represents. This is the solution to the original equation. |

Now your group will create a pennies-and-cups equation for another group to solve.

| Step 4 | Divide the pennies into two equal piles. If you have one left over, put it aside. Draw a large equal sign (or form one with two pencils) and place the penny stacks on opposite sides of it. |

| Step 5 | From the pile on one side of your equal sign, make three identical stacks, leaving at least a few pennies out of the stacks. Hide each stack under a paper cup. You should now have, on one side of your equal sign, 3 cups and some pennies. |

| Step 6 | On the other side you should have a pile of pennies. On both sides of the equal sign you have the same number of pennies, but on one side some of the pennies are hidden under cups. You can think of the two sides of the equal sign as being the two sides of a balance scale. Write an equation for this setup, using x to represent the number of pennies hidden under one cup. |

| Step 7 | Move to another group's setup. Look at their arrangement of pennies and cups, and write an equation for it. Solve the equation; that is, find how many pennies are under one cup without looking. When you're sure you know how many pennies are under each cup, you can look to check your answer. |

| Step 8 | Write a brief description of how you solved the equation. |

You could do problems like this on a balance scale as long as the weight of the cup is very small. But an actual balance scale can only model equations in which all the numbers involved are positive. Still, the idea of balancing equations can apply to equations involving negative numbers. Just remember, when you add any number to its opposite, you get 0. Think of negative and positive numbers as having opposite effects on a balance scale. You can remove 0 from either side of a balance-scale picture without affecting the balance. These three figures all represent 0:

$$1 + (-1) = 0$$

$$-3 + 3 = 0$$

$$2x + (-2x) = 0$$

EXAMPLE A | Draw a series of balance-scale pictures to solve the equation $6 = -2 + 4x$.

► **Solution** | The goal is to end up with a single x-cup on one side of the balance scale. One way to get rid of something on one side is to add its opposite to both sides.

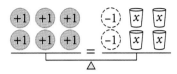

Here is the equation $6 = -2 + 4x$ solved by the balancing method:

Picture	Action taken	Equation
	Original equation.	$6 = -2 + 4x$
	Add 2 to both sides.	$6 + 2 = -2 + 2 + 4x$
	Remove the 0.	$8 = 4x$
	Divide both sides by 4.	$\dfrac{8}{4} = \dfrac{4x}{4}$
	Reduce.	$2 = x$ or $x = 2$

Balance-scale pictures can help you see what to do to solve an equation by the balancing method. But you won't need them once you get the idea of doing the same thing to both sides of an equation. And they're less useful if the numbers in the equation aren't "nice."

EXAMPLE B

Solve the equation $-31 = -50.25 + 1.55x$ using each method.

a. the balancing method

b. undoing operations

c. tracing on a calculator graph

d. zooming in on a calculator table

► **Solution**

a. the balancing method

$$-31 = -50.25 + 1.55x \qquad \text{Original equation.}$$

$$-31 + 50.25 = -50.25 + 50.25 + 1.55x \qquad \text{Add 50.25 to both sides.}$$

$$19.25 = 1.55x \qquad \text{Evaluate and remove the 0.}$$

$$\frac{19.25}{1.55} = \frac{1.55x}{1.55} \qquad \text{Divide both sides by 1.55.}$$

$$12.42 \approx x, \text{ or } x \approx 12.42 \qquad \text{Reduce.}$$

b. undoing operations

Start with -31.

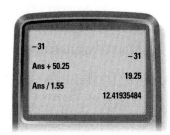

In parts a and b, if you convert the answer to a fraction, you get an exact solution of $\frac{385}{31}$.

c. tracing on a calculator graph

Enter the equation into Y1. Adjust your window settings and graph. Press TRACE and use the arrow keys to find the x-value for a y-value of -31. (See Example B in Lesson 4.6 to review this procedure.) The top screen shows that for a y-value of approximately -31.6 the x-value is 12.02.

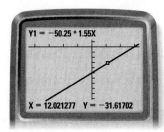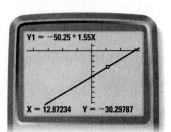

d. zooming in on a calculator table

To find a starting value for the table, use guess-and-check or a calculator graph to find an approximate answer. Then use the calculator table to find the answer to the desired accuracy. [► ▢ See **Calculator Note 3A** to review zooming in on a table. ◄]

Once you have determined a reasonable starting value, zoom in on a calculator table to find the answer using smaller and smaller values for the table increment.

You can also check your answer by using substitution.

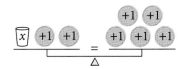

From Example B, you can see that each method has its advantages. The methods of balancing and undoing use the same process of working backward to get an exact solution. The two calculator methods are easy to use but usually give approximate solutions to the equation. You may prefer one method over others, depending on the equation you need to solve. If you are able to solve an equation using two or more different methods, you can check to see that each method gives the same result. With practice, you may develop symbolic solving methods of your own. Knowing a variety of methods, such as the balancing method and undoing, as well as the calculator methods, will improve your equation-solving skills, regardless of which method you prefer.

EXERCISES

You will need your calculator for problems **6, 8, 14,** and **15.**

▶ Practice Your Skills

1. Give the equation that each picture models.

a.

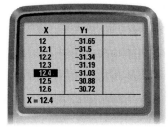

b.

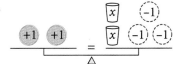

c.

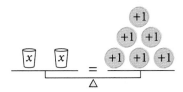

d.

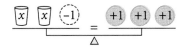

2. Copy and fill in the tables to solve the equations as in Example A.

Picture	Action taken	Equation
	Original equation.	
	Add 2 to both sides.	
	Divide both sides by 2.	$\dfrac{2x}{2} = \dfrac{6}{2}$
	Reduce.	

3. Give the next stage of the equation, matching the action taken, to reach the solution.

a. $0.1x + 12 = 2.2$

Original equation.

Subtract 12 from both sides.

Remove the 0 and subtract.

Divide both sides by 0.1.

b. $\dfrac{12 + 3.12x}{3} = -100$

Original equation.

Multiply both sides by 3.

Subtract 12 from both sides.

Remove the 0.

Divide both sides by 3.12.

4. For each original equation in problem 3, tell how you could solve the equation using different steps or using the same steps in a different order.

5. Solve these equations. Tell what action you take at each stage.

    a. $144x = 12$ **b.** $\frac{1}{6}x + 2 = 8$

▶ Reason and Apply

6. In the solution to the equation $-10 + 3x = 17$ shown below, some of the steps are left out.

$$-10 + 3x = 5$$
$$3x = 15$$
$$x = 5$$

    a. Describe the steps that transform the original equation into the second equation and the second equation into the third (the solution).

    b. Graph $Y_1 = -10 + 3x$ and $Y_2 = 5$, and trace to the lines' intersection. Write the coordinates of this point.

    c. Graph $Y_1 = 3x$ and $Y_2 = 15$, and trace to the lines' intersection. Write the coordinates of this point.

    d. Graph $Y_1 = x$ and $Y_2 = 5$, and trace to the lines' intersection. Write the coordinates of this point.

    e. What do you notice about your answers to 6b–d? Explain what this illustrates.

7. Solve the equation $4 - 1.2x = 12.4$ by using each method.

    a. balancing

    b. undoing

    c. tracing on a graph

    d. zooming in on a table

8. Write calculator routines that solve each equation by undoing. Give the solution to each one.

    a. $3 + 2x = 17$ **b.** $0.5x + 2.2 = 101.0$ **c.** $x + 307.2 = 2.1$

    d. $2(2x+2) = 7$ **e.** $\dfrac{4 + 0.01x}{6.2} - 6.2 = 0$

9. You can solve familiar formulas for a specific variable. For example, solving $A = lw$ for l you get

    $A = lw$ Original equation.

    $\dfrac{A}{w} = \dfrac{lw}{w}$ Divide both sides by w.

    $\dfrac{A}{w} = l$ Reduce.

You can also write $l = \frac{A}{w}$. Now try solving these formulas for the given variable.

    a. $C = 2\pi r$ for r **b.** $A = \frac{1}{2}(hb)$ for h **c.** $P = 2(l + w)$ for l

    d. $P = 4s$ for s **e.** $d = rt$ for t **f.** $A = \frac{1}{2}h(a + b)$ for h

10. Tell what number, if multiplied by the given number, gives 1.

    a. 12

    b. $\dfrac{1}{6}$

    c. 0.02

    d. $-\dfrac{1}{2}$

11. Tell what number, if added to the given number, gives 0.

    a. $\dfrac{1}{5}$

    b. 17

    c. -2.3

    d. $-x$

12. You can represent linear relationships with a graph, a table of values, an equation, or a rule stated in words. Here are two linear relationships. Give all the other ways to show each relationship.

    a.

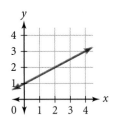

    b.

x	y
-2	2
-1.5	1.5
0	0
3	-3

▶ Review

13. **APPLICATION** Economy drapes for a certain size window cost $90. They have shallow pleats, and the width of the fabric is $2\frac{1}{4}$ times the window width. Luxury drapes of the same fabric for the same size window have deeper pleats. The width of the fabric is 3 times the window width. What price should the store manager ask for the luxury drapes?

14. Stella has decided to save to go on a trip at the end of her senior year of high school. She has decorated a glass jar and has put the $350 that she has saved so far into the jar. The first day of every month, she is planning to put $120 into the jar.

    a. Write a recursive routine that will generate a table showing the amount Stella will have in the jar each month for the next 12 months.

    b. The trip Stella wants to go on will cost $4,800. How many months will it take for her to save up enough to go on the trip?

    c. If Stella waited until September of her junior year to start saving, how much would she have to put in the jar every month in order to have saved enough by June of her senior year (assuming she still starts with $350)?

15. Run the LINES1 program on your calculator. [▶ 🖳] See **Calculator Note 4E** to learn how to use the LINES program. ◀] Sketch a graph of the randomly generated line on your paper. Use the trace function to locate the y-intercept and to determine the rate of change. When the calculator says you have the correct equation, write it under the graph. Repeat this program until you get three correct equations in a row.

16. The local bagel store sells a baker's dozen of bagels for $6.49, while the grocery store down the street sells a bag of 6 bagels for $2.50.

a. Copy and complete the tables showing the cost of bagels at the two stores.

Bagel Store

Bagels	13	26	39	52	65	78
Cost						

Grocery Store

Bagels	6	12	18	24	30	36	42	48	54	60
Cost										

b. Graph the information for each market on the same coordinate axes. Put bagels on the horizontal axis and cost on the vertical axis.

c. Find equations to describe the cost of bagels at each store.

d. How much does one bagel cost at each store? How do these cost values relate to the equations you wrote in 15c?

e. Looking at the graphs, how can you tell which store is the cheaper place to buy bagels?

f. Bernie and Buffy decided to use a recursive routine to complete the tables. Bernie used this routine for the bagel store:

6.49 (ENTER)

Ans · 2 (ENTER)

Buffy says that this routine isn't correct, even though it gives the correct answer for 13 and 26 bagels. Explain to Bernie what is wrong with his recursive routine. What routine should he use?

Activity Day

Modeling Data

Whenever measuring is involved in collecting data, you can expect some variation in the pattern of data points. Usually, you can't construct a mathematical model that fits the data exactly. But in general, the better a model fits, the more useful it is for making predictions or drawing conclusions from the data.

Activity
Tying Knots

You will need

- two pieces of rope of different lengths (around 1 m) and thickness
- a meterstick or a measuring tape

In this activity you'll explore the relationship between the number of knots in a rope and the length of the rope and write an equation to model the data.

Number of knots	Length of knotted rope (cm)
0	
1	
2	

Step 1 Choose one piece of rope and record its length in a table like the one shown. Tie 6 or 7 knots, remeasuring the rope after you tie each knot. As you measure, add data to complete a table like the one above.

Step 2 Graph your data, plotting the number of knots on the *x*-axis and the length of the knotted rope on the *y*-axis. What pattern does the data seem to form?

Step 3 What is the approximate rate of change for this data set? What is the real-world meaning of the rate of change? What factors have an effect on it?

Step 4 What is the *y*-intercept for the line that best models the data? What is its real-world meaning?

| Step 5 | Write an equation in intercept form for the line that you think best models the data. Graph your equation to check that it's a good fit. |

Now you'll make predictions and draw some conclusions from your data using the line model as a summary of the data.

Step 6	Use your equation to predict the length of your rope with 7 knots. What is the difference between the actual measurement of your rope with 7 knots and the length you predicted using your equation?
Step 7	Use your equation to predict the length of a rope with 17 knots. Explain the problems you might have in making or believing your prediction.
Step 8	What is the maximum number of knots that you can tie with your piece of rope? Explain your answer.
Step 9	Does your graph cross the x-axis? Explain the real-world meaning, if any, of the x-value of the intersection point.
Step 10	Substitute a value for y into the equation. What question does the equation ask? What is the answer?

| Step 11 | Repeat Steps 1–5 using a different piece of rope. Graph the data on the same pair of axes. |
| Step 12 | Compare the graphs of the lines of fit for both ropes. Give reasons for the differences in their y-intercepts, in their x-intercepts, and in their rates of change. |

IMPROVING YOUR **REASONING** SKILLS

There are 100 students and 100 lockers in a school hallway. All of the lockers are closed. The first student walks down the hallway and opens every locker. A second student closes every even numbered locker. The third student goes to every third locker and opens it if it is closed, or closes it if it is open. This pattern repeats so that the nth student leaves every nth locker the opposite of how it was before. After all 100 students have opened or closed the lockers, how many lockers are left open?

CHAPTER

4

REVIEW

You started this chapter by applying the rules for order of operations and learning how to enter numerical expressions into your calculator. You used the **distributive property** to write equivalent forms of **algebraic expressions.** You investigated **recursive sequences** by using their starting values and **rates of change** to write **recursive routines.** You saw how rates of change and starting values appear in plots.

In a walking investigation you observed, interpreted, and analyzed graphical representations of relationships between time and distance. What does the graph look like when you stand still? When you move away from or move toward the motion sensor? If you speed up or slow down? You identified real-world meanings of the *y*-intercept and the rate of change of a **linear relationship,** and used them to write the **equation** in the **intercept form** of $y = a + bx$. You learned the role of *b*, the coefficient of *x*. You explored relationships among graph sketches, tables, recursive rules, equations, and graphs.

Throughout the chapter you developed your equation-solving skills. You found **solutions** to equations by using an **undoing** process and by using a **balancing** process. You found approximate solutions by tracing calculator graphs and by zooming in on calculator tables. Finally, you learned how to model data that doesn't lie exactly on a line, and you used your model to predict inputs and outputs.

EXERCISES

You will need your calculator for problems **4, 6, 7,** and **8.**

1. Solve these equations. Give reasons for each step.

    a. $-x = 7$ **b.** $4.2 = -2x - 42.6$

2. These tables represent linear relationships. For each relationship, give the rate of change, the recursive rule, the *y*-intercept, and the equation in intercept form.

a.

x	y
0	3
1	4
2	5

b.

x	y
1	0.01
2	0.02
3	0.03

c.

x	y
−2	1
0	5
3	11

d.

x	y
−4	5
12	−3
2	2

3. Match these walking instructions with their graph sketches.

i.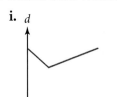

ii.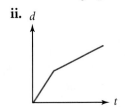

iii.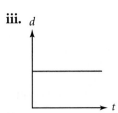

   a. The walker stands still.

   b. The walker takes a few steps toward the zero mark, then walks away.

   c. The walker steps away from the zero mark, stops, then continues more slowly in the same direction.

4. Graph each equation on your calculator, and trace to find the approximate y-value for the given x-value.

   a. $y = 1.21 - x$ when $x = 70.2$

   b. $y = 6.02 + 44.3x$ when $x = 96.7$

   c. $y = -0.06 + 0.313x$ when $x = 0.64$

   d. $y = 1183 - 2140x$ when $x = -111$

5. Write the equations for linear relationships that have these characteristics.

   a. The output value is equal to the input value.

   b. The output value is 3 less than the input value.

   c. The rate of change is 2.3 and the y-intercept is -4.3.

   d. The graph contains the points $(1, 1)$, $(2, 1)$, and $(3, 1)$.

6. On a recent trip to Detroit, Tom started from home, which is 12 miles from Traverse City. After 4 hours he had traveled 220 miles.

   a. Write a recursive routine to model Tom's distance from Traverse City during this trip. State at least two assumptions you're making.

   b. Use your recursive routine to determine his distance from Traverse City for each hour during the first 5 hours of the trip.

   c. What is the rate of change, and what does it mean in the context of this situation?

7. The profit for a small company depends on the number of bookcases it sells. One way to determine the profit is to use a recursive routine such as

   $\{0, -850\}$ **ENTER**

   $\{\text{Ans}(1) + 1, \text{Ans}(2) + 70\}$ **ENTER** ; **ENTER** , . . .

   a. Explain what the numbers and expressions 0, -850, $\text{Ans}(1)$, $\text{Ans}(1) + 1$, $\text{Ans}(2)$, and $\text{Ans}(2) + 70$ represent.

   b. Make a plot of this situation.

   c. When will the company begin to make a profit? Explain.

   d. Explain the relationship between the values -850 and 70 and your graph.

   e. Does it make sense to connect the points in the graph with a line? Explain why or why not.

8. A single section and a double section of a log fence are shown.

    a. How many additional logs are required each time the fence is increased by a single section?

    b. Copy and fill in the missing values in the table below.

Number of sections	1	2	3	4	. . .		. . .	50
Number of logs	4	7			. . .	91	. . .	

    c. Describe a recursive routine that relates the number of logs required to the number of sections.

    d. If each section is 3 meters long, what is the longest fence you can build with 217 logs?

9. The time-distance graph shows Carol walking at a steady rate. Her partner used a motion sensor to measure her distance from a given point.

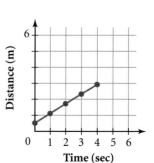

    a. According to the graph, how much time did Carol spend walking?

    b. Was Carol walking toward or away from the motion sensor? Explain your thinking.

    c. Approximately how far away from the motion sensor was she when she started walking?

    d. If you know Carol is 2.9 meters away from the motion sensor after 4 seconds, how fast was she walking?

    e. If the equipment will measure distances only up to 6 meters, how many seconds of data can be collected if Carol continues walking at the same rate?

    f. Looking only at the graph, how do you know that Carol was neither speeding up nor slowing down during her walk?

10. Suppose a new small-business computer system costs $5,400. Every year its value drops by $525.

    a. Define variables and write an equation modeling the value of the computer in any given year.

    b. What is the rate of change, and what does it mean in the context of the problem?

    c. What is the *y*-intercept, and what does it mean in the context of the problem?

    d. What is the *x*-intercept, and what does it mean in the context of the problem?

11. For each table, write a formula for list L_2 in terms of list L_1.

a.

List 1	List 2
0	−5.7
1	−3.4
2	−1.1
3	1.2
4	3.5
5	5.8

b.

List 1	List 2
−3	19
−1	3
0	−5
2	−21
5	−45
6	−53

c.

List 1	List 2
3	13.5
−2	11
−9	7.5
0	12
6	15
−5	9.5

12. Suppose Andrei and his younger brother are having a race. Because the younger brother can't run as fast, Andrei lets him start out 5 meters ahead. Andrei runs at a speed of 7.7 meters per second. His younger brother runs at 6.5 meters per second. The total length of the race is 50 meters.

a. Write an equation to find how long it will take Andrei to finish the race. Solve the equation to find the time.

b. Write an equation to find how long it will take Andrei's younger brother to finish the race. Solve the equation to find the time.

c. Who wins the race? How far ahead was the winner at the time he crossed the finish line?

13. Consider the equation $2(x − 6) = −5$.

a. Show two different ways you can solve the equation.

b. Show how you can check your result by substituting it into the original equation.

14. Solve each equation using the method of your choice. Then use a different method to verify your solution.

a. $14x = 63$

b. $−4.5x = 18.6$

c. $8 = 6 + 3x$

d. $5(x − 7) = 29$

e. $3(x − 5) + 8 = 12$

15. The equation $w = −38.3 + 1.4t$ approximates the wind chill temperatures in °F for a wind speed of 20 miles per hour. Find the actual temperature t for each given wind chill temperature w. Verify your results with another method.

a. $w = −40°$

b. $w = 15°$

c. $w = −23°$

d. $w = 0°$

TAKE ANOTHER LOOK

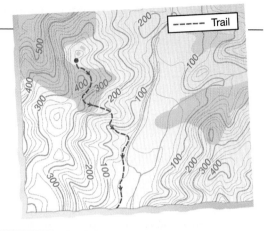

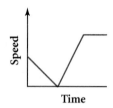 The picture at right is a **contour map** that reveals the character of the terrain. All points on an **isometric line** are the same height in feet above sea level. The graph below shows how the hiker's walking speed changes as she covers the terrain on the dotted-line trail shown on the map.

What quantities are changing in the graph and in the map?

How does each display reveal rate of change?

How could you measure distance on each display?

What would the graph sketch of this hike look like if distance were plotted on the vertical axis instead of speed?

What do these two displays tell you when you study them together?

Sediment layers form contour lines in the Grand Canyon.

IMPROVING YOUR **REASONING** SKILLS

Did these plants grow at the same rate? If not, which plant was tallest on Day 4? Which plant took the most time to reach 8 cm? Redraw the graphs so that you can compare their growth rates more easily.

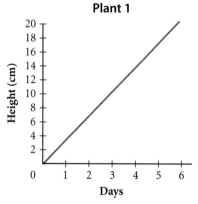

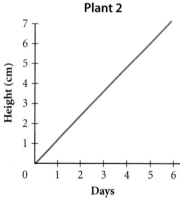

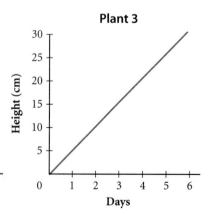

Assessing What You've Learned

GIVING A PRESENTATION

Making presentations is an important career skill. Most jobs require workers to share information, to help orient new coworkers, or to represent the employer to clients. Making a presentation to the class is a good way to develop your skill at organizing and delivering your ideas clearly and in an interesting way. Most teachers will tell you that they have learned more by trying to teach something than they did simply by studying it in school.

Here are some suggestions to make your presentation go well:

▶ Work with a group. Acting as a panel member might make you less nervous than giving a talk on your own. Be sure the role of each panel member is clear so that the work and the credit are equally shared.

▶ Choose the topic carefully. You can summarize the results of an investigation, do research for a project and present what you've learned and how it connects to the chapter, or give your own thinking on Take Another Look or Improving Your Reasoning Skills.

▶ Prepare thoroughly. Outline your presentation and think about what you have to say on each point. Decide how much detail to give, but don't try to memorize whole sentences. Illustrate your presentation with models, a poster, a handout, or overhead transparencies. Prepare these visual aids ahead of time and decide when to introduce them.

▶ Speak clearly. Practice talking loudly and clearly. Show your interest in the subject. Don't hide behind a poster or the projector. Look at the listeners when you talk.

Here are other ways to assess what you've learned:

ORGANIZE YOUR NOTEBOOK You might need to update it with examples of undoing and balancing to solve an equation, or with notes about how to trace a line or search a table to approximate the coordinates of the solution. Be sure you understand the meanings of important words like linear equation, rate of change, and intercept form.

WRITE IN YOUR JOURNAL What method for solving equations do you like better? Do you always remember to define variables before you graph or write an equation? How are you doing in algebra generally? What things don't you understand?

UPDATE YOUR PORTFOLIO Choose a piece of work you did in this chapter to add to your portfolio—your graph from the investigation On the Road Again (Lesson 4.4), your longest number trick with its undo operations, or your research on a project.

Fitting a Line to Data

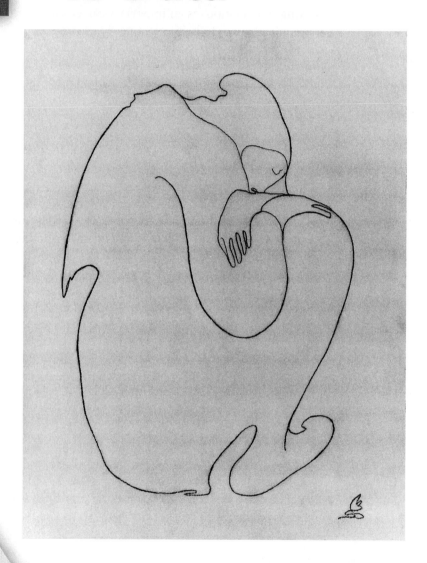

Artists, like mathematicians, use lines to summarize their observations. An artist's data include contour, texture, color, shape, motion, and balance. The American artist Romaine Brooks (1874–1970) reduced her entire set of observations into the lines you see in this pencil sketch titled *Departure*.

OBJECTIVES

In this chapter you will

- define and calculate slope
- write an equation that fits a set of real-world data
- review the intercept form of a linear equation
- learn the point-slope form of a linear equation
- recognize equivalent equations written in different forms

A Formula for Slope

The nearest thing to nothing that anything can be and still be something is zero.

ANONYMOUS

You have seen that the steepness of a line can be a graphical representation of a real-world rate of change like a car's speed, the number of calories burned with exercise, or a constant relating two units of measure. Often you can estimate the rate of change of a linear relationship just by looking at a graph of the line. Can you tell which line in the graph matches which equation?

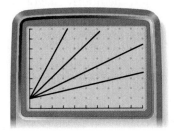

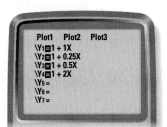

Slope is another word used to describe the steepness of a line or the rate of change of a linear relationship. In this investigation you will explore how to find the slope of a line from two points on the line.

Wayne Thiebaud's oil painting *Urban Downgrade, 20th and Noe* (1981) is an artistic representation of the steepness, or slope, of a street in San Francisco, California. Thiebaud is an American artist born in 1920.

Investigation
Points and Slope

You will need

• graph paper

Hector recently signed up with a limited-usage Internet provider. He knows that there is a flat monthly charge and an hourly rate for the number of hours he is connected during the month. The table shows the amount of time he spent using the Internet for the first three months and the total fee he was charged.

Step 1 Is there a linear relationship between the time in hours that Hector uses the Internet and his total fee in dollars? If so, why do you think such a relationship exists?

Step 2 Use the numbers in the table to find the rate in dollars per hour. Explain in words how you calculated this rate.

Internet Use

Month	Time (hr)	Total fee ($)
September	4	16.75
October	5	19.70
November	8	28.55

Step 3	Draw a pair of coordinate axes on graph paper. Use the x-axis for time in hours and the y-axis for total fee in dollars. Plot and label the three points the table of data represents. Draw a line through the three points. Does this line support your answer in Step 1?
Step 4	Choose two points on your graph. Use arrows to show how you could move from one point to the other using only one vertical move and one horizontal move. How long is each arrow? What is the real-world meaning of each length?
Step 5	How do the arrow lengths relate to the hourly rate that you found in Step 2? Use the arrow lengths to find the hourly rate of change, or slope, for this situation. What units should you apply to the number?

In Step 4, you used arrows to show the vertical change and the horizontal change when you moved from one point to another. The right triangle you created is called a **slope triangle.**

Step 6	Choose a different pair of points on your graph. Create a slope triangle between them and use it to find the slope of the line. How does this slope compare to your answers in Step 2 and Step 5?

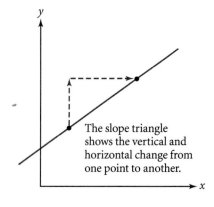

The slope triangle shows the vertical and horizontal change from one point to another.

Step 7	Think about what you have done with your slope triangles. How could you use the coordinates of any two points to find the vertical change and the horizontal change of each arrow? Write a single numerical expression using the coordinates of two points to show how you can calculate slope.
Step 8	Write a symbolic algebraic rule for finding the slope between any two points (x_1, y_1) and (x_2, y_2). The subscripts mean that these are two distinct points of the form (x, y).

A slope triangle helps you visualize slope by showing you the vertical change and the horizontal change from one point to another. These changes are also called the "change in y" (vertical) and the "change in x" (horizontal). The example will help you see how to work with positive and negative numbers in slope calculations.

The steps going up this pyramid (located in the ruins of the ancient Mayan city of Chichen Itza) are like slope triangles that define the slope of the pyramid's face.

EXAMPLE | Consider the line through the points (1, 7) and (6, 4).

a. Find the slope of the line.

b. Without graphing, verify that the point (4, 5.2) is also on that line.

c. Find the coordinates of another point on the same line.

▶ **Solution** | Plot the given points and draw the line between them.

a.

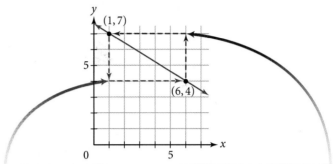

> If you move from (1, 7) to (6, 4), the change in y is -3 (down 3) and the change in x is $+5$ (right 5). The slope is $\frac{-3}{+5}$.

> If you move from (6, 4) to (1, 7), the change in y is $+3$ (up 3) and the change in x is -5 (left 5). The slope is $\frac{+3}{-5}$.

But $\frac{-3}{+5}$ is equivalent to $\frac{+3}{-5}$. You get the same slope, $-\frac{3}{5}$ or -0.6, no matter which point you start from. The slope triangles help you see this relationship more clearly.

Move to
(6, 4) from (1, 7).

$$\text{Slope} = \frac{4-7}{6-1} = \frac{-3}{5} = -\frac{3}{5} \quad \text{or}$$

Move to
(1, 7) from (6, 4).

$$\text{Slope} = \frac{7-4}{1-6} = \frac{3}{-5} = -\frac{3}{5}$$

b. The slope between any two points on the line will be the same. So, if the slope between the point (4, 5.2) and either of the original two points is -0.6, then the point is on the line. The slope between (4, 5.2) and (1, 7) is

$$\frac{7-5.2}{1-4} = \frac{1.8}{-3} = -\frac{1.8}{3} = -0.6$$

So the point (4, 5.2) is on the line.

c. You can find the coordinates of another point by adding the change in x and the change in y from any slope triangle on the line to a known point.

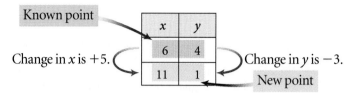

You could start with the point $(6, 4)$ and use

$$\frac{change\ in\ y}{change\ in\ x} = \frac{-3}{5}$$

This gives the new point $(6 + 5, 4 + (-3)) = (11, 1)$.

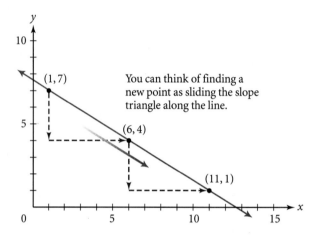

You can think of finding a new point as sliding the slope triangle along the line.

Try using point $(1, 7)$ and using

$$\frac{change\ in\ y}{change\ in\ x} = \frac{3}{-5}$$

to find another point. Try using either original point and using

$$\frac{change\ in\ y}{change\ in\ x} = \frac{-0.6}{1}$$

to find another point.

Slope is an extremely important concept in mathematics and in applications like medicine and engineering that rely on mathematics. You may encounter different ways of describing slope—for example, rise over run or vertical change over horizontal change. But you can always calculate the slope using this formula.

History
● CONNECTION ●

Slope is sometimes written $\frac{\Delta y}{\Delta x}$. The symbol Δ is the Greek capital letter delta. The use of Δ is linked to the history of calculus in the 18th century when it was used to mean "difference."

Slope Formula

The formula for the **slope** of the line passing through point 1 with coordinates (x_1, y_1) and point 2 with coordinates (x_2, y_2) is

$$slope = \frac{change\ in\ y}{change\ in\ x} = \frac{y_2 - y_1}{x_2 - x_1}$$

A line that goes up from left to right has a positive slope. A line that goes down from left to right has a negative slope.

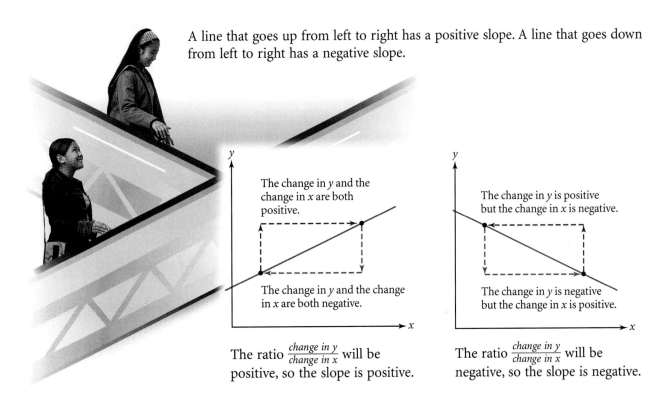

The change in y and the change in x are both positive.

The change in y and the change in x are both negative.

The ratio $\frac{change\ in\ y}{change\ in\ x}$ will be positive, so the slope is positive.

The change in y is positive but the change in x is negative.

The change in y is negative but the change in x is positive.

The ratio $\frac{change\ in\ y}{change\ in\ x}$ will be negative, so the slope is negative.

Horizontal lines have a slope of zero because they have no change in y. Vertical lines have no change in x. To calculate the slope of a vertical line, you would have to divide by zero, which is impossible—we say that the slope of a vertical line is undefined.

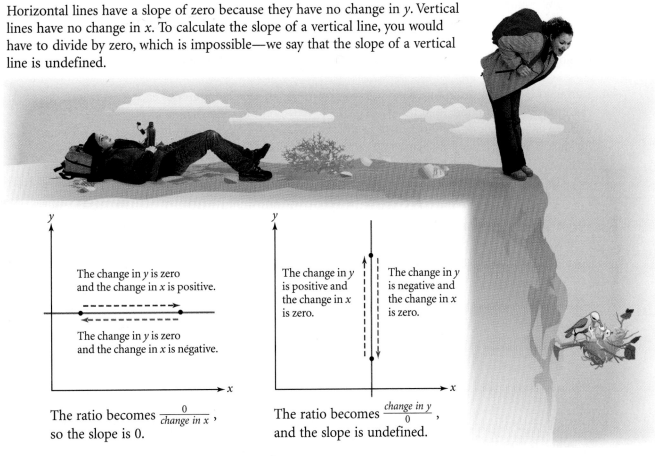

The change in y is zero and the change in x is positive.

The change in y is zero and the change in x is negative.

The change in y is positive and the change in x is zero.

The change in y is negative and the change in x is zero.

The ratio becomes $\frac{0}{change\ in\ x}$, so the slope is 0.

The ratio becomes $\frac{change\ in\ y}{0}$, and the slope is undefined.

As you work on the exercises, keep in mind that the slope of a line is the same as the rate of change of its equation. When a linear equation is written in intercept form, $y = a + bx$, which letter would represent the slope?

You will need your calculator for problems **4** and **14**.

Practice Your Skills

1. Find the slope of each line using a slope triangle or the slope formula.

a. 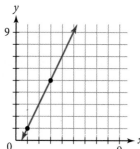   **b.** 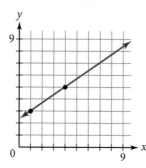   **c.**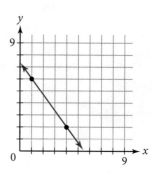

2. Find the slope of the line through each pair of points. Then name another point on the same line.

   a. $(2, 4), (4, 7)$ **b.** $(6, -1), (2, 5)$ **c.** $(-2, 4), (8, 4)$ **d.** $(1, -3), (9, 12)$

3. Given the slope of a line and one point on the line, name *two* other points on the same line. Then use the slope formula to check that the slope between each of the two new points and the given point is the same as the given slope.

   a. Slope $\frac{3}{1}$; point $(0, 4)$ **b.** Slope -5; point $(2, 8)$

   c. Slope $-\frac{3}{4}$; point $(8, 6)$ **d.** Slope 0.2; point $(5, 7)$

4. Run the LINES1 program five times. On your own paper, sketch a graph of each randomly generated line. Find the slope of the line by counting the change in y and the change in x on the grid, or trace the line for two points to use in the slope formula. Then find the y-intercept and write the equation of the line in intercept form. [▶ ▦ See **Calculator Note 4E** to learn how to use the LINES program. ◀]

Reason and Apply

5. Each table gives the coordinates of four points on a different line.

i.

x	y
4	−8
4	0
4	3
4	20

ii.

x	y
0	5
1	3
3	−1
4	−3

iii.

x	y
−4	−5
−3	−5
1	−5
4	−5

iv.

x	y
−4	−5
−2	−3.5
0	−2
4	1

   a. Without calculating, can you tell whether the slope of the line through each set of points is positive, negative, zero, or undefined? Explain how you can tell.

   b. Choose two points from each table and calculate the slope. Check that your answer is correct by calculating the slope with a different pair of points.

   c. Write an equation for each table of values.

6. Two lines have been graphed for you.

    a. How are the lines in the graph alike? How are they different?

    b. Which line matches the equation $y = -3 + \frac{2}{5}x$?

    c. What is the equation of the other line?

    d. How are the equations alike? How are they different?

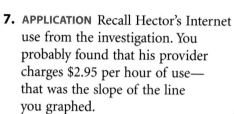

7. APPLICATION Recall Hector's Internet use from the investigation. You probably found that his provider charges $2.95 per hour of use—that was the slope of the line you graphed.

Internet Use

Month	Time (hr)	Total fee ($)
September	4	16.75
October	5	19.70
November	8	28.55

    a. Use the rate of change and the data in the table to find out how much the total fee is for 3 hours of use. How much is the total fee for 2 hours?

    b. Repeat the process in 7a to find out how much the total fee is for 0 hours of use. What is the real-world meaning of this number in this situation? (Look back at the investigation for help.)

    c. A mathematical model can be an equation, a graph, or a drawing that helps you better understand a real-world situation. Write a linear equation in intercept form that you can use to model this situation.

    d. Use your linear equation to find out how much the total fee is for 28 hours of use.

Many professions rely on models to understand real-world situations. For example, an architectural model like this one would be used to design a building. Mathematical models are another type of model that help people understand the real world.

8. APPLICATION A hot-air balloonist gathered the data in this table.

Hot-Air Balloon Height

Time (min)	Height (m)
0	14
2.2	80
3.4	116
4	134
4.6	152

a. What is the slope of the line through these points?

b. What are the units of the slope? What is the real-world meaning of the slope?

c. Write a linear equation in intercept form to model this situation.

d. What is the height of the balloon after 8 minutes?

e. During what time interval is the height less than or equal to 500 meters?

9. If a and c are the lengths of the vertical and horizontal segments and $(0, e)$ is the y-intercept, what is the equation of the line?

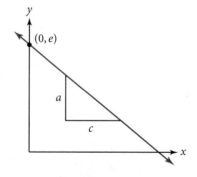

10. This line has a slope of 1. Graph it on your own paper.

a. Draw a slope triangle on your line. How do the change in y and the change in x compare?

b. Draw a line that is steeper than the given line. How do the change in y and the change in x compare? How does the numerical slope compare to that of the original line?

c. Draw a line that is less steep than the given line. How do the change in y and the change in x compare? How does the slope compare to the given line?

d. How would a line with a slope of -15 compare to your other lines? Explain your reasoning.

11. When you make a scatter plot of real-world data, you may see a linear pattern.

   a. Which line do you think "fits" each scatter plot? Think about slope and how the points are scattered. Explain how you chose your lines.

   i.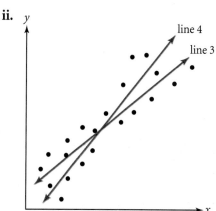

   ii.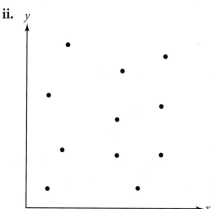

   b. Sketch each scatter plot on your own paper. Then draw a line that you think fits the data.

   i.

   c. List two features that you think are important for a line that fits data.

Review

12. Use the distributive property to rewrite each expression without using parentheses.

   a. $3(x - 2)$

   b. $-4(x - 5)$

   c. $-2(x + 8)$

13. Calista has five brothers. The mean of her brothers' ages is 10 years, and the median is 6 years. Create a data set that could represent the brothers' ages. Is this the only possible answer?

LESSON 5.1 A Formula for Slope **259**

14. Enter $\{-3, -1, 2, 8, 10\}$ into list L1 on your calculator.

    a. Write a rule for list L2 that adds 14 to each value in list L1 and then multiplies the results by 2.5. What are the values in list L2?

    b. Write a different rule for list L2 that produces the same results as the one you wrote in 14a but that doesn't use parentheses. Do you get the same results?

    c. Write a rule for list L3 that works backward and undoes the operations in list L2 to produce the values in list L1.

15. Convert each decimal number to a percent.

    a. 0.85   **b.** 1.50

    c. 0.065   **d.** 1.07

STEP RIGHT UP

How would it feel to climb a flight of stairs if every step was a little higher or lower than the previous one? The constant measure for treads and risers on most stairs keeps you from tripping. Have you noticed that the stairs outside some public buildings slow you down to a "ceremonial" pace? Or that little-used stairs to a cellar seem dangerously steep? Investigate the standards for stairs in various architectural settings and learn the reasons for their various slopes.

Your project should include

▶ Tread-and-riser data and slope calculations for several different stairways.

▶ The building codes or recommended standards in your area for home stairways. Is a range of slopes permitted? When are landings or railings required?

▶ Scale drawings for at least three different stairways.

After you've done your research, consider this question: Does a spiral staircase have a constant slope?

Slope triangles are like the steps of a staircase. This oil painting, *Bauhaus Stairway*, by the German artist Oskar Schlemmer (1888–1943) shows many slope triangles.

Bauhaus Stairway (1932) Oil on canvas, 63 7/8 × 45 in. The Museum of Modern Art, New York. Gift of Philip Johnson. Photograph © 2000 The Museum of Modern Art, New York

Writing a Linear Equation to Fit Data

When you plot real-world data, you will often see a linear pattern. However, the points will rarely fall exactly on a line. How can you tell if a particular line is a good model for the data? One of the simplest ways is to ask yourself if the line shows the general direction of the data and if there are about the same number of points above the line as below the line. If so, then the line will appear to "fit" the data, and we call it a **line of fit.**

Although these birds are not in a line, can you visualize a line that shows the flocks' general direction?

Sometimes one line will be a better model for your data than another. Each of these graphs shows a scatterplot of data points and possible lines of fit.

Graph A

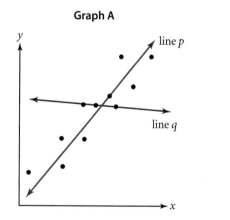

Graph B

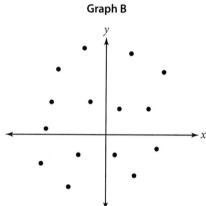

Graph C

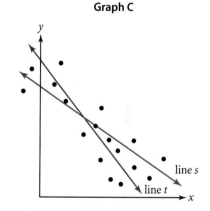

In Graph A, line *p* fits better because it shows the general direction of the data and there are the same number of points above the line as below the line. Although line *q* goes through several points, it does not show the direction of the data.

In Graph B, the data doesn't seem to have a pattern. No lines of fit are shown because you can't say that one line would fit the data better than another line would.

In Graph C, both lines show the general direction of the data and both lines have the same number of points above and below them. You could consider either line a line of fit. When making predictions, how would your calculations using the equation for line *s* differ from those using the equation for line *t*?

In the next investigation you will learn one method to find a possible line of fit.

Investigation
Beam Strength

How strong do the beams in a ceiling have to be? How do bridge engineers select beams to support traffic? In this investigation you will collect data and find a linear model to determine the strength of various "beams" made of spaghetti.

Step 1	Make two stacks of books of equal height. Punch holes on opposite sides of the cup and tie the string through the holes.
Step 2	Follow the procedure note for a beam made from one strand of spaghetti. Record the maximum load (the number of pennies) that this beam will support.
Step 3	Repeat Step 2 for beams made from two, three, four, five, and six strands of spaghetti.

> **Procedure Note**
>
> 1. Hang your cup at the center of your spaghetti beam.
> 2. Support the beam between the stacks of books so that it overlaps each stack by about 1 inch. Put another book on each stack to hold the beam in place.
> 3. Put pennies in the cup, one at a time, until the beam breaks.

Step 4	Plot your data on your calculator. Let x represent the number of strands of spaghetti, and let y represent the maximum load. Sketch the plot on paper too.
Step 5	Use a strand of spaghetti to visualize a line that you think fits the data on your sketch. Choose two points on the line. Note the coordinates of these points. Calculate the slope of the line between the two points.
Step 6	Use the slope, b, that you found in Step 5 to graph the equation $y = bx$ on your calculator. Why is this line parallel to the direction the points indicate? Is the line too low or too high to fit the data?
Step 7	Using the spaghetti strand on your sketch, estimate a good y-intercept, a, so that the equation $y = a + bx$ better fits your data. On your calculator, graph the equation $y = a + bx$ in place of $y = bx$. Adjust your estimate for a until you have a line of fit.
Step 8	In Step 5, everyone started with a visual model that went through two points. In your group, compare all final lines. Did everyone end up with the same line? Do you think a line of fit must go through at least two data points? Is any one line better than the others?

Your line is a model for the relationship between the number of strands of spaghetti in the beam and the load in pennies that the beam can support.

Step 9 | Explain the real-world meaning of the slope of your line.

Step 10 | Use your linear model to predict the number of spaghetti strands needed to support $5 worth of pennies.

Step 11 | Use your model to predict the maximum loads for beams made of 10 and 17 strands of spaghetti.

Step 12 | Some of your data points may be very close to your line, while others could be described as outliers. What could have caused these outliers?

Engineers conduct tests using procedures similar to the one you used in your investigation. The test results help them select the best materials and sizes for beams in buildings, bridges, and other forms of architecture.

Despite engineering tests, buildings can suffer damage during stress. This building in San Francisco, California, collapsed during an earthquake in October, 1989.

EXAMPLE

This table shows how many fat grams there are in some hamburgers sold by national chain restaurants.

Nutrition Facts

Burger	Saturated fat (g)	Total fat (g)
Burger King "Big King"	18	43
Burger King "Double Cheeseburger with Bacon"	18	39
Burger King "Whopper Jr."	8	24
Burger King "Whopper with Cheese"	16	46
Hardee's "The Works"	12	30
Jack in the Box "Jumbo Jack with Cheese"	14	40
McDonald's "Arch Deluxe with Bacon"	12	34
McDonald's "Big Mac"	10	28
Wendy's "Big Bacon Classic"	12	30
Wendy's "Single with Everything"	7	20

(*Consumer Reports*, Dec. 1997, pp. 12–13)

This is not a real hamburger but a ceramic sculpture. Can you imagine how much total fat this burger would have if it were real?

Hamburger (1983) by David Gilhooly, Collection of Harry W. and Mary Margaret Anderson, Photo by M. Lee Fatheree

a. Find a linear equation to model the data.

b. Tell the real-world meaning of the slope and intercept of your line.

c. Predict the total fat in a burger with 20 g of saturated fat.

d. Predict the saturated fat in a burger with 50 g of total fat.

▶ Solution

Draw a scatter plot of the data. Let x be the number of grams of saturated fat, and let y be the total number of grams of fat.

a. The scatter plot shows a linear pattern in the data. A line through the points (8, 24) and (18, 43) seems to show the direction of the data. Calculate the slope b of the line between these two points. Use (8, 24) as (x_1, y_1) and (18, 43) as (x_2, y_2).

$$b = \frac{y_2 - y_1}{x_2 - x_1} = \frac{43 - 24}{18 - 8} = \frac{19}{10} = 1.9$$

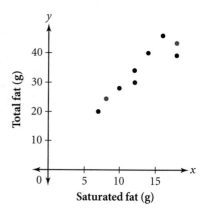

Substitute 1.9 for b in $y = bx$ to get

$$y = 1.9x$$

The equation $y = 1.9x$ shows the direction of the line, but it is too low.

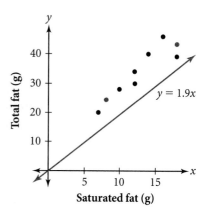

Estimate how far up to raise the line. The y-intercept is somewhere around 10. Adjust the intercept by tenths until you are satisfied. You may find that the equation

$$y = 9.4 + 1.9x$$

is a good model. Notice, however, that the line of fit doesn't have to go through any data points.

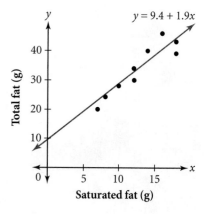

b. The y-intercept, 9.4, means that even without any saturated fat, a burger has about 9.4 grams of total fat. The slope, 1.9, means that for each additional gram of saturated fat there are an additional 1.9 grams of total fat.

c. Substitute 20 g of saturated fat for x in the equation.

$$y = 9.4 + 1.9x \qquad \text{Original equation.}$$
$$y = 9.4 + 1.9(20) \qquad \text{Substitute 20 for } x.$$
$$y = 47.4 \qquad \text{Multiply and add.}$$

The model predicts that there would be 47.4 g of total fat in a burger with 20 g of saturated fat.

d. Substitute 50 g of total fat for y in the equation.

$y = 9.4 + 1.9x$	Original equation.
$50 = 9.4 + 1.9x$	Substitute 50 for y.
$50 - 9.4 = 9.4 + 1.9x - 9.4$	Subtract 9.4 from both sides.
$40.6 = 1.9x$	Subtract.
$\dfrac{40.6}{1.9} = \dfrac{1.9x}{1.9}$	Divide both sides by 1.9.
$21.4 = x$	Reduce.

The model predicts that there would be about 21 g of saturated fat in a burger with 50 g of total fat.

Notice that you find the slope before the y-intercept when finding a line of fit. Because of the importance of slope, some mathematicians show it first. They use the **slope-intercept form** of a linear equation, often calling the slope m and the y-intercept b. This gives $y = mx + b$. Why is this equation equivalent to the intercept form that you have learned?

EXERCISES

You will need your calculator for problem **4.**

▶ Practice Your Skills

1. For each graph below, tell whether or not you think the line drawn is a good representation of the data. Explain your reasoning.

   a.
   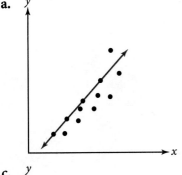

   b.
   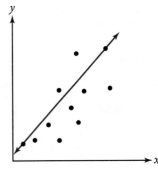

   c.
   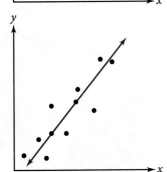

   d.
   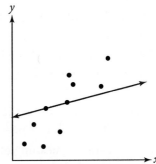

2. The line through the points $(0, 5)$ and $(4, 5)$ is horizontal. The equation of this line is $y = 5$ because the y-value of every point on it is 5. If a line goes through the points $(2, -6)$ and $(2, 8)$, what kind of line is it? What is its equation?

3. Write the equation of the line in each graph.

a.

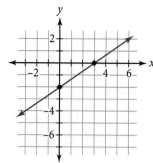

b.

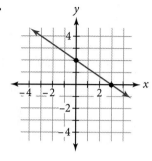

c.

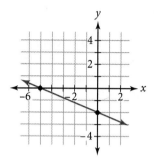

d.

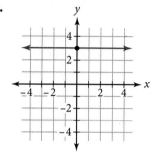

4. On Penny's 15th birthday, her grandmother gave her a large jar of quarters. Penny decided to continue to save quarters in the jar. Every few months she counts her quarters and records the number in a table like this one. Predict how many quarters she'll have on her 18th birthday.

Penny's Savings

Number of months x	3	5	8	12	15	19	22	26
Number of quarters y	270	275	376	420	602	684	800	830

a. Make a scatter plot of the data on your calculator. Is there a pattern?

b. Select two points through which a line of fit would pass. Find the slope of the line between these points.

c. What is the real-world meaning of the slope?

d. Use the slope you found in 4b to write an equation of the form $y = bx$. Graph this line on the scatter plot. What do you need to do to this line to better fit the data?

e. Estimate the y-intercept and write an equation of the form $y = a + bx$. Graph this new line.

f. What is the real-world meaning of the y-intercept?

g. Use your equation to predict how many quarters Penny will have on her 18th birthday.

Reason and Apply

5. Use the table to look for a relationship between a state's population and the number of members from that state in the House of Representatives.

Statistics for Some States

State	Estimated population (millions)	Number of members in House of Representatives	Number of members in Senate
Alabama	4.4	7	2
Indiana	5.9	10	2
Michigan	9.9	16	2
Mississippi	2.8	5	2
North Carolina	7.7	12	2
Oklahoma	3.4	6	2
Oregon	3.3	5	2
Tennessee	5.5	9	2
Utah	2.1	3	2
West Virginia	1.8	3	2

(*www.census.gov* and *www.house.gov*)

a. Which statement makes more sense: The population depends on the number of members in the House of Representatives, or the number of members in the House of Representatives depends on the population?

b. Based on your answer to 5a, define variables and make a scatter plot of the data.

c. Find the equation of a line of fit. What is the real-world meaning of the slope? What is the real-world meaning of the y-intercept?

d. California has an estimated population of 33 million. Use your equation to estimate the number of members California has in the House of Representatives.

e. Minnesota has eight members in the House of Representatives. Use your equation to estimate the population of Minnesota.

f. You might find that a direct variation equation of the form $y = bx$ fits your data. Is this a reasonable model for the data? Explain why or why not.

The United States Constitution gives each state representation in the House of Representatives in proportion to its population. In the Senate, each state has equal representation regardless of size. This photo shows a joint session of both the House and Senate.

6. Use the table in problem 5 to answer these questions.

a. Does the population of a state affect its number of members in the Senate?

b. Write an equation that models the number of senators from each state. Graph this equation on the same coordinate axes as 5c.

c. Describe the graph and explain why it looks this way.

7. Suppose your friend walks steadily away from you at a constant rate so that her distance at 2 seconds is 3.4 meters and her distance at 4.5 seconds is 4.4 meters. Let x represent time in seconds, and let y represent distance in meters.

  a. What is the slope of the line that models this situation?

  b. What is the y-intercept of this line? Explain how you found it.

  c. Write a linear equation in intercept form that models your friend's walk.

8. Suppose this line represents a walking situation in which you're using a motion sensor to measure distance. The x-axis shows time and the y-axis shows distance from the sensor.

  a. Is the slope positive, negative, zero, or undefined? Explain your reasoning.

  b. What is a real-world meaning for the y-intercept? For the x-intercept?

  c. If the line extended into Quadrant II, what could that mean? If the line extended into Quadrant IV?

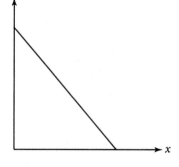

9. Find the equation of a line that

  a. Has a positive slope and a negative y-intercept.

  b. Has a negative slope and a y-intercept of zero.

  c. Passes through the points $(1, 7)$ and $(4, 10)$.

  d. Passes through the points $(-2, 10)$ and $(4, 10)$.

10. Each equation below represents a family of lines. Describe what the lines in each form have in common.

  a. $y = a + 3x$

  b. $y = 5 + bx$

  c. $y = a$

  d. $x = c$

Review

11. For each of these tables of x- and y-values, decide if the values indicate a direct variation, an inverse variation, or neither. Explain how you made your decision. If the values represent a direct or inverse variation, write an equation.

a.

x	y
-3	9
-1	1
-0.5	0.25
0.25	0.0625
7	49

b.

x	y
-20	-5
-8	-12.5
2	50
10	10
25	4

c.

x	y
0	0
-6	15
8	-20
-12	30
4	-10

d.

x	y
78	6
31.2	2.4
-145.6	-11.2
14.3	1.1
-44.2	-3.4

12. Show the steps to solve each equation. Then use your calculator to verify your solution.

    a. $8 - 12m = 17$ **b.** $2r + 7 = -24$ **c.** $-6 - 3w = 42$

13. Give the mean and median for each data set.

    a. {1, 2, 4, 7, 18, 20, 21, 21, 26, 31, 37, 45, 45, 47, 48}

    b. {30, 32, 33, 35, 39, 41, 42, 47, 72, 74}

    c. {107, 116, 120, 120, 138, 140, 145, 146, 147, 152, 155, 156, 179}

    d. {85, 91, 79, 86, 94, 90, 74, 87}

IMPROVING YOUR **VISUAL THINKING** SKILLS

The traditional Japanese abacus, or "soroban," is still widely used today. Each column shows a different place value—1, 10, 100, 1000, and so on. The four lower beads are moved up to represent the digits from 1 to 4. The fifth bead is moved down to show the digit 5. The digits 6 to 9 are shown with a combination of lower and upper beads. The first abacus below shows the number 6053.

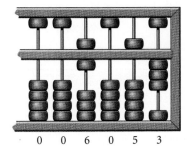

0 0 6 0 5 3

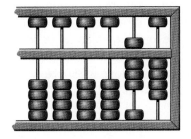

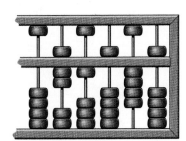

What numbers do the second and third abacuses show?

Sketch an abacus to show the number 27,059.

You can learn more about the abacus at www.keymath.com/DA .

Point-Slope Form of a Linear Equation

Success breeds confidence.

BERYL MARKHAM

So far you have worked with linear equations in intercept form, $y = a + bx$. When you know a line's slope and y-intercept, you can write its equation directly in intercept form. But what if you don't know the y-intercept? One method that you might remember from your homework is to work backward with the slope until you find the y-intercept. But you can also use the slope formula to find the equation of a line when you know the slope of the line and the coordinates of only one point on the line.

EXAMPLE

Since the time Beth was born, the population of her town has increased at a rate of approximately 850 people per year. On Beth's 9th birthday the total population was nearly 307,650. If this rate of growth continues, what will be the population on Beth's 16th birthday?

▶ Solution

Since the rate of change is approximately constant, a linear equation should model this population growth. Let x represent time in years since Beth's birth, and let y represent the population.

In the problem, you are given one point, (9, 307650). Any other point on the line will be in the form (x, y). So let (x, y) represent a second point on the line. You also know that the slope is 850. Now use the slope formula to find a linear equation.

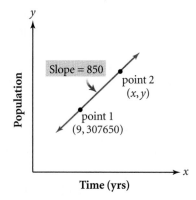

$$\frac{y_2 - y_1}{x_2 - x_1} = b \qquad \text{Slope formula.}$$

$$\frac{y - 307{,}650}{x - 9} = 850 \qquad \begin{array}{l}\text{Substitute the coordinates of the point (9, 307650)}\\ \text{for } (x_1, y_1), \text{ and the slope 850 for } b.\end{array}$$

Since we only know one point, we use (x, y) to represent any point.

$$(x - 9) \frac{y - 307{,}650}{(x - 9)} = 850(x - 9)$$ Multiply both sides by $(x - 9)$.

$$y - 307{,}650 = 850(x - 9)$$ Reduce the left side.

$$y - 307{,}650 + 307{,}650 = 307{,}650 + 850(x - 9)$$ Add 307,650 to both sides.

$$y = 307{,}650 + 850(x - 9)$$ Add.

The equation $y = 307{,}650 + 850(x - 9)$ is a linear equation that models the population growth. To find the population on Beth's 16th birthday, substitute 16 for x.

$$y = 307{,}650 + 850(x - 9)$$ Original equation.

$$y = 307{,}650 + 850(16 - 9)$$ Substitute 16 for x.

$$y = 313{,}600$$ Use order of operations.

The model equation predicts that the population on Beth's 16th birthday will be 313,600.

The equation $y = 307{,}650 + 850(x - 9)$ is a linear equation, but it is not in intercept form. This equation has its advantages too because you can clearly identify the slope and one point on the line. Do you see the slope of 850 and the point (9, 307650) within the equation? This form of a linear equation is appropriately called the **point-slope form.**

Point-Slope Form

If a line passes through the point (x_1, y_1) and has slope b, the **point-slope form** of the equation is

$$y = y_1 + b(x - x_1)$$

Investigation
The Point-Slope Form for Linear Equations

Silo and Jenny conducted an experiment in which Jenny walked at a constant rate. Unfortunately, Silo recorded only the data shown in this table.

Sorry Jenny, had to motor, but this should be enough info......
Peace, Silo

Elapsed time (sec) x	Distance to walker (m) y
3	4.6
6	2.8

Step 1 Find the slope of the line that represents this situation.

Step 2 Write a linear equation in point-slope form using the point (3, 4.6) and the slope you found in Step 1.

Step 3 Write another linear equation in point-slope form using the point (6, 2.8) and the slope you found in Step 1.

	Step 4	Enter the equation from Step 2 into Y₁ and the equation from Step 3 into Y₂ on your calculator, and graph both equations. What do you notice?
	Step 5	Look at a table of Y₁- and Y₂-values. What do you notice? What do you think the results mean?

Now that you have some practice at writing point-slope equations, try using a point-slope equation to fit data.

The table shows how the temperature of a pot of water changed over time as it was heated.

Water Temperature

Time (sec) x	Temperature (°C) y
24	25
36	30
49	35
62	40
76	45
89	50

Step 6	Define variables and plot the data on your calculator. Describe any patterns you notice.
Step 7	Choose a pair of points from the data. Find the slope of the line between your two points.
Step 8	Write an equation in point-slope form for a line that passes through your two points. Graph the line. Does your equation fit the data?

Step 9	Compare your graph to those of other members of your group. Does one graph show a line that is a better fit than the others?

If you look back at the investigation, you will notice that you found the point-slope form of a line even though you had only points to start with. This is possible because you can still use the point-slope form when you know two points on the line; there's just one additional step. What is it?

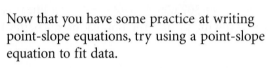

EXERCISES You will need your calculator for problems **3, 4, 5, 8, 9,** and **10.**

Practice Your Skills

1. Name the slope and one point on the line that each point-slope equation represents.

    a. $y = 3 + 4(x - 5)$ **b.** $y = 1.9 + 2(x + 3.1)$

    c. $y = -3.47(x - 7) - 2$ **d.** $y = 5 - 1.38(x - 2.5)$

2. Write an equation in point-slope form for a line, given its slope and one point that it passes through.

    a. Slope 3; point (2, 5)   **b.** Slope −5; point (1, −4)

3. A line passes through the points $(-2, -1)$ and $(5, 13)$.

   a. Find the slope of this line.

   b. Write an equation in point-slope form using the slope you found in 3a and the point $(-2, -1)$.

   c. Write an equation in point-slope form using the slope you found in 3a and the point $(5, 13)$.

   d. Verify that the equations in parts 3b and c are equivalent. Enter one equation into Y_1 and the other into Y_2 on your calculator, and compare their graphs and tables.

4. **APPLICATION** This table shows a linear relationship between actual temperature and approximate wind chill temperature when the wind speed is 20 miles per hour.

   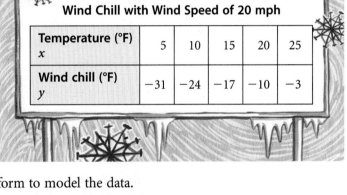

Temperature (°F) x	5	10	15	20	25
Wind chill (°F) y	-31	-24	-17	-10	-3

   Wind Chill with Wind Speed of 20 mph

   a. Find the rate of change of the data (the slope of the line).

   b. Choose one point and write an equation in point-slope form to model the data.

   c. Choose another point and write another equation in point-slope form to model the data.

   d. Verify that the two equations in 4b and c are equivalent. Enter one equation into Y_1 and the other into Y_2 on your calculator, and compare their graphs and tables.

   e. What is the wind chill temperature when the actual temperature is 0°F? What does this represent in the graph?

5. Play the BOWLING program at least four times. [▶ 🖳 See **Calculator Note 5A** for instructions on how to play the game. ◀] Each time you play, write down any equations you try and how many points you score.

Reason and Apply

6. The graph at right is made up of linear segments **a, b,** and **c.** Write an equation in point-slope form for the line that contains each segment.

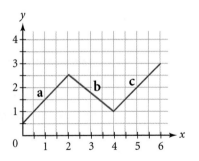

7. Look at quadrilateral $ABCD$.

   a. Write an equation in point-slope form for the line containing each segment in this quadrilateral. Check your equations by graphing them on your calculator.

   b. What is the same in the equations for the line through points A and D and the line through points B and C? What is different in these equations?

   c. What kind of figure does $ABCD$ appear to be? Do the results from 7b have anything to do with this?

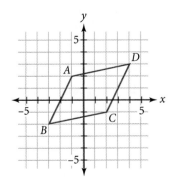

8. **APPLICATION** The table shows postal rates for first-class U.S. mail in the year 2001.

a. Make a scatter plot of the data. Describe any patterns you notice.

b. Find the slope of the line between any two points in the data. What is the real-world meaning of this slope?

c. Write a linear equation in point-slope form that models the data. Graph the equation to check that it fits your data points.

d. Use the equation you wrote in 8c to find the cost of mailing a 10-oz letter.

e. What would be the cost of mailing a 3.5-oz letter? A 9.1-oz letter? (Hint: Think about what the column header for the *x*-values means.)

f. The equation you found in 8c is useful for modeling this situation. Is the graph of this equation, a continuous line, a correct model for the situation? Explain why or why not.

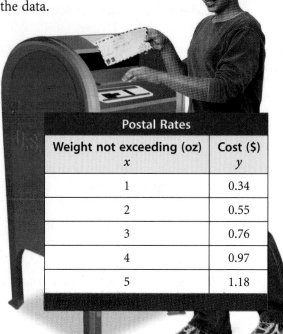

Postal Rates

Weight not exceeding (oz) x	Cost ($) y
1	0.34
2	0.55
3	0.76
4	0.97
5	1.18

9. **APPLICATION** The table below shows total fat grams and number of calories for some breakfast sandwiches sold by chain restaurants.

Nutrition Facts

Breakfast sandwich	Total fat (g) x	Calories y
Burger King Sausage, Egg, and Cheese Biscuit	43	620
Carl's Jr. Sunrise Sandwich	21	360
Hardee's Steak Biscuit	32	580
Jack in the Box Sourdough Breakfast Sandwich	24	450
McDonald's Sausage McMuffin with Egg	28	440
Subway Ham and Egg Breakfast Deli Sandwich	12	312
Taco Bell Country Breakfast Burrito	14	270
White Castle Sausage, Egg, and Cheese Breakfast Sandwich	25	340

(www.kenkuhl.com)

a. Make a scatter plot of the data. Describe any patterns you notice.

b. Select two points and find the equation of the line that passes through these two points in point-slope form. Graph the equation on the scatter plot.

c. According to your model, how many calories would you expect in a Hardee's Steak Biscuit with 32 grams of fat?

d. Does the actual data point representing the Hardee's Steak Biscuit lie above, on, or below the line you graphed in 9b? Explain what the point's location means.

e. Check each breakfast sandwich to find if its data point falls above, on, or below your line.

f. Based on your results for 9d–e, how well does your line fit the data?

g. If a sandwich has 0 grams of fat, how many calories does your equation predict? Does this answer make sense? Why or why not?

10. **APPLICATION** This table shows the amount of trash produced in the United States in 1980 and 1985.

   a. Let x represent the year, and let y represent the amount of trash in millions of tons for that year. Write an equation in point-slope form for the line passing through these two points.

   b. Plot the two data points and graph the equation you found in 10a.

   c. In 1990, 196 million tons of trash were produced in the United States. Plot this data point on the same graph you made in 10b. Do you think the linear equation you found in 10a is a good model for these data? Explain why or why not.

U.S. Trash Production

Year	Amount of trash (million tons)
1980	152
1985	164

(*The Universal Almanac*, 1995)

This table shows more data about the amount of trash produced in the United States.

   d. Add these data points to your graph. Adjust the window as necessary.

   e. Do you think the linear equation found in 10a is a good model for this larger data set? Explain why or why not.

   f. Find the equation of a better fitting line. You may find that you only need to adjust the slope value.

   g. Use your new equation from 10f to predict the amount of trash produced in 2000.

U.S. Trash Production

Year	Amount of trash (million tons)
1960	88
1965	103
1970	122
1975	128

(*The Universal Almanac*, 1995)

Review

11. **APPLICATION** The volume of a gas is directly proportional to its temperature in kelvins (K). The volume of this gas is 3.50 liters at 280 K.

   a. Find the volume of this gas when the temperature is 330 K.

   b. Find the temperature when the volume is 2.25 liters.

12. Insert operation signs, parentheses, or both into each string of numbers to create an expression equal to the answer given.

   a. 1   2   3   4   5   = 1   **b.** 1   2   3   4   5   = 3

   c. 1   2   3   4   5   = 5

13. Find the slope of the line through the first two points given. Assume the third point is also on the line and find the missing coordinate.

   a. $(-1, 5)$ and $(3, 1)$; $(5, \boxed{})$ **b.** $(2, -5)$ and $(2, -2)$; $(\boxed{}, 3)$

   c. $(-10, 22)$ and $(-2, 2)$; $(\boxed{}, -3)$

Equivalent Algebraic Equations

For the same line, there can be more than one equation in point-slope form. For example, you can write the equation for the line through the points (1, 2) and (2, 7) as $y = 2 + 5(x - 1)$ or $y = 7 + 5(x - 2)$ depending on which point you use. Actually, every line contains infinitely many points, so there are an infinite number of ways to write the equation in point-slope form. You can also write the equation of any line in intercept form, which looks different too. How can you tell when two equations actually represent the same line?

In this lesson you'll learn how to recognize equivalent equations by using mathematical properties and the rules for order of operations.

These self-portraits of the American pop artist Andy Warhol (1928–1987) are like equivalent equations. Each screen-printed image is the same as the next but Warhol's choice of colorization makes each look different.

Investigation
Equivalent Equations

Here are six different-looking equations in point-slope form.

a. $y = 3 - 2(x - 1)$ **d.** $y = 0 - 2(x - 2.5)$

b. $y = -5 - 2(x - 5)$ **e.** $y = 7 - 2(x + 1)$

c. $y = 9 - 2 (x + 2)$ **f.** $y = -9 - 2(x - 7)$

Step 1	Do the six equations represent the same line or different lines? Explain your reasoning.
Step 2	Divide these equations among the members of your group. Use the distributive property to rewrite the right side of each equation. You should get an equation in intercept form.
Step 3	Enter your point-slope equation into Y_1, and enter your intercept equation into Y_2. Check that the two equations have the same calculator graph or table. How does this show that the equations are equivalent?
Step 4	Now, as a group, compare your intercept equations. What do the results mean about the six equations?

Step 5 | As a group, explain how you can tell that an equation in point-slope form is equivalent to one in intercept form. Think about how you can do this graphically and symbolically.

Here are fifteen equations. They represent only four different lines.

a. $y = 2(x - 2.5)$

b. $y = 18 + 2(x - 8)$

c. $y = 52 - 6(x + 8)$

d. $y = -6 + 2(x + 4)$

e. $y = 21 - 6(x + 4)$

f. $y = -14 - 6(x - 3)$

g. $y = -10 + 2(x + 6)$

h. $6x + y = 4$

i. $y = 11 + 2(x - 8)$

j. $12x + 2y = -6$

k. $y = 2(x - 4) + 10$

l. $y = 15 - 2(10 - x)$

m. $y = 7 + 2(x - 6)$

n. $y = -6(x + 0.5)$

o. $y = -6(x + 2) + 16$

Step 6 | Test your answer to Step 5 by finding the intercept form of each equation and then grouping equivalent equations.

Step 7 | As a group, explain how you can tell that two equations in point-slope form are equivalent.

You have learned how to write linear equations in two different forms:

Intercept form $\qquad y = a + bx$

Point-slope form $\qquad y = y_1 + b(x - x_1)$

In the second part of the investigation, some of the equations had x and y on the same side, as in $12x + 2y = -6$. Equations like these are in **standard form.** Can you identify another equation in the investigation that is in standard form?

No matter what form you start with, you can always rewrite any linear equation in intercept form. Then it's easy to recognize equivalent equations. Let's review properties that help you change the form of an equation.

For any values of a, b, and c, these properties are true:

Distributive Property

$$a(b + c) = a(b) + a(c)$$ Example: $6(-2 + 3) = 6(-2) + 6(3)$

Commutative Property of Addition

$$a + b = b + a$$ Example: $3 + 4 = 4 + 3$

Commutative Property of Multiplication

$$ab = ba$$ Example: $\frac{1}{2} \cdot \frac{3}{4} = \frac{3}{4} \cdot \frac{1}{2}$

Associative Property of Addition

$$a + (b + c) = (a + b) + c$$ Example: $2 + (1.5 + 3) = (2 + 1.5) + 3$

Associative Property of Multiplication

$$a(bc) = (ab)c$$ Example: $4\left(\frac{1}{3} \cdot 6.3\right) = \left(4 \cdot \frac{1}{3}\right)6.3$

There are also the properties that you have used to solve equations by balancing.

Properties of Equality

Given $a = b$, for any number c,

$a + c = b + c$	addition property of equality
$a - c = b - c$	subtraction property of equality
$ac = bc$	multiplication property of equality
$\dfrac{a}{c} = \dfrac{b}{c} \; (c \neq 0)$	division property of equality

EXAMPLE A | Is the equation $y = 2 + 3(x - 1)$ equivalent to $6x - 2y = 2$?

▶ Solution | Use the properties to rewrite each equation in intercept form.

$y = 2 + 3(x - 1)$ Original equation.

$y = 2 + 3x - 3$ Distributive property (distribute 3 over $x - 1$).

$y = 2 - 3 + 3x$ Commutative property (swap $3x$ and -3).

$y = -1 + 3x$ Subtract $2 - 3$.

So the intercept form of the first equation is $y = -1 + 3x$.

$6x - 2y = 2$ Original equation.

$-2y = 2 - 6x$ Subtraction property (subtract $6x$ from both sides).

$y = -1 + 3x$ Division property (divide both sides by -2).

The intercept form of the second equation is also $y = -1 + 3x$. So both are equivalent. You can also check that the intercept form and the point-slope form of the equation are equivalent by verifying that they produce the same line graph and have the same table of values. Unfortunately, you cannot enter the standard form into your calculator.

One of the authors, Jerald Murdock, works with two students.

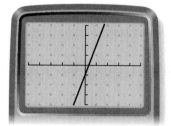

EXAMPLE B

Solve the equation $\frac{3x + 4}{6} - 5 = 7$. Identify the property of equality used in each step.

▶ **Solution**

$$\frac{3x + 4}{6} - 5 = 7$$ Original equation.

$$\frac{3x + 4}{6} = 12$$ Addition property (add 5 to both sides).

$$3x + 4 = 72$$ Multiplication property (multiply both sides by 6).

$$3x = 68$$ Subtraction property (subtract 4 from both sides).

$$x = 22\frac{2}{3}$$ Division property (divide both sides by 3).

EXERCISES

You will need your calculator for problems **1, 2, 9,** and **13.**

▶ **Practice Your Skills**

1. Is each pair of expressions equivalent? If they are not, change the second expression so that they are equivalent. Check your work on your calculator by comparing table values when you enter the equivalent expressions into Y_1 and Y_2.

   a. $3 - 3(x + 4)$ $3x - 9$

   b. $5 + 2(x - 2)$ $2x + 1$

   c. $5x - 3$ $2 + 5(x - 1)$

   d. $-2x - 8$ $-2(x - 4)$

2. Rewrite each equation in intercept form. Show your steps. Check your answer by using a calculator graph or table.

a. $y = 14 + 3(x - 5)$

b. $y = -5 - 2(x + 5)$

c. $6x + 2y = 24$

3. Solve each equation by balancing and tell which property you used for each step.

a. $3x = 12$

b. $-x - 45 = 47$

c. $x + 15 = 8$

d. $\frac{x}{4} = 28$

4. Solve each equation for x. Substitute your value into the original equation to check.

a. $35 = 3(x + 8)$

b. $\frac{15 - 3}{x - 4} = 10$

c. $4(2x - 5) - 12 = 16$

5. An equation of a line is $y = 25 - 2(x + 5)$.

a. Name the point used to write the point-slope equation.

b. Find x when y is 15.

Reason and Apply

6. In the expression $3x + 15$, the *greatest common factor* (GCF) of both $3x$ and 15 is 3. You can write the expression $3x + 15$ as $3(x + 5)$. This process, called **factoring,** is the reverse of distributing. Rewrite each expression by factoring out the GCF that will leave 1 as the coefficient of x. Use the distributive property to check your work.

a. $3x - 12$

b. $-5x + 20$

c. $32 + 4x$

d. $-7x - 28$

7. Consider the equation $y = 10 + 5x$ in intercept form.

a. Factor the right side of the equation.

b. Use the commutative property of addition to swap the addends inside the parentheses.

c. Your result should look similar to the point-slope form of the equation. What's missing? What is the value of this missing piece?

d. What point could you use to write the point-slope equation in 7c? What is special about this point?

8. In each set of three equations, two equations are equivalent. Find them and explain how you know they are equivalent.

a.
   i. $y = 14 - 2(x - 5)$
   ii. $y = 30 - 2(x + 3)$
   iii. $y = -12 + 2(x - 5)$

b.
   i. $y = -13 + 4(x + 2)$
   ii. $y = 10 + 3(x - 5)$
   iii. $y = -25 + 4(x + 5)$

c.
   i. $y = 5 + 5(x - 8)$
   ii. $y = 9 + 5(x + 8)$
   iii. $y = 94 + 5(x - 9)$

d.
   i. $y = -16 + 6(x + 5)$
   ii. $y = 8 + 6(x - 5)$
   iii. $y = 44 + 6(x - 5)$

9. The equation $3x + 2y = 6$ is in standard form.

   a. Find x when y is zero. Write your answer in the form (x, y). What is the significance of this point?

   b. Find y when x is zero. Write your answer in the form (x, y). What is the significance of this point?

   c. On graph paper, plot the points you found in 9a and b and draw the line through these points.

   d. Find the slope of the line you drew in 9c and write a linear equation in intercept form.

   e. On your calculator, graph the equation you wrote in 9d. Compare this graph to the one you drew on paper. Is the intercept equation equivalent to the standard-form equation? Explain why or why not.

   f. Symbolically show that the equation $3x + 2y = 6$ is equivalent to your equation from 9d.

10. A line has the equation $y = 4 - 4.2x$.

   a. Find the y-coordinate of the point on this line whose x-coordinate is 2.

   b. Use the point you found in 10a to write an equation in point-slope form.

   c. Find the x-coordinate of the point whose y-coordinate is 6.1.

   d. Use the point you found in 10c to write a different point-slope equation.

   e. Show that the point-slope equations you wrote in 10b and d are equivalent to the original equation in intercept form. Explain your procedure.

11. APPLICATION Dorine subscribes to an Internet service with a flat rate per month for up to 15 hours of use. For each hour over this limit, there is an additional per-hour fee. The table shows data about Dorine's first two bills.

Internet Use

Month	Logged on (hr)	Monthly fee ($)
January	20	15.20
February	23	17.75

   a. Define your variables and use the data in the table to write an equation in point-slope form that models Dorine's total fee.

   b. During March, Dorine was incorrectly charged $20 for being logged on for 25 hours. What should be her correct total fee?

   c. In April, Dorine was logged on for 14 hours. What was her total fee that month? Explain why you can't use your equation to answer this question. (Hint: Reread the problem carefully.)

   d. How many hours was Dorine logged on during a month when her fee was $23.70?

12. On Saturday morning, Avery took a hike in the hills near her house. The table shows the cumulative number of calories she burned from the time she went to sleep Friday night until she finished her hike.

  a. Write a point-slope equation of a line that fits the data.

  b. Rewrite your equation from 12a in intercept form.

  c. What are the real-world meanings of the slope and the y-intercept in this situation?

  d. Could you use the point-slope equation $y = 821 + 4.6(x - 60)$ to model this situation? Explain why or why not.

  e. What is the real-world meaning of the point used to write the equation in 12d?

Avery's Hike

Time spent hiking (min)	Cumulative number of calories burned
5	568
10	591
15	614
20	637

▶ Review

13. The table shows hourly compensation costs in 15 countries for 1975, 1985, and 1995. Use the list commands on your calculator to do this statistical analysis.

  a. Choose at least three countries and graph the hourly compensation costs for those countries over time. Write a paragraph describing the trends you notice and the conclusions you draw.

  b. Which of the 15 countries had the largest increase in compensation costs from 1975 to 1995? Which country had the least?

  c. Create three box plots that compare the compensation costs for the three years. Write a brief paragraph analyzing your graph.

Hourly Compensation Costs (in U.S. dollars) for Production Workers

Country	1975	1985	1995
Australia	5.62	8.20	15.05
Canada	5.96	10.94	16.04
Denmark	6.28	8.13	24.07
France	4.52	7.52	20.01
Germany	6.31	9.53	32.22
Hong Kong	0.76	1.73	4.82
Israel	2.25	4.06	10.54
Italy	4.67	7.63	16.21
Japan	3.00	6.34	23.82
Luxembourg	6.50	7.81	23.35
Mexico	1.47	1.59	1.51
Spain	2.53	4.66	12.88
Sri Lanka	0.28	0.28	0.48
Taiwan	0.40	1.50	5.92
United States	6.36	13.01	17.19

(*2000 New York Times Almanac*, p. 515)

The production workers are inspecting automobile bodies at an American factory.

14. Plot the points $(4, 2)$, $(1, 3.5)$, and $(10, -1)$ on graph paper. These points are on the same line, or collinear, so you can draw a line through them.

   a. Draw a slope triangle between $(4, 2)$ and $(1, 3.5)$, and calculate the slope from the change in y and the change in x.

   b. Draw another slope triangle between $(10, -1)$ and $(4, 2)$, and calculate the slope from the change in y and the change in x.

   c. Compare the slope triangles and the slopes you calculated. What do you notice?

   d. What would happen if you made a slope triangle between $(10, -1)$ and $(1, 3.5)$?

15. Show how to solve the equation $3.8 = 0.2(z + 6.2) - 5.4$ by using an undoing process to write an expression for z. Check your answer by substituting it into the original equation.

IMPROVING YOUR **GEOMETRY** SKILLS

Think about triangles drawn on the coordinate plane.

Draw a triangle that satisfies each of these sets of conditions. If it's not possible, tell why not.

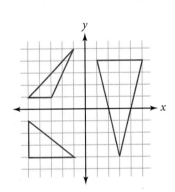

  1. a triangle with all three sides having positive slope

  2. a right triangle (one angle is 90°) with all three sides having negative slope

  3. an equilateral triangle (three equal sides) with one side having slope 0

  4. an obtuse triangle (one angle is greater than 90°) with one side having positive slope and the other two having negative slope

  5. an isosceles triangle (two equal sides) with all three sides having positive slope

  6. a right triangle with one side having undefined slope, one side having slope 0, and one side having slope 1

  7. a triangle with two sides having the same slope

LESSON
5.5

Writing Point-Slope Equations to Fit Data

To give an accurate description of what has never occurred is the proper occupation of the historian.

OSCAR WILDE

In this lesson you'll practice modeling data that has a linear pattern with the point-slope form of a linear equation. You may find that using the point-slope form is more efficient than using the intercept form because you don't have to first write a direct variation equation and then adjust it for the intercept.

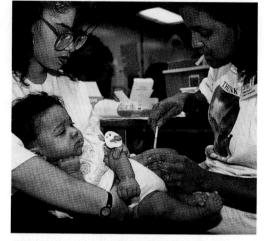

The development and improvement of vaccinations is one factor that has increased life expectancy over the decades.

Investigation
Life Expectancy

You will need

• graph paper

This table shows the relationship between the number of years a person might be expected to live and the year he or she was born. Life expectancy is a prediction that is very useful in professions like medicine and insurance.

Step 1	Choose one column of life expectancy data—female, male, or combined. Let *x* represent birth year, and let *y* represent life expectancy in years. Graph the data points.
Step 2	Choose two points on your graph so that a line through them closely reflects the pattern of all the points on the graph. Use the two points to write the equation of this line in point-slope form.
Step 3	Graph the line with your data points. Does it fit the data?
Step 4	Use your equation to predict the life expectancy of a person who will be born in 2022.

U.S. Life Expectancy at Birth

Birth year	Female	Male	Combined
1940	65.2	60.8	62.9
1950	71.1	65.6	68.2
1960	73.1	66.6	69.7
1970	74.7	67.1	70.8
1975	76.6	68.8	72.6
1980	77.5	70.0	73.7
1985	78.2	71.2	74.7
1990	78.8	71.8	75.4
1995	78.9	72.5	75.8
1998	79.4	73.9	76.7

(*2000 World Almanac*, p. 891)

Step 5	Compare your prediction from Step 4 to the prediction that another group made analyzing the same data. Are your predictions the same? Are they close? Explain why it's possible to make different predictions from the same data.

Step 6	Compare the slope of your line of fit to the slopes that other groups found working with different data sets. What does the slope for each data set tell you?
Step 7	As a class, select one line of fit that you think is the best model for each column of data—female, male, and combined. Graph all three lines on the same set of axes. Is it reasonable for the line representing the combined data to lie between the other two lines? Explain why or why not.
Step 8	How does the point-slope method of finding a line compare to the intercept-form method you learned about in Lesson 5.2? What are the strengths and weaknesses of each method?

You can summarize the point-slope method of fitting a line to the data like this: First, graph the data. Next, choose two points on a line that appears to show the direction of the data. Finally, write the equation of the line.

If you choose the two points well, you can write an equation that's a good model for the data. Still, two people choosing different points will get slightly different models from which they will make slightly different predictions. In the next lesson you'll learn a third method that will overcome these differences.

Each student will have a different impression of the artwork in this museum. Similarly, different people can have different impressions of a set of data; this can result in different mathematical models.

EXERCISES

You will need your calculator for problem **3, 4, 5, 6,** and **7.**

▶ Practice Your Skills

1. Write the point-slope form of the equation for each line graphed below.

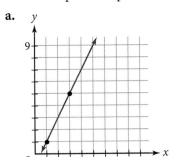 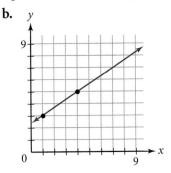 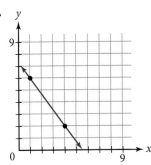

2. Look at each graph in problem 1 and estimate the y-intercept. Then convert your point-slope equations to intercept form. How well did you estimate?

3. Graph each linear equation on your calculator and name the x-intercept. Make an observation about the x-intercept of any equation in the form $y = b(x - x_1)$.

a. $y = 2(x - 3)$ **b.** $y = \frac{1}{3}(x + 4)$ **c.** $y = -1.5(x - 6)$

4. APPLICATION Carbon dioxide is one of several greenhouse gases that is emitted into the atmosphere from a variety of sources, including automobiles. The table shows the concentration of carbon dioxide (CO_2) in the atmosphere measured from the top of Mauna Loa volcano in Hawaii each January. The concentration of CO_2 is measured in parts per million (ppm).

CO$_2$ Concentration

Year	CO$_2$ (ppm)
1976	332
1978	335
1980	338
1982	341
1984	344
1986	346
1988	350
1990	354
1992	356
1994	358
1996	362

a. Define variables and write an equation in point-slope form that models the data.

b. Graph your equation to confirm that the line fits the data.

c. Use your equation to predict what the concentration of CO_2 will be in 2020.

d. What would be the x-intercept for your equation? Does its real-world meaning make sense? Explain why or why not.

Mauna Loa is the largest and most active volcano on Earth. Research on Mauna Loa has revealed a great deal about global changes in the atmosphere.

(Carbon Dioxide Information and Analysis Center)

Reason and Apply

5. APPLICATION Alex collected this table of data by using two thermometers simultaneously. Alex suspects that one or both of the thermometers are somewhat faulty.

Temperature Readings

Celsius (°C) x	Fahrenheit (°F) y
14.5	55.0
20.0	67.0
28.4	86.7
39.5	105.6
32.3	87.1
29.0	81.6
26.2	82.3
25.7	75.2
31.2	88.6

a. Graph the data.

b. Write an equation in point-slope form that models Alex's data.

c. Graph your equation to confirm that the line fits the data.

d. The freezing point of water is 0°C, which is equivalent to 32°F. The boiling point of water is 100°C, which is equivalent to 212°F. Use this information to write another equation in point-slope form that models the true relationship between the Celsius and Fahrenheit temperature scales.

e. Write the equations from 5b and d in intercept form. Are they equivalent?

f. Do you think that Alex's thermometers are faulty? Explain why or why not.

6. **APPLICATION** The table lists the concentration of dissolved oxygen (DO) in parts per million at various temperatures in degrees Celsius from a sample of lake water.

   a. Graph the data.

   b. Write an equation in point-slope form that models the data.

   c. Graph your equation to confirm that the line fits the data.

   d. Use your equation to predict the concentration of dissolved oxygen in parts per million when the water temperature is 2°C.

   e. Use your equation to predict the water temperature in degrees Celsius when the concentration of dissolved oxygen is 11 ppm.

Dissolved Oxygen

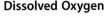

Temperature (°C) x	DO (ppm) y
17	8
15	9
13	11
16	10
11	14
13	11
10	14
8	14
6	16
7	13
8	14
4	17
5	15
9	13
6	16

Review

7. Rewrite each equation in intercept form. Show the steps to make the conversion. Check your answer with a calculator table or graph.

   a. $y = 3 + 2(x - 7)$

   b. $y = -11 + 41x + 28$

   c. $y = 5 - 6(x - 9)$

   d. $y = 4(7 - x) - 19$

8. Give the five-number summary for each data set.

   a. {1, 2, 4, 7, 18, 20, 21, 21, 26, 31, 37, 45, 45, 47, 48}

   b. {30, 32, 33, 35, 39, 41, 42, 47, 72, 74}

   c. {107, 116, 120, 120, 138, 140, 145, 146, 147, 152, 155, 156, 179}

   d. {85, 91, 79, 86, 94, 90, 74, 87}

9. **APPLICATION** Bryan has bought a box of biscuits for his dog, Anchor. Anchor always gets three biscuits a day. At the start of the 10th day after opening the box, Bryan counts 106 biscuits left. Let x represent the number of days after opening the box, and let y represent the number of biscuits left.

   a. In a graph of this situation, what is the slope?

   b. Write a point-slope equation that models the situation.

   c. When will the box be empty?

   d. What is the real-world meaning of the y-intercept?

10. Consider this expression:

    $$9\left(\frac{x - 11}{9} + 1\right) + 2$$

    a. Write in words the number trick the expression describes.

    b. Test the trick to be sure it works. Do you get the same result no matter what number you start with?

    c. What operation(s) make this trick work, and how?

More on Modeling

Several times in this chapter you have found the equation of a representative line to fit data. Making, analyzing, and using predictions based on equation models is important in the real world. For this reason it is often helpful and even important that different people arrive at the same model for a given set of data. For this to happen, each person must get the same slope and *y*-intercept. To do that, they have to follow the same systematic method.

When you can measure what you are talking about and express it in numbers, you know something about it.

LORD KELVIN

Investigation
Fire!!!!

In this investigation you will use a systematic method for finding a particular line of fit for data.

You will need

- a stopwatch
- a bucket
- graph paper

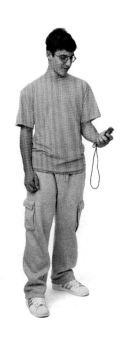

Step 1 | Line up in a bucket brigade. Record the number of people in the line. (See the procedure note.) Starting at one end of the line, pass the bucket as quickly as you can to the other end. Record the total passing time from picking up the bucket to setting it down at the very end.

Step 2 | Now have one or two people sit down and close up the gaps in the line. Repeat the bucket passing. Record the new number of people and the new passing time.

Step 3 | Continue the bucket brigade until you have collected 10 data points in the form (*number of people, passing time in seconds*).

Step 4	Let x represent the number of people, and let y represent time in seconds. Plot your data on graph paper.
Step 5	List the five-number summary for the x-values and the five-number summary for the y-values.
Step 6	What are the first-quartile (Q1) and third-quartile (Q3) values for the x-values in your data set? What are the Q1- and Q3-values for the y-values in your data set?
Step 7	On your graph, draw a horizontal box plot just below the x-axis using the five-number summary for the x-values. Draw a vertical box plot next to the y-axis using the five-number summary for the y-values. A sample graph is shown. Your data and graph will look different based on the data that you collect.
Step 8	Draw vertical lines from the Q1- and Q3-values on the x-axis box plot into the graph. Draw horizontal lines from the Q1- and Q3-values on the y-axis box plot into the graph. These lines should form a rectangle in the plot. Suppose we call the vertices of this rectangle **Q-points.** Do the Q-points have to be actual data points? Why or why not? Will everyone get the same Q-points?
Step 9	Draw the diagonal of this rectangle that shows the direction of the data. Extend this diagonal through the plot. Is the line a good fit for the data? Are any of the original data points on your line? If so, which ones?
Step 10	Find the coordinates of the two Q-points the line goes through and write a point-slope equation for the line.
Step 11	What are the real-world meanings of the slope and y-intercept of this model?
Step 12	What are the advantages and disadvantages of having a systematic procedure for finding a model for data?

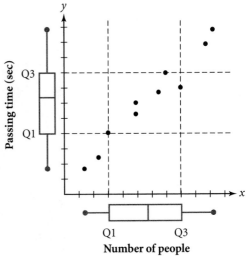

Step 13	Use your calculator to plot the data points, draw the vertical and horizontal lines, and find a line of fit by this method. [▶ ▤ See **Calculator Note 5B** for help on using the draw menu. ◀]

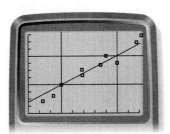

This method of finding the line of fit based on Q-points is more direct than the methods you used in Lessons 5.2 and 5.5. It is more systematic, too, because everyone will get the same points and the points themselves relate to measures of center in the upper and lower halves of the data set. Points that have some distance between them, but are not at the extremes of the data, are probably more reliable for locating the line of fit.

These students are collecting water samples. Their samples can be analyzed for many things, including dissolved oxygen.

EXAMPLE

The table lists the concentration of dissolved oxygen (DO) in parts per million at various temperatures in degrees Celsius from a sample of lake water. Find a line of fit based on Q-points for the data and use it to predict the temperature for water with only 4 ppm dissolved oxygen.

Dissolved Oxygen

Temperature (°C) x	DO (ppm) y	Temperature (°C) x	DO (ppm) y
17	8	8	14
16	10	8	14
15	9	7	13
13	11	6	16
13	11	6	16
11	14	5	15
10	14	4	17
9	13		

▶ **Solution**

The five-number summaries are

For temperature (x-values): 4, 6, 9, 13, 17

For dissolved oxygen (y-values): 8, 11, 14, 15, 17

The first-quartile and third-quartile values are

> For the x-values: Q1 = 6, Q3 = 13
> For the y-values: Q1 = 11, Q3 = 15

A sketch of the scatter plot shows the appropriate Q-points are (6, 15) and (13, 11). Note that (6, 15) is not actually one of the data points but (13, 11) is.

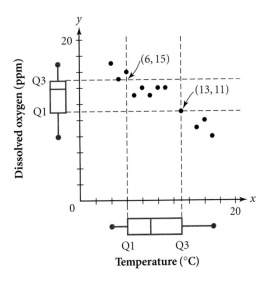

Calculating the slope between these two points you get

$$b = \frac{y_2 - y_1}{x_2 - x_1} = \frac{(11 - 15)}{(13 - 6)} = \frac{-4}{7} \approx -0.57$$

This means that if the temperature *rises* 1°C, the dissolved oxygen level *decreases* by 0.57 ppm. It also means that if the temperature *drops* 1°C, the dissolved oxygen level *increases* by 0.57 ppm.

Using the slope −0.57 and the coordinates of the point (6, 15) in the point-slope form gives

$$y = y_1 + (x - x_1)$$
$$y = 15 - 0.57(x - 6)$$

To find the temperature when the amount of dissolved oxygen is 4 ppm, substitute 4 for y in the equation and solve for x.

$y = 15 - 0.57(x - 6)$	Original equation.
$4 = 15 - 0.57(x - 6)$	Substitute 4 for y.
$-11 = -0.57(x - 6)$	Subtract 15 from both sides.
$19.3 \approx (x - 6)$	Divide both sides by –0.57.
$25.3 \approx x$	Add 6 to both sides.

At about 25°C, the water will have about 4 ppm dissolved oxygen.

If you go on to study statistics, you'll learn other systematic ways to find a line of fit for a data set, as well as how to find curves to model nonlinear patterns in data.

Practice Your Skills

1. **APPLICATION** This table shows that the traveling distances between some cities depend on how you travel.

Traveling Distances

From	To	Flying distance (mi)	Driving distance (mi)
Detroit, MI	Memphis, TN	623	756
St. Louis, MO	Minneapolis, MN	466	559
Dallas, TX	San Francisco, CA	1483	1765
Seattle, WA	Los Angeles, CA	959	1150
Washington, DC	Pittsburgh, PA	192	241
Philadelphia, PA	Indianapolis, IN	585	647
New Orleans, LA	Chicago, IL	833	947
Cleveland, OH	New York, NY	405	514
Birmingham, AL	Boston, MA	1052	1194
Denver, CO	Buffalo, NY	1370	1991
Kansas City, MO	Omaha, NE	166	204

a. What are the five-number summary values of the flying distances?

b. What are the five-number summary values of the driving distances?

c. Plot the data points. Let *x* represent flying distance in miles, and let *y* represent driving distance in miles.

d. Will the slope of the line through these points be positive or negative? Explain your reasoning.

e. Use the five-number summary values to draw a rectangle on the graph of the data. Name the two Q-points you should use for your line of fit.

f. Find the equation of the line and graph the line with your data points.

g. The flying distance from Louisville, Kentucky, to Miami, Florida, is 919 miles. Predict the driving distance from Louisville to Miami.

h. The driving distance from Phoenix, Arizona, to Salt Lake City, Utah, is 651 miles. Predict the flying distance from Phoenix to Salt Lake City.

2. APPLICATION Let x represent total fat in grams, and let y represent saturated fat in grams. Use the model $y = 10 + 0.5(x - 28)$ to predict

   a. The number of saturated fat grams for a hamburger with a total of 32 grams of fat.

   b. The total number of fat grams for a hamburger with 15 grams of saturated fat.

3. In a few sentences, describe the differences in the procedures for finding the Q-points for these two data sets.

   a.

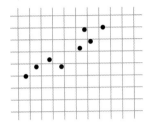

   b.

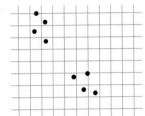

Reason and Apply

4. The table gives the winning times for the Olympic men's 400-meter dash.

   a. Define variables and find the line of fit based on Q-points for the data.

   b. Plot the data points and graph the equation of the model to verify that it is a good fit.

   c. What is the real-world meaning of the slope?

   d. Michael Johnson also won the 400-meter dash in the 2000 Olympic Games. Compare his actual winning time of 43.8 seconds with the winning time predicted by your model.

   e. Could you use this model to predict the winning time 100 years from now? Explain why or why not.

Men's 400-meter Dash

Year	Champion	Country	Time (sec)
1952	George Rhoden	Jamaica	45.9
1956	Charles Jenkins	United States	46.7
1960	Otis Davis	United States	44.9
1964	Mike Larrabee	United States	45.1
1968	Lee Evans	United States	43.9
1972	Vincent Mathews	United States	44.7
1976	Alberto Juantorena	Cuba	44.3
1980	Viktor Markin	USSR	44.6
1984	Alonzo Babers	United States	44.3
1988	Steve Lewis	United States	43.9
1992	Quincy Watts	United States	43.5
1996	Michael Johnson	United States	43.5

(2000 World Almanac, p. 917)

5. Create a data set that has Q-points at $(4, 28)$ and $(12, 47)$ so that only one of those two points is actually part of the data set.

6. Which linear equation below best fits the data at right? Explain your reasoning.

   i. $y = 1.3 + 0.18(x - 6)$

   ii. $y = 2.2 + 0.18(x - 6)$

   iii. $y = 1.3 - 0.18(x - 6)$

   iv. $y = 2.2 - 0.18(x - 6)$

Time (sec) x	Distance from motion sensor (m) y
2	2.8
6	2.2
8	1.7
9	1.5
11	1.3
14	0.9

7. At 2:00 P.M., elevator A passes the second floor of the Empire State Building going up. The table shows the floors and the times in seconds after 2:00.

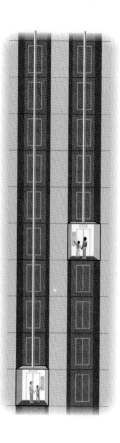

Floor x	2	4	6	8	10	12	14
Time after 2:00 (sec) y	0	1.3	2.5	3.8	5	6.3	7.5

  a. What is the line of fit based on Q-points for the data?

  b. Give a real-world meaning for the slope.

  c. About what time will this elevator pass the 60th floor if it makes no stops?

  d. Where will this elevator be at 2:00:45 if it makes no stops?

8. At 2:00 P.M., elevator B passes the 94th floor of the same building going down. The table shows the floors and the times in seconds after 2:00.

Floor x	94	92	90	88	86	84	80
Time after 2:00 (sec) y	0	1.3	2.5	3.8	5	6.3	8.6

  a. What is the line of fit based on Q-points for the data?

  b. Give a real-world meaning for the slope.

  c. About what time will this elevator pass the 10th floor if it makes no stops?

  d. Where will this elevator be at 2:00:34 if it makes no stops?

9. Think about the elevators in problems 7 and 8.

  a. Estimate when you expect that elevator A will pass elevator B if neither makes any stops.

  b. Calculate the actual time.

▶ **Review**

10. A 4-oz bottle of mustard costs $0.88, a 7.5-oz bottle costs $1.65, and an 18-oz bottle costs $3.99. Is the size of the mustard bottle directly proportional to the price? If so, show how you know. If not, suggest the change of one or two prices so that they will be directly proportional.

11. A car pulling a camper trailer is traveling from Sioux Falls, South Dakota, to Mt. Rushmore, which is near Rapid City, South Dakota. The car is traveling about 54 miles per hour, and it is about 370 miles from Sioux Falls to Mt. Rushmore.

  a. Write a recursive routine to create a table of values in the form

    (*time, distance from Mt. Rushmore*)

  for the relationship from 0 to 6 hours.

  b. Graph a scatter plot of your values using 1-hour time intervals.

  c. Draw a line through the points of your scatter plot. What is the real-world meaning of this line? What does the line represent that the points alone do not?

  d. What is the slope of the line? What is the real-world meaning of the slope?

e. When will the car be at the Wall Drug Store, which is 80 miles from Mt. Rushmore? Explain how you know.

f. When will the car arrive at Mt. Rushmore? Explain how you know.

12. Imagine that a classmate has been out of school for the past few days with the flu. Write him or her an e-mail describing how to convert an equation such as $y = 4 + 2(x - 3)$ from point-slope form to slope-intercept form. Be sure to include examples and explanations. End your note by telling your classmate how to find out if the two equations are equivalent.

Wall Drug is a landmark in South Dakota. The store's fame began during the 1930s—the Great Depression—when it offered free ice water to travelers.

project

STATE OF THE STATES

Many characteristics of a state vary with the size of the state's population. Some of these relationships are linear. The more people who live in a state, the more houses, cars, schools, and prisoners there are. A lot of data about the states is available on the Internet. You can link to a very useful site through www.keymath.com/DA .

Here are two scatter plots that show a comparison of the population of a state to two different characteristics of the state—number of prisoners and median household income. Which scatter plot shows a linear pattern?

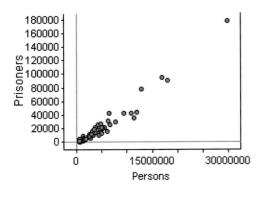 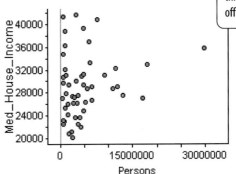

Your project should include

▶ Several scatter plots, investigating relationships between various pairs of characteristics for states.

▶ Lines of fit for your plotted data, their slopes and intercepts along with their real-world meanings (that is, if there appears to be a linear relationship).

▶ Explanations of why some relationships do not appear linear.

Fathom

Fathom comes with many data sets that contain information about the states, and you can easily download more information from web sites.

Use Fathom's movable line to "eyeball fit" lines through points and read off the slope.

LESSON
5.7

Applications of Modeling

In Lesson 5.6, you learned a systematic method, using quartile values, to find a line of fit for data points that appear to have a linear pattern. In this lesson, you'll contrast that method with ways you've used before and evaluate your results.

Investigation
What's My Line?

You will need

- graph paper
- a strand of spaghetti

These tables show how the diameter of the pupil of your eye changes with age. Your objective is to arrive at three linear models—the first by "eyeballing" using a pencil or a strand of spaghetti, the second by choosing two data points that show the direction of the data, and the third by finding Q-points and drawing the line through them. Finally, you'll analyze your models and use them to make a prediction.

> **Procedure Note**
>
> Two people in your group should work with the data in Table 1, and two people should work with the data in Table 2. Work as a group to discuss the answers to the questions in Steps 8, 10, and 13.

Table 1

Age (yrs)	Diameter of pupil in daylight (mm)
20	4.7
30	4.3
40	3.9
50	3.5
60	3.1
70	2.7
80	2.3

Table 2

Age (yrs)	Diameter of pupil at night (mm)
20	8.0
30	7.0
40	6.0
50	5.0
60	4.1
70	3.2
80	2.5

(John Lord, *Sizes*, 1995, p. 120)

First, you'll find a line of fit using an "eyeballing" method. Remember that the object of a linear model is to summarize or generalize the data.

Step 1 | Plot the data on graph paper. Lay a piece of spaghetti on the plot so that it crosses the *y*-axis and follows the direction of the data. Try not to focus on the points themselves, but on the general direction of the "cloud" of points.

Step 2 | Estimate the coordinate of the *y*-intercept. Locate a point with convenient coordinates along the strand. Use this information to write the equation of the line. Make a note of this equation.

Next, you'll find a line of fit by choosing "representative" data points.

Step 3 | Make a scatter plot of the data on your calculator. Choose two data points that you think show the direction of the data.

Step 4	Use the two points to write a linear equation in point-slope form. Make a note of this equation.

Next, you'll find a line of fit using Q-points.

Step 5	Use your calculator to get the five-number summaries of the x- and y-values. Draw a rectangle using the first- and third-quartile values for the x-values and the first- and third-quartile values for the y-values. Name the Q-points you should use for the data.
Step 6	Write the equation of the line of fit you can draw through your selected Q-points. Graph the equation to verify that it is the diagonal of the rectangle you drew on the plot. Make a note of this equation.

Finally, you'll compare the lines and their characteristics and decide which method has given you the best-fitting line.

Step 7	Compare the slopes of all three lines for each table. Do all these numbers have the same real-world meaning? What is it?
Step 8	What does the difference in slope values for Table 1 and Table 2 tell you about how your pupil diameter changes?
Step 9	Compare the y-intercepts of all three lines for each table. Do they all have the same real-world meaning? If so, what is it?
Step 10	At what ages would your models predict a pupil diameter of 4.5 mm? Show how to find this value symbolically.
Step 11	What is the effect of a small change in the y-intercept when you use the model to predict a value in the middle of the data set?
Step 12	What is the effect of a small change in the slope when you try to predict a y-value far from most of the given points?
Step 13	Discuss the pros and cons of each procedure you used to find a line of fit. Which method do you like best and why?

I like the "eyeballing" method because I can visualize the fit of the line before writing an equation.

Using two points and the point-slope method works best for me. It's nice to have definite points.

I prefer the certainty of using Q-points. I know my answer will be the same as anyone else's.

As you study more about finding models for data, you will also learn more about methods you can use to tell how well a model fits data. In this course, the emphasis will be on finding a reasonable model—even though it may not be the best-fitting model—so that you can use it to make reliable predictions.

EXERCISES

You will need your calculator for problem **6.**

Practice Your Skills

1. The equation of a line in point-slope form is $y = 6 - 3(x - 6)$.
   a. Name the point on this line that was used to write the equation.
   b. Name the point on this line with an x-coordinate of 5.
   c. Using the point you named in 1b, write another equation of the line in point-slope form.
   d. Write the equation of the line in intercept form.
   e. Find the coordinates of the x-intercept.

2. Solve each equation symbolically for x. Use another method to verify your solution.
   a. $3(x - 5) + 14 = 29$
   b. $\dfrac{8 - 13}{x + 5} = 2$
   c. $\dfrac{2(3 - x)}{4} - 8 = -7.75$
   d. $11 + \dfrac{6(x + 5)}{9} = 42$

3. Solve each equation for y.
   a. $2x + 5y = 18$
   b. $5x - 2y = -12$

Reason and Apply

4. **APPLICATION** This table shows winning distances for the Olympic men's discus throw.
   a. Define variables and find the line of fit based on Q-points for this data set. Give the real-world meanings of the slope and the y-intercept.
   b. In 1912, Armas Taipale of Finland threw the discus 45.21 m. What value does your model predict for that year? What is the difference in the two values?
   c. According to your model, what year might you expect the winning distance to pass 80 meters? Show how to find this value symbolically.

Men's Discus

Year	Champion	Country	Distance (m)
1952	Sim Iness	United States	55.03
1956	Al Oerter	United States	56.36
1960	Al Oerter	United States	59.18
1964	Al Oerter	United States	61.00
1968	Al Oerter	United States	64.78
1972	Ludvik Danek	Czechoslovakia	64.40
1976	Mac Wilkins	United States	67.50
1980	Viktor Rashchupkin	USSR	66.64
1984	Rolf Danneberg	West Germany	66.60
1988	Jürgen Schult	East Germany	68.82
1992	Romas Ubartas	Lithuania	65.12
1996	Lars Riedel	Germany	69.40
2000	Virgilijus Alelena	Lithuania	69.30

(*2000 World Almanac*, p. 919, and *www.olympics.com*)

5. **APPLICATION** The table shows the timetable for the Coast Starlight train from Seattle to Los Angeles.

Coast Starlight

Location	Distance from Los Angeles (mi)	Arrival time	Elapsed time from Seattle (min)
Kelso, WA	1252	12:52	172
Vancouver, WA	1213	13:35	205
Salem, OR	1150	15:45	275
Eugene, OR	1079	17:07	357
Sacramento, CA	552	6:30	1160
Emeryville, CA	468	8:30	1280
Salinas, CA	355	12:01	1491
Santa Barbara, CA	103	18:17	1867

a. Define variables and give the line of fit based on Q-points for this data set. Give the real-world meaning of the slope.

b. While riding the train, you pass a sign that says you are 200 miles from Los Angeles. What length of time does your model predict you have traveled?

c. The train comes to a stop after 10 hours (600 minutes). According to your model, how far are you from Los Angeles? Show how to find this value symbolically.

Before 1971, when Amtrak created the Coast Starlight, passengers had to ride three different trains to go from Seattle to Los Angeles.

▶ Review

6. A music teacher is using a catalog to purchase new strings for instruments. The number of violin, viola, cello, and bass strings, the type of string, and the cost are listed in the matrices below.

$$[A] = \begin{array}{c} \text{Violin} \\ \text{Viola} \\ \text{Cello} \\ \text{Bass} \end{array} \begin{array}{ccccc} E & A & D & G & C \\ \left[\begin{array}{ccccc} 6 & 8 & 6 & 1 & 0 \\ 0 & 4 & 8 & 3 & 1 \\ 0 & 6 & 5 & 4 & 4 \\ 1 & 1 & 2 & 2 & 0 \end{array}\right] \end{array} \qquad [B] = \begin{array}{c} E \\ A \\ D \\ G \\ C \end{array} \begin{array}{c} \text{Cost} \\ \left[\begin{array}{c} 8.50 \\ 12.25 \\ 16.50 \\ 18.75 \\ 22.25 \end{array}\right] \end{array}$$

a. Without using your calculator's matrix menu, find the cost of strings for each section. Explain how you found your answer.

b. Verify that your answer to 6a is correct by using your calculator to multiply the quantity matrix by the cost matrix.

c. Label the rows and columns of your answer to 6a. Explain what the entries in the matrix tell you.

d. Do you get a meaningful answer if you multiply $[B] \cdot [A]$? Why or why not?

7. A sample labeled "50 grains" weighs 3.24 grams on a balance. What is the conversion factor for grams to grains?

8. In a recent survey, visitors at several ski resorts preferred snowboarding over downhill skiing by a 5 to 4 ratio.

   a. What is the ratio of visitors who prefer snowboarding to all visitors?

   b. Imagine that you have a job in a rental shop at a ski resort. Using this information, what suggestions would you make about purchasing equipment for 1000 downhill skiers and snowboarders?

   c. What other information would you like to know about the survey to help you make accurate predictions?

IMPROVING YOUR **REASONING** SKILLS

Not all data sets form a linear pattern. Here is a set that doesn't. It relates speed and time for the same car trip made by several drivers. Plot the data and see if you recognize the shape. Once you do, write an equation whose graph shows this shape. Then adjust it, if necessary, to better show the shape of the data. Use your equation to predict how much time a driver who averages 45 mph would need for the trip and the average speed that would give a time of 70 minutes.

Speed and Time for the Same Trip

Speed (mph) x	Time (min) y	Speed (mph) x	Time (min) y
25	144	45	
26	137.6	50	72
30	120		70
34	106.1	55	65.5
36	99.7	56.5	63.7
37.4	96.4	60	59.8
40.5	89.3	62	58
42.2	85.4	65	55.5

Activity Day

Data Collection and Modeling

Here's your chance to take part in an extreme sport without the risk! In this activity you'll set up a bungee jump and collect data relating the distance a "jumper" falls and the number of rubber bands in the bungee cord. Then you model the data with an equation. Next you'll use your model to find the number of rubber bands you'd need in the cord for a near miss from a specific height.

Activity
The Toyland Bungee Jump

You will need

- a toy to serve as "jumper"
- a supply of rubber bands
- a tape measure

Step 1 | Make a bungee cord by attaching two rubber bands to your "jumper." (You may first need to make a harness by twisting a rubber band around the toy.)

Step 2 | Place your jumper on the edge of a table or another surface while holding the end of the bungee cord. Then let your jumper fall from the table. Use your tape measure to measure the maximum distance the jumper falls on the first plunge.

Step 3	Repeat this jump several times and find a mean value for the distance. Record the number of rubber bands (2) and the mean distance the jumper falls in a table like this one.

Number of rubber bands	2	4	5	6
Distance fallen				

Step 4	Add one or two rubber bands to the bungee cord and repeat the experiment. Record this new information.
Step 5	Continue to make bungee cords of different lengths, and measure the distance your jumper falls until you have at least seven pairs of data. When using long cords, you may need to move to a higher place to measure the falls.
Step 6	Define variables and make a scatter plot of the information from your table.
Step 7	Find the equation of a line of fit for your data. You may use any procedure, but be able to justify why your equation is a reasonable fit.

Step 8	The test! Decide on a good location for all the groups to conduct final bungee jumps from a particular height. Use your equation to determine the number of rubber bands you need in the cord to give your jumper the greatest thrill— falling as close as possible to the ground without touching. When you have determined the number of rubber bands, make the bungee cord and wait your turn to test your prediction.

Step 9	Write a group report for this activity. Follow this outline to produce a neat, organized, thorough, and accurate report. Any reader of your report should not need to have watched the activity to know what is going on.

Report Outline

A. Overview — Tell what the investigation was about, its purpose or objective.

B. Data collection — Describe the data you collected and how you collected it.

C. Data table — Use labels and units.

D. Graph — Show all data points. Use labels and units. Show the line of fit.

E. Model — Define your variables and give the equation.

F. Procedure — Tell how you found this equation and why you used this method.

G. Calculations — Show how you decided how many rubber bands to use in the final jump.

H. Results — Describe what happened on the final jump.

I. Conclusion — What problems did you have? What worked really well? If you could repeat the whole experiment, what would you do to improve it?

In Chapter 4, you learned how to write equations in intercept form, $y = a + bx$. In this chapter, you learned how to calculate **slope** using the slope formula $b = \frac{y_2 - y_1}{x_2 - x_1}$. You also used the slope formula to derive another form for a linear equation—the **point-slope form.** The point-slope form, $y = y_1 + b(x - x_1)$, is the equation of a line through point (x_1, y_1) with slope b. You learned that this form is very useful in real-world situations when the starting value is not on the y-axis.

You continued to investigate equivalent forms of expressions and equations using tables and graphs. You used the distributive property of multiplication over addition and the **commutative** and **associative** properties of addition and multiplication to write point-slope equations in intercept form.

You discovered how to use the first and third quartiles from the five-number summaries of x- and y-values in a data set to write a linear model for data based on **Q-points.**

EXERCISES

You will need your calculator for problems **3, 4,** and **9.**

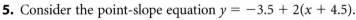

1. The slope of the line between $(2, 10)$ and $(x_2, 4)$ is -3. Find the value of x_2. Show your work.

2. Give the slope and the y-intercept for each equation.
   a. $y = -4 - 3x$ **b.** $2x + 7 = y$ **c.** $38x - 10y = 24$

3. Line a and line b are shown on the graph at right. Name the slope and the y-intercept, and write the equation for each line. Check your equations by graphing.

4. Write each equation in the form requested. Check your answers by graphing.
   a. Write $y = 13.6(x - 1902) + 158.2$ in intercept form.
   b. Write $y = -5.2x + 15$ in point-slope form using $x = 10$ as the first coordinate of the point.

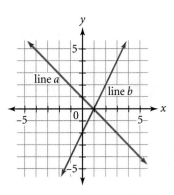

5. Consider the point-slope equation $y = -3.5 + 2(x + 4.5)$.
   a. Name the point used to write this equation.
   b. Write an equivalent equation in intercept form.
   c. Factor your answer to 5b and name the x-intercept.
   d. A point on the line has a y-coordinate of 16.5. Find the x-coordinate of this point and use this point to write an equivalent equation in point-slope form.
   e. Explain how you can verify that all four equations are equivalent.

6. Show all steps for a symbolic solution to each problem.

    a. $4 + 2.8x = 51$

    c. $11 + 3(x - 8) = 41$

    b. $38 - 0.35x = 27$

    d. $220 - 12.5(x - 6) = 470$

7. **APPLICATION** Suppose Karl bought a used car for $12,600. Each year its value is expected to decrease by $1,350.

    a. Write an equation modeling the value of the car over time. Let x represent the number of years Karl owns the car, and let y represent the value of the car in dollars.

    b. What is the slope, and what does it mean in the context of the problem?

    c. What is the y-intercept, and what does it mean in the context of the problem?

    d. What is the x-intercept, and what does it mean in the context of the problem?

8. Recall the data about heating a pot of water from the investigation in Lesson 5.3. A possible linear model relating the time x to the temperature y is $y = 30 + 0.375(x - 36)$.

    a. What equation could you solve to find how long it would take before the pot of water reaches 43°C?

    b. Find the approximate time indicated in 8a using a table or graph.

    c. Show a symbolic solution for your equation in 8a.

9. **APPLICATION** The table gives the winning heights for the Olympic women's high jump.

    a. Find the five-number summaries for the year and height data.

Women's High Jump

Year	Champion	Country	Height (m)
1952	Esther Brand	South Africa	1.67
1956	Mildred McDaniel	United States	1.76
1960	Iolanda Balas	Romania	1.85
1964	Iolanda Balas	Romania	1.90
1968	Miloslava Rezkova	Czechoslovakia	1.82
1972	Ulrike Meyfarth	West Germany	1.92
1976	Rosemarie Ackerman	East Germany	1.93
1980	Sara Simeoni	Italy	1.97
1984	Ulrike Meyfarth	West Germany	2.02
1988	Louise Ritter	United States	2.03
1992	Heike Henkel	Germany	2.02
1996	Stefka Kostadinova	Bulgaria	2.05
2000	Yelena Yelesina	Russia	2.01

Yelena Yelesina was the first Russian woman to win the Olympic high jump title.

(*2000 World Almanac*, p. 920, and *www.olympics.com*)

   b. Name the Q-points for this data set.

   c. Write an equation for the line through the Q-points.

   d. Graph the line and the data, and explain whether or not you think this line is a good model for the data pattern.

   e. Predict the winning height for the year 2012.

10. APPLICATION This table shows the federal minimum hourly wage for 1974–1997.

   a. Find the line of fit based on Q-points.

   b. Give the real-world meaning of the slope.

   c. Use your model to predict the minimum hourly wage for 2005.

   d. Predict when the minimum hourly wage would have been $1.00.

11. Explain how to find the equation of a line when you know

   a. The slope and the y-intercept.

   b. Two points on that line.

United States Minimum Wage

Year x	Hourly minimum y	Year x	Hourly minimum y
1974	$1.90	1980	$3.10
1975	$2.00	1981	$3.35
1976	$2.20	1990	$3.80
1977	$2.30	1991	$4.25
1978	$2.65	1996	$4.75
1979	$2.90	1997	$5.15

(Bureau of Labor Statistics, U.S. Department of Labor)

TAKE ANOTHER LOOK

Is rate of change the same as slope? For linear equations, you've seen that it is. But what about curves? You've studied inverse variations, whose equations have the form $y = \frac{k}{x}$. Let's look at the equation $y = \frac{12}{x}$ and its graph.

(3, 4) is a point on the curve. Let's choose another nearby point. Substituting 3.5 for x in the equation, you get $y \approx 3.4$. Using the points (3, 4) and (3.5, 3.4) in the formula for slope, you get

$$b = \frac{y_2 - y_1}{x_2 - x_1} = \frac{3.4 - 4}{3.5 - 3} = \frac{-0.6}{0.5} = -1.2$$

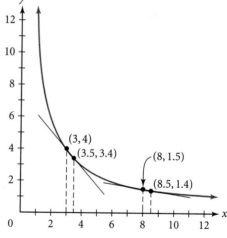

We can say that the *average* rate of change for $y = \frac{12}{x}$ on the interval $x = 3$ to $x = 3.5$ is -1.2. But -1.2 is not the "slope" of $y = \frac{12}{x}$. Instead, it is the slope of the *straight* line through the two points (3, 4) and (3.5, 3.4). Is the average rate of change on the x-interval from 3 to 3.25 the same as from 3.25 to 3.5?

Try points on the "wings" of the curve. For instance, (8, 1.5) is on the curve and so is (8.5, 1.4). Again, the y-coordinate is approximate. What is the average rate of change between these points? The x-interval is the same as for the points (3, 4) and (3.5, 3.4), but is the rate of change the same? What does this tell you? What *straight* line through (8, 1.5) has slope equal to the average rate of change on the interval $x = 8$ to $x = 8.5$?

Assessing What You've Learned

In each of the five chapters from Chapter 0 to Chapter 4, you were introduced to a different way to assess what you learned. Maybe you have tried all five ways—keeping a portfolio, writing in your journal, organizing your notebook, doing a performance assessment, and giving a presentation. Maybe you have tried just a couple of these methods. Probably, your teacher has adapted these ideas to suit the needs of your class.

By now, you should realize that assessment is not just giving and taking tests. In the working world, performance in some occupations can be measured in tests, but in all occupations, there is a need to communicate what you know to coworkers. In all jobs, workers demonstrate to their employer or to their clients, patients, or customers that they are skilled in their fields. They need to show they are creative and flexible enough to apply what they've learned in new situations. Assessing your own understanding and letting others assess what you know gives you practice in this important life skill. It also helps you develop good study habits, and that, in turn, will help you advance in school and give you the best possible opportunities in your work life. Keep that in mind as you try one or more of these suggestions.

 WRITE IN YOUR JOURNAL What have you enjoyed more in studying algebra—the numbers, symbols, graphs, and other abstract ways of describing relationships, or the concrete applications and examples that show how people use these ideas in the real world?

Do you find it interesting that a single linear relationship can be described in so many ways, or does that add confusion for you?

 ORGANIZE YOUR NOTEBOOK In your notebook, find examples of the different forms of a linear equation. Assign each form a color, and underline or highlight the equations of each type in the right color. Or create a table with the equation forms as the headings, and list several of each form in the columns.

 UPDATE YOUR PORTFOLIO Choose a piece of work from this chapter to add to your portfolio. Describe the work in a cover sheet, giving the objective, the result, and what you might have done differently.

 PERFORMANCE ASSESSMENT Show a classmate, a family member, or your teacher that you know how to solve an equation using the properties of equalities and other mathematical properties. Give your reasons for each step.

 GIVE A PRESENTATION Research a topic of interest to you that involves two kinds of numerical data. Present the data in a table, make a scatter plot, and describe the pattern of the points. If the data points show a linear pattern, tell how to find a line of fit for the data set and why that line is useful.

CHAPTER

6

Systems of Equations and Inequalities

Freshly painted umbrellas dry in the sun outside the Nagatsu factory in Kyushu, Japan. The sticks in their frames form intersecting lines like the graphs of linear equations.

OBJECTIVES

In this chapter you will

- learn to solve systems of linear equations
- solve systems using the substitution method
- solve systems using the elimination method
- solve systems using matrices
- graph inequalities in one and two variables
- solve systems of linear inequalities

LESSON 6.1

Solving Systems of Equations

In previous chapters you studied linear relationships in the contexts of elevators, wind chill, rope length, and walks. In this chapter you'll consider two or more linear equations together. A **system of equations** is a set of two or more equations that have variables in common. The common variables relate similar quantities. You can think of an equation as a condition imposed on one or more variables, and a system as several conditions imposed simultaneously.

When solving a system of equations, you look for a solution that makes each equation true. There are several strategies you can use. In this lesson you will solve systems using tables and graphs.

EXAMPLE

Edna leaves the trailhead at dawn to hike 12 miles toward the lake, where her friend Maria is camping. At the same time, Maria starts her hike toward the trailhead. Edna is walking uphill so she averages only 1.5 mi/hr, while Maria averages 2.5 mi/hr walking downhill. When and where will they meet?

    a. Define variables for time and for distance from the trailhead.

    b. Write a system of two equations to model this situation.

    c. Solve this system by creating a table and finding the values for the variables that make both equations true. Then locate this solution on a graph.

    d. Check your solution and explain its real-world meaning.

▶ **Solution**

    a. Let x represent the time in hours. Both women hike the same amount of time. Let y represent the distance in miles from the trailhead. When Edna and Maria meet they will both be the same distance from the trailhead, although they will have hiked different distances.

b. The system of equations that models this situation is grouped in a brace:

$$\begin{cases} y = 1.5x & \text{Edna's hike} \\ y = 12 - 2.5x & \text{Maria's hike} \end{cases}$$

Edna starts at the trailhead so she increases her distance from it as she hikes 1.5 mi/hr for x hours. Maria starts 12 miles from the trailhead and reduces her distance from it as she hikes 2.5 mi/hr for x hours.

c. Create a table from the equations. Fill in the times and calculate each distance. The table shows the x-value that gives equal y-values for both equations. When $x = 3$, both y-values are 4.5. So the solution is the ordered pair $(3, 4.5)$. We say that these values "satisfy" both equations.

Hiking Times and Distances

x	$y = 1.5x$	$y = 12 - 2.5x$
0	0	12
1	1.5	9.5
2	3	7
3	4.5	4.5
4	6	2
5	7.5	-0.5

On the graph this solution is the point where the two lines intersect. You can use the trace function on your calculator to approximate the coordinates of the solution point, though sometimes you'll get an exact answer.

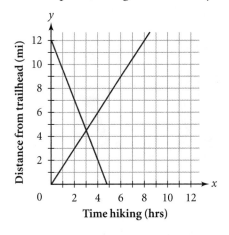

d. The coordinates $(3, 4.5)$ must satisfy both equations.

Edna	**Maria**	
$y = 1.5x$	$y = 12 - 2.5x$	Original equations.
$4.5 \stackrel{?}{=} 1.5(3)$	$4.5 \stackrel{?}{=} 12 - 2.5(3)$	Substitute 3 for the time x and 4.5 for the distance y into both equations.
$4.5 = 4.5$	$4.5 = 4.5$	We get true statements, so $(3, 4.5)$ is a solution for both equations.

So after hiking for 3 hours Edna and Maria meet on the trail 4.5 miles from the trailhead.

Investigation
Where Will They Meet?

You will need

- two motion sensors
- measuring tape or chalk to make a 6-meter line segment

In this investigation you'll solve a system of simultaneous equations to find the time and distance at which two walkers meet.

Suppose that two people begin walking in the same direction at different average speeds. The faster walker starts behind the slower walker. When and where will the faster walker overtake the slower walker?

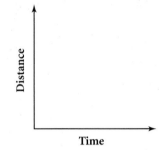

Step 1 | Draw a graph sketch showing both walks. Which line represents the faster walker?

Now act out the walk.

Step 2 | Mark a 6-meter segment at 1-meter intervals. In your group, designate walkers A and B, a timekeeper, and a recorder.

Step 3 | Practice these walks: Walker A starts at the 0.5-meter mark and walks toward the 6-meter mark at a speed of 1 m/sec. Walker B starts at the 2-meter mark and walks toward the 6-meter mark at 0.5 m/sec. When you can follow the walk instructions accurately, do the walk and record the solution.

Procedure Note

The timekeeper counts each second out loud. The recorder notes the time and position of the two walkers when they meet.

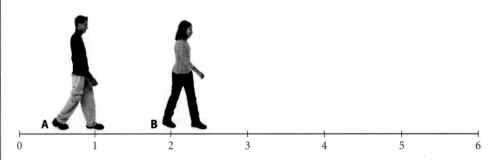

Step 4 | Graph points of the form (*time, distance*) for each walker. Then locate the point that you think represents the approximate time and distance from the zero mark when Walker A passes Walker B.

Next you'll model the walks with a system of equations.

Step 5 | Use the starting point and speed to write an equation for each walker.

Step 6 | Graph the two equations on the same set of axes as your data. Find the point of intersection of the two lines. Explain the real-world meaning of the coordinates of this point. Is the point you found by acting out the walks reasonably close to the point of intersection?

| Step 7 | Check that the coordinates of the point of intersection satisfy both of your equations. |

Next you'll consider what happens under different conditions.

Step 8	Suppose that Walker A walks even faster than 1 m/sec. How is the graph different? What happens to the point of intersection?
Step 9	Suppose that two people walk at the same speed and direction from different starting marks. What does this graph look like? What happens to the solution point?
Step 10	Suppose that two people walk at the same speed in the same direction from the same starting mark. What does this graph look like? How many points satisfy this system of equations?

Is it possible to draw two lines that intersect in two points? How many possible solutions do you think a linear system of two equations in two variables can have?

When you solve a system of two equations, you're finding a solution in the form (x, y) that makes both equations true. When you have a graph of two distinct linear equations the solution of the system is the point where the two lines intersect, if they cross at all. You can estimate these coordinates by tracing on the graph. To find the solution more precisely, zoom in on a table. In the next lesson you'll learn how to find the *exact* coordinates of the solution by working with the equations.

Dancers step between the parallel and intersecting sticks of a bamboo dance in Thailand.

EXERCISES

You will need your calculator for problems **3, 4, 6, 7, 10,** and **14.**

Practice Your Skills

1. Verify whether or not the given ordered pair is a solution to the system. If it is not a solution, explain why not.

   a. $(-15.6, 0.2)$
   $$\begin{cases} y = 47 + 3x \\ y = 8 + 0.5x \end{cases}$$

   b. $(-4, 23)$
   $$\begin{cases} y = 15 - 2x \\ y = 12 + x \end{cases}$$

   c. $(2, 12.3)$
   $$\begin{cases} y = 4.5 + 5x \\ y = 2.3 + 5x \end{cases}$$

2. Match each graph of a system of equations with its corresponding table values. The tick marks on each graph are one unit apart.

Graph of system	**Table values of system**

a.

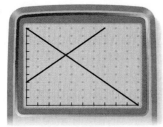

i.

X	Y₁	Y₂
1	4	1
2	4	2
3	4	3
4	4	4
5	4	5
6	4	6
7	4	7

X = 4

b.

ii.

X	Y₁	Y₂
−2	13	8
−1	11	7
0	9	6
1	7	5
2	5	4
3	3	3
4	1	2

X = 3

c.

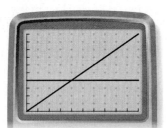

iii.

X	Y₁	Y₂
3	3.5	4
4	4	4.3333
5	4.5	4.6667
6	5	5
7	5.5	5.3333
8	6	5.6667
9	6.5	6

X = 6

d.

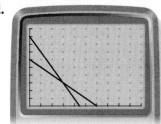

iv.

X	Y₁	Y₂
2	8	5
2.5	7.5	5.5
3	7	6
3.5	6.5	6.5
4	6	7
4.5	5.5	7.5
5	5	8

X = 3.5

3. Graph each system on your calculator using the window given. Use the trace function to find the point of intersection. Do you think the calculator is giving you approximate or exact solutions?

a. $[-18.8, 18.8, 5, -12.4, 12.4, 5]$
$$\begin{cases} y = 3 + 0.5x \\ y = -9 + 2x \end{cases}$$

b. $[-4.7, 4.7, 1, -3.1, 3.1, 1]$
$$\begin{cases} y = 4x - 5.5 \\ y = -3x + 5 \end{cases}$$

4. Use the calculator table function to find the solution to each system of equations. (In 4b, you'll need to solve the equations for y first.)

a. $y = 7 + 2.5x$
$y = 35.9 - 6x$

b. $2x + y = 9$
$3x + y = 16.3$

5. Solve the equations for y. Evaluate each equation for $x = 1$. Substitute these values for x and y into their original equations. What does this tell you?

a. $4x + 2y = 6$

b. $2x - 5y = 20$

Reason and Apply

6. **APPLICATION** Two friends start rival Internet companies in their homes. It costs Gizmo.kom $12,000 to set up the computers and buy the necessary office supplies. Advertisers pay Gizmo.kom $2.50 for each hit (each visit to the web site).

a. Define variables and write an equation to describe the profits for Gizmo.kom.

b. The profit equation for the rival company, Widget.kom, is $P = -5000 + 1.6N$. Explain possible real-world meanings of the numbers and variables in this equation, and tell why they're different from those in 6a.

c. Use a calculator table to find the N-value that gives approximately equal P-values for both equations.

d. Use your answer to 6c to select a viewing window, and graph both equations to display their intersection and all x- and y-intercepts.

e. What are the coordinates of the intersection point of the two graphs? Explain how you found this point and how accurate you think it is.

f. What is the real-world meaning of these coordinates?

7. **APPLICATION** After seeing her friends profit from their web sites in problem 6, Sally wants to start a third company, Gadget.kom, with the start-up costs of Widget.kom and the advertising rate of Gizmo.kom.

a. What is Sally's profit equation?

b. Graph the profit equations for Gadget.kom and Gizmo.kom.

c. What does the graph tell you about Sally's profits?

d. What can you learn about Sally's profits from the equations alone?

8. The total tuition for students at University College and State College consists of student fees plus costs per credit. Some classes have different credit values. The table shows the total tuition for programs with different numbers of credits at each college.

a. Write a system of equations that represents the relationship between credit hours and total tuition for each college.

b. Find the solution to this system of equations and check it.

c. Which method did you use to solve this system? Why?

d. What is the real-world meaning of the solution?

e. When is it cheaper to attend University College? State College?

Total Tuition

Credits	University College ($)	State College ($)
1	55	47
3	115	111
6	205	207
9	295	303
10	325	335
12	385	399

9. The high school band and drill team both practice on the football field. During one part of the routine, a drill team member walks from the 9-yard mark on the sideline at 1 yd/sec toward the 0-yard mark. At the same time, the tuba player walks from the 3-yard mark at 0.5 yd/sec in the opposite direction.

    a. Write a system of equations to describe this situation.

    b. Find the solution to this system and explain its meaning.

The marching band performs at half time during a football game at West Point.

10. In the equations $y = 45 - 0.054(x - 1962)$ and $y = 43.7 - 0.054(x - 1986)$, x represents the year and y represents the winning time in seconds. These are two ways to model the data for the men's 400-meter dash in the Olympics.

    a. Find the approximate winning time for the year 1972 given by each equation. What is the difference between the values?

    b. Find the approximate winning time for the year 2000 given by each equation. What is the difference between the values?

    c. Select an appropriate window and graph the two equations.

    d. Do you think these equations represent the same line? Explain your reasoning.

11. Consider the system of equations

$$\begin{cases} y = a + bx \\ y = 2 - 5x \end{cases}$$

Explain what values of a and b give this system

    a. exactly 1 solution   **b.** no solutions   **c.** infinitely many solutions

▶ Review

12. **APPLICATION** Hydroplanes are boats that move so fast that they skim the top of the water. The hydroplane, *Miss B*, qualified for the 2000 Columbia Cup race with a speed of 130.000 mi/hr. The hydroplane, *Club C*, qualified with a speed of 155.753 mi/hr. (*www.hydroracing.com, www.superior-racing.com, www.hydros.org*)

    a. How long will each hydroplane take to run a 5-lap race if one lap is 2.5 miles?

    b. Some boats limit the amount of fuel the motor burns to 4.3 gallons per minute. How much fuel will each boat use to run a 5-lap race?

c. Hydroplanes have a 50-gallon tank though generally only about 43 gallons are put in. The rest of the tank is filled with foam to prevent sloshing. How many miles can each hydroplane go on one 43-gallon tank of fuel?

d. Find each boat's fuel efficiency rate in miles per gallon.

13. Solve each equation using the method you like best. Then use a different method to check your solution.

a. $0.75x = 63.75$

b. $18.86 = -2.3x$

c. $6 = 12 - 2x$

d. $9 = 6(x - 2)$

e. $4(x + 5) - 8 = 18$

This hydroplane travels so fast that its image is blurred in the photo. Learn about hydroplane racing with a link at **www.keymath.com/DA** .

14. APPLICATION At the civic center, Johanna volunteers to design a model of the town and railroad as it looked 150 years ago. The model needs to fit on a piece of plywood that is 4 feet wide. The region that Johanna needs to show is about 0.1 mile wide. The director suggested that she choose a model train size from the scales in the table.

a. Complete the table. It is helpful to use calculator lists for this process.

b. Which scale best fits Johanna's model?

c. Express the scale of her model in inches to feet?

Train gauge (name of scale)	Scale (model:actual)	Represented width (ft)	Represented width (mi)
Z	1:220		
N	1:160		
HO	1:87		0.066
S	1:64		
O	1:48	192	
Maxi	1:32		
G	1:24		

15. Find each matrix sum and difference.

a. $\begin{bmatrix} 3 & -3 \\ -9 & 1 \end{bmatrix} + \begin{bmatrix} -2 & -8 \\ 3 & 7 \end{bmatrix}$

b. $\begin{bmatrix} 5 & 0 \\ 2 & 7 \end{bmatrix} - \begin{bmatrix} -8 & 1 \\ -5 & -1 \end{bmatrix}$

16. Solve each equation for y.

a. $y + 2 = 5x$

b. $5y = 4 - 7x$

c. $2y - 6x = 3$

LESSON 6.2

Solving Systems of Equations Using Substitution

Graphing systems and comparing their table values are good ways to see solutions. However, it's not always easy to find a good graphing window or the right settings for a table. Also, the solutions you find are often only approximations. To find exact solutions, you'll need to work with the equations themselves. One way is called the **substitution method.**

EXAMPLE A

On a rural highway a police officer sees a motorist run a red light at 50 miles per hour, and begins pursuit. At the instant the police officer passes through the intersection at 60 mph, the motorist is 0.2 mile down the road. When and where will the officer catch the motorist?

a. Write a system of equations in two variables to model this situation.

b. Solve this system by the substitution method and check the solution.

c. Explain the real-world meaning of the solution.

▶ **Solution**

a. Let t represent the time in hours, with $t = 0$ being the instant the officer passes through the intersection. Let d represent the distance in miles from the intersection. Then the system of equations is

$$\begin{cases} d = 0.2 + 50t \\ d = 60t \end{cases}$$

The first equation represents the motorist, who is already 0.2 mile away when the timing begins. The second equation represents the officer.

b. When the officer catches the motorist, they will both be the same distance from the intersection. At this time, both equations will have the same d-value. So you can replace d in one equation with an equivalent expression for d that you find from the other equation. Substituting $60t$ for d into $d = 0.2 + 50t$ gives the new equation:

$$\begin{cases} d = 0.2 + 50t \\ d = 60t \end{cases} \longrightarrow 60t = 0.2 + 50t$$

There is now one equation to solve. Notice that the variable t occurs on both sides of the equal sign and d has been dropped out. Now you use the balancing method to find the solution.

$60t = 0.2 + 50t$	New equation.
$60t - 50t = 0.2 + 50t - 50t$	Subtract 50t from both sides of the equation.
$10t = 0.2$	Evaluate.
$\dfrac{10t}{10} = \dfrac{0.2}{10}$	Divide both sides of the equation by 10.
$t = 0.02$	Reduce and divide.

To find the other half of the solution, substitute 0.02 for t into one of the original equations.

$$t = \boxed{(0.02)}$$

$$d = 0.2 + 50t \qquad\qquad d = 60t$$
$$d = 0.2 + 50(0.02) \quad \text{or} \quad d = 60(0.02)$$

If both equations give the same d-value, 1.2 in this case, then you have the correct solution.

c. The solution is the only ordered pair of values, (0.02, 1.2), that works in both equations. The police officer will catch the motorist 1.2 miles from the intersection in 0.02 hour, which is 1 minute 12 seconds after passing through the intersection.

The calculator screen shows the system of equations from the example in the window [0, 0.04, 0.01, −1, 3, 1]. It is difficult to guess at these window settings because the two lines have very similar slopes and close y-intercepts. But the substitution method helps you find the exact solution no matter how difficult it is to set windows or tables. Once you have the exact solution, it is much easier to find a good window to display it.

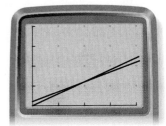

Investigation
All Tied Up

You will need

- two ropes of different thickness, both about 1 meter long
- a meterstick or measuring tape
- a 9-meter-long thin rope (optional)
- a 10-meter-long thick rope (optional)

In this investigation you'll work with rope lengths and predict how many knots it would take in each rope to make a thicker rope the same length as a thinner one.

First you'll collect data and write equations.

Step 1   Measure the length of the thinner rope without any knots. Then tie a knot and measure the length of the rope again. Continue tying knots until no more can be tied. Knots should be of the same kind, size, and tightness. Record the data for number of knots and length of rope in a table.

Step 2   Define variables and write an equation in intercept form to model the data you collected in Step 1. What are the slope and y-intercept, and how do they relate to the rope?

Step 3   Repeat Steps 1 and 2 for the thicker rope.

Step 4   Suppose you have a 9-meter-long thin rope and a 10-meter-long thick rope. Write a system of equations that gives the length of each rope depending on the number of knots tied.

Next you'll analyze the system to find a meaningful solution.

Step 5	Solve this system of equations using the substitution method.
Step 6	Select an appropriate window setting and graph this system of equations. Estimate coordinates for the point of intersection to check your solution. Compare this solution with the one from Step 5.
Step 7	Explain the real-world meaning of the solution to the system of equations.
Step 8	What happens to the graph of the system if the two ropes have the same thickness? The same length?

So far in this chapter, you've seen equations only in intercept form. In other words, they are already solved for the output variable y. This form makes it easy to use the substitution method: You can simply set the two expressions in x equal to each other because they are both equal to y. Sometimes you have to put the equations into intercept form before substituting. In the next example, you'll have to change an equation in standard form to the intercept form.

EXAMPLE B

A molecule of hexane, C_6H_{14}, has six carbon atoms and fourteen hydrogen atoms. Its molecular weight in grams per mole, the sum of the atomic weights of carbon and hydrogen, is 86.178. The molecular weight of octane, C_8H_{18}, is 114.232 grams per mole. Octane has eight carbon atoms and eighteen hydrogen atoms per molecule. Find the atomic weights of carbon and hydrogen.

hexane

● carbon
○ hydrogen

a. Define variables and write a system of linear equations in the standard form $ax + by = c$ for these molecular weights.

b. Put one equation in the intercept form $y = a + bx$ and substitute into the other equation to find a solution.

octane

c. Check your solution in the original equations.

▶ **Solution**

a. Let c represent the atomic weight of carbon in grams per mole. Let h represent the atomic weight of hydrogen in grams per mole. Since the molecular weight of the compounds is the sum of the atomic weights of carbon and hydrogen, we have the system

$$\begin{cases} 6c + 14h = 86.178 & \text{hexane's molecular weight} \\ 8c + 18h = 114.232 & \text{octane's molecular weight} \end{cases}$$

b. Use the balancing method to put one of the equations in intercept form. You can solve for either c or h.

Let's choose to solve the octane equation for c.

$$8c + 18h = 114.232 \qquad \text{Original equation.}$$

$$8c + 18h - 18h = 114.232 - 18h \qquad \text{Subtract } 18h \text{ from both sides.}$$

$$8c = 114.232 - 18h \qquad \text{Evaluate.}$$

$$\frac{8c}{8} = \frac{114.232}{8} - \frac{18h}{8} \qquad \text{Divide every term on both sides by 8.}$$

$$c = 14.279 - 2.25h \qquad \text{Reduce and divide.}$$

Substitute this expression for c into the other equation, $6c + 14h = 86.178$, and solve for h:

$$6c + 14h = 86.178 \qquad \text{Original equation.}$$

$$\mathbf{6(14.279 - 2.25h)} + 14h = 86.178 \qquad \text{Substitute } 14.279 - 2.25h \text{ for } c.$$

$$85.674 - 13.5h + 14h = 86.178 \qquad \text{Distribute the 6 through the ( ).}$$

$$85.674 + 0.5h = 86.178 \qquad \text{Evaluate.}$$

$$85.674 - 85.674 + 0.5h = 86.178 - 85.674 \qquad \text{Subtract 85.674 from both sides.}$$

$$0.5h = 0.504 \qquad \text{Evaluate.}$$

$$\frac{0.5h}{0.5} = \frac{0.504}{0.5} \qquad \text{Divide both sides by 0.5.}$$

$$h = 1.008 \qquad \text{Evaluate.}$$

To find the corresponding c-value, substitute 1.008 for h in the equation.

$$c = 14.279 - 2.25h \qquad \text{The equation in intercept form.}$$

$$= 14.279 - 2.25(1.008) \qquad \text{Substitute 1.008 for } h.$$

$$= 14.279 - 2.268 \qquad \text{Multiply.}$$

$$c = 12.011 \qquad \text{Subtract.}$$

c. The solution to the system is (12.011, 1.008). So the atomic weight of carbon is 12.011 grams per mole and the atomic weight of hydrogen is 1.008 grams per mole. Check your answers by substituting them into the original equations.

$$8c + 18h = 114.232$$

$$8(12.011) + 18(1.008) \overset{?}{=} 114.232$$

$$96.088 + 18.144 \overset{?}{=} 114.232$$

$$114.232 = 114.232$$

$$6c + 14h = 86.178$$

$$6(12.011) + 14(1.008) \overset{?}{=} 86.178$$

$$72.066 + 14.112 \overset{?}{=} 86.178$$

$$86.178 = 86.178$$

Since we get true statements for both equations, the solution checks.

There are many ways to solve systems using the substitution method. You can set expressions equal to one another, or solve for one of the variables and substitute the expression you get into the other equation. Both ways are examples of **symbolic manipulation,** which simply means that you are working with the properties you have used in the balancing method and "undoing" to keep sides of the equation equal. It does not matter which equation or variable you work with first, but you must always substitute the resulting expression into the *other* equation to find a solution. When you solve a system of equations using the substitution method, you can always find an exact solution, not just its approximate coordinates.

EXERCISES

You will need your calculator for problems **12** and **14.**

▶ Practice Your Skills

1. The system of equations

$$\begin{cases} d = 1.5t \\ d = 12 - 2.5t \end{cases}$$

describes the distance of two hikers, Edna and Maria, from the example in Lesson 6.1. By setting the expressions of the right sides of the equations equal to each other, you can find the time when Edna and Maria meet. Explain what happens in Stages 3 and 5 of the substitution process.

$d = 12 - 2.5t$	1. Original equation.
$1.5t = 12 - 2.5t$	2. Substitute $1.5t$ for d.
$1.5t + 2.5t = 12 - 2.5t + 2.5t$	3. _____
$4t = 12$	4. Evaluate.
$\dfrac{4t}{4} = \dfrac{12}{4}$	5. _____
$t = 3$	6. Reduce.

2. Check that each ordered pair is a solution to each system. If the pair is not a solution point, explain why not.

   a. $(-2, 34)$
   $$\begin{cases} y = 38 + 2x \\ y = -21 - 0.5x \end{cases}$$

   b. $(4.25, 19.25)$
   $$\begin{cases} y = 32 - 3x \\ y = 15 + x \end{cases}$$

   c. $(2, 12.3)$
   $$\begin{cases} y = 2.3 + 3.2x \\ y = 5.9 + 3.2x \end{cases}$$

3. Solve each equation by symbolic manipulation.

   a. $14 + 2x = 4 - 3x$
   b. $7 - 2y = -3 - y$
   c. $5d = 9 + 2d$
   d. $12 + t = 4t$

4. Solve the system of equations using the substitution method, and check your solution.

   $$\begin{cases} y = 25 + 30x \\ y = 15 + 32x \end{cases}$$

5. Substitute $4 - 3x$ for y. Then rewrite each expression in terms of one variable.

    a. $5x + 2y$ **b.** $7x - 2y$

6. Solve each system of equations by substitution, and check your solution.

    a. $\begin{cases} y = 4 - 3x \\ y = 2x - 1 \end{cases}$ **b.** $\begin{cases} 2x - 2y = 4 \\ x + 3y = 1 \end{cases}$

▶ Reason and Apply

7. APPLICATION This system of equations models the profits of two home-based Internet companies.

$$\begin{cases} P = -12000 + 2.5N \\ P = -5000 + 1.6N \end{cases}$$

The variable P represents profit in dollars, and N represents hits to the company's web site.

    a. Use the substitution method to find an exact solution.

    b. Is an approximate or exact solution more meaningful in this model?

8. Consider the system of equations

$$\begin{cases} P = -5000 + 2.5N \\ P = -12000 + 2.5N \end{cases}$$

from problem 6 in Lesson 6.1.

    a. What does the graph tell you about this system?

    b. How can you find the answer to 8a using the substitution method?

9. APPLICATION The manager of a movie theater wants to know the number of adults and children who go to the movies. The theater charges $8 for each adult ticket and $4 for each child ticket. At a showing where 200 tickets were sold, the theater collected $1304.

    a. Let the variable A represent the number of adult tickets and C represent the number of child tickets. Write an equation for the total number of tickets sold.

    b. Write an equation showing the total cost of the tickets.

    c. Use your equations from 9a and b to write a system whose solution represents the number of adult and child tickets sold. Solve this system by symbolic manipulation.

10. Students in an algebra class did an experiment similar to the investigation Where Will They Meet? from Lesson 6.1. They had two walkers start at opposite ends of the marked length and walk toward each other. They wrote the system

$$\begin{cases} d = 8.5 - 0.5t \\ d = 2.5 + 0.75t \end{cases}$$

   a. What real-world information does the system tell you?

   b. Use the substitution method to solve this system.

   c. What is the real-world meaning of the solution you found in 10b?

11. The table at right gives the equations that model the three vehicles' distances in the investigation On the Road Again from Lesson 4.4.

   The variable d represents the distance in miles from Flint and t represents time in minutes, with $t = 0$ being the instant all three vehicles start traveling.

   For each event described in 11a–c, write a system of equations, solve using the substitution method, and explain the real-world meaning of your solution.

Distance from Flint

Equation	Vehicle
$d = 220 - 1.2t$	minivan
$d = 35 + 0.8t$	sports car
$d = 1.1t$	pickup truck

   a. The pickup truck passes the sports car.

   b. The minivan meets the pickup truck.

   c. The minivan meets the sports car.

   d. Write and solve an equation to find when the minivan is twice as far from Flint as the sports car.

   e. How do the solutions that you found using symbolic manipulation in 11a through d compare with those you found using recursive routines in the investigation in Lesson 4.4?

12. **APPLICATION** The table below shows the winning times for the women's 400-meter dash in the Olympics since 1964.

Women's 400-meter Dash

Year	Champion	Country	Time (sec)
1964	Betty Cuthbert	Australia	52.00
1968	Colette Besson	France	52.00
1972	Monika Zehrt	East Germany	51.08
1976	Irena-Szewinska Kirszenstein	Poland	49.29
1980	Marita Koch	East Germany	48.88
1984	Valerie Brisco-Hooks	United States	48.83
1988	Olga Bryzgina	U.S.S.R	48.65
1992	Marie-José Perec	France	48.83
1996	Marie-José Perec	France	48.25
2000	Cathy Freeman	Australia	49.11

(*2000 World Almanac*, p. 920, and *www.olympics.com*)

   a. Find a line of fit based on Q-points for the data in the table.

Cathy Freeman won the gold medal in the 400-meter dash at the 2000 Olympics in Sydney, Australia.

b. A possible equation from the data for the men's 400-meter dash in Lesson 5.6 is $y = 45 - 0.054(x - 1962)$. Write and solve a system of equations whose solution tells you when the men and the women will have equal winning times for this Olympic event.

c. Select an appropriate window to graph this system and its solution.

d. Discuss the reasonableness of this model and the solution.

▶ Review

13. You and your family are visiting Seattle and take the elevator to the observation deck of the Space Needle. The observation deck is 520 feet high while the needle itself is 605 feet high. The elevator travels at a constant speed, and it takes 43 seconds to travel from the base at 0 feet to the observation deck.

a. What is the slope of the graph of this situation?

b. If the elevator could go all the way to the top, how long would it take to get there?

c. If a rider got on the elevator at the restaurant at the 100-foot level, what equation models her ride to the observation deck?

14. Do each calculation by hand, and then check your results with a calculator. Express your answers as fractions.

a. $3 - \dfrac{5}{6}$

b. $\dfrac{1}{4} + \dfrac{5}{12}$

c. $\dfrac{3}{4} \cdot \dfrac{2}{9}$

d. $\dfrac{1}{5} + \dfrac{2}{3} + \dfrac{3}{4}$

The Space Needle, shown here in the city skyline, was built for the 1962 Seattle World's Fair.

15. Match each matrix multiplication with its answer.

a. $\begin{bmatrix} 8 & -2 \\ 1 & 9 \end{bmatrix} \times \begin{bmatrix} 3 & 8 \\ -1 & -4 \end{bmatrix}$

b. $\begin{bmatrix} 24 & -16 \\ -1 & -36 \end{bmatrix} \times \begin{bmatrix} 1 & 0 \\ 0 & 1 \end{bmatrix}$

c. $\begin{bmatrix} 6 & 8 \\ -7 & -1 \end{bmatrix} \times \begin{bmatrix} 2 \\ 3 \end{bmatrix}$

i. $\begin{bmatrix} 26 & 72 \\ -6 & -28 \end{bmatrix}$

ii. $\begin{bmatrix} 36 \\ -17 \end{bmatrix}$

iii. $\begin{bmatrix} 24 & -16 \\ -1 & -36 \end{bmatrix}$

Solving Systems of Equations Using Elimination

I happen to feel that the degree of a person's intelligence is directly reflected by the number of conflicting attitudes she can bring to bear on the same topic.

LISA ALTHER

When you add equal quantities to each side of an equation, the resulting equation has the same solution as the original.

$$y - 7 = 12$$
$$+ \quad 7 = 7$$
$$\overline{\quad y \quad = 19}$$

Original equations.
Add equal quantities to both sides.
The resulting equations are true and have the same solutions as the originals.

$$3x - 5y = 9$$
$$+ \quad\quad 5y = 5y$$
$$\overline{3x \quad\quad = 9 + 5y}$$

In the same way, when you add two quantities that are equal, c and d, to two other quantities that are equal, a and b, the resulting expressions are equal.

$$a = b$$
$$+\, c = d$$
$$\overline{a + c = b + d}$$

Original equation.
Add equal quantities.
The resulting equation is true and has the same solutions as the originals.

The **elimination method** makes use of this fact to solve systems of linear equations.

EXAMPLE A

J. P. is thinking of two numbers, but he won't say what they are. He tells you that the sum of the two numbers is 163 and that their difference is 33. Find the two numbers.

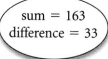

sum = 163
difference = 33

a. Write a system of equations for the sum and difference of these numbers.

b. Use the elimination method to solve this system.

▶ Solution

a. Let f and s represent the first and second numbers, respectively. Then the system is

$$\begin{cases} f + s = 163 \\ f - s = 33 \end{cases}$$

The first equation describes the sum, and the second describes the difference.

b. Note that adding the equations eliminates the variable s. Then solve for f.

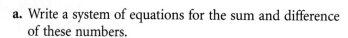

$$f + s = 163$$
$$\underline{f - s = 33}$$
$$2f = 196$$
$$f = 98$$

Original equations.

Add.
Divide both sides by 2.

So the first number is 98. To find the second number, substitute 98 for f into one of the original equations:

$$98 + s = 163 \qquad \text{or} \qquad 98 - s = 33$$

Either way, the second number is 65. Check that your solutions are correct.

$$f + s = 163 \qquad\qquad f - s = 33$$
$$98 + 65 \stackrel{?}{=} 163 \qquad\qquad 98 - 65 \stackrel{?}{=} 33$$
$$163 = 163 \qquad\qquad 33 = 33$$

Adding the two equations quickly leads to a solution because the resulting equation has only one variable. The other variable was eliminated! You won't always have coefficients that add to 0. You'll need another strategy for the elimination method to work.

Investigation
Paper Clips and Pennies

You will need

- three paper clips
- several pennies
- an 8.5-by-11-inch sheet of paper

In this investigation you'll create a system of equations by using paper clips and pennies as variables.

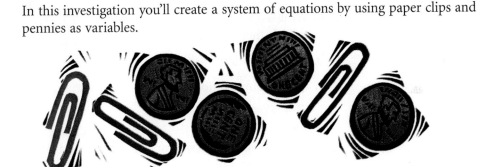

Step 1	Lay one paper clip along the long side of the paper. Then add enough pennies to complete the 11-inch length.
Step 2	Use C for the length of one paper clip and P for the diameter of one penny. Write an equation in standard form showing your results.
Step 3	Now you'll write the other equation for the system. Lay two paper clips along the shorter edge of your paper, and then add pennies to complete the 8.5-inch length.
Step 4	Using the same variables as in Step 2, write an equation to record your results for the shorter side.
Step 5	In this system the equations from Steps 2 and 4 have different coefficients for each variable. What can you do to one equation so that the variable C is eliminated when you add both equations?
Step 6	Use your answer to Step 5 to set up the addition of two equations. Once you eliminate the variable C, use the balancing method to solve for P.
Step 7	Substitute the value for P into one of the original equations to find C.
Step 8	Check that your solution satisfies both equations.
Step 9	Describe at least one other way to solve this system by elimination.
Step 10	Explain the real-world meaning of the solution. Describe other experiments in measuring that you can solve using a system of equations.

The goal of the elimination method is to get one of the variables to have a coefficient of 0 when you add the two equations. Sometimes you must first multiply one or both of the equations by some convenient number before you combine them. If you start with additive inverses such as s and $-s$ in Example A, then you can simply add the equations.

EXAMPLE B

The makers of FastBreak breakfast cereal offer a basketball in a promotional giveaway. There are two ways to get it. In Offer A you earn the basketball with 2 UPC symbols and $7. With Offer B you need 10 UPC symbols and $4. FastBreak has collected 1234 UPC symbols and $1,405. How many basketballs must it send out?

a. Write a system of equations that models this problem.

b. Use elimination to solve this system.

c. Check your solution.

▶ Solution

It is helpful to organize the information in a table first. Let a represent the number of basketballs sent out in Offer A, and let b represent the number of basketballs sent out in Offer B.

Basketball Giveaway

	Offer A (per basketball)	Offer B (per basketball)	Total
UPC symbols	2	10	1234
Dollars	7	4	1405

a. These equations describe the number of UPC symbols and the amount of money collected:

UPC symbols $\qquad 2a + 10b = 1234$

dollars $\qquad 7a + 4b\ = 1405$

b. To eliminate a from the resulting equation you must make its coefficients additive inverses, that is, numbers with opposite signs. If you multiply the UPC equation by 7 and the dollars equation by -2, then you get two new equations set up for elimination.

$$7(2a + 10b) = 7(1234) \quad \rightarrow \quad 14a + 70b = 8638 \qquad \text{Multiply both sides by 7.}$$
$$-2(7a + 4b) = -2(1405) \rightarrow \underline{-14a - 8b\ = -2810} \qquad \text{Multiply both sides by } -2.$$
$$62b = 5828 \qquad \text{Add the equations.}$$
$$\frac{62b}{62} = \frac{5828}{62} \qquad \text{Divide both sides by 62.}$$
$$b = 94 \qquad \text{Reduce.}$$

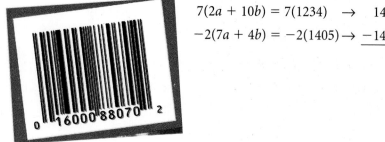

To find the value of a, you could substitute 94 for b in one of your original equations and solve for a, as you did in the previous lesson. Or you could go back to the original equations and use elimination on b. If you multiply the UPC equation by -2 and the dollars equation by 5, then you get two equations set up to eliminate b.

$$-2(2a + 10b) = -2(1234) \rightarrow -4a - 20b = -2468 \qquad \text{Multiply both sides by } -2.$$
$$5(7a + 4b) = 5(1405) \qquad \rightarrow \underline{35a + 20b = 7025} \qquad \text{Multiply both sides by 5.}$$
$$31a = 4557 \qquad \text{Add the equations.}$$
$$\frac{31a}{31} = \frac{4557}{31} \qquad \text{Divide both sides by 31.}$$
$$a = 147 \qquad \text{Reduce.}$$

c. Substitute 147 for a and 94 for b in the original equations.

$$2a + 10b \stackrel{?}{=} 1234$$
$$2(147) + 10(94) \stackrel{?}{=} 1234$$
$$294 + 940 \stackrel{?}{=} 1234$$
$$1234 = 1234$$

$$7a + 4b = 1405$$
$$7(147) + 4(94) \stackrel{?}{=} 1405$$
$$1029 + 376 \stackrel{?}{=} 1405$$
$$1405 = 1405$$

So there are 147 basketballs from Offer A and 94 from Offer B. FastBreak must send out 241 basketballs.

Since there is no single right order to the steps in solving a system of equations, you can start by choosing a variable that's easy to eliminate. You can use both elimination and substitution if that's easiest. Always check your solution in the original system.

EXERCISES

You will need your calculator for problem **9.**

Practice Your Skills

1. Consider the equation $5x + 2y = 10$.

  a. Solve the equation for y and sketch the graph.

  b. Multiply the equation $5x + 2y = 10$ by 3, and then solve for y. How does the graph of this equation compare with the graph of the original equation? Explain your answer.

2. Use the equation $5x - 2y = 10$ to find the missing coordinates of each point.

  a. $(6, a)$ **b.** $(-4, b)$

  c. $(c, 25)$ **d.** $(d, -5)$

3. Solve each system of equations by elimination. Show your work.

  a. $\begin{cases} 6x + 5y = -20 \\ -6x - 10y = 25 \end{cases}$ **b.** $\begin{cases} 5x - 4y = 23 \\ 7x + 8y = 5 \end{cases}$

4. Consider this system of equations:

$$\begin{cases} 2x - 5y = 12 \\ 6x - 15y = 36 \end{cases}$$

    a. By what number can you multiply which equation to eliminate the x-term when you combine the equations by addition? Do this multiplication.

    b. What is the sum of these equations?

    c. Solve each equation for y and graph the system.

    d. How does multiplying both sides of an equation affect the graph?

5. Consider this system of equations:

$$\begin{cases} 3x + 7y = -8 \\ 5x + 8y = -6 \end{cases}$$

In 5a and b, tell how you can eliminate each variable when you combine the equations by addition.

    a. the x-term   **b.** the y-term

Reason and Apply

6. List the different ways you have learned to solve the system. Then choose one method and find the solution.

$$\begin{cases} 3x + 7y = -8 \\ 5x + 8y = -6 \end{cases}$$

7. Solve each system using the elimination method.

    a. $\begin{cases} 2x + y = 10 \\ 5x - y = 18 \end{cases}$ **b.** $\begin{cases} 3x + 5y = 4 \\ 3x + 7y = 2 \end{cases}$ **c.** $\begin{cases} 2x + 9y = -15 \\ 5x + 9y = -24 \end{cases}$

8. In 8a–c, solve each equation for y and sketch a graph of the result on the same set of axes.

    a. $x - 2y = 6$

    b. $3x + 4y = 8$

    c. Graph the equation you get from adding the original two equations in 8a and b.

    d. What does the graph tell you?

9. Refer to this system from Example A to answer each question.

$$\begin{cases} x + y = 163 \\ x - y = 33 \end{cases}$$

    a. Solve each equation for y and enter each equation into your calculator. Use the window [0, 150, 10, 0, 150, 10] to graph this system.

    b. Use the elimination method to find the y-value of the solution. Enter the resulting equation into Y₃ and add it to your graph from 9a.

    c. Use elimination to find the x-value of the solution. Draw a vertical line on the graph to represent the equation you found in 9b.

    d. Describe what you notice about the four lines on your screen and explain why this happens.

10. Part of Adam's homework paper is missing. If (5, 2) is the only solution to the system shown, write a possible equation that completes the system.

$$2x + y = 12$$
$$4x$$

11. Anisha turned in this quiz in her algebra class.

Anisha _____ Score _____

Solve this system:
$$y = x - 5$$
$$3y + 2x = 5$$

Solution:
$$3(x - 5) + 2x = 5$$
$$3x - 15 + 2x = 5$$
$$5x = 20$$
$$x = 4$$

a. What method did she use?

b. Is her solution correct?

12. The school's photographer took pictures of couples at this year's prom. She charged $3.25 for wallet-size pictures and $10.50 for portrait-size pictures.

a. Write a system of equations representing the fact that Crystal and Dan bought a total of 10 pictures for $61.50.

b. Solve this system and explain what your answer means.

13. **APPLICATION** Automobile companies advertise two rates for fuel mileage. City mileage is the rate of fuel consumption for driving in stop-and-go traffic. Highway mileage is the rate for driving at higher speeds for long periods of time.

Cynthia's new car gets 17 mi/gal in the city and 25 mi/gal on the highway. She drove 220 miles on 11 gallons of gas.

a. Define variables and write a system of equations for the gallons burned at each mileage rate.

b. Solve this system and explain the meaning of the solution.

c. Find the number of city miles and highway miles Cynthia drove.

d. Check your answers.

▶ Review

14. For each pair of fractions, name a fraction that lies between them.

a. $\frac{1}{2}$ and $\frac{3}{4}$ **b.** $\frac{2}{3}$ and $\frac{7}{8}$ **c.** $-\frac{1}{4}$ and $-\frac{1}{5}$ **d.** $\frac{7}{11}$ and $\frac{5}{6}$

e. Describe a strategy for naming a fraction between any two fractions.

15. APPLICATION When you go up a mountain, the temperature drops about 4 degrees Fahrenheit for every 1000 feet you ascend.

a. While climbing a trail on Mt. McKinley in Alaska, Marsha intended to record the elevation and temperature at three locations. Complete the table for her.

Marsha's Climb

	Elevation (ft)	Temperature (°F)
Start	4,300	78
Rest station		64
Highest point	11,900	

b. Write an equation to model the relationship between elevation and temperature. Explain the meaning of the slope and y-intercept.

c. Mt. McKinley is 20,320 feet high. On the day Marsha was climbing, how cold was it at the summit?

This mountain climber is ascending Mt. McKinley in Denali National Park, Alaska.

16. Write an equation in point-slope form using the given information.

a. A line whose slope is -2 that passes through the point $(5, -3)$.

b. A line whose slope is 2.5 that passes through the point $(-3, 7)$.

17. Students are randomly assigned to a locker in a school where even numbered lockers are on the bottom and odd numbered lockers are on top. There is a total of 800 lockers. However, due to construction mess, the bottom row of lockers is blocked off under lockers numbered 201 to 261. Those lockers have been removed from the system. The odd numbered lockers are still available.

a. What is the probability of being assigned a locker on the top row? What is the probability of being assigned a locker on the bottom row?

b. In the situation described, what is a trial? What is an outcome?

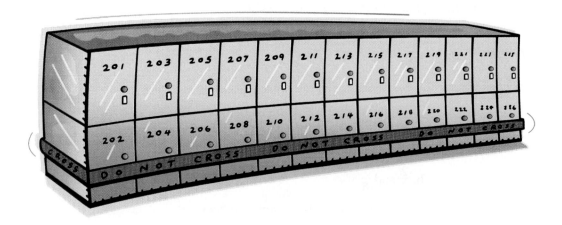

<table><tr><th>LESSON</th><th>6.4</th></tr></table>

Solving Systems of Equations Using Matrices

In Lesson 1.8, you learned how to enter, display, and use matrices to organize and analyze data. In this lesson you will use matrices to solve systems of equations.

The essence of mathematics is not to make simple things complicated but to make complicated things simple.
STANLEY GUDDER

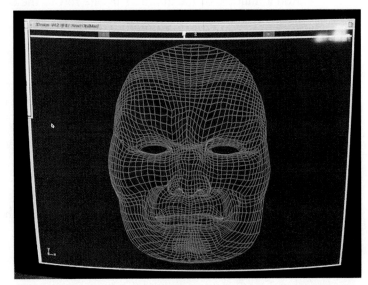

Software that renders 3-D computer-generated images uses matrices to organize data. This program graphs thousands of points and lines to draw the contours of a person's face.

If you look only at the numerals in a system of equations in standard form $ax + by = c$—that is, the coefficients of both variables and the constant terms—you have a matrix with two rows and three columns. If you have a system with both equations in standard form $ax + by = c$, you can write a matrix for the system:

$$\begin{cases} 5x + 3y = -1 \\ 2x - 6y = 50 \end{cases} \qquad \begin{bmatrix} 5 & 3 & -1 \\ 2 & -6 & 50 \end{bmatrix}$$

The numerals in the first equation match the numerals in the first row, and the numerals in the second equation match the numerals in the second row. But what does the solution look like in a matrix? The solution to the system above is $(4, -7)$, or $x = 4$ and $y = -7$. You want the rows of the solution matrix to represent the equations. So you can rewrite each equation to get the numerals for each row of the solution matrix:

$$\begin{matrix} x = 4 \\ y = -7 \end{matrix} \rightarrow \begin{matrix} x + 0y = 4 \\ 0x + y = -7 \end{matrix} \rightarrow \begin{bmatrix} 1 & 0 & 4 \\ 0 & 1 & -7 \end{bmatrix}$$

In the elimination method, you combined equations and multiplied them by numbers. In much the same way, you can modify the rows of a matrix by performing **row operations** on each number in those rows.

Row Operations in a Matrix

▶ Multiply (or divide) all numbers in a row by a nonzero number.
▶ Add all numbers in a row to corresponding numbers in another row.
▶ Add a multiple of the numbers in one row to the corresponding numbers in another row.
▶ Exchange two rows.

You can do these operations on the rows of a matrix to change the starting matrix into the solution matrix. The general strategy is to get a diagonal of 1's in the matrix with 0's above and below, like this:

$$\begin{bmatrix} 1 & 0 & a \\ 0 & 1 & b \end{bmatrix}$$

The ordered pair (a, b) is the solution, if one exists, to the system.

EXAMPLE A   Solve this system of equations using matrices:

$$\begin{cases} x - 2y = 3 \\ 3x + y = 23 \end{cases}$$

▶ Solution   Copy the numerals from each equation into each row of the matrix. Then use row operations to transform it into the solution matrix.

$$\begin{cases} x - 2y = 3 \\ 3x + y = 23 \end{cases} \longrightarrow \begin{bmatrix} 1 & -2 & 3 \\ 3 & 1 & 23 \end{bmatrix}$$

Add -3 times row 1 to row 2.

$$\begin{array}{llrrr} -3 \text{ times row 1} & \to & -3 & 6 & -9 \\ + \text{ row 2} & \to & 3 & 1 & 23 \\ \hline \text{New row 2} & \to & 0 & 7 & 14 \end{array} \qquad \begin{bmatrix} 1 & -2 & 3 \\ 0 & 7 & 14 \end{bmatrix}$$

Divide row 2 by 7.

$$\begin{bmatrix} 1 & -2 & 3 \\ 0 & 1 & 2 \end{bmatrix}$$

Add 2 times row 2 to row 1.

$$\begin{array}{llrrr} 2 \text{ times row 2} & \to & 0 & 2 & 4 \\ + \text{ row 1} & \to & +1 & -2 & 3 \\ \hline \text{New row 1} & \to & 1 & 0 & 7 \end{array} \qquad \begin{bmatrix} 1 & 0 & 7 \\ 0 & 1 & 2 \end{bmatrix}$$

Use the solution matrix to write the equations:

$$\begin{bmatrix} 1 & 0 & 7 \\ 0 & 1 & 2 \end{bmatrix} \rightarrow \begin{array}{ll} 1x + 0y = 7 & \text{or} \quad x = 7 \\ 0x + 1y = 2 & \text{or} \quad y = 2 \end{array}$$

The solution to the system is (7, 2).

Investigation
Diagonalization

In this investigation you will see how to combine row operations in your solution process.

Consider the system of equations

$$\begin{cases} 2x + y = 11 \\ 6x - 5y = 9 \end{cases}$$

Step 1   Write the matrix for this system. What does the first row contain? The second row?

Step 2   Describe how to use row operations to get a 0 as the first entry in the second row. Write this matrix.

Step 3   Next, get a 1 as the second number in the second row of your matrix from Step 2.

Step 4   Use row operations on the matrix from Step 3 to get a 0 as the second number in row 1.

Step 5   Next, get a 1 as the first number of row 1 of your matrix from Step 4. Tell what this matrix means, and give the solution to the system.

Step 6   Check your solution by solving the system with another method.

Look at the first three rules for Row Operations in a Matrix. How do they correspond to steps in the elimination process?

History
● CONNECTION ●

German mathematician Carl Friedrich Gauss (1777–1855) made many contributions to mathematics. He developed the elementary row operations on matrices. In his honor the process of solving systems with matrices is sometimes called "Gaussian elimination."

Matrices are useful for solving systems involving large numbers. Here is another example.

EXAMPLE B

On Friday, 3247 people attended the county fair. The entrance fee for an adult was $5, and for a child 12 or under the fee was $3. The fair collected a total of $14,273. How many of the total attendees were adults and how many were children?

▶ **Solution**

Use A for the number of adults attending the fair and C for the number of children attending. Use these variables to write a system of equations and solve it using matrices. The attendance is the number of adults and children at the fair. So the first equation is $A + C = 3247$. The fair collected $5A$ dollars for A adults and $3C$ dollars for C children in attendance. The total collected is $5A + 3C$, so the second equation is $5A + 3C = 14273$.

With one equation describing attendance at the fair, and another describing ticket money collected, the system is

$$\begin{cases} A + C = 3247 \\ 5A + 3C = 14273 \end{cases} \longrightarrow \begin{bmatrix} 1 & 1 & 3247 \\ 5 & 3 & 14273 \end{bmatrix}$$

Use row operations to find the solution.

Add -5 times row 1 to row 2 to get new row 2. $\begin{bmatrix} 1 & 1 & 3247 \\ 0 & -2 & -1962 \end{bmatrix}$

Divide row 2 by -2. $\begin{bmatrix} 1 & 1 & 3247 \\ 0 & 1 & 981 \end{bmatrix}$

Add -1 times row 2 to row 1 to get new row 1. $\begin{bmatrix} 1 & 0 & 2266 \\ 0 & 1 & 981 \end{bmatrix}$

The final matrix shows that $A = 2266$ and $C = 981$. So there were 2266 adults and 981 children at the fair on Friday.

To check this solution, substitute 2266 for A and 981 for C into the original equations.

$$A + C = 3247 \qquad\qquad 5A + 3C = 14273$$
$$2266 + 981 \stackrel{?}{=} 3247 \qquad 5(2266) + 3(981) \stackrel{?}{=} 14273$$
$$3247 = 3247 \qquad\qquad 11330 + 2943 \stackrel{?}{=} 14273$$
$$14273 = 14273$$

These are true statements, so the solution checks. With row operations on matrices, you now have five methods to solve systems of linear equations. Like elimination and substitution, row operations on matrices give exact solutions. With practice, you will develop a sense of when it is easiest to use each solution method. The form of the equation often makes some methods easier to use than others. If an equation is solved for y, then it is easiest to use the substitution method. If the equations are in standard form, then it is easiest to solve by elimination or by using matrices.

EXERCISES

You will need your calculator for problem **8.**

▶ Practice Your Skills

1. Write a system of equations whose matrix is

a. $\begin{bmatrix} 2 & 1.5 & 12.75 \\ -3 & 4 & 9 \end{bmatrix}$

b. $\begin{bmatrix} \frac{1}{2} & 0 & \frac{1}{2} \\ -1 & 2 & 0 \end{bmatrix}$

c. $\begin{bmatrix} 2 & 3 & 1 \\ 0 & 2 & 0 \end{bmatrix}$

2. Write the matrix for each system.

a. $\begin{cases} x + 4y = 3 \\ -x + 2y = 9 \end{cases}$

b. $\begin{cases} 7x - y = 3 \\ 0.1x - 2.1y = 3 \end{cases}$

c. $\begin{cases} x + y = 3 \\ x + y = 6 \end{cases}$

3. Write each solution matrix as an ordered pair.

a. $\begin{bmatrix} 1 & 0 & 8.5 \\ 0 & 1 & 2.8 \end{bmatrix}$

b. $\begin{bmatrix} 1 & 0 & \frac{1}{2} \\ 0 & 1 & \frac{13}{16} \end{bmatrix}$

c. $\begin{bmatrix} 1 & 0 & 0 \\ 0 & 1 & 0 \end{bmatrix}$

4. Use row operations to transform the matrix $\begin{bmatrix} 4.2 & 0 & 12.6 \\ 0 & -1 & 5.25 \end{bmatrix}$ into the form
$\begin{bmatrix} 1 & 0 & a \\ 0 & 1 & b \end{bmatrix}$

Write the solution as an ordered pair.

5. Consider the system

$\begin{cases} y = 7 - 3x \\ y = 11 - 2(x - 5) \end{cases}$

a. Convert each equation to the standard form $ax + by = c$.

b. Write a matrix for the system.

▶ Reason and Apply

6. Give the missing description or matrix for each step of the process below. Give the solution as an ordered pair.

Description	Matrix
The matrix for $\begin{cases} 3x + 2y = 28.9 \\ 8x + 5y = 74.6 \end{cases}$	$\begin{bmatrix} 3 & 2 & 28.9 \\ 8 & 5 & 74.6 \end{bmatrix}$
Add 8 times row 1 to -3 times row 2 and put the result in row 2.	$\begin{bmatrix} & & \\ & & \end{bmatrix}$
	$\begin{bmatrix} 3 & 0 & 14.1 \\ 0 & 1 & 7.4 \end{bmatrix}$
	$\begin{bmatrix} 1 & 0 & 4.7 \\ 0 & 1 & 7.4 \end{bmatrix}$

7. APPLICATION Each day Sal prepares a large basket of self-serve tortilla chips in his restaurant. On Monday, 40 adult patrons and 15 child patrons ate 10.8 kg of chips. On Tuesday, 35 adult patrons and 22 child patrons ate 12.29 kg of chips. Sal wants to know whether adults or children eat more chips on average.

a. Organize the information into a table.

b. Define variables and write the system of equations showing the average amount of chips the adults and the children eat each day.

c. Write a matrix for the system.

d. Solve the system by transforming the matrix into the solution matrix $\begin{bmatrix} 1 & 0 & a \\ 0 & 1 & b \end{bmatrix}$.

e. Write a sentence that describes the real-world meaning of the solution to the system.

8. Your calculator probably has built-in row operations to transform a matrix into its solution form. Transform this matrix using row operations on your calculator. [▶ 🖳 See **Calculator Note 6B.** ◀]

$$\begin{bmatrix} 8 & 7 & -1 \\ 3 & -1 & -4 \end{bmatrix}$$

9. APPLICATION Zoe must ship 532 tubas and 284 kettledrums from her warehouse to a store across the country. A truck rental company offers two sizes of trucks. A small truck will hold 5 tubas and 7 kettledrums. A large truck will hold 12 tubas and 4 kettledrums. If she wants to fill each truck so that the cargo won't shift, how many small and large trucks should she rent?

a. Define variables and write a system of equations to find the number of small trucks and the number of large trucks Zoe needs to ship the instruments. Write one equation for each instrument.

b. Write a matrix that represents the system.

c. Perform row operations to transform the matrix into a solution matrix.

d. Write a sentence describing the real-world meaning of the solution.

10. APPLICATION Will is baking a new kind of bread. He has two different kinds of flour. Flour X is enriched with 0.12 mg of calcium per gram; Flour Y is enriched with 0.04 mg of calcium per gram. Each loaf has 300 grams of flour, and Will wants each loaf to have 30 mg of calcium. How much of each type of flour should he use for each loaf?

a. Write a system of equations that relates the number of grams of Flour X and the number of grams of Flour Y. (Hint: Write one equation for the total grams of flour and one for the total grams of calcium.)

b. Write a matrix for the system.

c. Find the solution matrix.

d. Explain the real-world meaning of the solution.

11. APPLICATION On Monday a group of students started on a three-day bicycle tour covering a total of 286 km. On Tuesday they cycled 7 km less than on Monday. On Wednesday they traveled 24 km less than on Tuesday.

   a. Write a system of three linear equations representing this trip. Use m, t, and w to represent the distances in kilometers they cycled on Monday, Tuesday, and Wednesday, respectively. Write each equation in the form $am + bt + cw = d$.

   b. Write a 3 × 4 matrix to model this system of equations. Describe what the rows and columns of your matrix represent.

   c. List and describe a sequence of matrix row operations that will produce a matrix of the form

$$\begin{bmatrix} 1 & 0 & 0 & ? \\ 0 & 1 & 0 & ? \\ 0 & 0 & 1 & ? \end{bmatrix}$$

   d. What is the solution to this problem?

Review

12. These matrices show the cost of a 1-day ticket and a 3-day ticket for an adult, a teen, and a child at two amusement parks, Tivoli and Hill.

1-day ticket

	Tivoli	Hill
Adult	31	28
Teen	26	24
Child	21	16

3-day ticket

	Tivoli	Hill
Adult	72	65
Teen	55	55
Child	45	35

   a. Write a matrix equation displaying the difference in cost between a 3-day ticket and a 1-day ticket.

   b. Which type of ticket is the better deal and why?

   c. Which type of ticket should you buy if you are in the park for only 2 days?

13. Write a recursive sequence for the y-coordinates of the points shown on each graph. On each graph one tick mark represents 1 unit.

a.

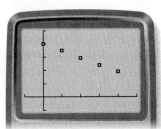

b.

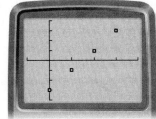

c.

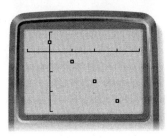

d.

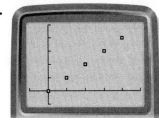

14. At the Coffee Stop you can buy a mug for $25 and then pay only $0.75 per hot drink.

    a. What is the slope of the equation that models the total cost of refills? What is the real-world meaning of the slope?

    b. Use the point (33, 49.75) to write an equation in point-slope form that models this situation.

    c. Rewrite your equation in intercept form. What is the real-world meaning of the y-intercept?

15. Over 2000 years ago the Chinese developed column equation matrices as a method to solve linear equations. The numerals of each equation are arranged in columns instead of rows. Use the biancheng (translated as "multiply throughout") and zhichu (translated as "direct reduction") rules of operation to solve the system.

Rules of Operation

    1. Biancheng: Multiply the numerals of the left column by the numeral at the top of the right column.

    2. Zhichu: Subtract the right column from the resulting left column repeatedly until you get a 0 at the top.

For example, represent this system as a column equation matrix.

$$\begin{cases} 2x + y = 11 \\ 6x - 5y = 9 \end{cases} \rightarrow \begin{bmatrix} 2 & 6 \\ 1 & -5 \\ 11 & 9 \end{bmatrix}$$

Biancheng: Multiply the first column by 6 (highest top row numeral).

$$2(6) \quad \rightarrow \quad 12$$
$$1(6) \quad \rightarrow \quad 6$$
$$11(6) \quad \rightarrow \quad 66$$

Zhichu: Subtract the right column from the left column twice.

$$12 \;-\; 6 \;-\; 6 \;\rightarrow\; 0$$
$$6 \;-\; -5 \;-\; -5 \;\rightarrow\; 16$$
$$66 \;-\; 9 \;-\; 9 \;\rightarrow\; 48$$

Write a new equation and solve for y.

$$16y = 48 \quad \text{or} \quad y = 3$$

Substitute and solve for x.

$$2x + 3 = 11 \quad \text{or} \quad x = 4$$

Use a Chinese column equation matrix to solve the system

$$\begin{cases} x - 2y = 3 \\ 3x + y = 23 \end{cases}$$

(Jean-Claude Martzloff, *A History of Chinese Mathematics*, 1997, pp. 252–254; Lǐ Yǎn and Dù Shírán, *Chinese Mathematics, a Concise History*, 1987, pp. 46–48)

Inequalities in One Variable

Drink at least six glasses of water a day. Store milk at temperatures below 40°F. Eat snacks with fewer than 20 calories. Spend at most $10 for a gift. These are a few examples of inequalities in everyday life. In this lesson you will analyze situations involving inequalities in one variable and learn how to find and graph their solutions.

An **inequality** is a statement that one quantity is less than or greater than another. You write inequalities using these symbols:

| less than | < | less than or equal to | ≤ |
| greater than | > | greater than or equal to | ≥ |

Sometimes you need to translate everyday language into the phrases you see in the table above. Here are some examples.

History
CONNECTION

Thomas Harriot (1560–1621) introduced the symbols of inequality < and >. Pierre Bouguer first used the symbols ≤ and ≥ about a century later.

(Florian Cajori, *A History of Mathematics*, 1985)

Everyday phrase	Translation	Inequality
at least six glasses	The number of glasses is greater than or equal to 6.	$g \geq 6$
below 40°	The temperature is less than 40°.	$t < 40$
fewer than 20 calories	The number of calories is less than 20.	$c < 20$
at most $10	The price of the gift is less than or equal to $10.	$p \leq 10$
between 35°and 120°	35° is less than the temperature and the temperature is less than 120°.	$35 < t < 120$

You solve inequalities very much like you solve equations. You use the same strategies—adding or subtracting the same quantity to both sides, multiplying both sides by the same number or expression, and so on. However, there is one exception you need to remember when solving inequalities. You will explore this exception in the investigation.

Investigation
Toe the Line

You will need
- chalk or measuring tape to mark a segment

In this investigation you will analyze properties of inequalities and discover some interesting results.

First you'll act out operations on a number line.

Procedure Note
The announcer calls out operations for Walkers A and B. The walkers perform operations on their numbers by walking to the resulting values on the number line. The recorder logs the position of each walker after each operation.

Step 1 In your group choose an announcer, a recorder, and two walkers. The two walkers make a number line on the ground with marks from −10 to 10. The announcer and recorder make a table with these column headings and twelve rows. The operations to use as row headings are Add 2. Subtract 3. Add −2. Subtract −4. Multiply by 2. Subtract 7. Multiply by −3. Add 5. Divide by −4. Subtract 2. Multiply by −1.

Operation	Walker A's position	Inequality symbol	Walker B's position
Starting number	2		4

Step 2 As a trial, act out the first operation in the table: Walker A simply stands at 2 on the number line, and Walker B stands at 4.

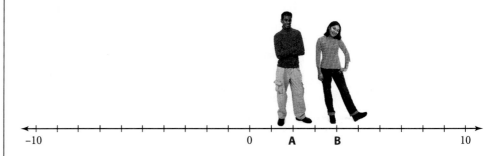

Enter the inequality symbol in the table that describes the relative position of Walkers A and B on the number line. Be sure you have written a true inequality.

Step 3 Call out the operations. After the walkers calculate their new numbers, record the operation and walkers' positions in the next row.

Step 4 As a group, discuss which inequality symbol to enter into each cell of the third column.

Next you'll analyze what each operation does to the inequality.

Step 5 What happens to the walkers' relative positions on the number line when the operation adds or subtracts a positive number? A negative number? Does anything happen to the direction of the inequality symbol?

Step 6 What happens to the walkers' relative positions on the number line when the operation multiplies or divides by a positive number? Does anything happen to the inequality symbol?

Step 7	What happens to the walkers' relative positions on the number line when the operation multiplies or divides by a negative number? Does the inequality symbol change directions?
Step 8	Which operations on an inequality reverse the inequality symbol? Does it make any difference which numbers you use? Consider fractions and decimals as well as integers.
Step 9	Check your findings about the effects of adding, subtracting, multiplying, and dividing by the same number on both sides of an inequality by creating your own table of operations and walkers' positions.

In square dancing a caller tells the dancers which steps to take. Their maneuvers depend on their relative positions.

This example will show you how to graph solutions to inequalities.

EXAMPLE A

Graph each inequality on a number line.

a. $t > 5$

b. $x \leq -1$

c. $-2 \leq x < 4$

▶ **Solution**

a. Any number greater than 5 satisfies this inequality. So $5.0001 > 5$, $7\frac{1}{2} > 5$, and $1,000,000 > 5$ are all true statements. You show this by drawing an arrow through the values that are greater than 5.

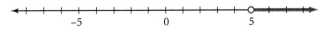

The open circle at 5 excludes 5 from the solutions because $5 > 5$ is not a true statement.

b. The inequality $x \leq -1$ reads, "x is less than or equal to -1." The solid circle at -1 includes the value -1 in the solutions because $-1 \leq -1$ is a true statement.

c. This statement is a **compound inequality.** It says that -2 is less than or equal to x and that x is less than 4. So the graph includes all values that are greater than or equal to -2 and less than 4. The solid circle at -2 includes -2 in the solutions because $-2 \leq -2$ is true. The open circle at 4 excludes 4 from the solutions because $4 < 4$ is not true.

When you graph inequalities, always label 0 on the number line as a point of reference.

EXAMPLE B

Erin says, "I lose 15 minutes of sleep every time the dog barks. Last night I got less than 5 hours of sleep. I usually sleep 8 hours." Find the number of times Erin woke up.

To solve the problem, let x represent the number of times Erin woke up, and write an inequality.

Graph your solutions.

Solve the inequality.

▶ Solution

The number of hours Erin slept is 8 hours, minus $\frac{1}{4}$ hour times x, the number of times she woke up. The total is less than 5 hours. So the inequality is $8 - 0.25x < 5$.

Solve the inequality for x. Remember to reverse the inequality symbol if you multiply or divide by a negative number.

$8 - 0.25x < 5$	Original inequality.
$8 - 0.25x - 8 < 5 - 8$	Subtract 8 from both sides of the inequality.
$-0.25x < -3$	Evaluate.
$\dfrac{-0.25x}{-0.25} > \dfrac{-3}{-0.25}$	Divide both sides by -0.25, and reverse the inequality symbol.
$x > 12$	Divide.

The dog woke her up more than 12 times. The graph of the solutions is

Any number to the right of 12 on the number line satisfies the inequality. The open circle at 12 shows that 12 itself is not a solution.

Working with inequalities is very much like working with equations. An equation shows a balance between two quantities, but an inequality shows an imbalance. The important thing to remember is that multiplying and dividing both sides of an equation by a negative number tips the scales in the opposite direction.

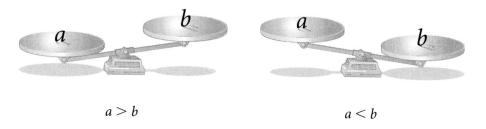

$a > b$ $a < b$

EXERCISES

You will need your calculator for problem **14.**

Practice Your Skills

1. Tell what operation on the first inequality gives the second one, and give the answer using the correct inequality symbol.

   a. $3 < 7$
   $4 \cdot 3 \ \Box \ 7 \cdot 4$

   b. $5 \leq 12$
   $-3 \cdot 5 \ \Box \ 12 \cdot -3$

   c. $-4 \geq x$
   $-4 + -10 \ \Box \ x + -10$

   d. $b + 3 > 15$
   $b + 3 - 8 \ \Box \ 15 - 8$

   e. $24d < 32$
   $\dfrac{24d}{3} \ \Box \ \dfrac{32}{3}$

   f. $24x \leq 32$
   $\dfrac{24x}{-3} \ \Box \ \dfrac{32}{-3}$

2. Find three values of the variable that satisfy each inequality.

   a. $5 + 2a > 21$

   b. $7 - 3b < 28$

   c. $-11.6 + 2.5c < 8.2$

   d. $4.7 - 3.25d > -25.3$

3. Give the inequality graphed on each number line.

   a.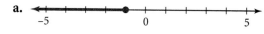

   b.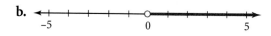

   c.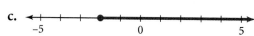

   d.

   e.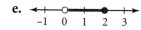

4. Translate each phrase into symbols.

   a. 3 is more than x

   b. y is at least -2

   c. z is no more than 12

   d. n is not greater than 7

5. Solve each equation for y.

   a. $3x + 4y = 5.2$

   b. $3(y - 5) = 2x$

▶ Reason and Apply

6. Solve each inequality and show your work.

   a. $4.1 + 3.2x > 18$

   b. $7.2 - 2.1b < 4.4$

   c. $7 - 2(x - 3) \geq 25$

   d. $11.5 + 4.5(x + 1.8) \leq x$

7. Solve each inequality and graph the solutions on a number line.

   a. $3x - 2 \leq 7$

   b. $4 - x > 6$

   c. $3 + 2x \geq -3$

   d. $10 \leq 2(5 - 3x)$

8. Ezra received $50 from his grandparents for his birthday. He makes $7.50 each week for odd jobs he does around the neighborhood. Since his birthday, he has saved more than enough to buy the $120 gift he wants to buy for his parents' 20th wedding anniversary. How many weeks ago was his birthday?

9. For each graph, tell what operation moves the two points in the inequality to their new positions. Write the new inequality, stating the position of the red dot first.

   a. $1 < 2$

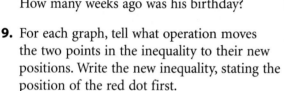

   b. $6 > 2$

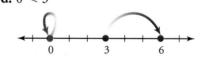

   c. $-1 < 1$

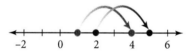

   d. $0 < 3$

10. Tell whether each inequality is true or false for the given value.

   a. $x - 14 < 9, x = 5$

   b. $3x \geq 51, x = 7$

   c. $2x - 3 < 7, x = 5$

   d. $4(x - 6) \geq 18, x = 12$

11. Solve each inequality. Explain the meaning of the result. On a number line, graph the values of x that make the original inequality true.

   a. $2x - 3 > 5x - 3x + 3$

   b. $-2.2(5x + 3) \geq -11x - 15$

12. You read the inequality symbols $<$, $\leq$, $>$, and $\geq$ as "is less than," "is less than or equal to," "is greater than," and "is greater than or equal to," respectively. But you describe everyday situations with different expressions. Identify the variable in each statement and give the inequality to describe each situation.

a. I'll spend no more than $30 on CDs this month.

b. You must be at least 48 inches tall to go on this ride.

c. Three or more people make a carpool.

d. No one under age 17 will be admitted without a parent or guardian.

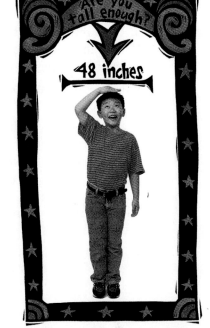

Review

13. List the order in which you would perform these operations to get the correct answer.

a. $72 - 12 \cdot 3.2 = 33.6$ **b.** $2 + 1.5\,(3 - 5^2) = -31$

c. $21 \div 7 - 6 \div 2 = 0$

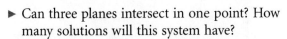

IMPROVING YOUR **VISUAL THINKING** SKILLS

In this chapter you have seen three possible outcomes for a system of two equations in two variables. If one solution exists, it is the point of intersection. If no solution exists, the lines are parallel and there is no point of intersection. If infinitely many solutions exist, the two lines overlap.

But what do the solutions look like in a system of three linear equations in three unknowns? An equation like $3x + 2y = 12$ is a line, but an equation in three variables is a plane. Consider the graph of $3x + 2y + 6z = 12$. Imagine the x-axis coming out of the page. The shaded triangle indicates the part of the solution plane whose coordinates are all positive. The complete plane is infinite.

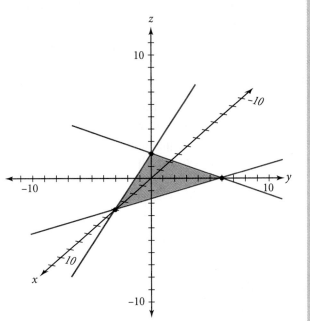

If you have two more planar equations, you have a system of three equations in three variables. There will be three planes on the graph. So the solutions to this system are where the planes intersect, if they do at all. Visualize how three planes could intersect to answer these questions.

▶ Can three planes intersect in one point? How many solutions will this system have?

▶ If a system has infinitely many solutions, are all three equations the same plane?

▶ If the system has no solutions, are the planes parallel?

14. The table shows the 2001 U.S. postal rates for letters, flats, and parcels.

U.S. Postal Rates

Weight	Rate
First ounce or fraction of an ounce	$0.34
Each additional ounce or fraction	$0.21

(*http://new.usps.com*)

a. Use a recursive routine to create a table that shows the cost of sending letters weighing from 0 to 11 ounces.

b. Use 1-ounce units on the horizontal axis to plot the postal costs.

c. Kasey has drawn a line through the points on her graph. What real-world meaning does this line have? Is a line useful in this situation? Why or why not?

d. What is the cost of sending a 10.5-ounce parcel?

15. Use the distributive property to rewrite each expression without using parentheses.

a. $-2(x + 8)$ **b.** $4(0.75 - y)$

c. $-(z - 5)$

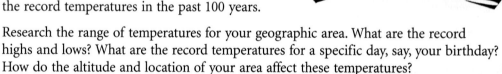

TEMPERATURES

Temperatures for your city vary depending on the time of day, season, and its location. Weather reports give the daily high and low temperatures and often compare them with the record temperatures in the past 100 years.

Research the range of temperatures for your geographic area. What are the record highs and lows? What are the record temperatures for a specific day, say, your birthday? How do the altitude and location of your area affect these temperatures?

Compare your results to temperatures on the moon. Research the temperatures of other planets such as Venus, Mars, and Pluto. What factors affect these data sets? Are the temperatures given in degrees Fahrenheit or degrees Celsius? Be sure to convert all data to the same units before comparing. Describe your findings with inequalities and graphs in a paper or give a presentation.

Your project should include

▶ Your hometown high and low temperatures.

▶ Algebraic expressions with compound inequalities.

▶ Clearly labeled graphs.

Some people think it may be possible to live on another planet or moon someday. Based on your findings, what do you think?

This view from the Apollo II spacecraft shows Earth above the Lunar terrain.

Graphing Inequalities in Two Variables

In Lesson 6.5, you learned to graph inequalities in one variable on a number line. However, some situations, such as the number of points a football team scores by touchdowns and field goals, require more than one variable. In this lesson you will learn to graph inequalities in two variables on the coordinate plane.

You have graphed many equations like $y = 1 + 0.5x$. In the following investigation you will learn how to graph inequalities such as $y < 1 + 0.5x$ and $y > 1 + 0.5x$.

Investigation
Graphing Inequalities

You will need

- the worksheet Graphing Inequalities Grids

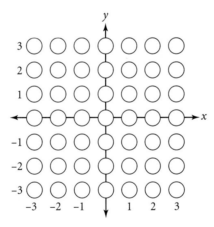

First you'll make a graph from one of four statements.

 i. $y \,\square\, 1 + 0.5x$

 ii. $y \,\square\, -1 - 2x$

 iii. $y \,\square\, 1 - 0.5x$

 iv. $y \,\square\, 1 - 2x$

Step 1 | Each member of the group should choose a different statement from above.

Step 2 | Use the coordinates of each point shown with a circle to test the statement you selected. Fill in each circle with the relational symbol, $<$, $>$, or $=$, that makes the statement true. For example, to test the upper left point in statement i, substitute $(-3, 3)$ for (x, y) as follows:

$$y \;\square\; 1 + 0.5x$$ 　　　　　　　　Original statement.

$$3 \;\square\; 1 + 0.5(-3)$$ 　　　　　　Substitute 3 for y and -3 for x.

$$3 > -0.5$$ 　　　　　　　　　　Evaluate and choose the symbol.

Place a ">" in the upper left circle because this symbol makes statement i true.

Step 3 | Repeat Step 2 for your statement. Work down each column to fill all 49 circles with one of the three symbols.

Next you'll analyze the results of your graph.

Step 4 | What do you notice about the circles filled with the equal sign? Tell any other patterns you see.

Step 5 | Test a point with fractional or decimal coordinates that is not represented by a circle on the grid. Compare your result with the symbols on the same side of the line of equal signs as your point.

Step 6 | Draw a set of xy-axes, with scales from -3 to 3 on each axis. Under the graph write your statement with the less than symbol $<$. Shade the region of points that makes your statement true. If the points on the line make an inequality true, draw a solid line through them. If not, draw a dashed line. Repeat this step for each of the remaining symbols ($>, \leq, \geq, =$).

Finally, you'll draw general conclusions by comparing graphs in your group.

Step 7 | Compare your graphs with those of others in your group. What graphs require a solid line? A dashed line?

Step 8 | What graphs require shading? Shading above the line? Below the line?

Step 9 | Discuss how to check the graph of an inequality with one point.

The graph of the solutions to a single inequality is called a **half-plane** because it includes all the points in the coordinate plane that fall on one side of the boundary line.

EXAMPLE | Graph the inequality $2x - 3y > 3$, and check to see whether each point is part of the solution.

　　i. $(3, -2)$ 　　　　**ii.** $(3, 1)$ 　　　　**iii.** $(-1, 2)$ 　　　　**iv.** $(-2, -3)$

▶ **Solution** | To graph the inequality, first solve it for y:

$$2x - 3y > 3$$ 　　　　　　　　Original inequality.

$$-3y > 3 - 2x$$ 　　　　　　　Subtract $2x$ from both sides.

$$y < -1 + \frac{2}{3}x$$ 　　　　　　Divide both sides by -3 and reverse the inequality symbol.

Graph the line $y = -1 + \frac{2}{3}x$ with a dashed line to indicate that points on the line are not part of the solution to the inequality. Because the inequality in y is less than the expression in x on the right side, shade the region *below* the line. Points in this region will have y-values that are less than the expression in x.

If you plot the given points, you'll see that the points that satisfy the inequality lie in the shaded part of the plane.

To check numerically whether the given points satisfy the inequality, substitute the x- and y-values from each given coordinate pair for x and y in the inequality $2x - 3y > 3$, and enter the inequality into your calculator. When you press (ENTER) you'll see 1 if the inequality is true or 0 if the inequality is false, as shown on the calculator screen.

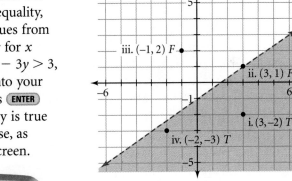

	2(3) – 3(–2) > 3	
		1
	2(3) – 3(1) > 3	
		0
	2(–1) – 3(2) > 3	
		0
	2(–2) – 3(–3) > 3	
		1

i. $2(3) - 3(-2) > 3$ $\longrightarrow$ $12 > 3$ $\longrightarrow$ True
ii. $2(3) - 3(1) > 3$ $\longrightarrow$ $3 > 3$ $\longrightarrow$ False
iii. $2(-1) - 3(2) > 3$ $\longrightarrow$ $-8 > 3$ $\longrightarrow$ False
iv. $2(-2) - 3(-3) > 3$ $\longrightarrow$ $5 > 3$ $\longrightarrow$ True

[▶ 🖳 See **Calculator Note 6C** to test inequalities. ◀]

Graphing Inequalities

▶ Draw a broken or dashed line on the boundary for inequalities with $>$ or $<$.
▶ Draw a solid line on the boundary for inequalities with $\geq$ or $\leq$.
▶ To graph inequalities of the form $y <$ or $y \leq$ shade below the boundary line.
▶ To graph inequalities of the form $y >$ or $y \geq$ shade above the boundary line.

[▶ 🖳 See **Calculator Note 6D** to graph inequalities in two variables on your calculator. ◀]

▶ **Practice Your Skills**

1. Match each graph with an inequality.

   a. $y \leq 3 + 2x$ **b.** $y \leq 2 + 3x$ **c.** $2x + 3y \leq 6$ **d.** $2x + 3y \geq 6$

   i.
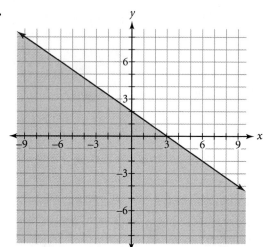

   ii.
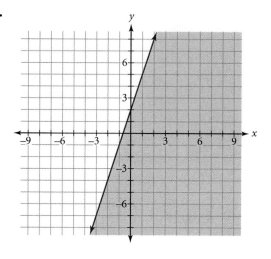

   iii.
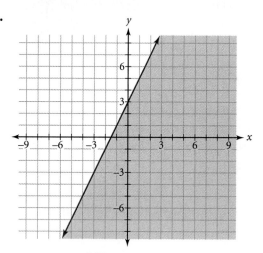

   iv.
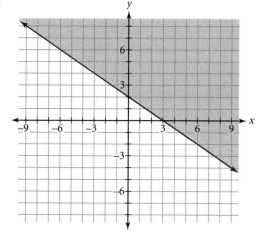

2. Solve each inequality for y.

   a. $84x + 7y \geq 70$ **b.** $4.8x - 0.12y < 7.2$

3. Sketch each inequality on a number line.

   a. $x \leq -5$ **b.** $x > 2.5$ **c.** $-3 \leq x \leq 3$ **d.** $-1 \leq x < 2$

4. Consider the inequality $y < 2 - 0.5x$.

   a. Graph the boundary line for the inequality on axes with scales from -6 to 6 on each axis.

   b. Determine whether each given point satisfies $y < 2 - 0.5x$. Plot the point on the graph you drew in 4a. Label the point T (true) if it is part of the solution or F (false) if it is not part of the solution region.

        i. $(1, 2)$ **ii.** $(4, 0)$ **iii.** $(2, -3)$ **iv.** $(-2, -1)$

   c. Use your results from 4b to shade the half-plane that represents the inequality.

5. Consider the inequality $y \geq 1 + 2x$.

a. Graph the boundary line for the inequality on axes with scales from -6 to 6 on each axis.

b. Determine whether each given point satisfies $y \geq 1 + 2x$. Plot the point on the graph you drew in 5a, and label the point T (true) if it is part of the solution or F (false) if it is not part of the solution.

    i. $(-2, 2)$ **ii.** $(3, 2)$ **iii.** $(-1, -1)$ **iv.** $(-4, -3)$

c. Use your results from 5b to shade the half-plane that represents the inequality.

▶ Reason and Apply

6. Sketch each inequality.

    a. $y \leq -3 + x$

    b. $y > -2 - 1.5x$

    c. $2x - y \geq 4$

7. Write the inequality for each graph.

a.

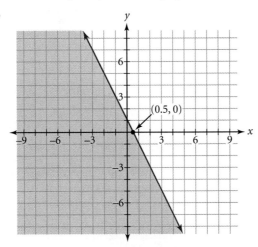

b.

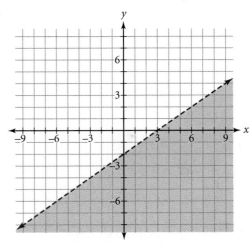

c.

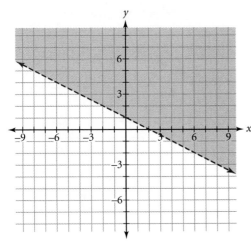

d.

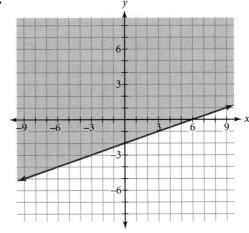

8. Which of these points lie in the shaded region of each graph?

$A(4, 3)$ $B(1, -2)$ $C(5, -4)$ $D(2, 0)$ $E(0, 5)$

$F(4, -7)$ $G(-2, 3)$ $H(1, 8)$ $I(-1, -4)$ $J(3, 11)$

i.

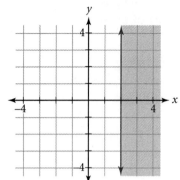

ii.

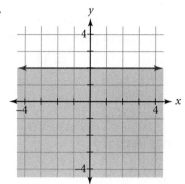

iii.

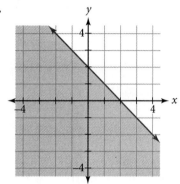

iv.

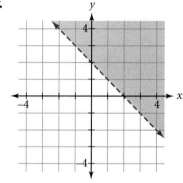

v.

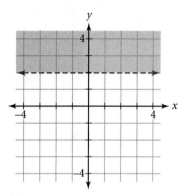

vi.

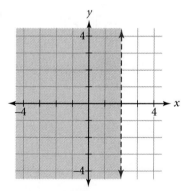

vii.

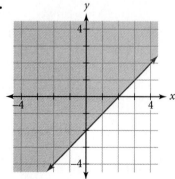

viii.

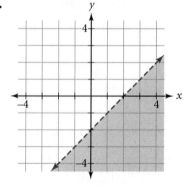

9. Sketch each inequality on coordinate axes.

     a. $y < 4$ **b.** $x \leq -3$ **c.** $y \geq -1$ **d.** $x > 3$

10. **APPLICATION** The total number of points from a combination of one-point free throws, *F*, and two-point shots, *S*, is less than 84 points.

   a. Write an inequality to represent this situation.

   b. Write the equation for the boundary line of this situation.

   c. Graph this inequality. Label the horizontal axis two-point shots and the vertical axis free throws. Show the scale you used on the axes.

   d. On your graph, indicate three possible combinations of free throws and two-point shots that give a point total of 50. Name the coordinates of these points.

11. Graph the inequalities in problems 4 and 5 on your calculator. [▶ ▢ See **Calculator Note 6D** to review graphing inequalities on your calculator. ◀]

Raul Acosta plays wheelchair basketball for the Eastern Paralyzed Veterans Association in New Jersey. The sport started in 1946 and is governed worldwide by the International Wheelchair Basketball Federation.

Review

12. In social studies, Zach studies minimum wages of the past 60 years. He finds this set of data on the Internet.

   a. Graph the data from the table on the same set of axes. Use one color for minimum wage and another for 1998 dollars.

   b. Which is better represented by a line, the hourly minimum wage or the dollar value?

   c. Find the line of fit based on Q-points for the data points of the form (*year, 1998 dollars*) that are best modeled by a line.

   d. Graph the equation in 12c to verify that it is a good fit.

   e. What is the real-world meaning of the slope? How does it compare with the 1998 dollars graph?

13. Ellie was talking with her grandmother about a trip she took this summer. Ellie made the trip in 2.5 hours traveling at 65 mph. Ellie's grandmother remembers that she made the same trip in about 6 hours when she was Ellie's age.

   a. What speed was Ellie's grandmother traveling when she made the trip?

   b. Explain how this is an application of inverse variation.

14. Solve each equation for *y*.

   a. $7x - 3y = 22$

   b. $5x + 4y = -12$

Year	Minimum wage	1998 Equivalent dollars
1938	0.25	2.89
1939	0.30	3.52
1945	0.40	3.62
1950	0.75	5.07
1956	1.00	5.99
1961	1.25	6.27
1967	1.40	6.83
1968	1.60	7.49
1974	2.00	6.61
1975	2.10	6.36
1976	2.30	6.59
1978	2.65	6.63
1979	2.90	6.51
1980	3.10	6.13
1981	3.35	6.01
1990	3.50	4.74
1991	4.25	5.09
1996	4.75	4.93
1997	5.15	5.23

(www.dol.gov)

Systems of Inequalities

You learned that the solution to a system of two linear equations, if there is exactly one solution, it is the coordinates of the point where the two lines intersect. In this lesson you'll learn about **systems of inequalities** and their solutions. Unlike the graphs of linear equations, the graphs of linear inequalities don't intersect in a single point, as you'll see in the examples and investigation in this lesson.

Translucent sheets of blue, red, and yellow intersect to form overlapping regions of new colors—orange, green, and purple.

EXAMPLE A

Graph the system of inequalities

$$\begin{cases} y \le -2 + \dfrac{3}{2}x \\ y > 1 - x \end{cases}$$

Graph the boundary lines and shade the half-planes. Indicate the solution area as the darkest region.

▶ Solution

First, determine if the boundary lines are solid or dashed. Graph $y = -2 + \frac{3}{2}x$ with a solid line because points on the line satisfy the inequality. Graph $y = 1 - x$ with a dotted line because its points do not satisfy the inequality.

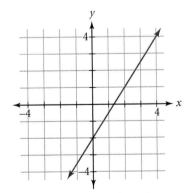 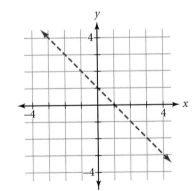 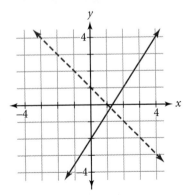

Shade the half-plane below the solid line $y = -2 + \frac{3}{2}x$ because its inequality has the "less than or equal to" symbol, $\le$. Shade above the dotted line $y = 1 - x$ because its inequality has the "greater than" symbol, $>$. Use different colors or patterns to distinguish each area shaded.

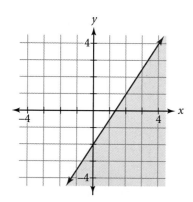 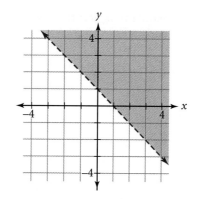 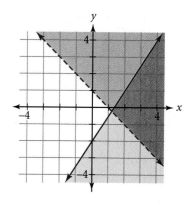

Each shaded area indicates the region of points that satisfy each inequality. The overlapping area bounded by $y \leq -2 + \frac{3}{2}x$ and $y > 1 - x$ satisfies both. The points that lie in both half-planes are the solutions to the system of inequalities.

EXAMPLE B

A cereal company is including a chance to win a $1,000 scholarship in each box of cereal. In this promotional campaign, it will give away one scholarship each month, regardless of the number of boxes sold. Because the cereal is priced differently at various locations, the profit from a single box is between $0.47 and $1.10. Graph the expected profit, given the initial cost of the scholarship, for up to 5000 boxes sold in a month. Show the solution region on a graph. Is it possible to sell 3000 boxes and make a profit of $1,000?

▶ Solution

Write a system of inequalities to model this situation. The lowest profit per box is $0.47. So $0.47x$ is the profit when x boxes are sold. Subtract $1,000 for the scholarship given each month. So the profit y is at least $0.47x - 1000$ dollars for x boxes sold. This is given by the inequality

$$y \geq -1000 + 0.47x$$

Likewise, if the maximum profit is $1.10 per box, then the profit is at most $1.1x - 1000$ dollars. So the second inequality is

$$y \leq -1000 + 1.1x$$

The profit during each month is given by the system

$$\begin{cases} y \geq -1000 + 0.47x \\ y \leq -1000 + 1.1x \end{cases}$$

Each inequality is graphed for up to 5000 boxes on separate axes below.

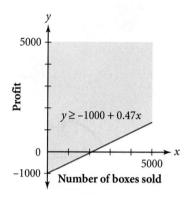

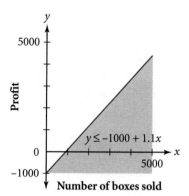

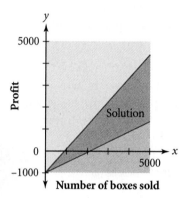

The possible profits are in the region where the two half-planes overlap.
[▶ 🖳 See **Calculator Note 6D** to graph systems of inequalities on your calculator. ◀]

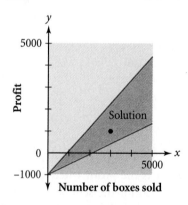

To see if it is possible to make \$1000 when 3000 boxes are sold, plot the point (3000, 1000) on the graph. Since the point is in the solution region, the coordinates satisfy both inequalities.

$$y \geq -1000 + 0.47x \qquad\qquad y \leq -1000 + 1.1x$$
$$1000 \,\square\, -1000 + 0.47(3000) \quad \text{and} \quad 1000 \,\square\, -1000 + 1.1(3000)$$
$$1000 \geq 410 \qquad\qquad\qquad 1000 \leq 2300$$

Both inequalities are true, so it is possible to sell 3000 boxes and make \$1000.

The inequalities in a system are often called **constraints.** In Example B, the inequalities model constraints on the possible profits in the situation. In the following investigation you'll learn about another application that has constraints.

Investigation
A "Typical" Envelope

The U.S. Postal Service imposes several constraints on the acceptable sizes for an envelope. One constraint is that the ratio of length to width must be less than or equal to 2.5, and another is that this ratio must be greater than or equal to 1.3.

Step 1	Define variables and write an inequality for each constraint.
Step 2	Solve each inequality for the variable representing length. Decide whether or not you have to reverse directions on the inequality symbols. Then write a system of inequalities to describe the Postal Service's constraints on envelope sizes.
Step 3	Decide on appropriate scales for each axis and label a set of axes. Decide if you should draw the boundaries of the system with solid or dashed lines. Graph each inequality on the same set of axes. Shade each half-plane with a different color or pattern.
Step 4	Where on the graph are the solutions to the system of inequalities? Discuss how to check that your answer is correct.

Step 5	Decide if each envelope satisfies the constraints by locating the corresponding point on your graph.

a. 5 in. by 8 in.

b. 3 in. by 3 in.

c. 2.5 in. by 7.5 in.

d. 5.5 in. by 7.5 in.

Step 6	Do the coordinates of the origin satisfy this system of inequalities? Explain the real-world meaning of this point. What constraints can you add to more realistically model the Postal Service's acceptable envelope sizes? How do these additions affect the graph?

With enough constraints the solution to a system of inequalities might resemble a geometric shape or polygon. No matter how small the region, there are infinitely many points that satisfy the system. Even a line may contain solutions to a system of inequalities. Of course, if no solutions exist, there is no solution region at all.

EXERCISES

Practice Your Skills

1. Match each system of inequalities with its graph.

    a. $\begin{cases} y < 3 \\ x \geq 2 \end{cases}$

    b. $\begin{cases} y > 2 + x \\ y > 1 - x \end{cases}$

    c. $\begin{cases} 2x - y \leq 6 \\ 3x + 2y \geq 12 \end{cases}$

    i.
    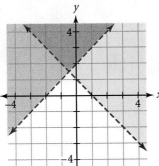

    ii.
    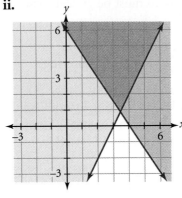

    iii.
    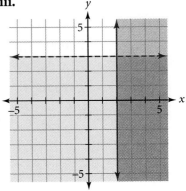

2. Here is the graph of this system of inequalities:

    $$\begin{cases} y > x \\ y > 2 - \dfrac{1}{2}x \end{cases}$$

    Is each point listed a solution to the system?
    Explain why or why not.

    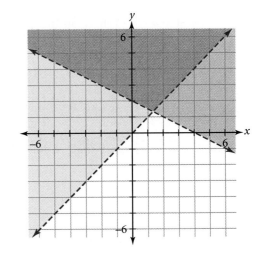

    a. $(1, 2)$

    b. $(3, 2)$

    c. $\left(\dfrac{4}{3}, \dfrac{4}{3} \right)$

    d. $(5, -3)$

3. Consider these two inequalities together as a system.

    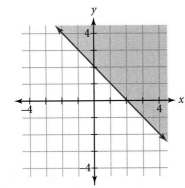

    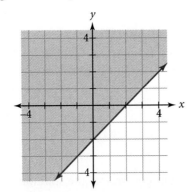

    a. Name the inequality pictured in each graph.
    b. Sketch a graph showing the solution to this system.

4. Sketch a graph showing the solution to each system.

    a. $y \leq 2$

        $x < 2$

    b. $x + y \leq 4$

        $x - y \leq 4$

5. Write a system of inequalities for the solution shown on the graph.

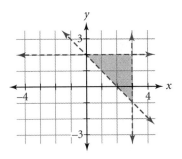

▶ Reason and Apply

6. **APPLICATION** The cereal company from Example B decides to raise the scholarship amount to $1,250. It also lowers the cereal's price so that the expected profit from a single box is between $0.40 and $1.00.

    a. Write the inequalities to represent this new situation.

    b. Graph the expected revenue for up to 5000 boxes sold in a month.

7. On Kids' Night, every adult admitted into a restaurant must be escorted by at least one child. The restaurant has a maximum seating capacity of 75 people.

    a. Write a system of inequalities to represent the constraints in this situation.

    b. Graph the solution. Is it possible for 50 children to escort 10 adults to the restaurant?

    c. Why would the restaurant reconsider the rules for Kids' Night? Add a new constraint to address these concerns. Draw a graph of the new solution.

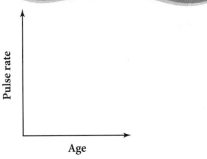

8. **APPLICATION** The American College of Sports Medicine considers age as one factor when it recommends low and high heart rates during workout sessions. For safe and efficient training, your heart rate should be between 55% and 90% of the maximum heart rate level. The maximum heart rate is calculated by subtracting a person's age from 220 beats per minute.

    a. Define variables and write an equation relating age and maximum heart rate during workouts.

    b. Write a system of inequalities to represent the recommended high and low heart rates during a workout.

    c. Graph the solution to show a region of safe and efficient heart rates for people of any age.

    d. What constraints should you add to limit your region to show the safe and efficient heart rates for people between the ages of 14 and 40?

    e. Graph the new solution for 8d.

9. Write two inequalities that describe the shaded area below. Assume that the boundaries are solid lines and that each grid mark represents 1 unit.

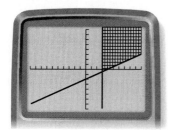

10. Write a system of inequalities to describe the shaded area on the graph at right. Write each slope as a fraction.

11. Graph this system of inequalities on the same set of axes. Describe the shape of the region.

$$\begin{cases} y \le 4 + \frac{2}{3}(x - 1) \\ y \le 6 - \frac{2}{3}(x - 4) \\ y \ge -17 + 3x \\ y \ge 1 \\ y \ge 7 - 3x \end{cases}$$

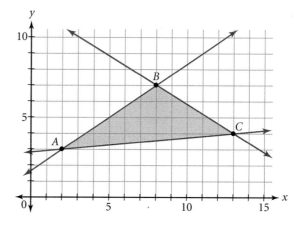

12. Write a system of inequalities that defines each shaded region in the parallelogram on the graph at right.

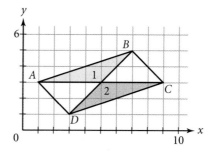

▶ Review

13. APPLICATION Manuel has a sales job at a local furniture store. Once a year, on Employees' Day, every item in the store is 15% off regular price. In addition, salespeople get to take home 25% commission on the items they sell as a bonus.

a. A loft bed with a built-in desk and closet usually costs $839. What will it cost on Employee's Day?

b. At the end of the day, Manuel's bonus is $239.45. How many dollars worth of merchandise did he sell?

14. Hugh went for a hike in Colorado that went by windmills that generate electricity. He decided to figure out how tall they are by using proportions. Hugh knows that he is about 6 feet tall, and he measured his shadow as being 4 feet long. The shadow cast by the windmill was about 175 feet long. How tall is the windmill?

15. Think about this number trick.

    a. Layla got a final number of 4. What was her original number?

    b. Robert got a final answer of 10. What was his original number?

    c. Write an algebraic expression to represent this sequence of operations. Let x represent the number you start with.

x	——
Ans · 3	——
Ans + 12	——
Ans ÷ 5	——
Ans − 1.4	——
Ans · 10	——
Ans − 10	——
Ans ÷ 6	——

16. Solve each system of equations by using a symbolic method. Check that your solutions are correct.

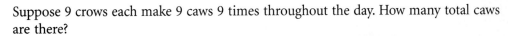

    a. $\begin{cases} y = 4x - 3 \\ y = 2x + 9 \end{cases}$ **b.** $\begin{cases} 3x - 4y = -2 \\ -2x + 3y = 1 \end{cases}$

IMPROVING YOUR **REASONING** SKILLS

Suppose 9 crows each make 9 caws 9 times throughout the day. How many total caws are there?

Suppose 99 crows make 99 caws 99 separate times in one day. Now how many caws are there?

Answer the question again for 999 crows making 999 caws 999 times. If you continue this pattern of problems, at what number does your calculator round the answer? What is the exact number of caws in this case?

Write the answers to the first three questions and look for a pattern. Use it to find how many caws there are when the number is 99,999. With 86,400 seconds in a day, this means that each crow makes more than one caw per second every hour!

CHAPTER 6 REVIEW

In this chapter you learned to model many situations with a **system of equations** in two variables. You learned that systems of linear equations can have 0, 1, or infinitely many solutions. You used tables, used graphs, and solved symbolically to find the solutions of systems. You discovered that the methods of **elimination, substitution,** and **row operations** on a matrix allow you to find exact solutions to problems, not just the approximations of graphs and tables.

Then you analyzed situations involving **inequalities** and discovered how to find their solutions using graphs, tables, and **symbolic manipulation.** The graph of an inequality in one variable is a part of a number line, and the graph of a linear inequality in two variables is a shaded **half-plane** that contains points whose coordinates make the inequality true. You learned that a **compound inequality** is the combination of two inequalities.

You discovered how to use inequalities to define **constraints** that limit the solution possibilities in real-world applications. You learned how to graph a **system of inequalities.**

EXERCISES

1. Lines *a* and *b* at right form a system of equations. Write the equations of the lines and find the point of intersection.

2. Find the point where the graphs of the equations intersect. Check your answer.
$$\begin{cases} 3x - 2y = 10 \\ x + 2y = 6 \end{cases}$$

3. Graph this system of equations, and find the solution point.
$$\begin{cases} y = 5 - 0.5(x - 3) \\ y = -4 + 1.5(x + 2) \end{cases}$$

4. Show the steps involved in solving this system symbolically by the substitution method. Explain each step.
$$\begin{cases} y = 16 + 4.3(x - 5) \\ y = -7 + 4.2x \end{cases}$$

5. Complete each sentence.

a. A system of two linear equations has no solution if . . .

b. A system of two linear equations has infinitely many solutions if . . .

c. A system of two linear equations has exactly one solution if . . .

6. Name the inequality that each graph represents.

a.

b.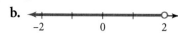

c.

7. Solve the inequality $5 \leq 2 - 3x$ for x and graph the solution on a number line.

8. Write a system of inequalities to describe this shaded area.

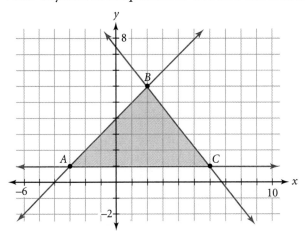

9. **APPLICATION** Harold cuts lawns after school. He has a problem on Wednesdays when he cuts Mr. Fleming's lawn. His lawn mower has two speeds—at the higher speed he can get the job done quickly, but he always runs out of gas; at the lower speed he has plenty of gas, but it seems to take forever to get the job done. So he has collected this information.

- On Monday he cut a 15-meter-by-12-meter lawn at the higher speed in 18 minutes. He used a half tank of gas, or 0.6 liter.
- On Tuesday he cut a 20-meter-by-14-meter lawn at the lower speed in 40 minutes. He used a half tank of gas.
- Mr. Fleming's lawn measures 22 m by 18 m.

a. At the higher speed, how many square meters of lawn can Harold cut per minute?

b. At the lower speed, how many square meters of lawn can Harold cut per minute?

c. If Harold decides to cut Mr. Fleming's lawn using the higher speed for 10 minutes and the lower speed for 8 minutes, will he finish the job?

d. Let h represent the number of minutes cutting at higher speed, and let l represent the number of minutes cutting at lower speed. Write an equation that models completion of Mr. Fleming's lawn.

e. At the higher speed, how much gas does the lawn mower use in liters per minute?

f. At the lower speed, how much gas does the lawn mower use in liters per minute?

g. If Harold starts with a full tank, will he have enough gas to use the higher speed for 10 minutes and the lower speed for 8 minutes?

h. Write an equation in terms of h and l that has Harold use all of his gas.

i. Using the equations from 9d and h, solve the system.

j. What is the real-world meaning of the solution?

10. Use row operations to find the solution matrix for this system.

$$\begin{cases} 7x + 3y = -45 \\ x + 6y = -51 \end{cases}$$

TAKE ANOTHER LOOK

Businesses use systems of equations and inequalities to determine how to maximize profits. A process called **linear programming** applies the concepts of constraints, points of intersection, and algebraic expressions to solve this very real application problem. Here is one example.

A company manufactures scooters and skateboards. The factory has the capacity to make at most 6000 scooters in one day, and the factory can make at most 8000 skateboards in one day. However, the factory can produce a combination of no more than 10,000 scooters and skateboards together. Define variables and write a system of three inequalities to describe these constraints. Label a set of axes and graph the solution. This is called a **feasible region.** What do the points in this shaded region represent? Find the points of intersection at the corners of this region.

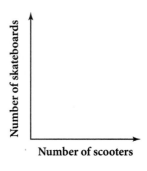

The company makes a profit of $15 per scooter and $10 per skateboard. How many of each should the company make to maximize its profits? To answer this question, use the variables defined earlier to write an expression to find the total profit the company makes from scooters and skateboards. Then substitute the coordinates of several points from the feasible region including the points of intersection. For example, if the company makes 5000 scooters and 5000 skateboards, substitute 5000 for x and 5000 for y into your expression to find the profit. Which point gave you the greatest profit? Be sure to try the points of intersection.

Professional skateboarder Tony Hawk performs a trick at the X Games.

Assessing What You've Learned

 ORGANIZE YOUR NOTEBOOK Update your notebook with an example, investigation, or exercise that demonstrates each solution method for a system of equations. Add one problem that demonstrates each concept—inequalities in one variable, inequalities in two variables, and systems of inequalities.

 WRITE IN YOUR JOURNAL Add to your journal by answering one of these prompts:

▶ You have learned five methods to find a solution for a system of equations. Which method do you like best? Which one is the most challenging to you? What are the advantages and disadvantages of each method?

▶ You have now studied more than half of this book. What mathematical skills in the previous chapters were most crucial to your success in this chapter? Which concepts are your strengths and weaknesses?

▶ Describe in writing the difference between an inequality in one variable and an inequality in two variables. How do the graphs of the solutions differ? Compare these to the graph of a system of inequalities.

 UPDATE YOUR PORTFOLIO Choose your best graph of a system of inequalities from this chapter to add to your portfolio. Redraw the graph with a clearly labeled set of axes. Use color to highlight each inequality and its half-plane. Indicate the solution region with a visually pleasing design or pattern.

 GIVE A PRESENTATION Write your own word problem for a system of equations or inequalities. Choose a setting that is meaningful to you or that you wish to know more about, and write a problem to model the situation. It can be about winning times for Olympic events, the point where two objects meet while traveling, percent mixture problems, or something new you created. Solve the problem using one of the methods you learned in this chapter. Make a poster of the problem and its solution, and present it to the class. Work with a partner or in a group.

Exponents and Exponential Models

OBJECTIVES

In this chapter you will

- write recursive routines for nonlinear sequences
- learn an equation for exponential growth or decrease
- use properties of exponents to rewrite expressions
- write numbers in scientific notation
- model real-world data with exponential equations

This "Chinese Horse" is part of a prehistoric cave painting in Lascaux, France. Scientific methods that use equations with exponents have determined that parts of the Lascaux cave paintings are more than 15,000 years old. For archaeologists, dating ancient artifacts helps them understand how civilizations evolved. Drawings and pieces of art help them understand what existed at that time and what was important to the civilization. You will see that exponents are useful in many other real-world settings too.

LESSON
7.1

Slow buds the pink dawn
like a rose
From out night's gray and
cloudy sheath
Softly and still it grows
and grows
Petal by petal, leaf by leaf

SUSAN COOLIDGE

Recursive Routines

Have you ever noticed that it doesn't take very long for a cup of steaming hot chocolate to cool to sipping temperature? If so, then you've also noticed that it stays about the same temperature for a long time. Have you ever left food in your locker? It might look fine for several days, then suddenly some mold appears and a few days later it's covered with mold. The same mathematical principle describes both of these situations. Yet these patterns are different from the linear patterns you saw in rising elevators and shortening ropes— you modeled those situations with repeated addition or subtraction. Now you'll investigate a different type of pattern, a pattern seen in a population that increases very rapidly.

Investigation
Bocks, Bugs, Everywhere Bugs

You will need

• graph paper

Imagine that a bug population has invaded your classroom. One day you notice 16 bugs. Every day new bugs hatch, increasing the population by 50% each week. So, in the first week the population increases by 8 bugs.

Step 1 In a table like this one, record the total number of bugs at the end of each week for 4 weeks.

Bug Invasion

Weeks elapsed	Total number of bugs	Increase in number of bugs (rate of change per week)	Ratio of this week's total to last week's total
Start (0)	16	/////////	/////////
1		8	
2			

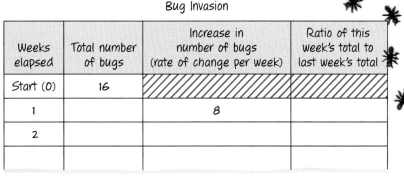

Step 2 The increase in the number of bugs each week is the population's rate of change per week. Calculate each rate of change and record it in your table. Does the rate of change show a linear pattern? Why or why not?

Step 3 Let *x* represent the number of weeks elapsed, and let *y* represent the total number of bugs. Graph the data using (0, 16) for the first point. Connect the points with line segments and describe how the slope changes from point to point.

Step 4	Calculate the ratio of the number of bugs each week to the number of bugs the previous week and record it in the table. For example, divide the population after 1 week has elapsed by the population when 0 weeks have elapsed. Repeat this process to complete your table. How do these ratios compare? Explain what the ratios tell you about the bug population growth.
Step 5	What is the *constant multiplier* for the bug population?
Step 6	Model the population growth by writing a recursive routine that shows the growing number of bugs. [▶ 🖳 See **Calculator Note 4B** to review recursive routines. ◀] Describe what each part of this calculator command does.
Step 7	By pressing (ENTER) a few times, check that your recursive routine gives the sequence of values in your table (in the column "Total number of bugs"). Use the routine to find the bug population at the end of weeks 5 to 8.
Step 8	What is the bug population after 20 weeks have elapsed? After 30 weeks have elapsed? What happens in the long run?

In the investigation, you found that repeated multiplication is the key to growth of the bug population. Populations of people, animals, and even bacteria show similar growth patterns. Many decreasing patterns, like cooling liquids and decay of substances, can also be described with repeated multiplication.

EXAMPLE

Maria has saved $10,000 from her part-time job and wants to invest it for college. She is considering two options. Plan A guarantees a payment, or return, of $550 each year. Plan B grows by 5% each year. With each plan, what would Maria's new balance be after 5 years? After 10 years?

▶ **Solution**

With plan A, Maria's investment would grow by $550 each year.

Year	Current balance	+	Return	=	New balance
1	10,000	+	550	=	10,550
2	10,550	+	550	=	11,100
3	11,100	+	550	=	11,650

A recursive routine to do this on your calculator is

{0, 10000} (ENTER)

{Ans(1)+1, Ans(2)+550} (ENTER)

(ENTER), (ENTER), . . .

After 5 years the new balance is $12,750. After 10 years it is $15,500.

```
{3 11650}
{4 12200}
{5 12750}
{6 13300}
{7 13850}
{8 14400}
{9 14950}
{10 15500}
```

With plan B, money earns *interest* each year. The amount of interest is 5% of the current balance. To find the new balance at the end of the first year, add the interest to the current balance. Notice that there is a factor of 10,000 in both the current balance and the interest. You can apply the distributive property to write the expression for the new balance in **factored form.**

Year	Current balance	+	Interest (balance × interest rate)	=	New balance (factored form)
1	10,000	+	10,000 × 0.05	=	10,000(1 + 0.05), or 10,500
2	10,500	+	10,500 × 0.05	=	10,500(1 + 0.05), or 11,025
3	11,025	+	11,025 × 0.05	=	11,025(1 + 0.05), or about 11,576

In the first year the balance grows by $500 to $10,500. To find the new balance for the next year, you need to add 5% of $10,500 to the current $10,500 balance in the account.

Each year the balance grows by 5%. To find each new balance, you use the constant multiplier 1 + 0.05, or 1.05.

You can generate the sequence of balances from year to year on your calculator using this recursive routine:

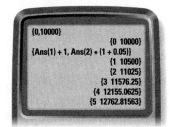

{0, 10000} (ENTER)

{Ans(1) + 1, Ans(2) · (1 + 0.05)} (ENTER)

(ENTER), (ENTER), . . .

The calculator screen shows the sequence of new balances after the first 5 years. Notice that the balance grows by a larger amount each year. That's because each year you're finding a percent of a larger current balance than in the previous year. After 5 years the new balance is about $12,763. After 10 years it is about $16,289.

A graph illustrates how the investment plans compare. Given enough time, the balance from plan B, which is growing by a constant percent, will always outgrow the balance from plan A, which has only a constant amount added to it. After 20 years you see an even more significant difference: $26,533 compared to $21,000.

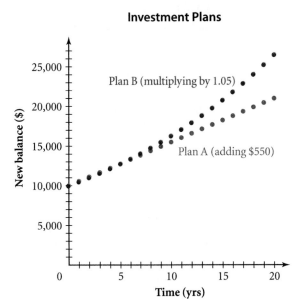

Investment Plans

It is helpful to think of a constant multiplier, like 1.05 in the example, as a sum. The plus sign in $1 + 0.05$ shows that the pattern increases and 0.05 is the percent growth per year, written as a decimal. When a balance or population decreases, say, by 15% during a given time period, you write the constant multiplier as a difference, for example, $1 - 0.15$. The subtraction sign shows that the pattern decreases and 0.15 is the percent decrease per time period, written as a decimal.

Constant multipliers can be positive or negative. These two sequences have the same starting value, but one has a multiplier of 2 and the other has a multiplier of -2.

$3, 6, 12, 24, 48, \ldots$

$3, -6, 12, -24, 48, \ldots$

How does the negative multiplier affect the sequence?

EXERCISES

You will need your calculator for problems **2, 3, 5, 6, 7, 9,** and **10.**

Practice Your Skills

1. Give the starting value and constant multiplier for each sequence. Then find the 7th term of the sequence.

   a. $16, 20, 25, 31.25, \ldots$

   b. $27, 18, 12, 8, \ldots$

2. Use a recursive routine to find the first six terms of a sequence that starts with 100 and has a constant multiplier of -1.6.

3. Use a recursive routine to find the first five terms of a sequence that starts with 72 and increases by 40% with each term. (Hint: Identify the constant multiplier first.)

4. Use the distributive property to rewrite each expression in an equivalent form. For example, you can write $500(1 + 0.05)$ as $500 + 500(0.05)$.

   a. $75 + 75(0.02)$

   b. $1000 - 1000(0.18)$

   c. $P + Pr$

   d. $75(1 - 0.02)$

   e. $80(1 - 0.24)$

   f. $A(1 - r)$

5. You may remember from Chapter 0 that the geometric pattern below is the beginning of a fractal called the *Sierpiński triangle*. In this example, the beginning triangle, Stage 0, has a total shaded area of 32 square units. Write a recursive routine that generates the sequence of shaded areas in the pattern. Then use your routine to find the shaded area in Stage 2.

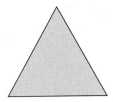

Stage 0
Area = 32 square units

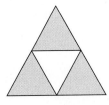

Stage 1
Area = 24 square units

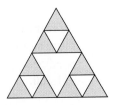

Stage 2

Reason and Apply

6. **APPLICATION** Toward the end of the year, to make room for next year's models, a car dealer may decide to drop prices on this year's models. Imagine a car that has a sticker price of $20,000. The dealer lowers the price by 4% each week until the car sells.

   a. Write a recursive routine to generate the sequence of decreasing prices.

   b. Find the 5th term and explain what your answer means in this situation.

   c. If the cost of the car to the dealer was $10,000, how many weeks will pass before its sale price will produce no profit for the dealer?

7. Ima Shivring took a cup of hot cocoa outdoors where the temperature was 0°F. When she stepped outside the cocoa was 115°F. The temperature in the cup dropped by 3% each minute.

   a. Write a recursive routine to generate the sequence representing the temperature of the cocoa each minute.

   b. How many minutes does it take for the cocoa to cool to less than 80°F?

8. Recall the six expressions in problem 4. Imagine that each expression represents the value of an antique increasing or decreasing in value per year. For each expression, identify whether it represents an increasing or decreasing situation, give the starting value, and give the percent of increase or decrease per year.

9. **APPLICATION** Health care expenditures in the United States exceeded $1 trillion in the mid-1990s and are expected to exceed $2 trillion before 2010. Many elderly and disabled persons rely on Medicare benefits to help cover health care costs. According to the *1998 Wall Street Journal Almanac*, Medicare expenditures were only $6.9 billion in 1970.

   a. Assume Medicare spending has increased by 14.2% per year since 1970. Write a recursive routine to generate the sequence of increasing Medicare spending.

   b. Use your recursive routine to find the missing table values. Round to the nearest $0.1 billion.

Medicare Spending

Year	1970	1975	1980	1985	1990	1995	2000	2005
Elapsed time (yrs) x	0	5	10	15	20	25	30	35
Spending ($ billion) y	6.9							

c. Plot the data points from your table and draw a smooth curve through them.

d. What does the shape of the curve suggest about Medicare spending? Do you think this is a realistic model?

10. **APPLICATION** The advertisement for a super-duper bouncing ball says it rebounds to 85% of the height from which it is dropped.

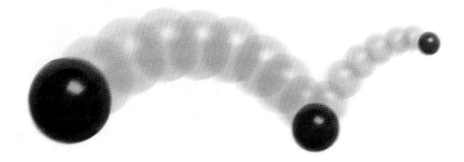

a. If the ball is dropped from a starting height of 2 m, how high should it rebound on the first bounce?

b. Write a recursive routine to generate the sequence of heights for the ball when it is dropped from a height of 2 m.

c. How high should the ball rebound on the sixth bounce?

d. If the ball is dropped from a height of 10 ft, how high should it rebound on the tenth bounce? (Hint: Modify your recursive routine in 10b.)

e. When the ball is dropped from a height of 10 ft, how many times will it bounce before the rebound height is less than 0.5 ft?

f. A collection of super-duper bouncing balls was tested. Each ball was dropped from a height of 2 m. The table shows the height of the first rebound for eight different balls. Do you think the advertisement's claim that the ball rebounds to 85% of the original height is fair? Explain your thinking.

Balls Dropped from 2 m

Ball number	1	2	3	4	5	6	7	8
Height of rebound (m)	1.68	1.67	1.69	1.78	1.64	1.68	1.66	1.8

11. Grace manages a local charity. A wealthy benefactor has offered two options for making a donation over the next year. One option is to give $50 now and $25 each month after that. The second option is to give $1 now and twice that amount next month; each month afterward the benefactor would give twice the amount given the month before.

a. Determine how much Grace's charity would receive each month under each option. Use a table to show the values over the course of one year.

b. Use another table to record the total amount Grace's charity will have received after each month.

c. Let x represent the number of the month (1 to 12), and let y represent the total amount Grace received after each month. On the same coordinate axes, graph the data for both options. How do the graphs compare?

d. Which option should Grace choose? Why?

▶ Review

12. Write an equation in point-slope form for a line with slope -1.2 that goes through the point $(600, 0)$. Find the y-intercept.

13. Match the recursive routine to the equation.

  a. $y = 3x + 7$
  b. $y = -3x + 7$
  c. $y = 7x + 3$
  d. $y = -7x + 3$

  i. Start with 7, then apply the rule Ans $+ 3$.
  ii. Start with 3, then apply the rule Ans $+ 7$.
  iii. Start with 7, then apply the rule Ans $- 3$.
  iv. Start with 3, then apply the rule Ans $- 7$.

14. **APPLICATION** A long-distance phone company offers two calling plans. The first plan costs \$12 per month and offers 60 minutes free per month; additional minutes cost 5¢ per minute. The second plan costs only \$10 a month and offers 50 minutes free per month; additional minutes cost 9¢ per minute.

  a. Define variables and write an equation for the first plan if you use it for 60 minutes or less.

  b. Write an equation for the first plan if you use it for more than 60 minutes.

  c. Write two equations for the second plan similar to those you wrote in 14a and b. Explain what each equation represents.

  d. Sydney frequently calls her cousin in Australia from her hometown in Kentucky and talks for a long time, generally about 150 minutes per month. How much would each plan cost for her? Which plan should she choose?

  e. Louis makes only a few long-distance calls since most of his friends and family members live nearby. How much would each plan cost for Louis if he averages 55 minutes of long-distance calls per month? Which plan should he choose?

  f. For how many minutes of use will the cost of the plans be the same? How can you decide which of these two long-distance plans is better for a new subscriber?

IMPROVING YOUR **GEOMETRY** SKILLS

Use 16 toothpicks to make this pattern. Then remove 4 toothpicks so that you have 4 congruent triangles.

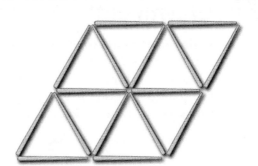

Exponential Equations

Recursive routines are useful for seeing how a sequence develops and generating the first few terms. But, as you learned in Chapter 4, if you're looking for the 50th term, you'll have to do many calculations to find your answer. For most of the sequences in Chapter 4, you found that the graphs of the points formed a linear pattern, so you learned how to write the equation of a line.

Recursive routines with a constant multiplier create a different type of increasing or decreasing pattern. In this lesson you'll discover the connection between these recursive routines and exponents. Then, with a new type of equation, you'll be able to find any term in a sequence based on a constant multiplier without having to find all the terms before it.

This sculpture, *Door to Door* (1995), was created by Filipino artist José Tence Ruiz (b 1956) from wood, cardboard, and other materials. The decreasing size of the boxes suggests an exponential pattern.

Investigation
Growth of the Koch Curve

You will need

- the worksheet Growth of the Koch Curve

In this investigation you will look for patterns in the growth of a fractal. You may remember the *Koch curve* from Chapter 0. Here you will think about the relationship between the length of the Koch curve and the repeated multiplication you studied in Lesson 7.1. Stage 0 of the Koch curve is already drawn on the worksheet. It is a segment 27 units long.

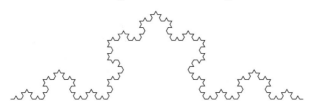

| Step 1 | Draw the Stage 1 figure below the Stage 0 figure. The first segment is drawn for you on the worksheet. As shown here, the Stage 1 figure has four segments, each $\frac{1}{3}$ the length of the Stage 0 segment. |

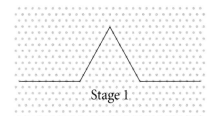

Stage 1

| Step 2 | Determine the total length at Stage 1 and record it in a table like this. |

Stage number	Total length (units)	Ratio of this stage's length to previous stage's length
0	27	////////////
1		
2		
3		

| Step 3 | Draw the Stage 2 and Stage 3 figures of the fractal. Again, the first segment for each stage is drawn for you. Record the total length at each stage. |

| Step 4 | Find the ratio of the total length at any stage to the total length at the previous stage. What is the constant multiplier? |

| Step 5 | Use your constant multiplier from Step 4 to predict the total lengths of this fractal at Stages 4 and 5. |

| Step 6 | How many times do you multiply the original length at Stage 0 by the constant multiplier to get the length at Stage 2? Write an expression that calculates the length at Stage 2. |

| Step 7 | How many times do you multiply the length at Stage 0 by the constant multiplier to get the length at Stage 3? Write an expression that calculates the length at Stage 3. |

| Step 8 | If your expressions in Steps 6 and 7 do not use exponents, rewrite them so that they do. |

| Step 9 | Use an exponent to write an expression that predicts the total length of the Stage 5 figure. Evaluate this expression using your calculator. Is the result the same as you predicted in Step 5? |

| Step 10 | Let x represent the stage number, and let y represent the total length. Write an equation to model the total length of this fractal at any stage. Graph your equation and check that the calculator table contains the same values as your table. |

| Step 11 | What does the graph tell you about the growth of the Koch curve? |

A recursive routine that uses a constant multiplier represents a pattern that increases or decreases by a constant ratio or a constant percent. Because exponents are another way of writing repeated multiplication, you can use exponents to model these patterns. In the investigation you discovered how to calculate the length of the Koch curve at any stage by using this equation:

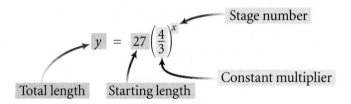

Equations like this are called **exponential equations** because a variable, in this case x, appears in the exponent.

When you write out a repeated multiplication expression to show each factor, it is written in **expanded form.** When you show a repeated multiplication expression with an exponent, it is in **exponential form** and the factor being multiplied is called the **base.**

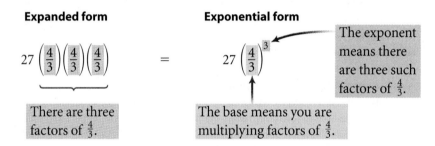

EXAMPLE A

Write each expression in exponential form.

a. $(5)(5)(5)(5)(5)(5)$

b. $3(3)(2)(2)(2)(2)(2)(2)(2)(2)$

c. The balance of a savings account that was opened 7 years ago with $200 earning 2.5% interest per year.

▶ **Solution**

a. 5^6

b. There are two factors of 3 and nine factors of 2, so you write

$$3^2 \cdot 2^9$$

You can't combine 3^2 and 2^9 any further because they have different bases.

c. There will be 7 factors of $(1 + 0.025)$ multiplied by the starting value of $200, so you write

$$200(1 + 0.025)^7$$

EXAMPLE B

Seth deposits $200 in a savings account. The account pays 5% annual interest. Assuming that he makes no more deposits and no withdrawals, calculate his new balance after 10 years.

▶ **Solution**

The interest represents a 5% rate of growth per year, so the constant multiplier is $(1 + 0.05)$. Now find an equation that you can use to find the new balance after any number of years by considering these yearly calculations and results.

	Expanded form	Exponential form	New balance
Starting balance:	$200		= $200.00
After 1 year:	$200(1 + 0.05)$	$= \$200(1 + 0.05)^1$	= $210.00
After 2 years:	$200(1 + 0.05)(1 + 0.05)$	$= \$200(1 + 0.05)^2$	= $220.50
After 3 years:	$200(1 + 0.05)(1 + 0.05)(1 + 0.05)$	$= \$200(1 + 0.05)^3$	= $231.53
After x years:	$200(1 + 0.05)(1 + 0.05) \ldots (1 + 0.05)$	$= \$200(1 + 0.05)^x$	

You can now use the equation $y = 200(1 + 0.05)^x$, where x represents time in years and y represents the new balance in dollars, to find the new balance after 10 years.

$$y = 200(1 + 0.5)^x \qquad \text{Original equation.}$$

$$y = 200(1 + 0.05)^{10} \qquad \text{Substitute 10 for } x.$$

$$y \approx 325.78 \qquad \text{Use your calculator to evaluate the exponential expression.}$$

The new balance after 10 years would be $325.78.

Amounts that increase by a constant percent, like the savings account in the example, have **exponential growth.**

Exponential Growth

Any constant percent growth can be modeled by the exponential equation

$$y = A(1 + r)^x$$

A is the starting value, r is the rate of growth written as a positive decimal or fraction, x is the number of time periods elapsed, and y is the final value.

You can model amounts that decrease by a constant percent with a similar equation. What would need to change in the exponential equation to show a constant percent decrease?

EXERCISES

You will need your calculator for problems **2, 5, 7, 8, 9, 11, 12,** and **13.**

▶ **Practice Your Skills**

1. Rewrite each expression with exponents.

 a. $(7)(7)(7)(7)(7)(7)(7)(7)$

 b. $(3)(3)(3)(3)(5)(5)(5)(5)(5)$

 c. $(1 + 0.12)(1 + 0.12)(1 + 0.12)(1 + 0.12)$

2. A bacteria culture grows at a rate of 20% each day. There are 450 bacteria today. How many will there be

 a. Tomorrow?

 b. One week from now?

A technician puts bacteria in several petri dishes of agar. Agar is a gelatin-like substance made from algae. The agar holds the bacteria in place on the petri dish and provides nutrients for growth of the bacteria.

3. Match each equation with a table of values.

 a. $y = 4(2)^x$ **b.** $y = 4(0.5)^x$ **c.** $y = 2(4)^x$ **d.** $y = 2(0.25)^x$

i.

x	y
0	2
1	0.5
2	0.12
3	0.03

ii.

x	y
0	4
1	8
2	16
3	32

iii.

x	y
0	4
1	2
2	1
3	0.5

iv.

x	y
0	2
1	8
2	32
3	128

4. Match each recursive routine with the equation that gives the same values.

 a. 1.05 `ENTER`
   Ans · (0.95) `ENTER`

 b. 1.05 `ENTER`
   Ans + Ans · 0.05 `ENTER`

 c. 0.95 `ENTER`
   Ans · (1 + 0.05) `ENTER`

 d. 0.95 `ENTER`
   Ans · (1 − 0.05) `ENTER`

 i. $y = 0.95(1.05)^x$

 ii. $y = 1.05(1 + 0.05)^x$

 iii. $y = 0.95(0.95)^x$

 iv. $y = 1.05(1 − 0.05)^x$

5. For each table, find the value of the constants A and r such that $y = A \cdot r^x$.
(Hint: Enter your equation into Y_1 on your calculator. Then see if a table of values matches the table in the book.)

a.

x	y
0	1.2
1	2.4
2	4.8
3	9.6
4	19.2

b.

x	y
0	500
2	20
3	4
5	0.16
7	0.0064

c.

x	y
3	8
1	50
5	1.28
2	20
7	0.2048

6. The equation $y = 500(1 + 0.04)^x$ models the amount of money in a savings account that earns annual interest. Explain what each number and variable in this expression means.

7. Run the calculator program INOUTEXP and play the easy-level game five times. Each time you play, write down the input and output values you were given and the exponential equation that models those values. [▶ 🖵 See **Calculator Note 7A** for instructions on running the program INOUTEXP. ◀]

▶ Reason and Apply

8. APPLICATION A credit card account is essentially a loan. A constant percent interest is added to the balance. Stanley buys $100 worth of groceries with his credit card. The balance then grows by 4.5% interest each month. How much will he owe if he makes no payments in 4 months? Write the expression you used to do this calculation in expanded form and also in exponential form.

9. APPLICATION Phil purchases a used truck for $11,500. The value of the truck is expected to decrease by 20% each year. (A decrease in monetary value over time is sometimes called *depreciation*.)

a. Find the truck's value after 1 year.

b. Write a recursive routine that generates the value of the truck after each year.

c. Create a table showing the value of the truck when Phil purchases it and after each of the next 4 years.

d. Write an equation in the form $y = A(1 - r)^x$ to calculate the value, y, of the truck after x years.

e. Graph the equation from 9d, showing the value of the truck up to an age of 10 years.

Many people, like these ranch hands in Montana, rely on a truck for work and leisure.

10. Draw a "starting" line segment 2 cm long on a sheet of paper.

  a. Draw a segment 3 times as long as the starting segment. How long is this segment?

  b. Draw a segment 3 times as long as the segment in 10a. How long is this segment?

  c. Use the starting length and an exponent to write an expression that gives the length in centimeters of the next segment you would draw.

  d. Use the starting length and an exponent to write an expression that gives the length in centimeters of the longest segment you could draw on a 100-meter soccer field.

11. Run the calculator program INOUTEXP and play the medium- or difficult-level game five times. Each time you play, write down the input and output values you were given and the exponential equation that models those values. [▶ 🔲 See **Calculator Note 7A** for instructions on running the program INOUTEXP. ◀]

12. Fold a sheet of paper in half. You should have two layers. Fold it in half again so that there are four layers. Do this as many times as you can. Make a table and record the number of folds and number of layers.

  a. As you fold the paper in half each time, what happens to the number of layers?

  b. Estimate the number of folds you would have to make before you have about the same number of layers as the number of sheets in this textbook.

  c. Calculate the answer for 12b. You may use a recursive routine, the graph or table of an equation, or a trial-and-error method.

Origami is the Japanese art of paper folding.

13. **APPLICATION** Phil's friend Shawna buys an antique car for $5,000. She estimates that it will increase in value (*appreciate*) by 5% a year.

  a. Write an equation to calculate the value, y, of Shawna's car after x years.

  b. Simultaneously graph the equation in 13a and the equation you found in 9d. Where do the two graphs intersect? What is the meaning of this point of intersection?

14. Invent a situation that could be modeled by each equation below. Sketch a graph of each equation, and describe similarities and differences between the two models.

$$y = 400 + 20x$$
$$y = 400(1 + 0.05)^x$$

15. Consider the recursive routine

  {0, 100} (ENTER)

  {Ans(1)+1, Ans(2) · (1−0.035)} (ENTER)

  a. Invent a situation that this routine could model.

  b. Create a problem related to your situation. Carefully describe the meaning of the numbers in your problem.

  c. Use an exponential equation to solve your problem.

16. Look at this "step" pattern. In the first figure, which has one step, each side of the block is 1 cm long.

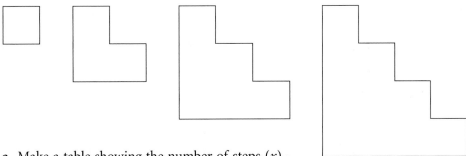

  a. Make a table showing the number of steps (x) and the perimeter (y) of each figure.

  b. On a graph, plot the coordinates your table represents.

  c. Write an equation that relates the perimeter of these figures to the number of steps.

  d. Use your equation to predict the perimeter of a figure with 47 steps.

  e. Is there a figure with a perimeter of 74 cm? If so, how many steps does it have? If not, why not?

AUTOMOBILE DEPRECIATION

Cars usually lose value as they get older. Dealers and buyers may rely on books or Internet resources to help them find out how much a used car is worth. But many people don't understand what type of math is used to make these judgments.

Choose a model of automobile that has been manufactured for several years. Research the new-car value now. Then research how much the same model would be worth now if it were manufactured last year, the year before that, and so on. Does your data show a pattern? Can you write an equation that models your data?

Your project should include

▶ Data for at least 10 consecutive years.

▶ A scatter plot comparing age and value.

▶ The rate of change (if your data appears linear) or the rate of depreciation as a percent (if your data looks exponential).

▶ An equation that fits your data.

▶ A summary of your procedures and findings; include how you collected your data and how well your equation fits the data.

You might want to ask a local auto dealership how it determines a car's value. Does it use the same rate of depreciation for all cars? And how do special features, like a custom stereo, affect the value?

Multiplication and Exponents

*Growth for the sake of
growth is the ideology
of the cancer cell.*

EDWARD ABBEY

In Lesson 7.2, you learned that the exponential expression $200(1 + 0.05)^3$ can model a situation with a starting value of 200 and a rate of growth of 5% over three time periods. How would you change the expression to model five time periods, seven time periods, or more? In this lesson you will explore that question and discover how the answer is related to a rule for showing multiplication with exponents.

Every year, the population of the United States increases. This photo shows Grand Central Station in New York City, which is the most populated U.S. city.

Social Science
• CONNECTION •

The U.S. Bureau of the Census only collects population information every 10 years. It uses mathematical models, like exponential equations, to make population predictions between census years.

Suppose the population of a town is 12,800 and the town's population grows at a rate of 2.5% each year.

An expression for the population 3 years from now is $12{,}800(1 + 0.025)^3$. To represent one more year, you can write the expression $12{,}800(1 + 0.025)^4$. You can also think about the growth from 3 years to 4 years recursively. Because the rate of growth is constant, multiply the expression for 3 years by one more constant multiplier to get $12{,}800(1 + 0.025)^3 \cdot (1 + 0.025)^1$.

This means that

$$12{,}800(1 + 0.025)^3 \cdot (1 + 0.025)^1 = 12{,}800(1 + 0.025)^4$$

Both methods make sense and both evaluate to the same result.

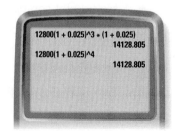

So you can advance exponential growth one time period either by multiplying the previous amount by the base (the constant multiplier) or by increasing the exponent by one. Every time you increase by one the number of times the base is used as a factor, the exponent increases by one. But what happens when you want to advance the growth by more than one time period? In the next investigation, you will discover a shortcut for multiplying exponential expressions.

Investigation
Moving Ahead

Step 1 | Rewrite each product below in expanded form and then rewrite it in exponential form with a single base.

 a. $3^4 \cdot 3^2$

 b. $x^3 \cdot x^5$

 c. $(1 + 0.05)^2 \cdot (1 + 0.05)^4$

 d. $10^3 \cdot 10^6$

Step 2 | Compare the exponents in each final expression you got in Step 1 to the exponents in the original product. Describe a way to find the exponents in the final expression without using expanded form.

Step 3 | Generalize your observations in Step 2 by filling in the blank.

$b^m \cdot b^n = b^\square$

Step 4 | Apply what you have discovered about multiplying expressions with exponents.

 a. The number of bugs in a colony after 5 weeks is $16(1 + 0.5)^5$. What does the expression $16(1 + 0.5)^5 \cdot (1 + 0.5)^3$ mean in this situation? Rewrite the expression with a single exponent.

All ants live in colonies.

 b. The depreciating value of a truck after 7 years is $11{,}500(1 - 0.2)^7$. What does the expression $11{,}500(1 - 0.2)^7 \cdot (1 - 0.2)^2$ mean in this situation? Rewrite the expression with a single exponent.

 c. The expression $A(1 + r)^n$ can model n time periods of exponential growth. What does the expression $A(1 + r)^{n+m}$ model?

Step 5 | How does looking ahead in time with an exponential model relate to multiplying expressions with exponents?

This investigation has helped you to discover the **multiplication property of exponents.**

Multiplication Property of Exponents

For any nonzero value of b and any integer values of m and n,

$$b^m \cdot b^n = b^{m+n}$$

This property is very handy in rewriting exponential expressions. However, you can add exponents to multiply numbers only when the bases are the same.

EXAMPLE A Cal and Al got different answers when asked to write $3^4 \cdot 2^2$ in another exponential form. Who was right and why?

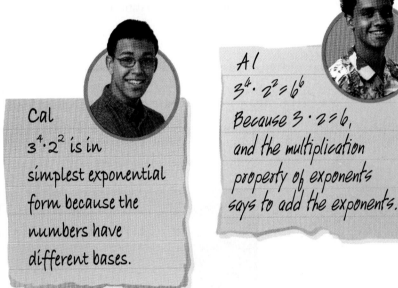

Cal
$3^4 \cdot 2^2$ is in simplest exponential form because the numbers have different bases.

Al
$3^4 \cdot 2^2 = 6^6$
Because $3 \cdot 2 = 6$, and the multiplication property of exponents says to add the exponents.

▶ **Solution** Rewrite the original expression in expanded form

$$3 \cdot 3 \cdot 3 \cdot 3 \cdot 2 \cdot 2 \;=\; 3^4 \cdot 2^2$$

4 factors of 3   2 factors of 2

The factors are not all the same, so the multiplication property of exponents does not allow you to write this expression with a single exponent. Cal was right. Use your calculator to check that $3^4 \cdot 2^2$ and 6^6 are not equivalent.

EXAMPLE B Rewrite each expression with a single exponent.

a. $(4^5)^2$

b. $(x^3)^4$

c. $(5^m)^n$

d. $(xy)^3$

▶ **Solution**

a. Here, a number with an exponent has another exponent. You can say that 4^5 is **raised to the power** of 2. Begin by writing $(4^5)^2$ as two factors of 4^5.

$$\left(4^5\right)^2 = 4^5 \cdot 4^5 = 4^{5+5} = 4^{10}$$

There is a total of $5 \cdot 2$ or 10 factors of 4.

b. $(x^3)^4 = x^3 \cdot x^3 \cdot x^3 \cdot x^3 = x^{3+3+3+3} = x^{12}$

There is a total of $3 \cdot 4$ or 12 factors of x.

c. Based on parts a and b, when you raise an exponential expression to a power, you multiply the exponents.

$$(5^m)^n = 5^{mn}$$

d. Here, a product is raised to a power. Begin by writing $(xy)^3$ as 3 factors of xy.

$$(xy)^3 = xy \cdot xy \cdot xy = x \cdot x \cdot x \cdot y \cdot y \cdot y = x^3 y^3$$

Do you remember which property allows you to write $xy \cdot xy \cdot xy$ as $x \cdot x \cdot x \cdot y \cdot y \cdot y$?

This example has illustrated two more properties of exponents.

Power Properties of Exponents

For any nonzero values of a and b and any integer values of m and n,

$$(b^m)^n = b^{mn}$$
$$(ab)^n = a^n b^n$$

EXERCISES

You will need your calculator for problems **1, 8, 9, 12,** and **13.**

▶ Practice Your Skills

1. Use the properties of exponents to rewrite each expression. Use your calculator to check that your expression is equivalent to the original expression. [▶ ◻ See **Calculator Note 7B** to learn how to check equivalent expressions. ◀]

   a. $(5)(x)(x)(x)(x)$ **b.** $3x^4 \cdot 5x^6$ **c.** $4x^7 \cdot 2x^3$ **d.** $(-2x^2)(x^2 + x^4)$

2. Write each expression in expanded form. Then rewrite the product in exponential form.

   a. $3^5 \cdot 3^8$ **b.** $7^3 \cdot 7^4$ **c.** $x^6 \cdot x^2$ **d.** $y^8 y^5$ **e.** $x^2 y^4 \cdot xy^3$

3. Rewrite each expression with a single exponent.

   a. $(3^5)^8$ **b.** $(7^3)^4$ **c.** $(x^6)^2$ **d.** $(y^8)^5$

4. Use the properties of exponents to rewrite each expression.

   a. $(rt)^2$ **b.** $(x^2 y)^3$ **c.** $(4x)^5$ **d.** $(2x^4 y^2 z^5)^3$

Reason and Apply

5. An algebra class had this problem on a quiz: "Find the value of $2x^2$ when $x = 3$."
Two students reasoned differently.

    Student 1   Two times three is six. Six squared is thirty-six.
    Student 2   Three squared is nine. Two times nine is eighteen.

    Who was correct? Explain why.

6. Match expressions from this list that are equivalent but written in different
exponential forms. There can be multiple matches.

    a. $(4x^4)(3x)$ **b.** $(8x^2)(3x^2)$ **c.** $(12x)(4x)$ **d.** $(6x^3)(2x^2)$

    e. $12x^6$ **f.** $24x^4$ **g.** $12x^5$ **h.** $48x^2$

7. Evaluate each expression in problem 6 using an x-value of 4.7.

8. Use the properties of exponents to rewrite each expression. Use your calculator to
check that your expression is equivalent to the original expression. [▶  See **Calculator
Note 7B** to learn how to check equivalent expressions. ◀]

    a. $3x^2 \cdot 2x^4$ **b.** $5x^2y^3 \cdot 4x^4y^5$ **c.** $2x^2 \cdot 3x^3y^4$ **d.** $x^3 \cdot 4x^4$

9. Cal and Al's teacher asked them, "What do you get when you square
negative 5?" Al said, "Negative five times negative five is positive
twenty-five." Cal replied, "My calculator says negative twenty-five.
Doesn't my calculator know how to do exponents?" Experiment
with your calculator and see if you can find a way for Cal to get
the correct answer.

10. Evaluate $2x^2 + 3x + 1$ for each x-value.

    a. $x = 3$ **b.** $x = 5$ **c.** $x = -2$ **d.** $x = 0$

11. The properties you learned in this section involve adding and multiplying
exponents, and applying an exponent to more than one factor.

    a. Write and solve a problem that requires adding exponents.

    b. Write and solve a problem that requires multiplying exponents.

    c. Write and solve a problem that requires applying an exponent to two factors.

    d. Write a few sentences describing when to
add exponents, when to multiply exponents,
and when to apply an exponent to more
than one factor.

12. **APPLICATION** Lara buys a $500 sofa at a
furniture store. She buys the sofa with a
new credit card that charges 1.5% interest
per month, with an offer for "no payments
for a year."

    a. What balance will Lara's credit card bill
show after 6 months? Write an
exponential expression and evaluate it.

b. How much much total interest will be added after 6 months?

c. What balance will Lara's credit card bill show after 12 months? Write an exponential expression and evaluate it.

d. How much more interest will be added between 6 and 12 months?

e. Explain why more interest builds up between 6 and 12 months than between 0 and 6 months.

13. Use the distributive property and the properties of exponents to write an equivalent expression without parentheses. Use your calculator to check your answers, as you did in problem 1.

a. $x(x^3 + x^4)$ **b.** $(-2x^2)(x^2 + x^4)$ **c.** $2.5x^{4.2}(6.8x^{3.3} + 3.4x^{4.2})$

14. Write an equivalent expression in the form $a \cdot b^n$.

a. $3x \cdot 5x^3$ **b.** $x \cdot x^5$ **c.** $2x^3 \cdot 2x^3$

d. $3.5(x + 0.15)^4 \cdot (x + 0.15)^2$ **e.** $(2x^3)^3$ **f.** $[3(x + 0.05)^3]^2$

► Review

15. Jack Frost started a snow-shoveling business. He spent $47 on a new shovel and gloves. Jack plans to charge $4.50 for every sidewalk he shovels.

     a. Write an expression for Jack's profit from shoveling x sidewalks. (Hint: Don't forget his expenses.)

     b. Write and solve an inequality to find how many sidewalks Jack must shovel before he makes enough money to pay for his equipment.

     c. How many sidewalks must Jack shovel before he makes enough money to buy a $100 used lawn mower for his summer business? Write and solve an inequality to find out.

16. Find the equation of the line that passes through $(2.2, 4.7)$ and $(6.8, -3.9)$.

17. Solve each system.

     a. $\begin{cases} y = 7.3 + 2.5(x - 8) \\ y = 4.4 - 1.5(x - 2.9) \end{cases}$ **b.** $\begin{cases} 2x + 5y = 10 \\ 3x - 3y = 7 \end{cases}$

Scientific Notation for Large Numbers

In fact, everything that can be known has number, for it is not possible to conceive of or to know anything that has not.

PHILOLAUS

Did you know that there are approximately 75,000 genes in each human cell and more than 50 trillion cells in the human body? This means that 75,000 · 50,000,000,000,000 is a low estimate of the number of genes in your body!

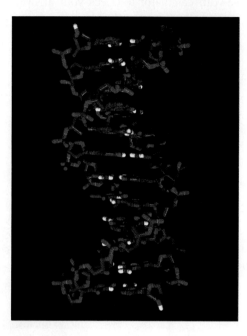

Whether you use paper and pencil, an old-fashioned slide rule, or your calculator, exponents are useful when you work with very large numbers. For example, instead of writing 3,750,000,000,000,000,000 genes, scientists write this number more compactly as 3.75×10^{18}. This compact method of writing numbers is called **scientific notation.** You will learn how to use this notation for large numbers—numbers far from 0 on a number line.

This is a computer model of a DNA strand. Many strands of DNA combine to form the genetic information in each cell.

Investigation
A Scientific Quandary

Consider these two lists of numbers.

In scientific notation	Not in scientific notation
3.4×10^5	27×10^4
7.04×10^3	120,000,000
6.023×10^{17}	42.682×10^{29}
8×10^1	4.2×12^6
1.6×10^2	$4^2 \times 10^2$

Step 1 | Classify each of these numbers as in scientific notation or not. If a number is not in scientific notation, tell why not.

    a. 4.7×10^3 **b.** 32×10^5 **c.** $2^4 \times 10^6$

    d. 1.107×10^{13} **e.** 0.28×10^{11}

Step 2 | Define what it means for a number to be in scientific notation.

Use your calculator's scientific notation mode to help you figure out how to convert standard notation to scientific notation and vice versa.

| Step 3 | Set your calculator to scientific notation mode. [▶ 🖳 See **Calculator Note 7C.** ◀] |

| Step 4 | Enter the number 5000 and press `ENTER`. Your calculator will display its version of 5×10^3. Use a table to record the standard notation for this number, 5000, and the equivalent scientific notation. |

| Step 5 | Repeat Step 4 for these numbers: |

a. 250   **b.** −5,530

c. 14,000   **d.** 7,000,000

e. 18   **f.** −470,000

| Step 6 | In scientific notation, how is the exponent on the 10 related to the number in standard notation? How are the digits before the 10 related to the number in standard notation? If the number in standard notation is negative, how does that show up in scientific notation? |

| Step 7 | Write a set of instructions for converting 415,000,000 from standard notation to scientific notation. |

| Step 8 | Write a set of instructions for converting 6.4×10^5 from scientific notation to standard notation. |

Physicist Suzanne Willis repairs a particle detector at Fermi National Accelerator Lab in Batavia, Illinois. When working with the physics of atomic particles, physicists need scientific notation to write quantities such as 2 trillion electron volts.

A number in scientific notation has the form $a \times 10^n$ where $1 \le a < 10$ or $-10 < a \le -1$ and n is an integer. In other words, the number is written as a number with one nonzero digit to the left of the decimal point multiplied by a power of 10. The number of digits to the right of the decimal point in a depends on the degree of accuracy you want to show.

EXAMPLE

Meredith is doing a report on stars and wants an estimate for the total number of stars in the universe. She reads that astronomers estimate there are at least 125 billion galaxies in the universe. An encyclopedia says that the Milky Way, Earth's galaxy, is estimated to contain more than 100 billion stars. Estimate the total number of stars in the universe. Give your answer in scientific notation.

Maria Mitchell (1818–1889) was the first professional woman astronomer in the United States.

▶ Solution

One billion is 1,000,000,000, or 10^9. Write the numbers in the example using powers of 10 and multiply them.

$$(125 \times 10^9)(100 \times 10^9)$$

125 billion (galaxies) times 100 billion (stars per galaxy).

$$125 \times 100 \times 10^9 \times 10^9$$

Regroup using the associative and commutative properties of multiplication.

$$125 \times 10^2 \times 10^9 \times 10^9$$

Express 100 as 10^2.

$$125 \times 10^{20}$$

Use the multiplication property of exponents.

Since 125 is greater than 10, the answer is not yet in scientific notation.

$$1.25 \times 10^2 \times 10^{20}$$

Convert 125 to scientific notation.

$$1.25 \times 10^{22}$$

Use the multiplication property of exponents.

So the universe contains more than 1.25×10^{22} stars.

Notice in this example that you used exponential expressions that were not in scientific notation. Numbers like 125 billion, 100×10^{18}, or 0.03×10^{12} can come up in calculations and sometimes these numbers make comparisons easier. Scientific notation is one of several ways to write large numbers consistently.

History
● ━━ **CONNECTION** ●

A slide rule is a mechanical device that uses a scale related to exponential notation. Slide rules were widely used for calculating with large numbers until electronic calculators became readily available in the 1970s.

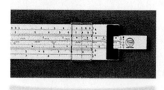

EXERCISES

You will need your calculator for problems **5, 8, 10,** and **14.**

▶ Practice Your Skills

1. Write each number in scientific notation.
   a. 34,000,000,000
   b. −2,100,000
   c. 10,060

2. Write each number in standard notation.
   a. 7.4×10^4
   b. -2.134×10^6
   c. 4.01×10^3

3. Use the properties of exponents to rewrite each expression.
   a. $3x^5(4x)$
   b. $y^8(7y^8)$
   c. $b^4(2b^2 + b)$
   d. $2x(5x^3 - 3x)$

4. Use the properties of exponents to rewrite each expression.
   a. $3x^2 \cdot 4x^3$
   b. $(3y^3)^4$
   c. $2x^3(5x^4)^2$
   d. $(3m^2n^3)^3$

5. Owen insists on reading his calculator's display as "three point five to the seventh." Bethany tells him that he should read it as "three point five times ten to the seventh." He says, "They are the same thing. Why say all those extra words?" Write Owen and Bethany's expressions in expanded form and evaluate each to show Owen why they are not the same thing.

▶ Reason and Apply

6. There are approximately 5.58×10^{21} atoms in a gram of silver. How many atoms are there in 3 kilograms of silver? Express your answer in scientific notation.

7. Because the number of molecules in a given amount of a compound is usually a very large number, scientists often work with a quantity called a mole. One mole is about 6.02×10^{23} molecules.

   a. A liter of water has about 55.5 moles of H_2O. How many molecules is this? Write your answer in scientific notation.

   b. How many molecules are in 6.02×10^{23} moles of a compound? Write your answer in scientific notation.

8. Write each number in scientific notation. How does your calculator show each answer?

   a. 250

   b. 7,420,000,000,000

   c. -18

9. Cal and Al were assigned this multiplication problem for homework:

The number of molecules in one mole is called Avogadro's number. The number is named after the Italian chemist and physicist Amadeo Avogadro (1776–1856).

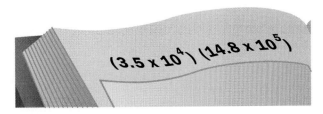

$$(3.5 \times 10^4)(14.8 \times 10^5)$$

Cal got an answer of 51.8×10^9, and Al got 5.18×10^{10}.

   a. Are Cal's and Al's answers equivalent? Explain why or why not.

   b. Whose answer is in scientific notation?

   c. Find another exponential expression equivalent to Cal's and Al's answers.

   d. Explain how you can rewrite a number, such as 432.5×10^3, in scientific notation.

10. Consider these multiplication expressions.

   i. $(2 \times 10^5)(3 \times 10^8)$ ii. $(4.1 \times 10^3)(2 \times 10^5)$

   a. Set your calculator in scientific notation mode and multiply each expression.

   b. Explain how you could do the multiplication in 10a without using a calculator.

   c. Find the product $(4 \times 10^5)(6 \times 10^7)$ and write it in scientific notation without using your calculator.

11. Americans make almost 2 billion telephone calls each day. (www.britannica.com)

   a. Write this number in standard notation and in scientific notation.

   b. How many phone calls do Americans make in one year? (Assume that there are 365 days in a year.) Write your answer in scientific notation.

12. On average a person sheds 1 million dead skin cells every 40 minutes. (*The World in One Day,* 1997, p. 16)

    a. How many dead skin cells does a person shed in an hour? Write your answer in scientific notation.

    b. How many dead skin cells does a person shed in a year? (Assume that there are 365 days in a year.) Write your answer in scientific notation.

13. A *light-year* is the distance light can travel in one year. This distance is approximately 9,460 billion kilometers. The Milky Way galaxy is estimated to be about 100,000 light-years in diameter.

Dead skin cells are one of the components of dust.

    a. Write both distances in scientific notation.

    b. Find the diameter of the Milky Way in kilometers. Use scientific notation.

    c. Scientists estimate the diameter of the earth is greater than 1.27×10^4 km. How many times larger is the diameter of the Milky Way?

▶ Review

14. **APPLICATION** The exponential equation $P = 3.8(1 + 0.017)^t$ approximates Australia's annual population (in millions) since 1900.

    a. Explain the real-world meaning of each number and variable in the equation.

    b. What interval of t-values will give information up to the current year?

    c. Graph $P = 3.8(1 + 0.017)^t$ over the time interval you named in 14b.

    d. What population does the model predict for the year 1950?

    e. Use your answer from 14d to predict today's population.

15. Graph $y \leq -2(x - 5)$.

IMPROVING YOUR **REASONING** SKILLS

The *Jinkōki* (2000, Wasan Institute, p. 146) tells this ancient Japanese problem:

> A breeding pair of rats produced 12 baby rats (6 female and 6 male) in January. There were 14 rats at that time. In February, each female-male pair of rats again bred 12 baby rats. The total number of rats was then 98. In this way, each month, the parents, their children, their grandchildren, and so forth, breed 12 baby rats each. How many rats would there be at the end of one year?

Solve this problem using an exponential model. If you use your calculator, you will get an answer in scientific notation that doesn't show all the digits of the answer. Can you devise a way to find the "missing" digits?

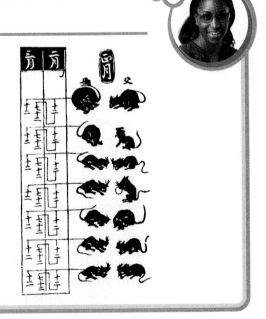

Looking Back with Exponents

The eye that directs a needle in the delicate meshes of embroidery, will equally well bisect a star with the spider web of the micrometer.

MARIA MITCHELL

You've learned that looking ahead in time to predict future growth with an exponential model is related to the multiplication property of exponents. In this lesson you'll discover a rule for dividing expressions with exponents. Then you'll see how dividing expressions with exponents is like looking *back* in time.

Investigation
The Division Property of Exponents

Step 1 Write the numerator and the denominator of each quotient in expanded form. Then reduce to eliminate common factors. Rewrite the factors that remain with exponents.

    a. $\dfrac{5^9}{5^6}$ **b.** $\dfrac{3^3 \cdot 5^3}{3 \cdot 5^2}$ **c.** $\dfrac{4^4 x^6}{4^2 x^3}$

Step 2 Compare the exponents in each final expression you got in Step 1 to the exponents in the original quotient. Describe a way to find the exponents in the final expression without using expanded form.

Step 3 Use your method from Step 2 to rewrite this expression so that it is not a fraction. You can leave $\frac{0.08}{12}$ as a fraction.

$$\dfrac{5^{15}\left(1 + \dfrac{0.08}{12}\right)^{24}}{5^{11}\left(1 + \dfrac{0.08}{12}\right)^{18}}$$

Recall that exponential growth is related to repeated multiplication. When you look ahead in time you multiply by more constant multipliers, or increase the exponent. To look back in time you will need to undo some of the constant multipliers, or divide.

Step 4	Apply what you have discovered about dividing expressions with exponents.

a. After 7 years the balance in a savings account is $500(1 + 0.04)^7$. What does the expression $\frac{500(1 + 0.04)^7}{(1 + 0.04)^3}$ mean in this situation? Rewrite this expression with a single exponent.

b. After 9 years of depreciation, the value of a car is $21,300(1 - 0.12)^9$. What does the expression $\frac{21,300(1 - 0.12)^9}{(1 - 0.12)^5}$ mean in this situation? Rewrite this expression with a single exponent.

c. After 5 weeks the population of a bug colony is $32(1 + 0.50)^5$. Write a division expression to show the population 2 weeks earlier. Rewrite your expression with a single exponent.

d. The expression $A(1 + r)^n$ can model n time periods of exponential growth. What expression models the growth m time periods earlier?

Step 5	How does looking back in time with an exponential model relate to dividing expressions with exponents?

Expanded form helps you understand many properties of exponents. It also helps you understand how the properties work together.

EXAMPLE A | Use the properties of exponents to rewrite each expression.

a. $\dfrac{6x^9}{5x^4}$

b. $\dfrac{(3x^2)(8x^4)}{-4x^3}$

c. $\dfrac{7.5 \times 10^8}{1.5 \times 10^3}$

▶ **Solution** | **a.**

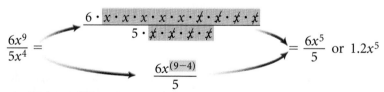

Use expanded form and reduce.

$$\frac{6x^9}{5x^4} = \frac{6 \cdot x \cdot x \cdot x \cdot x \cdot x \cdot \cancel{x} \cdot \cancel{x} \cdot \cancel{x} \cdot \cancel{x}}{5 \cdot \cancel{x} \cdot \cancel{x} \cdot \cancel{x} \cdot \cancel{x}} = \frac{6x^5}{5} \text{ or } 1.2x^5$$

$$\frac{6x^{(9-4)}}{5}$$

In expanded form, 4 factors of x are removed in the numerator and denominator. That leaves $9 - 4$, or 5, factors of x in the numerator.

b.

Use expanded form and reduce.

$$\frac{(3x^2)(8x^4)}{-4x^3} = \frac{3 \cdot 8 \cdot x^2 \cdot x^4}{-4 \cdot x^3} = \frac{3 \cdot 8}{-4} \cdot \frac{x \cdot x \cdot x \cdot \cancel{x} \cdot \cancel{x} \cdot \cancel{x}}{\cancel{x} \cdot \cancel{x} \cdot \cancel{x}} = -6x^3$$

$$\frac{3 \cdot 8}{-4} \cdot x^{(2+4)-3}$$

In expanded form, 2 factors of x are combined with 4 factors of x in the numerator. Then 3 factors of x are removed in the numerator and denominator. That leaves $(2 + 4) - 3$, or 3, factors of x in the numerator.

c.

$$\frac{7.5 \times 10^8}{1.5 \times 10^3} = \frac{7.5}{1.5} \times \frac{10 \cdot 10 \cdot 10 \cdot 10 \cdot 10 \cdot \cancel{10} \cdot \cancel{10} \cdot \cancel{10}}{\cancel{10} \cdot \cancel{10} \cdot \cancel{10}} = 5.0 \times 10^5$$

$$\frac{7.5}{1.5} \times 10^{(8-3)}$$

So, division involving scientific notation can be done just like any other expression with exponents.

The investigation and example have introduced the **division property of exponents.**

Division Property of Exponents

For any nonzero value of b and any integer values of m and n,

$$\frac{b^n}{b^m} = b^{n-m}$$

The division property of exponents lets you divide expressions with exponents simply by subtracting the exponents.

EXAMPLE B

Six years ago, Anne bought a van for $18,500 for her flower delivery service. Based on the prices of similar used vans, she estimates a rate of depreciation of 9% per year.

a. How much is the van worth now?

b. How much was it worth last year?

c. How much was it worth 2 years ago?

▶ **Solution**

a. Right now the value of the van has been decreasing for 6 years. The original price was $18,500, and the rate of depreciation as a decimal is 0.09.

$$A(1 - r)^x = 18,500(1 - 0.09)^6 \approx 10,505.58$$

The van is currently worth $10,505.58.

b. A year ago the van was 5 years old. One approach is to use 5 as the exponent.

$$18,500(1 - 0.09)^5 \approx 11,544.59$$

Another approach is to undo the multiplication in part a by using division.

$$\frac{18,500(1 - 0.12)^6}{(1 - 0.12)} = 18,500(1 - 0.12)^5$$

The numerator on the left side of this equation represents the starting value multiplied by 6 factors of the constant multiplier $(1 - 0.09)$. Dividing by the constant multiplier once leaves you with an expression representing 5 years of exponential depreciation. Either way, the exponent is decreased by 1. The van was worth $11,544.59 last year.

c. To find the value 2 years ago, decrease the exponent in part a by 2.

$$18,500(1 - 0.09)^{(6-2)} = 18,500(1 - 0.09)^4 \approx 12,686.37$$

Subtracting 2 from the exponent gives the same result as undoing two multiplications. The van was worth \$12,686.37 two years ago.

EXERCISES

You will need your calculator for problems **4, 6, 8, 9, 10, 11, 12, 13, 14,** and **15.**

▶ Practice Your Skills

1. Eliminate factors equivalent to 1 and rewrite the right side of this equation.

$$\frac{x^5 y^4}{x^2 y^3} = \frac{x \cdot x \cdot x \cdot x \cdot x \cdot y \cdot y \cdot y \cdot y}{x \cdot x \cdot y \cdot y \cdot y}$$

2. Use the properties of exponents to rewrite each expression.

   a. $\dfrac{7^{12}}{7^4}$ **b.** $\dfrac{x^{11}}{x^5}$ **c.** $\dfrac{12x^5}{3x^2}$ **d.** $\dfrac{7x^6 y^3}{14x^3 y}$

3. Cal says that $\frac{3^6}{3^2}$ equals 1^4 because you divide the 3's and subtract the exponents. Al knows Cal is incorrect, but he doesn't know how to explain it. Write an explanation so that Cal will understand why he is wrong and how to get the correct answer.

4. APPLICATION Webster owns a set of antique dining-room furniture that has been in his family for many years. The historical society tells him that furniture similar to his has been appreciating in value at 10% per year for the last 20 years and that his furniture could be worth \$10,000 now.

   a. Which letter in the equation $y = A(1 + r)^x$ could represent the value of the furniture 20 years ago when it started appreciating?

   b. Substitute the other given information into the equation $y = A(1 + r)^x$.

   c. Solve your equation in 4b to find out how much Webster's furniture was worth 20 years ago. Show your work.

5. Use the properties of exponents to rewrite each expression.

   a. $(2x)^3 \cdot (3x^2)^4$ **b.** $\dfrac{(5x)^7}{(5x)^5}$ **c.** $\dfrac{(2x)^5}{-8x^3}$ **d.** $(4x^2 y^5) \cdot (-3xy^3)^3$

▶ Reason and Apply

6. The earth is 1.5×10^{11} meters from the sun. Light travels at a speed of 3×10^8 meters per second. How long does it take light to travel from the sun to the earth? Answer to the nearest minute.

7. APPLICATION Population density is the number of people per square mile. That is, if the population of a country were spread out evenly across an entire nation, the population density would be the number of people in each square mile.

  a. In 2000, the population of Mexico was about 1.0×10^8. Mexico has a land area of about 7.6×10^5 square miles. What was the population density of Mexico in 2000? *(2001 World Almanac, p. 822)*

  b. In 2000, the population of Japan was about 1.3×10^8. Japan has a land area of about 1.5×10^5 square miles. What was the population density of Japan in 2000? *(2001 World Almanac, p. 803)*

  c. How did the population densities of Mexico and Japan compare in 2000?

8. APPLICATION Eight months ago, Tori's parents put $5,000 into a savings account that earns 0.25% interest per month. Now, her dentist has suggested that she get braces.

  a. If Tori's parents use the money in their savings account, how much do they have?

  b. If Tori's dentist had suggested braces 3 months ago, how much money would have been in her parents' savings account?

  c. Tori's dentist says she can probably wait up to 2 months before having the braces fitted. How much will be in her parents' savings account if she waits?

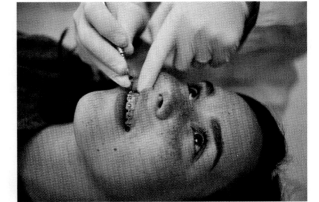

Orthodontic treatment can cost between $4,000 and $6,000 depending on the extent of the procedure. It is estimated that 5 million people were treated by orthodontists in the U.S. in 2000.

9. APPLICATION During its early stages, a disease can spread exponentially as those already infected come in contact with others. Assume that the number of people infected by a disease approximately triples every day. At one point in time there are 864 people who are infected. How many days earlier had fewer than 20 people been infected? Show at least two different methods for solving this problem.

10. The population of a city has been growing at a rate of 2% for the last 5 years. The population is now 120,000. Find the population 5 years ago.

11. **APPLICATION** In the course of a mammal's lifetime its heart beats about 800 million times, regardless of the mammal's size or weight. (This excludes humans.)

a. An elephant's heart beats approximately 25 times a minute. How many years would you expect an elephant to live? Use scientific notation to calculate your answer.

b. A pygmy shrew's heart beats approximately 1150 times a minute. How many years would you expect a pygmy shrew to live?

c. If this relationship were true for humans, how many years would you expect a human being with a heart rate of 60 bpm to live?

Pygmy shrews may be the world's smallest mammal, as small as 5 cm from nose to tail.

12. More than 57,000 tons of cotton are produced in the world each day. It takes about 8 ounces of cotton to make a T-shirt. The population of the United States in 2000 was estimated to be more than 275 million. If all the available cotton were used to make T-shirts, how many T-shirts could have been manufactured every day for each person in the United States in 2000? Write your answer in scientific notation. (*http://cotton.net*)

13. Each day bees sip the nectar from approximately 3 trillion flowers to make 3300 tons of honey. How many flowers does it take to make 8 ounces of honey? Write your answer in scientific notation. (*The World in One Day*, 1997, p. 21)

Review

14. On his birthday Jon figured out that he was 441,504,000 seconds old. Find Jon's age in years. (Assume that there are 365 days per year.)

15. Halley is doing a report on the solar system and wants to make models of the sun and the planets showing relative size. She decides that Pluto, the smallest planet, should have a model diameter of 2 cm.

a. Using the table, find the diameters of the other models she would have to make.

b. What advice would you give Halley on her project?

Size of Planets and Sun

Planet	Diameter (mi)
Mercury	3.1×10^3
Venus	7.5×10^3
Earth	7.9×10^3
Mars	4.2×10^3
Jupiter	8.8×10^4
Saturn	7.1×10^4
Uranus	5.2×10^4
Neptune	3.1×10^4
Pluto	1.5×10^3
Sun	8.64×10^5

LESSON 7.6

Zero and Negative Exponents

It is not knowledge which is dangerous, but the poor use of it.

HROTSWITHA

Have you noticed that so far in this chapter the exponents have been positive integers? In this lesson you will learn what a zero or a negative integer means as an exponent.

x^0 x^{-2}

Investigation
More Exponents

Step 1 Use the division property of exponents to rewrite each of these expressions with a single exponent.

a. $\dfrac{y^7}{y^2}$ **b.** $\dfrac{3^2}{3^4}$ **c.** $\dfrac{7^4}{7^4}$ **d.** $\dfrac{2}{2^5}$ **e.** $\dfrac{x^3}{x^6}$

f. $\dfrac{z^8}{z}$ **g.** $\dfrac{2^3}{2^3}$ **h.** $\dfrac{x^5}{x^5}$ **i.** $\dfrac{m^6}{m^3}$ **j.** $\dfrac{5^3}{5^5}$

Some of your answers in Step 1 should have positive exponents, some should have negative exponents, and some should have an exponent of zero.

Step 2 How can you tell what type of exponent will result simply by looking at the original expression?

Step 3 Go back to the expressions in Step 1 that resulted in a negative exponent. Write each in expanded form. Then reduce them.

Step 4 Compare your answers from Step 3 and Step 1. Tell what a base raised to a negative exponent means.

Step 5 Go back to the expressions in Step 1 that resulted in an exponent of zero. Write each in expanded form. Then reduce them.

Step 6 Compare your answers from Step 5 and Step 1. Tell what a base raised to an exponent of zero means.

Step 7	Use what you have learned about negative exponents to rewrite each of these expressions with positive exponents and only one fraction bar.

a. $\dfrac{5^{-2}}{1}$ b. $\dfrac{1}{3^{-8}}$ c. $\dfrac{4x^{-2}}{z^2y^{-5}}$

Step 8	In one or two sentences, explain how to rewrite a fraction with a negative exponent in the numerator or denominator as a fraction with positive exponents.

This table supports what you have learned about negative exponents and exponents of zero. To go down either column of the table, you divide by 3. Notice that each time you divide, the exponent decreases by 1. (Likewise, to go up either column of the table, you multiply by 3 and the exponent increases by 1.) In order to continue the pattern, 3^0 must have the value 1. As the exponents become negative, the base 3 appears in the denominator with a positive exponent.

$3^1 \div 3 = \dfrac{3^1}{3^1} = 3^{(1-1)} = 3^0$

Exponential form	Fraction form
3^3	27
3^2	9
3^1	3
3^0	1
3^{-1}	$\dfrac{1}{3}$
3^{-2}	$\dfrac{1}{9}$
3^{-3}	$\dfrac{1}{27}$

$3 \div 3 = \dfrac{3}{3} = 1$

$3^{-1} \div 3 = \dfrac{3^{-1}}{3^1} = 3^{(-1-1)} = 3^{-2}$

$\dfrac{1}{3} \div 3 = \dfrac{1}{3} \cdot \dfrac{1}{3} = \dfrac{1}{3^2} = \dfrac{1}{9}$

Negative Exponents and Exponents of Zero

For any nonzero value of b and for any integer value of n,

$$b^{-n} = \dfrac{1}{b^n} \quad \text{and} \quad \dfrac{1}{b^{-n}} = b^n$$

$$b^0 = 1$$

EXAMPLE A

Use the properties of exponents to rewrite each expression without a fraction bar.

a. $\dfrac{3^5}{4^7}$

b. $\dfrac{25}{x^8}$

c. $\dfrac{5^{-3}}{2^{-8}}$

d. $\dfrac{3(17)^8}{17^8}$

► **Solution**

a. $\dfrac{3^5}{4^7} = 3^5 \cdot \dfrac{1}{4^7}$ 　　　　　Think of the original expression as having two separate factors.

　　$= 3^5 \cdot 4^{-7}$ 　　　　　Use the definition of negative exponents.

b. $\dfrac{25}{x^8} = 25 \cdot \dfrac{1}{x^8} = 25 \cdot x^{-8} = 25x^{-8}$

c. $\dfrac{5^{-3}}{2^{-8}} = 5^{-3} \cdot \dfrac{1}{2^{-8}} = 5^{-3} \cdot 2^8$

d. $\dfrac{3(17)^8}{17^8} = 3 \cdot 17^0$ 　　　　　Use the division property of exponents.

　　$= 3 \cdot 1$ 　　　　　Use the definition of an exponent of zero.

　　$= 3$ 　　　　　Multiply.

You can also use negative exponents to look back in time with increasing or decreasing exponential situations.

EXAMPLE B

Solomon bought a used car for $5,600. He estimates that it has been decreasing in value by 15% each year.

a. If his estimate of the rate of depreciation is correct, how much was the car worth 3 years ago?

b. If the car is 7 years old, what was the original price of the car?

► **Solution**

a. You can solve this problem by considering $5,600 to be the starting value and then looking back 3 years.

$y = A(1 - r)^x$ 　　　　　The general form of the equation.

$y = 5,600(1 - 0.15)^{-3}$ 　　　　　Substitute the given information in the equation. -3 means you look back 3 years.

$y \approx 9{,}118.66$

The value of the car 3 years ago was approximately $9,118.66.

b. The original price is the value of the car 7 years ago.

$$y = 5{,}600(1 - 0.15)^{-7}$$

$$y \approx 17{,}468.50$$

The original price was approximately $17,468.50.

You can also use negative exponents to write numbers close to 0 in scientific notation. Just as positive powers of 10 help you rewrite numbers with lots of zeros, negative powers of 10 help you rewrite numbers with lots of zeros between the decimal point and a nonzero digit.

EXAMPLE C

These particle tracks show the paths of particles like protons, electrons, and mesons during a nuclear reaction.

Convert each number to standard notation from scientific notation, or vice versa.

a. A pi meson, an unstable particle released in a nuclear reaction, "lives" only 0.000000026 second.

b. The number 6.67×10^{-11} is the gravitational constant in the metric system used to calculate the gravitational attraction between two objects that have given masses and are a given distance apart.

c. The mass of an electron is 9.1×10^{-31} kilogram.

▶ Solution

a. $0.000000026 = \dfrac{2.6}{100{,}000{,}000} = \dfrac{2.6}{10^8} = 2.6 \times 10^{-8}$

Notice that the decimal point in the original number was moved to the right eight places to get a number between 1 and 10, in this case, 2.6. To undo that, you must multiply 2.6 by 10^{-8}.

b. $6.67 \times 10^{-11} = \dfrac{6.67}{10^{11}} = \dfrac{6.67}{100{,}000{,}000{,}000} = 0.0000000000667$

Multiplying 6.67 by 10^{-11} moves the decimal point 11 places to the left, requiring 10 zeros after the decimal point—the first move of the decimal point changes 6.67 to 0.667.

c. Generalize the method in part b. To write 9.1×10^{-31} in standard notation you move the decimal point 31 places to the left, requiring 30 zeros after the decimal point.

$$9.1 \times 10^{-31} = 0.00000000000000000000000000000091$$

▶ Practice Your Skills

1. Rewrite each expression using only positive exponents.

   a. 2^{-3} **b.** 5^{-2} **c.** 1.35×10^{-4}

2. Insert the appropriate symbol ($<$, $=$, or $>$) between each pair of numbers.

   a. $6.35 \times 10^5 \ \square \ 63.5 \times 10^4$ **b.** $-5.24 \times 10^{-7} \ \square \ -5.2 \times 10^{-7}$

   c. $2.674 \times 10^{-5} \ \square \ 2.674 \times 10^{-6}$ **d.** $-2.7 \times 10^{-4} \ \square \ -2.8 \times 10^{-3}$

3. Find the exponent of 10 that you need for scientific notation.

   a. $0.0000412 = 4.12 \times 10^{\square}$ **b.** $46 \times 10^{-5} = 4.6 \times 10^{\square}$ **c.** $0.00046 = 4.6 \times 10^{\square}$

4. The population of a town is currently 45,647. It has been growing at a rate of about 2.8% per year.

   a. Write an expression of the form $45,647(1 + 0.028)^x$ for the current population.

   b. What does the expression $45,647(1 + 0.028)^{-12}$ represent in this situation?

   c. Write and evaluate an expression for the population 8 years ago.

   d. Write expressions without negative exponents that are equivalent to the exponential expressions from 4b and c.

5. Juan says that 6^{-3} is the same as -6^3. Write an explanation of how Juan should interpret 6^{-3}, then show him how each expression results in a different value.

▶ Reason and Apply

6. Use the properties of exponents to rewrite each expression without negative exponents.

   a. $(2x^3)^2(3x^4)$ **b.** $(5x^4)^2(2x^2)$

   c. $3(2x)^3(3x)^{-2}$ **d.** $\left(\dfrac{2x^4}{3x}\right)^3$

7. **APPLICATION** Suppose the annual rate of inflation is about 4%. This means that the cost of an item increases by about 4% each year. Write and evaluate an exponential expression to find the answers to these questions.

   a. If a piano costs $3,500 today, what did it cost 4 years ago?

   b. If a vacuum cleaner costs $250 today, what did the same model cost 3 years ago?

   c. If tickets to a college basketball game cost $25 today, what did they cost 5 years ago?

   d. The median price of a house in the Midwest United States in June 2000 was $126,800. What was the median price 30 years ago?

   (*National Association of Realtors*)

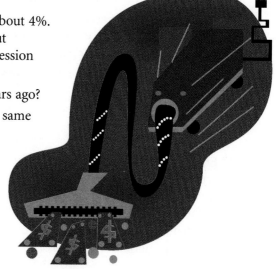

8. APPLICATION The population of Japan in 2000 was about 1.3×10^8. Japan has a land area of about 1.5×10^5 square miles. (*2001 World Almanac*, p. 803)

   a. On average, how much land in square miles is there per person? (Note: This is a different problem from the one you may have solved in Lesson 7.5.)

   b. Convert your answer from 8a to square feet per person.

9. Decide whether each statement is true or false. Use expanded form to show either that the statement is true or what the correct statement should be.

   a. $(2^3)^2 = 2^6$ **b.** $(3^3)^4 = 3^7$

   c. $(10^{-2})^4 = -10^8$ **d.** $(5^{-3})^{-4} = 5^{12}$

10. A large ball of string originally held 1 mile of string. Abigail cut off a piece of string one-tenth of that length. Barbara then cut a piece of string that was one-tenth as long as the piece Abigail had cut. Cruz came along and cut a piece that was one-tenth the length of what Barbara had cut.

   a. Write each length of string in miles in scientific notation.

   b. If the process continues, how long a piece will the next person, Damien, cut off?

   c. Do any of the people have a piece of string too short to use as a shoelace? (Hint: Would you use inches, feet, or miles to measure a shoelace?)

11. Suppose $36(1 + 0.5)^4$ represents the number of bacteria cells in a sample after 4 hours of growth at a rate of 50% per hour. Write an exponential expression for the number of cells 6 hours earlier.

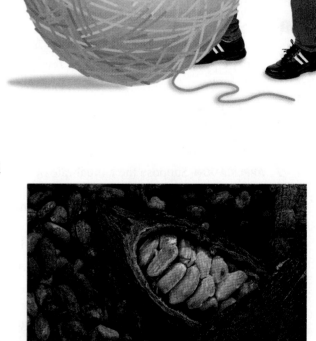

12. APPLICATION Camila received a $1,200 prize for one of her essays. She decides to invest $1,000 of it for college. Her bank offers two options. The first is a regular savings account that pays 2.5% interest every 6 months. The second is a certificate of deposit that pays 5% interest each year.

   a. With the savings account, how much would Camila have after 1 year? After 2 years?

   b. With the certificate of deposit, how much would Camila have after 1 year? After 2 years?

   c. Explain why you get different results for 12a and b.

13. Enough chocolate is produced in the world each day to make 600,000,000 chocolate bars. If you use an estimate of 6,000,000,000 people as the world population, how many chocolate bars is this per person? Write your answer in scientific notation. (*The World in One Day*, 1997, p. 20)

A major ingredient of chocolate is cocoa. Cocoa beans grow in large pods.

▶ Review

14. Find the solution for each system.

a. $\begin{cases} y = 3x - 5 \\ y = -2.5x + 9 \end{cases}$

b. $\begin{cases} y = 2(x - 4) + 15 \\ y = 15(x + 5) - 12 \end{cases}$

15. Set your calculator in scientific notation mode for this problem.

a. Use your calculator to do each division.

i. $\dfrac{8 \times 10^8}{2 \times 10^3}$ **ii.** $\dfrac{9.3 \times 10^{13}}{3 \times 10^3}$ **iii.** $\dfrac{4.84 \times 10^9}{4 \times 10^4}$ **iv.** $\dfrac{6.2 \times 10^4}{3.1 \times 10^8}$

b. Describe how you could do the calculations in 15a without using a calculator.

c. Find the answer to the quotient $\dfrac{4.8 \times 10^7}{8 \times 10^2}$ without using your calculator.

IMPROVING YOUR **REASONING** SKILLS

You have learned about scientific notation in this chapter. There is another convention for writing numbers called **engineering notation.**

Engineering notation	Not in engineering notation
2.5×10^9	2500×10^3
630×10^{-3}	630×10^{-2}
12×10^0	1.5×10^5
400×10^3	0.4×10^6
10.8×10^6	1.08×10^7

1. Write a definition for engineering notation based on the numbers in the lists. If your calculator has an engineering notation mode, you can enter more numbers to help support your definition.

2. Convert these numbers to engineering notation.

a. 78,000,000   **b.** 9,450

c. 130,000,000,000   **d.** 0.0034

e. 0.31   **f.** 1.4×10^8

3. You may have seen these symbols used as shorthand for numbers:

n ("nano," or times $\frac{1}{1,000,000,000}$),

μ ("micro," or times $\frac{1}{1,000,000}$),

k ("kilo," or times 1,000),

M ("mega," or times 1,000,000),

G ("giga," or times 1,000,000,000).

Explain how engineering notation is related to these symbols.

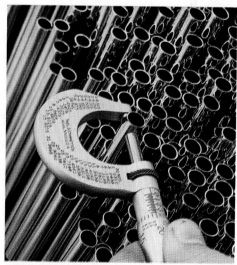

This tool, a micrometer, is used to accurately measure very small distances. Measurements made with it may be recorded in engineering notation.

Fitting Exponential Models to Data

In broken mathematics
We estimate our prize
Vast—in its fading ratio
To our penurious eyes!

EMILY DICKINSON

Victoria Julian has been collecting data on changes in median house prices in her area over the past 10 years. She plans to buy a house 5 years from now and wants to know how much money she needs to save each month toward the down payment. How can she make an intelligent prediction of what a house might cost in the future? What assumptions will she have to make?

Sale pending

Charming Mock Tudor on 1/2 acre

6 bedrms, 3 baths, stone fireplace, marquetry floors throughout, 2-car garage, huge landscaped lot. Best offer.

Lovely Westside Bungalow
Newly remodeled, 2 bedroom, 1.5 bath, off-street parking, laundry, porch, hdwd floors and fireplace, charming back patio

Affordable Brownstone
- 3 floors
- 3 bed/2 bath
- spacious
- charming details
- street parking
- needs only minor repairs
- walk to shops and park
- accepting offers starting Wed

In the real world, situations like population growth, price inflation, and the decay of substances often tend to approximate an exponential pattern over time. With an appropriate exponential model, you can sometimes predict what might happen in the future.

In Chapter 5, you learned about fitting linear models to data. In this lesson you'll learn how to find an exponential model to fit data.

Investigation
Radioactive Decay

You will need

- a paper plate
- a protractor
- a supply of small counters

The particles that make up an atom of some elements, like uranium, are unstable. Over a period of time specific to the element, the particles will change so that the atom eventually will become a different element. This process is called **radioactive decay.**

In this investigation, your counters represent atoms of a radioactive substance. Draw an angle from the center of your plate, as illustrated. Counters that fall inside the angle represent atoms that have decayed.

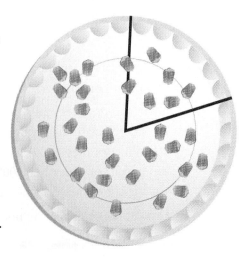

Step 1	Count the number of counters. Record this in a table as the number of "atoms" after 0 years of decay. Pick up all of the counters.
Step 2	Drop the counters on the plate. Count and remove the counters that fall inside the angle—these atoms have decayed. Subtract from the previous value and record the number remaining after 1 year of decay. Pick up the remaining counters.
Step 3	Repeat Step 2 until you have fewer than ten atoms that have not decayed. Each drop will represent another year of decay. Record the number of atoms remaining each time.

Procedure Note

Create a procedure for dropping counters randomly and evenly on the plate. Practice your method until you think you have a good technique. Make a plan for handling counters that fall on the lines of your angle and those that miss the plate—they need to be accounted for too.

Step 4	Let x represent elapsed time in years, and let y represent the number of atoms remaining. Make a scatter plot of the data. What do you notice about the graph?
Step 5	Calculate the ratios of atoms remaining between successive years. That is, divide the number of atoms after 1 year by the number of atoms after 0 years; then divide the number of atoms after 2 years by the number of atoms after 1 year; and so on. How do the ratios compare?
Step 6	Choose one representative ratio. Explain how and why you made your choice.
Step 7	At what rate did your atoms decay?
Step 8	Write an exponential equation that models the relationship between time elapsed and the number of atoms remaining.
Step 9	Graph the equation with the scatter plot. How well does it fit the data?
Step 10	If the equation does not fit well, which values could you try to adjust to give a better fit? Record your final equation when you are satisfied.
Step 11	Measure the angle on your plate. Describe a connection between your angle and the numbers in your equation.
Step 12	Create a table of x- and y-values on your calculator. Is the starting value at year 0 the same as the data you collected? Are any of the subsequent values the same as your data? Explain why you could find differences.

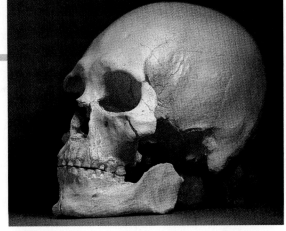

Archaeologists can approximate the age of artifacts with *carbon dating*. This process uses the rate of radioactive decay of carbon-14. Carbon is found in all living things, so the amount left in a bone, for example, is an indicator of the bone's age. This is a plastic casting of a skull found in 1997 in Richland, Washington. Carbon dating has dated the skull as 9200 years old.

The steps of finding an equation in this investigation provide a good method for finding an exponential equation that models data that displays an exponential pattern, either increasing or decreasing. These situations are often generated recursively by multiplying by a constant ratio. Thinking of the constant multiplier in the form $1 + r$ or $1 - r$ leads to these familiar equations:

$$y = A(1 + r)^x$$
$$y = A(1 - r)^x$$

You can then fine-tune the fit of your model by slightly adjusting the values of A and r.

EXAMPLE

Every musical note has an associated frequency measured in hertz (Hz), or vibrations per second. The table shows the approximate frequencies of the notes in the octave from middle C up to the next C on a piano. (In this scale, E# is the same as F and B# is the same as C.)

Piano Notes

Note name	Note above middle C	Frequency (Hz)
Middle C	0	262
C#	1	277
D	2	294
D#	3	311
E	4	330
F	5	349
F#	6	370
G	7	392
G#	8	415
A	9	440
A#	10	466
B	11	494
C above middle C	12	523

The arrangement of strings in a piano shows an exponential-like curve.

a. Find a model that fits the data.

b. Use the model to find the frequency of the note two octaves above middle C (note 24).

c. Find the note with a frequency of 600 Hz.

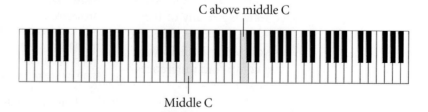

C above middle C

Middle C

a. Let x represent the note number above middle C, and let y represent the frequency. A scatter plot shows the exponential-like pattern. To find the exponential model, first calculate the ratios between successive data points. The mean of the ratios is 1.0593. So the frequency of the notes increases by about 5.93% each time you move up one note on the keyboard. The starting frequency is 262 Hz. So an equation is

$$y = 262(1 + 0.0593)^x$$

The graph shows a very good fit.

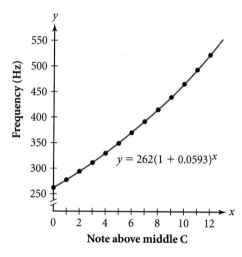

b. To find the frequency of the C two octaves above middle C (note 24), substitute 24 for x in the model.

$$y = 262(1 + 0.0593)^{24} \approx 1044$$

By this model, the frequency of note 24 is 1044 Hz.

c. To find the note with a frequency of 600 Hz, substitute 600 for y in the model.

$$600 = 262(1 + 0.0593)^x$$

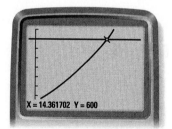

Enter $262(1 + 0.0593)^x$ into Y_1 and 600 into Y_2 on your calculator. Graph both equations and trace to approximate the intersection point. Or you could look at a table to see where $Y_1 = Y_2$.

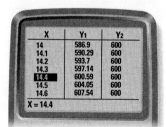

Both the graph and the table show an x-value between 14 and 15. The 14th note above middle C is a D and the 15th note is a D#. Since the piano notes correspond only to whole numbers, you cannot make a note with a frequency of 600 Hz on this piano.

Music

●────**CONNECTION**●

Before the 17th century, there were many ways to tune an instrument. The most popular, developed by the ancient Greek philosopher Pythagoras, used different tuning ratios between each pair of adjacent notes. This made some scales, like the scale of C, sound good but others, like the scale of A-flat, sound bad. Modern Western tuning now uses *even temperament*, based on an equal tuning ratio between adjacent notes, which leads to an exponential model.

If you wanted to find the frequency of notes below middle C, you would need to use negative values for x. The frequencies found using this equation will be fairly accurate because the data fits the equation so well. If the piano were very out of tune, the equation probably would not fit so nicely, and the model might be less valuable for predicting.

Practice Your Skills

1. Rewrite each value as either $1 + r$ or $1 - r$. Then state the rate of increase or decrease as a percent.

   a. 1.15   **b.** 1.08   **c.** 0.76

   d. 0.998   **e.** 2.5

2. Use the equation $y = 47(1 - 0.12)^x$ to answer each question.

   a. Does this equation model an increasing or decreasing pattern?

   b. What is the rate of increase or decrease?

   c. What is the y-value when x is 13?

   d. What happens to the y-values as the x-values get very large?

3. Write an equation to model the growth of an initial deposit of $250 in a savings account that pays 4.25% annual interest. Let B represent the balance in the account, and let t represent the number of years the money has been in the account.

4. Use the properties of exponents to rewrite each expression with only positive exponents.

   a. $4x^3 \cdot (3x^5)^3$
   b. $\dfrac{60x^8y^4}{15x^3y}$
   c. $3^2 \cdot 2^3$
   d. $\dfrac{(8x^3)^2}{(4x^2)^3}$

   e. $x^{-3}y^4$
   f. $(2x)^{-3}$
   g. $2x^{-3}$
   h. $\dfrac{2x^{-4}}{(3y^2)^{-3}}$

Reason and Apply

5. Mya placed a cup of hot water in a freezer. Then she recorded the temperature of the water each minute.

Water Temperature

Time (min) x	0	1	2	3	4	5	6	7	8	9	10
Temperature (°C) y	47	45	43	41.5	40	38.5	37	35.5	34	33	31.5

   a. Find the ratios between successive temperatures.

   b. Find the mean of the ratios in 5a.

   c. Write the ratio from 5b in the form $1 - r$.

   d. Use your answer from 5c and the starting temperature to write an equation in the form $y = A(1 - r)^x$.

   e. Graph your equation with a scatter plot of the data. Adjust the values of A or r until you get a satisfactory fit.

   f. Use your equation to predict how long it will take for the water temperature to drop below 5°C.

6. In science class Phylis used a light sensor to measure the intensity of light (in lumens per square meter, or lux) that passes through layers of colored plastic. The table below shows her readings.

Light Experiment

Number of layers	0	1	2	3	4	5	6
Intensity of light (lux)	431	316	233	174	128	98	73

a. Write an exponential equation to model Phylis's data. Let x represent the number of layers, and let y represent the intensity of light in lux.

b. What does your r-value represent?

c. If Phylis's sensor cannot register readings below 30 lux, how many layers can she add before the sensor stops registering?

7. APPLICATION Recall Victoria from the opening of this lesson. She has collected this table of data on median house prices for her area.

a. Define variables and find an exponential equation to model Victoria's data.

b. Victoria plans to buy a house five years from now. What median price should she expect then?

c. Victoria plans to make a down payment of 10% of the purchase price. Based on your answer to 7b, how much money will she need for her down payment?

d. If Victoria saves the same amount each year for the next 5 years (without interest), how much will she need to save each month for her down payment?

Median House Prices

Year	Years since 1990	Median price ($)
1990	0	85,000
1991	1	95,000
1992	2	95,400
1993	3	101,250
1994	4	107,000
1995	5	114,000
1996	6	120,250
1997	7	127,580
1998	8	135,500
1999	9	144,000

8. The equation $y = 262(1 + 0.0593)^x$ models the frequency in hertz of various notes on the piano, with middle C considered as note 0. The average human ear can detect frequencies between 20 and 20,000 hertz. If a piano keyboard were extended, the highest and lowest notes audible to the average human ear would be how far above and below middle C?

9. Very small amounts of time much less than a second have special names. Some of these names may be familiar to you, such as a millisecond, or 0.001 second. Have you heard of a nanosecond or a microsecond? A nanosecond is 1×10^{-9} second, and a microsecond is 1×10^{-6} second. How many nanoseconds are in a microsecond?

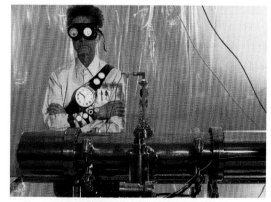

This is Jim Gray, keeper of the NBS-4 atomic clock. Much more accurate than mechanical clocks or watches, some atomic clocks may gain or lose less than a microsecond each year.

10. Suppose that on Sunday you see 32 mosquitoes in your room. On Monday you count 48 mosquitoes. On Tuesday there are 72 mosquitoes. Assume that the population will continue to grow exponentially.

 a. What is the percent rate of growth?

 b. Write an equation that models the number of mosquitoes, y, after x days.

 c. Graph your equation and use it to find the number of mosquitoes after 5 days. After 2 weeks. After 4 weeks.

 d. Name at least one real-life factor that would cause the population of mosquitoes not to grow exponentially.

11. Many stories in children's literature involve magic pots. An Italian variation goes something like this: A woman puts a pot of water on the stove to boil. She says some special words, and the pot is filled with pasta. Then she says another set of special words, and the pot stops filling up.

Suppose someone overhears the first words, takes the pot, and starts it in its pasta-creating mode. Two liters of pasta are created. Then the pot continues to create more pasta because the impostor doesn't know the second set of words. The volume continues to increase 50% per minute.

 a. Write an equation that models the amount of pasta in liters, y, after x minutes.

 b. How much pasta will there be after 30 seconds?

 c. How much pasta will there be after 10 minutes?

 d. How long, to the nearest second, will it be until the entire house, which can hold 450,000 liters, is full of pasta?

12. In this problem you will explore the equation $y = 10(1 - 0.25)^x$.

 a. Find y for some large positive values of x, such as 100, 500, and 1000. What happens to y as x gets larger and larger?

 b. The calculator will say y is 0 when x equals 10,000. Is this correct? Explain why or why not.

 c. Find y for some large negative values of x, such as -100, -500, and -1000. What happens to y as x moves farther and farther from 0 in the negative direction?

▶ Review

13. These polygons are similar. Find the lengths of the three unknown sides.

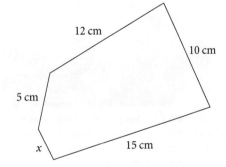

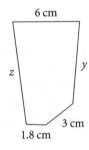

14. One of the most famous formulas in science is

$$E = mc^2$$

The formula describes the relationship between mass (m, measured in kilograms) and energy (E, measured in joules) and shows how they can be converted from one to the other. The variable c is the speed of light, 3×10^8 meters per second. How much energy could be created from a 5-kilogram bowling ball? Express your answer in scientific notation.

James Joule (1818–1889) was one of the first scientists to study how energy was related to heat. At the time of his experiments, many scientists thought heat was a gas that seeped in and out of objects. The SI (metric) unit of energy was named in his honor.

project

MOORE'S LAW

In 1965 Gordon Moore, the co-founder of Intel Corporation, observed that the number of transistors on a computer chip doubled approximately every 2 years. Since a computer processor's speed and power are proportional to the number of transistors on it, computers should get twice as powerful every 2 years.

Has "Moore's Law" come true since 1965? Research technical specifications for various computer processors and find an exponential model that relates time and number of transistors. You can research data in magazines or at www.keymath.com/DA . How many years or months has it taken for computers to double in power? At what rate has the power of computers increased each year?

Your project should include

▶ A scatter plot of your data.

▶ An exponential equation that models the data and an explanation of each number and variable in your equation.

▶ A report summarizing your findings.

You may want to research news items that give recent projections and see if computer chip manufacturers are continuing to meet or exceed Moore's Law. You may also want to research other theories on computer production that examine variables such as purchase price or equipment required for production.

Fathom™

With Fathom you can easily graph an exponential equation through data points. You can use a slider to make small adjustments in your equation until it fits. You can graph multiple models each with its own slider to compare different exponential equations.

Activity Day

Decreasing Exponential Models and Half-Life

In Lesson 7.7, you learned that data can sometimes be modeled using the exponential equation $y = A(1 - r)^x$. In this lesson you will do an experiment, write an equation that models the decreasing exponential pattern, and find the **half-life**—the amount of time needed for a substance or activity to decrease to one-half its starting value. To find the half-life, approximate the value of x that makes y equal to $\frac{1}{2} \cdot A$.

In the previous investigation, if your plate was marked with a 72° angle and you started with 200 "atoms," a model for the data could be $y = 200(1 - 0.20)^x$. This is because the ratio of the angle to the whole plate is $\frac{72}{360}$, or 0.20. To determine the half-life of your atoms, you would need to find out how many drops you would

Technology
CONNECTION

You can see simulations of atomic half-life with a link at **www.keymath.com/DA** .

expect to do before you had 100 atoms remaining. Hence, you could solve the equation $100 = 200(1 - 0.20)^x$ for x using a graph or a calculator table. The x-value in this situation is approximately 3, which means your atoms have a half-life of about 3 years.

Activity
Bouncing and Swinging

There are two experiments described in this activity. Each group should choose at least one, collect and analyze data, and prepare a presentation of results.

You will need

- a motion sensor
- a meterstick
- a ball
- string
- a soda can half-filled with water

Step 1 | Select one of these two experiments.

Experiment 1: Ball Bounce

You will drop a ball from a height of about 1 meter and measure its rebound height for at least 6 bounces. You can collect data "by eye" using a meterstick, or you can use a motion sensor. [▶ 🖥 See **Calculator Note 7D** for a program to use with your motion sensor. ◀] If you use a motion sensor, hold it $\frac{1}{2}$ meter above the ball and collect data for about 8 seconds; trace the resulting scatter plot of data points to find the maximum rebound heights.

Experiment 2: Pendulum Swing

Resting position

Make a pendulum with a soda can half-filled with water tied to at least one meter of string—use the pull tab on the can to connect it to the string. Pull the can back about $\frac{1}{2}$ meter from its resting position and then release it. Measure how far the can swings from the resting position for several swings. You can collect data "by eye" using a meterstick (you may have to collect data for every fifth swing in this case), or you can use a motion sensor. [▶ 🖥 See **Calculator Note 7D** for a program to use with your motion sensor. ◀] If you use a motion sensor, position it 1 meter from the can along the path of the swing; the program will collect the maximum distance from the resting position for 30 swings.

Step 2 | Set up your experiment and collect data. Based on your results, you might want to modify your setup and repeat your data collection.

Step 3 | Define variables and make a scatter plot of your data on your calculator. (If you used a motion sensor, you should have this already.) Sketch the plot on paper. Does the graph show an exponential pattern?

Step 4 | Find an equation of the form $y = A(1 - r)^x$ that models your data. Graph this equation with your scatter plot and adjust the values if a better fit is needed.

Step 5 | Find the half-life of your data. Explain what the half-life means for the situation in your experiment. (Read p. 414 to review the calculation of half-life.)

Step 6 | Find the y-value after 1 half-life, 2 half-lives, and 3 half-lives. How do these values compare?

Step 7 | Write a summary of your results. Include descriptions of how you found your exponential model, what the rate r means in your equation, and how you found the half-life. You might want to include ways you could improve your setup and data collection.

In the real world, eventually your ball will stop bouncing or your pendulum will stop swinging. Your exponential model, however, will never reach a y-value of zero. Remember that any mathematical model is, at best, an approximation and will therefore have limitations.

7

REVIEW

You started this chapter by creating sequences that increase or decrease when you multiply each term by a constant factor. Repeated multiplication causes the rate of change between successive terms to increase or decrease. So the graphs of these sequences curve, getting steeper and steeper or less and less steep. You then discovered that **exponential equations** model these sequences, in which the constant multiplier is the **base** and the number of the term in the sequence is the exponent.

By writing exponential expressions in both **expanded form** and **exponential form**, you learned the **multiplication, division,** and **power properties of exponents,** and you explored the meanings of zero and negative exponents. You applied these properties to **scientific notation,** a way to express numbers with powers of 10.

When modeling data, you can often use an equation to make predictions. You now have two kinds of models for real-world data—a linear equation and an exponential equation. You can model many real-world quantities that increase as **exponential growth** with an equation in the form $y = A(1 + r)^x$. You can model many quantities that decrease, like **radioactive decay,** with an equation in the form $y = A(1 - r)^x$.

EXERCISES

You will need your calculator for problems **2, 3, 5, 7, 8, 10, 13, 19,** and **20.**

1. Write each number in exponential form with a base of 3.

   a. 81
   b. 27
   c. 9
   d. $\frac{1}{3}$
   e. $\frac{1}{9}$
   f. 1

2. Use the properties of exponents to rewrite each expression. Your final answer should have only positive exponents. Use calculator tables to check that your expression is equivalent to the original equation.

   a. $\dfrac{x \cdot x \cdot x}{x}$
   b. $2x^{-1}$
   c. $\dfrac{6.273x^8}{5.1x^3}$
   d. 3^{-x}

   e. $3x^0$
   f. $x^{2.1} \cdot x^{5.6}$
   g. $(3^4)^x$
   h. $\dfrac{1}{x^{-2}}$

3. Consider this exponential equation:

   $$y = 300(1 - 0.15)^x$$

   a. Invent a real-world situation that you can model with this equation. Give the meaning of 300 and of 0.15 in your situation.

   b. What would the inequality $75 \le 300(1 - 0.15)^x$ mean for your situation in 3a?

   c. Find all integer values of x such that $75 \le 300(1 - 0.15)^x$.

4. Proaga says, "Three to the power of zero must be zero. An exponent tells you how many times to multiply the base, and if you multiply zero times you would have nothing!" Give her a convincing argument as to why 3^0 equals 1.

5. For each table, find the value of the constants A and r such that $y = A(1 + r)^x$ or $y = A(1 - r)^x$. Then use your equations to find the missing values.

a.

x	y
0	200
1	280
2	392
3	548.8
4	768.32
5	1075.648
6	

b.

x	y
−2	
−1	
0	850
1	722.5
2	614.125
3	522.00625
4	

6. Convert each number from scientific notation to standard notation, or vice versa.

a. -2.4×10^6

b. 3.25×10^{-4}

c. 37,140,000,000

d. 0.00000008011

7. A person blinks about 9365 times a day. Each blink lasts about 0.15 second. If one person lives 72 years, how many years will be spent with his or her eyes closed while blinking? Write your answer in scientific notation.

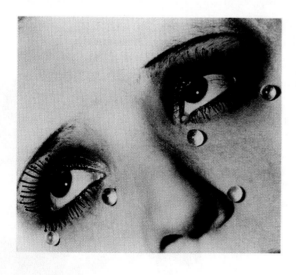

One of the purposes of blinking is to spread tears over the eye. The American photographer Man Ray (1890–1976) is well-known for this photo titled *Glass Tear*.

8. **APPLICATION** In 1995, a can of soda cost 75¢ in a vending machine. If prices increase about 3% per year, in what year will the cost first exceed $2?

9. Classify each equation as true or false. If false, explain why and change the right side of the equation to make it true.

   a. $(3x^2)^3 = 9x^6$

   b. $3^2 \cdot 2^3 = 6^5$

   c. $2x^{-2} = \dfrac{1}{2x^2}$

   d. $\left(\dfrac{x^2}{y^3}\right)^3 = \dfrac{x^5}{y^6}$

10. APPLICATION A pendulum is pulled back 80 centimeters horizontally from its resting position and then released. The maximum distance of the swings from the resting position is recorded for 5 minutes.

Pendulum Swings

Time elapsed (min)	0	1	2	3	4	5
Maximum distance from the resting position (cm)	80	66	55	46	38	32

   a. Define variables and write an equation that models the maximum distance of the swing after each minute.

   b. What is the maximum distance from the resting position after 9 minutes?

   c. After how many minutes will the maximum distance from the resting position be less than 5 centimeters?

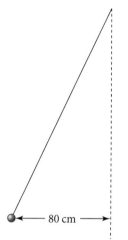

← 80 cm →

MIXED REVIEW

11. Three sisters went shopping for T-shirts and sweatshirts at an outlet store. They paid $6 for each T-shirt and $10 for each sweatshirt. They bought 12 shirts in all and the total cost was $88.

   a. Define variables and write a system of equations to represent this situation.

   b. Solve the system symbolically. How many shirts of each kind did the sisters buy?

12. Mr. Lee's science class received a collection of praying mantises from a local entomologist. The students have been measuring the mantises' lengths in centimeters.

Praying Mantis Length (cm)

5.6, 9.4, 1.7, 3.4, 5.3, 6.2, 8.2, 2.1, 5.3, 2.6, 5.6, 2.6, 5.4, 12.1, 5.3, 2.2, 4.8, 9.8

   a. Organize the data in a stem plot.

   b. What is the range of the data?

   c. What are the measures of center? Which do you think best represents the data? Explain your thinking.

Praying mantises use their front legs to capture other insects.

13. Write a recursive routine to generate each sequence. Then use your routine to find the 10th term of the sequence.

    a. 21, 17, 13, 9, . . .   **b.** −5, 15, −45, 135, . . .   **c.** 2, 9, 16, 23, . . .

14. APPLICATION Chad polled the 9th graders and 12th graders at his school to ask about their plans for post–high school education. He got the following results.

Post–High School Plans

Plans	Number of 9th graders	Number of 12th graders
College or university	126	212
Junior college	88	122
Technical school	64	98
Travel	92	142
Work full time	132	78
Undecided	260	46
Total	762	698

  a. Make two circle graphs to represent the data that Chad collected.

  b. What is the percent of increase or decrease in students planning to attend a college or university from 9th to 12th grade?

  c. What is the percent of increase or decrease in students planning to work full time from 9th to 12th grade?

  d. Marta attends another high school in Chad's town. She wants to use Chad's information to predict how many 12th graders from a class of 520 will travel after graduation. Explain how she can do this.

15. Angie has some guests visiting from Italy, and they are planning to drive from her house to Washington, D.C. Angie wants her friends to understand how far they will need to drive.

  a. Write a direct variation equation to convert miles to kilometers.

  b. Angie's house is about 250 miles from Washington, D.C. How many kilometers will her friends have to drive?

  c. The hotel her friends are staying at says it is 2 miles from the Washington Monument. How far is that in kilometers?

  d. The Washington Monument is taller than 555 feet. How tall is this in meters?

The Washington Monument was built between 1848 and 1884. It was the tallest structure until the Eiffel Tower was built in 1889.

16. Sketch a graph showing these three inequalities on the same coordinate axes. What shape do you get?

$$\begin{cases} x > -2 \\ y \geq x \\ y \leq 4 \end{cases}$$

17. Solve each equation using the method of your choice. Then use a different method to verify your answer.

   a. $-2(4-d) + 3 = -13$

   b. $42 - 7(d - 8) = 7$

   c. $0.5(d - 2) - 3 = -10$

18. Find the slope and y-intercept of the line through each pair of points.

   a. $(3, 2), (1, -3)$

   b. $(-1, 4), (-1, 8)$

   c. $(-11, 3), (-9, 2)$

19. Does someone use the same amount of soap each day when he or she showers? Rex Boggs of Glenmore State High School in Queensland, Australia, decided to find out. He collected data for three weeks.

   a. Make a scatter plot and find a linear equation that fits the data. Use any method that you prefer.

   b. Write your equation in intercept form.

   c. Using your equation, after how many days would Rex's soap weigh 34 grams?

   d. How would your equation in 19b change if Rex starts with a family size soap bar that weighs 200 grams? Write an equation for a family size soap bar. (Assume the soap is used at the same rate.)

   e. If Rex starts with a family size soap bar, how much will it weigh after 20 days have elapsed?

Soap Usage

Number of days elapsed	Weight of bar of soap (grams)
0	124
1	121
4	103
5	96
6	90
7	84
8	78
9	71
11	58
12	50
17	27
19	16
20	12
21	8
22	6

(*www.maths.uq.oz.au*)

20. APPLICATION When Anton started his career as an assistant manager of a tutoring center, his salary was $18,500 per year. He was told that he would get a 2.25% raise every year. Anton has now been with the company for 3 years.

a. Find Anton's current salary in whole dollars.

b. What will Anton's salary be in another 5 years?

c. Anton's coworker Kobra has been with the company for 10 years. She can't remember her starting salary but knows that she got the same 2.25% raise for each of her first 8 years and then got a 3.5% raise during each of the last 2 years. She now makes $23,039. Write and solve an equation to find Kobra's starting salary.

TAKE ANOTHER LOOK

Scientific notation gives scientists and mathematicians one way to express extremely large and extremely small numbers. Sometimes scientists focus on only the power of 10 to describe size or quantity, calling this the **order of magnitude.**

Consider that the average distance from the earth to the sun is 9.29×10^7 miles. Unless a scientist is going to calculate with this figure, she may simply say the distance in miles from the earth to the sun is *on the order of 10^7*. By stating only the power of 10, what range of values is the scientist including?

Order of magnitude is also used to compare numbers. Suppose a sample of bacteria grows from several hundred to several thousand cells overnight. How many times larger is the sample now? A scientist may say the number of cells in the sample *increased by one order of magnitude,* because $\frac{10^3}{10^2}$ equals 10^1. What would the scientist say when the sample grows from several hundred cells to several hundred thousand cells? What fraction of cells would remain if the sample *decreased* by two orders of magnitude? (Note: The units must be equal to compare orders of magnitude.)

Think about the relative size of our universe as you answer these questions:

1. Explain what it means for the typical size of a cell in meters to be on the order of 10^{-6}.

2. Explain what it means for the length of a cow in meters to be on the order of 10^0.

3. The distance in meters from the earth to the nearest star (other than the sun) is on the order of 10^{17}. Is it correct to compare the distance from the earth to the sun and the distance from the earth to the nearest star as an increase by 10 orders of magnitude, because $\frac{10^{17}}{10^7}$ equals 10^{10}?

4. The diameter in meters of the Milky Way galaxy is 10^{20}. Describe the increase in order of magnitude between the size of a cell and the size of the galaxy.

When something increases 100%, should it be described as an increase in order of magnitude? Give an example to support your conclusion.

Assessing What You've Learned

 WRITE IN YOUR JOURNAL Add to your journal by considering one of these prompts.

▶ Why is scientific notation convenient for writing extremely large or extremely small numbers? Are there numbers that you find to be less convenient to write in scientific notation? Does scientific notation help you to understand why our standard number system is called a "base 10" system?

▶ Compare and contrast linear and exponential data. How do the graphs differ? If you weren't specifically told to find either a linear or exponential equation to fit a graph of data, how would you decide which to try? How do the methods of fitting linear and exponential models compare?

 PERFORMANCE ASSESSMENT Show a classmate, a family member, or your teacher that you know how to find an exponential model in the form $y = A(1 + r)^x$. You may want to go back and use the data sets from Lesson 7.7 or Lesson 7.8, or use data that you have collected from a project. Explain why you think the data is exponential, and when and why you would want to adjust the value of A or r.

 GIVE A PRESENTATION Review the properties of exponents that you learned in this chapter. Think about the techniques you have used to remember these properties, or ask your peers, teachers, or family members how they remember these properties. Prepare a presentation for your class and demonstrate the memory methods you have learned. Your presentation will help your classmates remember the properties of exponents too!

Functions

The musician in
The Lute Player by an
unknown artist called the
Master of the Half Figures
plays her lute while reading
sheet music. When music is
composed or transcribed it is written
on a staff as notes in standard notation
or as numbers in tablature. Playing music
from notation and writing notation from
music are very much like the relationships
between input and output in
mathematical functions.

OBJECTIVES

In this chapter you will

- learn the mathematics of
  code breaking
- use the vertical line test
- graph functions of real-
  world situations
- learn about function
  notation and vocabulary
- learn the absolute value
  and squaring functions

Secret Codes

The study of secret codes is called **cryptography.** Early examples of codes go back 4000 years to Egypt. Writing messages in code plays an important role in history and in technology. Today you can find applications of codes at ATMs, in communications, and on the Internet.

Cryptography is an intellectual battle between the code-maker and the code-breaker.

SIMON SINGH

The Rosetta Stone, found near Rashid, Egypt, in 1799, bears inscriptions in Greek, Egyptian hieroglyphics, and demotic (everyday) Egyptian. Having these three versions of the same text helped language researchers "break the code" of hieroglyphics.

In this investigation you will learn some of the mathematics behind secret codes.

Investigation
TFDSFU DPEFT

You will need

- the worksheet Coding Grid

The table below shows that the letter A is coded into the letter Q, and the letter B is coded into R, and so on. It also shows that the letter U is coded into the letter K. This code is an example of a letter-shift code. Can you see why? How would you use the code to write a message?

Original input	A	B	C	D	E	F	G	H	I	J	K	L	M	N	O	P	Q	R	S	T	U	V	W	X	Y	Z
Coded output	Q	R	S	T	U	V	W	X	Y	Z	A	B	C	D	E	F	G	H	I	J	K	L	M	N	O	P

You can also represent the code with a grid. Note that the input letters run across (horizontally). To code a letter, look for the shaded square directly above it. Then find the coded output by looking across to the letters that run up (vertically).

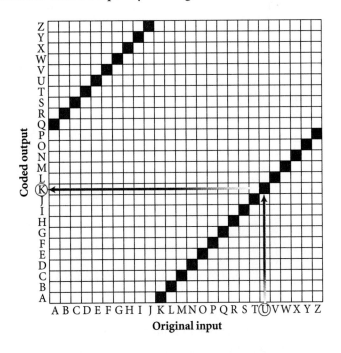

Step 1	Use the coding grid to write a two-word or three-word message.
Step 2	Think of the letters A through Z as the numbers 1 through 26. Write two different rules to describe this code. Write one rule that adds a constant number to a letter's position in the alphabet. Write a second rule that subtracts a constant number from the letter's position. Explain how to code letters that are shifted to a position which is less than 1 or greater than 26.
Step 3	Exchange your coded message with a partner. Use this grid or one of your rules to decode each other's messages.

Next you'll invent your own letter-shift code.

Step 4	Create a new code by writing a rule that shifts letters a certain specified number of places. Put the code on a grid like the one shown above. Do not allow your partner to see this grid.
Step 5	Use your new grid to code the same message you wrote in Step 1.
Step 6	Exchange your newly coded message with your partner. Use it along with the message in the first code to try to figure out each other's new codes. Write a rule or create a coding grid to represent your partner's new code.
Step 7	Compare your grid to your classmates' new grids. In what ways are the grids the same? How are they different? For any one grid, how many coded outputs are possible for one input letter? How many ways are there to decode any one letter in a coded message?

Here is another new code.

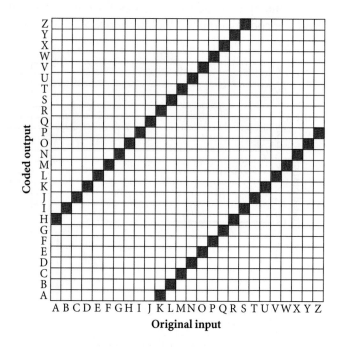

Coded output (vertical axis): A B C D E F G H I J K L M N O P Q R S T U V W X Y Z

Original input (horizontal axis): A B C D E F G H I J K L M N O P Q R S T U V W X Y Z

Step 8 | Use the grid above to send a new two- or three-word message to your partner. Exchange and decode each other's messages.

Step 9 | Did your partner successfully decode your message? Why or why not?

Step 10 | How is the new grid in Step 8 different from the grid in Step 1? Code the word FUNCTION to help you answer this question.

Step 11 | Which grid makes it easier to decode messages? Which coded output letters are difficult to decode into their original input letters?

Step 12 | Create a new coding scheme by shading squares that don't touch each other on the grid. Make the grid so that there is exactly one output for each input. How is it similar to the grid in Step 1? How is it different?

Letter-shift codes are relationships. Codes that have exactly one output letter for every input letter are examples of **functions.** The set of values that are inputs for a function is called its **domain.** In the investigation, the domain is the set of all letters of the alphabet. The **range** of a function is the set of all its possible output values for these codes. The range happens to be all the letters of the alphabet as well. But often the domain contains many values different from those in the range. Here is an example.

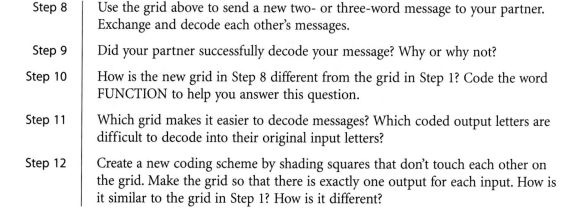

Domain	A	B	C	D	E	F	G	H	I	J	K	L	M
Range	65	66	67	68	69	70	71	72	73	74	75	76	77

Domain	N	O	P	Q	R	S	T	U	V	W	X	Y	Z
Range	78	79	80	81	82	83	84	85	86	87	88	89	90

Computers store letters as numbers. In the preceding example, the letter A is coded as the number 65, B as 66, and so on. In this case, the domain is the letters of the alphabet, but the range is the set of whole numbers from 65 through 90. The table on the opposite page shows how each letter is represented in this code.

Notice that each letter in the domain matches no more than one number in the range. This is what makes the code a function.

EXAMPLE

Tell whether or not each table of values represents a function. Give the domain and range of each relationship.

Table A

Input	Output
1	2
2	4
3	6

Table B

Input	1	0	1
Output	1	2	5

Table C

Input	1	2	3	4	5	6
Output	0	0	0	0	0	0

▶ **Solution**

To be a function, each input must have exactly one output. It is helpful to use arrows to show which input value matches which output value.

Table A

Each input value matches one output value. So this relationship is a function. The domain is {1, 2, 3}, and the range is {2, 4, 6}.

Input: 1   2   3
↓   ↓   ↓
Output: 2   4   6

Table B

The input value 1 has two outputs, 1 and 5. This relationship is not a function because there is an input value with more than one output value. The domain is {0, 1}, and the range is {1, 2, 5}.

Input: 0   1

Output: 1   2   5

Table C

Each input value matches exactly one output value, 0. So this relationship is a function, even though all the inputs have the same output. The domain is {1, 2, 3, 4, 5, 6}, and the range is {0}.

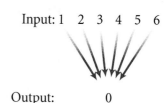

Input: 1   2   3   4   5   6

Output: 0

You can represent a function with a table, a graph, an equation, symbols, a diagram, or even a written rule or description. Many of the relationships you have studied in this book are functions. You will revisit some of them as you learn more about functions in this chapter.

EXERCISES

▶ Practice Your Skills

1. Use this table to code each word.

Input	A	B	C	D	E	F	G	H	I	J	K	L	M	N	O	P	Q	R	S	T	U	V	W	X	Y	Z
Coded output	B	C	D	E	F	G	H	I	J	K	L	M	N	O	P	Q	R	S	T	U	V	W	X	Y	Z	A

a. RANGE   **b.** DOMAIN   **c.** TABLE   **d.** GRAPH

2. Use the grid at right to decode each word.
  a. SXZED
  b. YEDZED
  c. BOVKDSYXCRSZ
  d. BEVO

3. The title of the investigation, TFDSFU DPEFT, is the output of a one–letter–shift code.
  a. Decode TFDSFU DPEFT.
  b. Write the rule or create the coding grid for the code.

4. Use the coding grid below to answer 4a–c.
  a. What are the possible input values?
  b. What are the possible output values?
  c. Is this code a function? Explain why or why not.

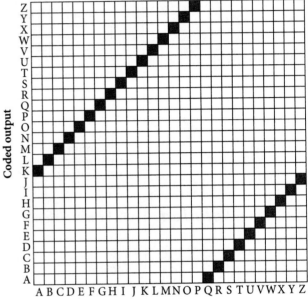

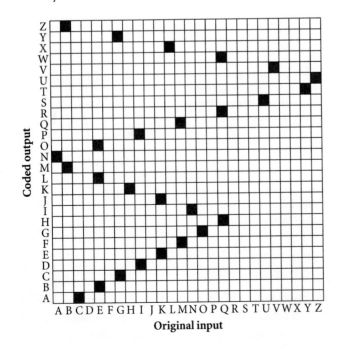

5. The table converts standard time to military time.

Standard time (A.M.)	1:00	2:00	3:00	4:00	5:00	6:00	7:00	8:00	9:00	10:00	11:00	12:00
Military time	0100	0200	0300	0400	0500	0600	0700	0800	0900	1000	1100	1200

Standard time (P.M.)	1:00	2:00	3:00	4:00	5:00	6:00	7:00	8:00	9:00	10:00	11:00	12:00
Military time	1300	1400	1500	1600	1700	1800	1900	2000	2100	2200	2300	2400

a. Describe the input.

b. Describe the output.

c. Does the table represent a function? Explain why or why not.

Reason and Apply

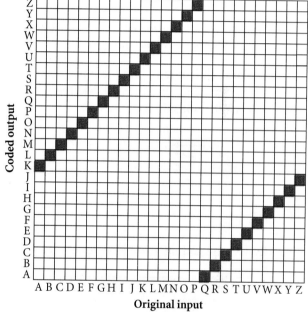

6. Use the letter-shift grid at right to

a. Find the output when the input is W.

b. Find the input when the output is W.

c. Code a Q.

d. Decode a K.

7. APPLICATION Think of the letters A through Z as the numbers 1 through 26.

a. Enter the position for each letter in the word FUNCTIONS into list L_1. Use your calculator to shift the letters 9 places to the right in the alphabet. Store the results in list L_2.

b. What must you do to some of these numbers before coding them all into letters? What are the numbers in list L_2 after you do this?

c. Use the results from 7b to code the word FUNCTIONS.

d. Plot pairs of the form (*input position, output position*) for this code.

e. If you design a different letter-shift code, what letter-shift values should you avoid so that FUNCTIONS is not coded as the same word?

8. Sylvana creates a code that doubles the position number of each letter in the alphabet. Then she subtracts 26 from the new positions that do not correspond to a letter in the alphabet. She stores the input values in list L_1 and the output values in list L_2.

a. What numbers are in list L_1?

b. What numbers are in list L_2?

c. Plot Sylvana's code.

d. Will she have difficulty coding or decoding messages?

9. Use the coding grid at right to decode CEOKEQC into a word.

10. Here is a corner of a coding grid.

a. Does each input letter code to a single output? Does each output letter decode to a single input?

b. If this code were a function, which would be made easier, coding or decoding?

c. How would you change this grid to make the other part of coding in 10b easier?

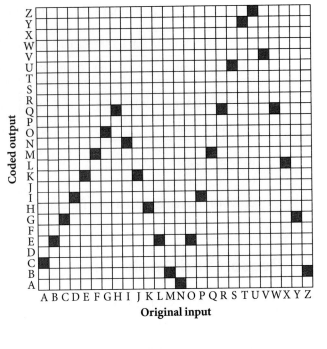

11. For each diagram, give the input values and output values and then tell whether or not each relationship is a function.

a.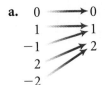

b.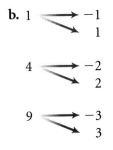

12. Use the coding grid below to answer 12a–c.

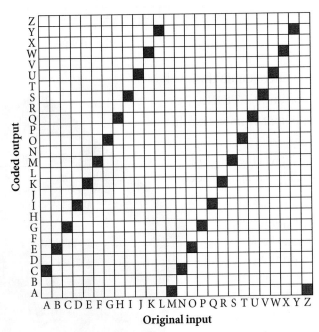

a. Write a rule for this coding grid.

b. Code the word CODE.

c. Can you decode the word SPY? Explain why or why not.

13. The grid at right shows an ancient Hebrew code called "atbash."

(*http://all.net/books/ip/Chap2-1.html*)

  a. Create a rule for the atbash code.

  b. Is this code a function? Explain why or why not.

  c. Use the atbash code to code your name.

14. If you know that SHOFJEWHQFXO is the study of coding and decoding, what is the rule for breaking this code? What is the original message?

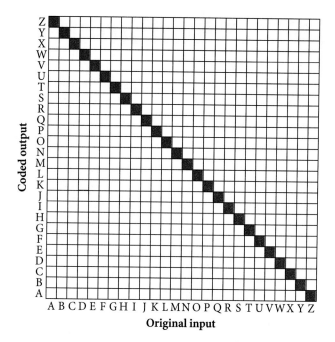

Review

15. If possible, perform the indicated operation.

  a. $(4b^3)(7b^3)$

  b. $4a^3 + 7a^3$

  c. $\dfrac{7c^3}{4c^3}$

  d. $4(7d^3)^3$

  e. $2a^3 + 3b^2$

  f. $(2x^3)(2x^2)$

16. If 1 calorie is 4.1868 joules, then how many calories is 470 joules?

COMPUTER NUMBER SYSTEMS

A computer stores alphanumeric symbols—letters, digits, and special characters—as a sequence of 1's and 0's in its memory. These numbers are called **binary numbers,** or base 2 numbers, because they contain only two digits—1 and 0. The number system that people use is a base 10 decimal system because it contains the ten digits from 0 through 9. How does a computer store 10 numerical digits, 26 letters, and several other characters using only two digits?

Research the binary number system and its use in computer memory. Are there other number systems that computers also use? How do computers convert letters into numbers? Is there a standard code that most systems follow?

Your project should include

▶ Sample conversions of base 10 numbers to binary numbers, and vice versa.

▶ A table that shows how to code letters and special characters.

Functions and Graphs

In Lesson 8.1, you learned that you can write rules for some of the coding grids. You can also write rules, often in the form of equations, to transform numbers into other numbers. One simple example is "Add one to each number." You can represent this rule with a table, an equation, a graph, or even a diagram.

Table

Input x	Output y
7	8
−47	−46
10.28	11.28
x	x + 1

Equation

$y = x + 1$

Graph

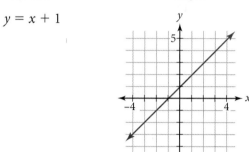

Diagram

Domain		Range
−47	⟶	−46
7	⟶	8
10.28	⟶	11.28

This rule turns 7 into 8, −47 into −46, 10.28 into 11 28, and x into x + 1.

When you explored relationships in previous chapters, you used recursive routines, graphs, and equations to relate input and output data. To tell whether a relationship between input and output data is also a function, there is a test that you can apply to the relationship's graph on the xy-plane.

The Spanish painter Pablo Picasso (1881–1973) was one of the originators of the art movement Cubism. Cubists were interested in creating a new visual language, translating realism into a different way of seeing.

Investigation
Testing for Functions

Here are three relationships, each in different form.

Relationship A

Input x	Output y
3	7
31	63
4.7	10.4
0	1
−11	−21
51	103

Relationship B

$y = 47(1.10)^x$

Relationship C

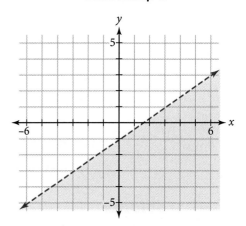

Step 1 | The table in Relationship A gives a few input and output values for a function. Make a graph of the complete function and write an equation. Use these three different expressions of the same relationship to answer Steps 2–5.

Step 2 | Enter possible input values into list L1 and the corresponding output values into list L2. Select an appropriate window and plot the points on your calculator. What is the complete domain and range of the graph?

Step 3 | Next, enter an equation for the relationship into Y1. Graph the equation onto your point plot. What set of values can be entered into x? What is the resulting set of values for y?

Step 4 | Examine your table and graph. Does each input value have exactly one output value?

Step 5 | Move a vertical line, such as the edge of a ruler, from side to side on your graph. Count the number of intersections of your vertical line at each x-value to see if the graph represents a function. How does this result help you tell whether the relationship is a function?

Step 6 | Consider the equation in Relationship B. Express it as a graph and as a table. Use these three different expressions of Relationship B to again answer Steps 2–5.

Step 7 | Consider the graph in Relationship C. Express this relationship one other way and repeat Steps 2–5.

A function is a relationship between input and output values. Each input has exactly one output. The **vertical line test** helps you determine if a relationship is a function. If all possible vertical lines cross the graph once or not at all, then the graph represents a function. The graph does not represent a function if you can draw even one vertical line that crosses the graph two or more times.

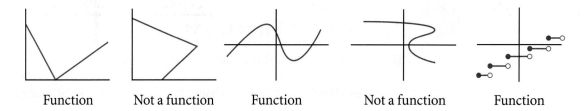

Function   Not a function   Function   Not a function   Function

You have learned many forms of linear equations. In the example, you will see whether or not all lines represent functions.

EXAMPLE

Name the form of each linear equation and use a graph to explain why it is or is not a function.

a. $y = 1 - 3x$ **b.** $y = 0.5x + 2$ **c.** $y = \frac{3}{4}x$ **d.** $2x + 3y = 6$

e. $y = 5 + 2(x - 8)$ **f.** $y = 7$ **g.** $x = 9$

▶ Solution

Each equation is written in one of the forms you have learned in this course. If you graph the equations, you can see that all of them except the graph for part g pass the vertical line test. So all the equations represent functions except for the one in part g.

a. This equation is in the intercept form $y = a + bx$.

b. This equation is in the slope-intercept form $y = mx + b$.

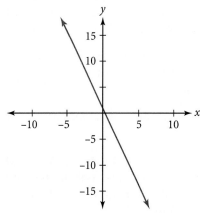

c. This equation is a direct variation in the form $y = kx$.

d. This equation is in the standard form $ax + by = c$.

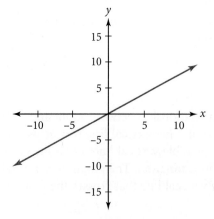

e. This equation is in the point-slope form $y = y_1 + b(x - x_1)$.

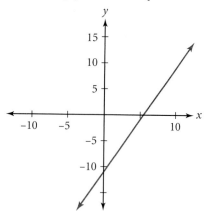

f. This equation is a horizontal line in the form $y = k$.

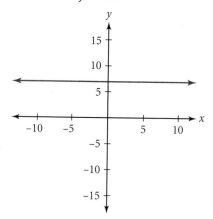

g. This equation is a vertical line in the form $x = k$.

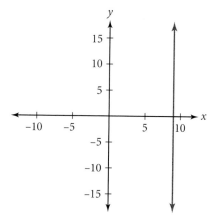

The graph of $x = 9$ fails the vertical line test. In fact, all vertical lines fail the vertical line test. If you rewrite the equation as $0y + x = 9$, you'll see that you can match infinitely many output values of y to one input value of x. (No matter what value you substitute for y, the factor 0 will drop the y-term out of the equation.) So this equation does not represent a function.

As you work more with functions, you will be able to tell if a relationship is a function without having to consider its graph on the xy-plane. If the graph is shown, use the vertical line test. Otherwise, see if there is more than one output value for any single input value.

Carpenters use a tool called a level to determine if support beams are truly vertical.

You will need your calculator for problem **15.**

Practice Your Skills

1. Use the equations to find the missing entries in each table.

a. $y = 4.2 + 0.8x$

Input x	Output y
−4	
−1	
1.5	
6.4	
9	

b. $y = 1.2 − 0.8x$

Domain x	Range y
−4	
−1	
2.4	
	−7.6
	−10

2. On the same set of axes, plot the points in the table and graph the equation in problem 1a.

3. On the same set of axes, plot the points in the table and graph the equation in problem 1b.

4. Use the tables and graphs in problems 1–3 to tell whether or not the relationships in problem 1 are functions.

Reason and Apply

5. The graph at right describes another student's distance from you. What are the walking instructions for the graph? Does it represent a function?

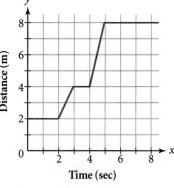

6. Find whether or not each graph below represents a function. Does it pass the vertical line test?

a.

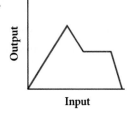

Output

Input

b.

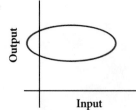

Output

Input

c.

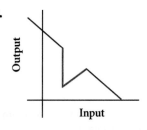

Output

Input

d.

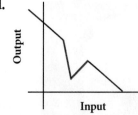

Output

Input

7. Does each relationship of the form (*input, output*) represent a function? If the relationship does not represent a function, find an example of one input that has two or more outputs.

  a. (*city, area code*)

  b. (*person, birth date*)

  c. (*last name, first name*)

  d. (*state, capital*)

8. The graphs of seven walks showing distance from a motion sensor are shown.

  i.
  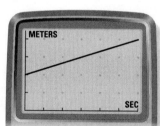

  ii.
  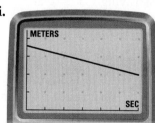

  iii.
  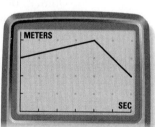

  iv.
  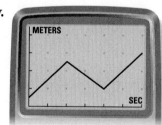

  v.
  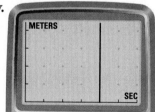

  vi.
  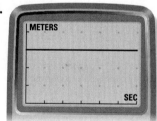

  vii.
  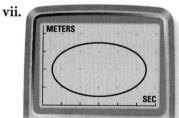

  a. Which graphs represent functions?

  b. For which graphs is it not possible to write walking instructions?

  c. What conclusion can you make?

9. Find whether or not each table of x- and y-values represents a function. Explain your reasoning.

a.

Input x	Output y
0	5
1	7
3	10
7	9
5	7
4	5
3	8

b.

Input x	Output y
3	7
4	9
8	4
5	5
9	3
11	9
7	6

c.

Input x	Output y
2	8
3	11
5	12
7	3
9	5
8	7
4	11

10. On graph paper, draw a graph that is a function and has these three properties:

   ▶ Domain of x-values satisfying $-3 \leq x \leq 5$
   ▶ Range of y-values satisfying $-4 \leq y \leq 4$
   ▶ Includes the points $(-2, 3)$ and $(3, -2)$

11. On graph paper, draw a graph that is *not* a function and has these three properties:

   ▶ Domain of x-values satisfying $-3 \leq x \leq 5$
   ▶ Range of y-values satisfying $-4 \leq y \leq 4$
   ▶ Includes the points $(-2, 3)$ and $(3, -2)$

12. Complete the table of values for each equation. Let x represent the input values, and let y represent the output values. Graph the points and find whether or not the equation describes a function. Explain your reasoning.

a. $x - 3y = 5$

x	2		-4		0	
y		1		-2		0

b. $y = 2x^2 + 1$

x	-2	3	0	-3	-1	
y						9

c. $x + y^2 = 2$

x	-7				-2	2
y		1	-2	-3		

d. $x + 2y = 4x$

x						
y						

13. Identify all numbers in the domain and range of each graph.

a.

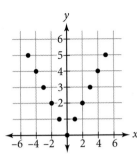

b.

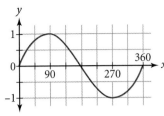

c.

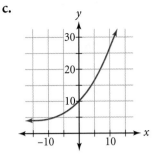

14. Which letters of the alphabet pass the vertical line test?

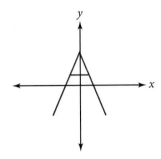

Review

15. If x represents actual temperature and y represents wind chill temperature, the equation

$$y = -52 + 1.6x$$

approximates the wind chill temperatures for a wind speed of 40 miles per hour. Enter this equation into Y_1 on your calculator and find the requested x- and y-values.

a. What x-value gives a y-value of $-15°$? Explain how you use the calculator table function to find this answer.

b. Enter

$$y = -15$$

into Y_2 on your calculator. Graph both equations. Explain how to use the graph to answer 15a.

16. Show how you can use an undoing process to solve these equations. (Hint for 16b: First, invert both fractions.)

a. $\dfrac{4(x - 7) - 8}{3} = 20$

b. $\dfrac{4.5}{x - 3} = \dfrac{2}{3}$

Graphs of Real-World Situations

Like pictures, graphs communicate a lot of information. So you need to be able to draw and make sense of graphs. In previous chapters you learned to interpret bar graphs, circle graphs, histograms, and box plots. Then you learned to graph data from recursive routines and equations. Some graphs were lines and others were curves.

In this lesson you will apply many of those ideas as you begin to explore the graphs of functions. You will learn how to draw and interpret graphs of some real-world situations.

Frida Kahlo (1907–1954), a Mexican artist, is well known for her fascinating self-portraits. After surviving a bus accident, she had 32 surgical operations, painting many works from her bed. The contrasts between *Self-Portrait with Changuito* (left) and *Self-Portrait with Dog Ixcmintli and Sun* (right) reflect differences in the output of her work at different times in her life.

EXAMPLE

This graph shows the depth of the water in a leaky swimming pool. Tell what quantities are varying and how they are related. Give possible real-world events in your explanation.

▶ **Solution**

The graph shows that the water level or depth changes over a 15-hour time period. At the beginning, when no time has passed, $t = 0$, and the water in the pool is 2 feet deep, so $d = 2$. During the first 6-hour interval ($0 \leq t \leq 6$), the water level drops. The leak seems to get worse as time passes. When $t = 6$ and $d = 1$, it seems that someone starts to refill the pool. The water level rises for the next 5 hours, during the interval $6 \leq t \leq 11$. At $t = 11$, the water reaches its highest level at just above 3 feet, so $d = 3$. At the

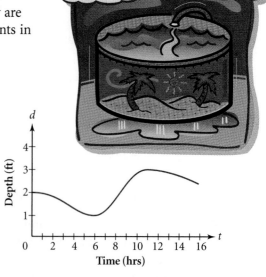

11-hour mark, the in-flowing water must have been turned off. The pool still has a leak, so the water level starts to drop again.

In the example the depth of the water is a function of time. That is, the depth depends on how much time has passed. So, in this case, depth is called the **dependent variable.** Time is the **independent variable.** When you draw a graph, put the independent variable on the horizontal axis and put the dependent variable on the vertical axis.

On the graph of this function, you can see domain values that are possible for the independent variable. In the example the domain is the set of all instants of time from 0 through 15 hours. You express this interval as $0 \leq x \leq 15$, where x is the independent variable representing time.

You can also see the values that are possible for the dependent variable. In the example the range is the set of all numbers from 1 through about 3.3. You express this as $1 \leq y \leq 3.3$, where y is the dependent variable representing the depth of the water. Notice that the lowest value for the range does not have to be the starting value when x is zero.

Investigation
Identify the Variables

The people in the school gymnasium before a volleyball game consist of the players, coaches, and the people working the event. Slowly the fans arrive for the match. Just before the first game, the people are coming in as fast as the tickets can be sold. After the match is over, most of the parents and fans leave. Then more students arrive for the after-game dance. Most of the students leave after an hour. The people that remain are the ones who have been working at the gym all night long.

Step 1	Define an independent variable for this situation. Decide what units of measurement you will use for this variable.
Step 2	What are reasonable values for the domain? Are they positive or negative numbers? Whole numbers or decimals?
Step 3	Define the dependent variable for this situation. What are the units of measurement?
Step 4	What are reasonable values for the range? Are they positive or negative numbers? Whole numbers or decimals?
Step 5	Draw and label a graph for this situation.

Now you will create a graph to tell a story.

Step 6	Choose a scenario you wish to describe from the list shown.
	a. The phone company offers a long distance calling plan.
	b. The Egyptians built pyramids from bricks and stone.
	c. A school club plans a fundraiser.
	d. The temperature you feel depends on the wind chill factor.
	e. An elevator moves in a building.
Step 7	Repeat Steps 1–5 for your scenario. Draw and label your graph on a separate piece of paper.
Step 8	Exchange your graphs with students from another group. Ask them to tell a story based on your graph. What are the strengths and weaknesses of using graphs to communicate information?

The relationship between the independent variable and the dependent variable is not always one of cause and effect. In many situations, time is the independent variable. It is the independent variable in the graphs of many walks, in problems of population growth and decay, and in several relationships of the form (*time, distance*). But time does not cause a walker's distance to change or a population to grow. People do that.

The values of the range depend on the domain. If you know the value of the independent variable, you can determine the corresponding value of the dependent variable. You do this every time you locate a point on the graph of a function.

EXERCISES

▶ Practice Your Skills

1. The graph shows the amount of medicine in the bloodstream after an injection of 500 milligrams of medication.

   a. Describe what happens to the level of medication over a 10-hour period.

   b. Identify the independent and dependent variables for this situation.

   c. Identify the domain and range for this situation.

   d. Define variables and write an exponential equation that corresponds to this graph.

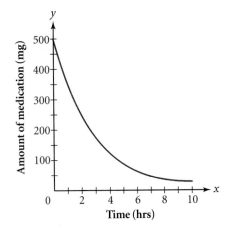

2. Sketch a reasonable graph and label the axes for each situation described. Write a few sentences explaining each graph.

   a. The more students who help to decorate for the homecoming dance, the less time it will take to decorate.

   b. The more you charge for T-shirts, the fewer T-shirts you will sell.

   c. The more you spend on advertising, the more product you will sell.

3. For each relationship, identify the independent variable and the dependent variable.

   a. the weight of your dog and the reading on the scale

   b. the amount of time you spend in an airplane and the distance between your departure and your destination

   c. the number of times you dip a wick into hot wax and the diameter of a handmade candle

4. Match each description with its most likely graph.

   a. the wind chill factor for a wind speed of 35 mph

   b. the amount of a radioactive substance over time

   c. the height of an elevator relative to floor number

   d. the population of a city over time

   e. instructions for a walk

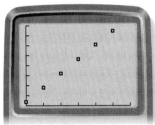

Graph 1

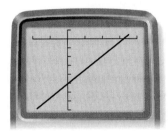

Graph 2

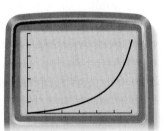

Graph 3

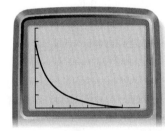

Graph 4

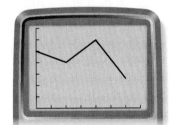

Graph 5

5. Pair the words that best go together: dependent, distance, x, independent, y, input, output, time.

Reason and Apply

6. The graph describes a function of another student's distance from you.

   a. What is the domain of this function?

   b. What is the range of this function?

   c. Explain what $(0, 2)$ means in this situation.

   d. Find the missing coordinate in each ordered pair.

     $(3.5, y)$ $(5, y)$ $(x, 3)$

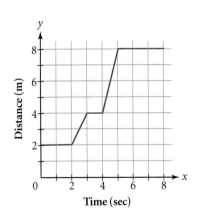

7. APPLICATION The drawing shows a side view of a swimming pool that is being filled. The water enters the pool at a constant rate. Sketch a graph of your interpretation of the relationship between depth and time as the pool is being filled. Explain your graph.

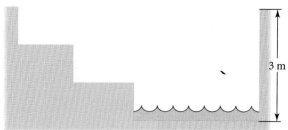

3 m

8. For each situation, identify the independent and dependent variables and sketch a reasonable graph. Write a few sentences explaining your graph.

   a. the height of the grass in a yard over the summer

   b. the number of buses needed to take different numbers of students on a field trip

   c. the height of a baseball as time passes after it is hit

   d. the number of dice thrown and the probability of getting at least one 6

   e. the area inside a wire coat hanger and the size of the angle formed at the center of the base

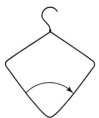

9. Name a reasonable domain and range for each graph you sketched in problem 8. Explain your answers.

10. The graph pictures a 100-meter dash between Erica and Eileen.

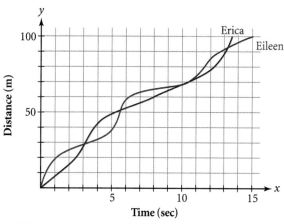

   a. Who won the race and in how many seconds? Explain.

   b. Who was ahead at the 60-meter mark?

   c. At what approximate times were the runners tied?

   d. When was Eileen in the lead?

11. Sketch a graph and describe a reasonable scenario for each statement.

　　a. a domain for the independent variable, *time*, of 0 to 8 seconds and a range for the dependent variable, *velocity*, of $\{0, 2, -2\}$ meters per second

　　b. your speed while you are riding or driving in a car following a school bus

　　c. the height of a basketball during a free throw shot

12. What does the graph of the function that relates *time* and *distance* look like for problem 11a? Explain your graph.

13. The radius, r, and height, h, of a 1-liter (1000-cm^3) cylindrical container vary according to the equation $1000 = \pi r^2 h$.

　　a. Complete this statement: As the can becomes wider, the height . . .

　　b. Complete this statement: As the can becomes narrower, the height . . .

　　c. Solve the equation for h.

　　d. Sketch a graph of (r, h), assuming r and h are measured in centimeters.

▶ Review

14. It is possible for two different functions to have coordinates in common.

　　a. Write the equation of a line through the points $(0, 8)$ and $(1, 10)$.

　　b. Write an exponential equation of a function whose graph goes through the points $(0, 8)$ and $(1, 10)$.

15. Solve each equation for x using any method. Use another method to check your answer.

　　a. $\dfrac{2x - 4}{3} + 7 = 4$

　　b. $\dfrac{5(3 - x)}{-2} = -17.5$

　　c. $\dfrac{2}{x - 1} = 3$

16. Solve each equation for y.

　　a. $3y = 15x$

　　b. $y + 2 = |x - 3|$

　　c. $4y - x = 20$

Function Notation

Every function defines a relationship between an input (independent) variable and an output (dependent) variable. **Function notation** uses parentheses to name the input or independent variable for the function. For instance, $y = f(x)$, which you read as "y equals f of x," says "y is a function of x" or "y depends on x." (In function notation, the parentheses do *not* mean multiplication.)

Albert Einstein writes mathematical notation in a lecture to scientists in 1931.

You can show some functions with an equation. For example, the equation $y = 2x + 4$ represents a function, so you can write it as $f(x) = 2x + 4$. The notation $f(3)$ tells you to substitute 3 for x in the equation $y = 2x + 4$. So $f(3) = 2(3) + 4$. The value of $f(x)$ when $x = 3$ is 8. By itself, f is the name of the function. In this case, its rule is $2x + 4$.

Not all functions are expressed as equations. The graph shows a new function, $f(x)$. No rule or equation is given, but you can still use function notation to find output values. For example, on the graph below, the point at $x = 4$ has the coordinates $(4, f(4))$ or $(4, 1)$. The value of y when x is 4 is $f(4)$. So $f(4) = 1$. Check that $f(2)$ is 4. What is the value of $f(6)$? Can you find two x-values for which $f(x) = 1$?

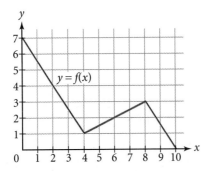

In the next investigation you will learn more about using function notation with graphs.

Investigation
A Graphic Message

You will need

• the worksheet
A Graphic Message

In this investigation you will apply function notation to learn the identity of the mathematician who introduced functions.

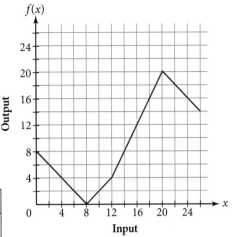

Step 1 Describe the domain and range of the function f in the graph.

Step 2 Use the graph to find each function value in the table. Then do the indicated operations.

Notation	Value
$f(3)$	
$f(18) + f(3)$	
$f(5) \cdot f(4)$	
$f(15) \div f(6)$	
$f(20) - f(10)$	

Step 3 Use the rules for the order of operations to evaluate the expressions that involve function values. Do the operations inside parentheses first. Then find the function values before doing the remaining operations. Write your answers in a table.

Notation	Value
$f(0) + f(1) - 3$	
$5 \cdot f(9)$	
x when $f(x) = 10$	
$f(9 + 8)$	
$\dfrac{f(17) + f(10)}{2}$	
$f(8 \cdot 3) - 5 \cdot f(11)$	
$f(4 \cdot 5 - 1)$	
$f(12)$	

Step 4 Think of the numbers 1 through 26 as the letters A through Z. Find the letters that match your answers to Step 2 to learn the mathematician's last name. Find the letters that match your answers to Step 3 to discover the first name.

The mathematician whose name you decoded invented much of the mathematical notation in use today. In the example that follows, you will practice function notation with an equation.

EXAMPLE

You can use the function $f(x) = \frac{9}{5}x + 32$ to find the temperature $f(x)$ in degrees Fahrenheit for any given temperature x in degrees Celsius. Find the specified value.

a. $f(15)$

b. $f(-10)$

c. $f(5)$

d. x when $f(x) = -4$

▶ **Solution**

In each case, substitute the value in parentheses for x in the function.

a. $f(15) = \frac{9}{5}(15) + 32$

$f(15) = 27 + 32$

$f(15) = 59$

b. $f(-10) = \frac{9}{5}(-10) + 32$

$f(-10) = -18 + 32$

$f(-10) = 14$

c. $f(5) = \frac{9}{5}(5) + 32$

$f(5) = 9 + 32$

$f(5) = 41$

d. $-4 = \frac{9}{5}x + 32$

$-36 = \frac{9}{5}x$

$-20 = x$

[▶ 🔲 See **Calculator Note 8A** to learn how to evaluate functions on your calculator. ◀]

Some calculators use the notation $Y_1(x)$ instead of $f(x)$. The function depends on the equation you have entered into Y_1. Other calculators allow you to directly define the function f as an expression.

EXERCISES

You will need your calculator for problems **1, 2,** and **9.**

▶ **Practice Your Skills**

1. Find each function value for $f(x) = 3x + 2$ and $g(x) = x^2 - 1$ without using your calculator. Then enter the equation for $f(x)$ into Y_1 and the equation for $g(x)$ into Y_2. Use function notation on your calculator to check your answers. [▶ 🔲 See **Calculator Note 8A** to review function notation on your calculator. ◀]

a. $f(3)$ **b.** $f(-4)$ **c.** $g(5)$ **d.** $g(-3)$

2. Find the y-coordinate corresponding to each x-coordinate if the functions are $f(x) = -2x - 5$ and $g(x) = 3.75(2.5)^x$. Check your answers with your calculator.

a. $f(6)$ **b.** $f(0)$ **c.** $g(2)$ **d.** $g(-2)$

3. Use the graph of $y = f(x)$ at right to answer each question.

a. What is the value of $f(4)$?

b. What is the value of $f(6)$?

c. For what value or values does $f(x) = 2$?

d. For what value or values does $f(x) = 1$?

e. How many x-values make the statement $f(x) = 0.5$ true?

f. For what x-values is $f(x)$ greater than 2?

g. What are the domain and range shown on the graph?

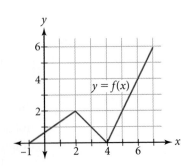

4. APPLICATION The graph of the function $y = f(x)$ below shows the temperature y outside at different times x over a 24-hour period.

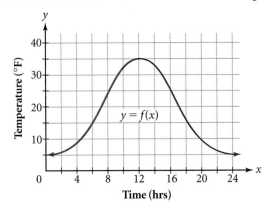

a. What are the dependent and independent variables?

b. What are the domain and range shown on the graph?

c. Use function notation to represent the temperature at 10 hours.

d. Use function notation to represent the time at which the temperature is 10°F.

5. Use function notation to write the equation for a line through each pair of points.

a. $(0, 5)$ and $(1, 12)$ b. $(1, 5)$ and $(2, 12)$

Reason and Apply

6. The function $f(x)$ gives the lake level over the past year, with x measured in days and y, that is, the $f(x)$ values, measured in inches above last year's mean height.

a. What is the real-world meaning of $f(60)$?

b. What is the real-world meaning of $f(x) = -3$?

c. What is an interpretation of $f(x) = f(150)$?

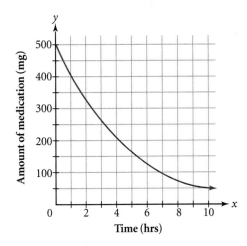

7. The graph shows part of the function $f(x) = 500(0.80)^x$.

a. What is the dependent variable and what are its units?

b. What is the independent variable and what are its units?

c. What part of the domain is pictured? What is the domain of the function?

d. What part of the range is pictured? What is the range of the function?

e. What is $f(0)$ for this graph?

f. Find the value of x when $f(x) = 200$.

8. APPLICATION A bacteria population decreases at the rate of 8.5% an hour. There are 650 bacteria present at the start. The time it takes for the population to decrease to half its original size is called its half-life.

a. Write an equation and graph the function that describes this population decay.

b. Graph the line that represents half the starting amount of bacteria. Find the point of intersection of this line with the population decay function.

c. What is the real-world meaning of your answer in 8b?

9. Many of the commands in your calculator are programmed as functions. Try each command several times with a variety of inputs. Describe the allowable input and corresponding output of each command. If you think the command is a function, describe its domain and range. [▶ 🖳 See **Calculator Note 8B** to access and use these commands. ◀]

a. the square (x^2) command

b. the square root $\left(\sqrt{}\right)$ command

c. the sum of a list command

d. the random command

10. Use the function $f(x) = \frac{5}{9}(x - 32)$ to convert temperatures in degrees Fahrenheit (x-values) to temperatures in degrees Celsius ($f(x)$-values) and vice versa.

a. 72°F

b. −10°F

c. 20°C

d. −5°C

11. The graphs of $f(x)$ and $g(x)$ below show two different aspects of an object dropped straight down from the Tower of Pisa.

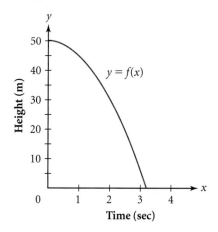 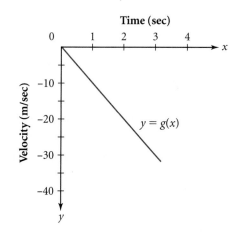

Answer these questions for the graph of each function.

a. What are the dependent and independent variables?

b. What are the domain and range?

c. Describe a real-world sequence of events for each graph.

d. About how far does the ball drop in the 1st second (from $x = 0$ to $x = 1$)?

e. About how far does the ball drop in the 2nd second (from $x = 1$ to $x = 2$)?

f. At what speed does the object hit the ground?

12. Could this set of ordered pairs represent a function? If so, what are its domain and range values?

$(-2, 3), (3, -2), (1, 3), (0, -2)$

13. Could this set of ordered pairs represent a function? Explain your reasoning.

$(3, -2), (-2, 3), (3, 1), (-2, 0)$

14. Use the graph of $f(x)$ at right to evaluate each expression. Write your answers as a number sequence. Then think of the numbers 1 through 26 as the letters A through Z to decode a message.

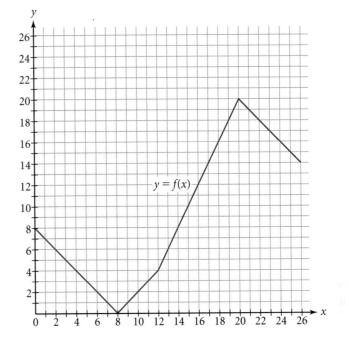

a. $f(8) + 6$

b. $f(20) + 1$

c. the sum of two x-values that give $f(x) = 8$

d. $f(0) - 4$

e. $f(7)$

f. x when x is an integer and $f(x) = 15$

g. $f(18) + f(5)$

h. (the sum of two x-values that give $f(x) = 16$) $\div 42$

i. (x when $f(x) = 12$) $- 8$

j. $\dfrac{f(25)}{3}$

k. $f(7) + f(8)$

l. the largest domain value $-$ the largest range value $- 2$

Review

15. Write each expression in exponential form without using negative exponents.

a. $(2^3)^{-3}$ **b.** $(5^2)^5$ **c.** $(2^4 3^2)^3$ **d.** $(3^2 5^3)^{-4}$

16. Find the slope for the line through each pair of points.

a. $(1, 3)$ and $(-2, 6)$

b. $(-4, -5)$ and $(7, 0)$

c. $(-3, 6)$ and $(9, 6)$

17. Solve each equation.

a. $2x - 5 = 7x + 15$

b. $3(x + 6) = 12 - 5x$

c. $\dfrac{7(8 - x)}{4} = x + 3$

Interpreting Graphs

News reports and advertisements use graphs. So do science articles and political debates. To "read" a graph, you have to understand how the quantities in the graph relate to each other, how they make the graph go up or down or level off.

The function values in a graph can change at a constant rate or at a varying rate as the *x*-values of a function increase steadily. A function is **linear** if, as *x* increases at a constant rate, the function values change at a constant rate. Graphs A and D on the opposite page show linear functions. Here is another linear function. What is the constant rate of change for the function values in the graph below?

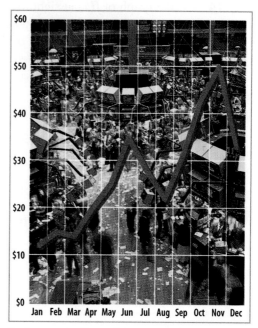

Traders on the floor of the New York Stock Exchange use graphs to show stock prices.

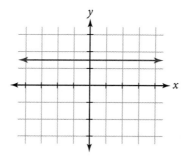

A function is **nonlinear** if, as *x* changes at a constant rate, the function values change at a varying rate. Graphs B, C, E, and F on the opposite page show nonlinear functions. Here is another nonlinear function.

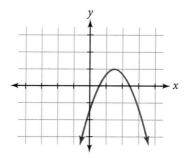

The function shown in the graph above rises, peaks, and then falls.

Many graphs do not show the whole domain and the whole range of the function. Often, the function whose graph you're looking at has limitless domain and range. The graph below cannot show the whole domain of the function, but it *can* show the whole range.

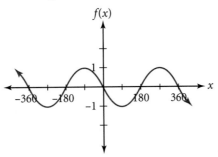

In the investigation that follows, you'll discover another aspect of functions and use their graphs to describe real-world situations.

Investigation
Matching Up

First, you'll consider the concepts of increasing and decreasing functions.

Step 1 | These are graphs of *increasing functions*. What do the three graphs have in common? How would you describe the rate of change in each?

Graph A

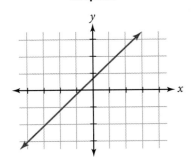

Graph B

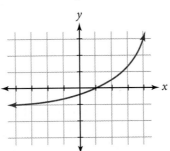

Graph C

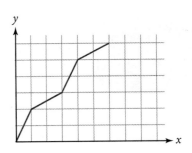

Step 2 | These are graphs of *decreasing functions*. What do these three graphs have in common? How are they different from those in Step 1? How would you describe the rate of change in these graphs?

Graph D

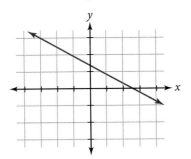

Graph E

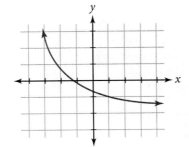

Graph F

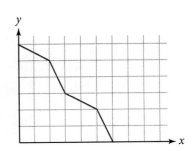

In Steps 3–5 you'll use this vocabulary to find and describe four of the graphs that match each of these real-world situations.

Situation A During the first few years, the number of deer on the island increased by a steady percentage. As food became less plentiful, the growth rate started slowing down. Now, the number of births and deaths is about the same.

Situation B In the Northern Hemisphere the amount of daylight increases slowly from January through February, faster until mid-May, and then slowly until the maximum in June. Then it decreases slowly though July, faster from August until mid-November, and then slowly until the year's end.

Situation C If you have a fixed amount of fencing, the width of your rectangular garden determines its area. If the width is very short, the garden won't have much area. As the width increases, the area also increases. The area increases more slowly until it reaches a maximum. As the width continues to increase, the area becomes smaller more quickly until it is zero.

Situation D Your cup of tea is very hot. The difference between the tea temperature and the room temperature decreases quickly at first as the tea starts to cool to room temperature. But when the two temperatures are close together, the cooling rate slows down. It actually takes a long time for the tea to finally reach room temperature.

Step 3 In Situation A decide which quantities are varying. Also decide which variable is independent and which is dependent.

Step 4 Match and describe the graph that best fits the situation. Write a description of the function and its graph using the words *linear, nonlinear, increasing, decreasing, rate of change, maximum* or *greatest value,* and *minimum* or *least value.* Tell why you think the graph and your description match the situation.

Step 5 Repeat Steps 3–4 for the other three situations.

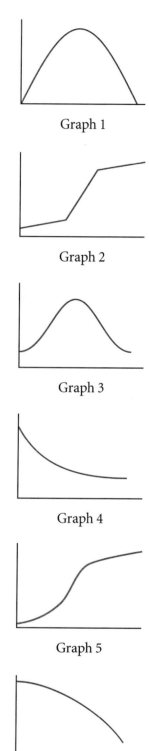

Graph 1

Graph 2

Graph 3

Graph 4

Graph 5

Graph 6

In the investigation you learned how to describe real-world situations with graphs and some function vocabulary. A function is **increasing** when the variables change in the same way—that is, the *y*-values *grow* when reading the graph from left to right. A function is **decreasing** when the variables change in different directions— that is, the *y*-values *drop* when reading the graph from left to right.

Situations C and D in the investigation represent **continuous** functions because there are no breaks in the domain or range. Many functions that are not continuous involve quantities that are counted or measured in whole numbers—for instance, people, cars, or stories of a building. In the investigation you have already seen two functions like this—the number of deer in Situation A and the number of days in Situation B. These are called **discrete** functions. When graphing the amount of daylight for every day of the year, the graph should really be a set of 365 points, as in the graph below. There is no value for day 47.35. Likewise, there may not be a day with exactly 11 hours 1 minute of daylight. But it's easier to draw this relationship as a smooth curve than to plot 365 points.

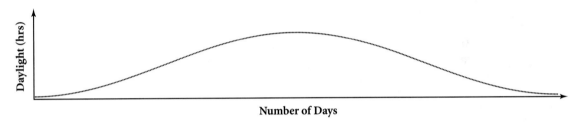

Number of Days

Sometimes it is useful to name a part of the domain for which a function has a certain characteristic.

EXAMPLE

Describe this graph, telling how the quantities in the graph relate to each other.

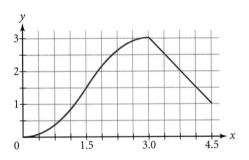

▶ **Solution**

Use the intervals marked on the *x*-axis to help you discuss where the function is increasing or decreasing and where it is linear or nonlinear.

On the interval $0 \le x < 3.0$, the function is nonlinear and increasing. As *x* increases steadily, *y* changes at a varying rate, so the graph is nonlinear. When read from left to right, the graph rises. So the *y*-values grow and the function is increasing.

On the interval $3.0 \le x < 4.5$, the function appears linear and is decreasing. As both *x* and *y* appear to change at a constant rate on the graph, the function is linear. When read from left to right, the graph falls. So the *y*-values drop and the function is decreasing.

EXERCISES

▶ Practice Your Skills

1. Sketch a graph of a continuous function to fit each description.
   a. always increasing with a faster and faster rate of change
   b. decreasing with a slower and slower rate of change, then increasing with a slower and slower rate of change
   c. linear and decreasing
   d. decreasing with a faster and faster rate of change

2. Write an inequality for each interval in 2a–e. Include the least point in each interval and exclude the greatest point in each interval.

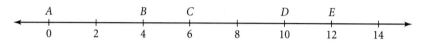

   a. *A* to *B*   b. *B* to *C*   c. *B* to *D*   d. *C* to *E*   e. *A* to *E*

3. Describe each of these discrete function graphs using the words *increasing, decreasing, linear, nonlinear,* and *rate of change.*

   a. 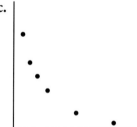   b.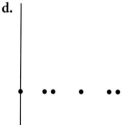

   c.   d.

4. Sketch a discrete function graph to fit each description.
   a. always increasing with a slower and slower rate of change
   b. linear with a constant rate of change equal to zero
   c. linear and decreasing
   d. decreasing with a faster and faster rate of change

Reason and Apply

5. A turtle crawls steadily from its pond across the lawn. Then a small dog picks up the turtle and runs with it across the lawn. The dog slows down and finally drops the turtle. The turtle rests for a few minutes after this excitement. Then a young boy comes along, picks up the turtle, and slowly carries it back to the pond. Which of the graphs describes the turtle's distance from the pond?

Graph A

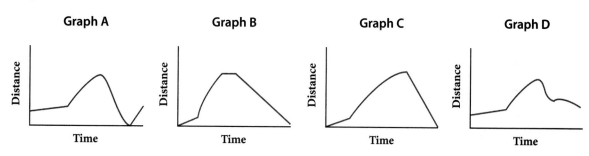

Graph B

Graph C

Graph D

6. The graphs show the distance of a person from a fixed point for a 4-second interval. Answer both questions for each graph.

    i. Is the person moving toward or away from the point?
    ii. Is the person speeding up, slowing down, or moving at a constant speed?

a.

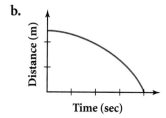

b.

c.

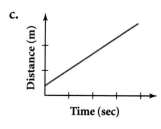

d.

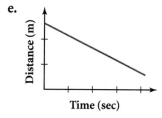

e.

f.

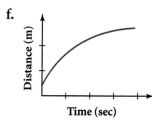

7. Which of the graphs most realistically shows the relationship between the number of volunteers and the amount of time required to clean up the trash on the school's campus? Explain your choice. If none of the choices seems correct, sketch your own graph to answer the question.

a.
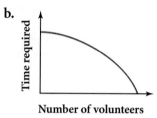

b.
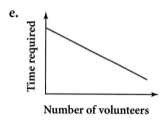

c.
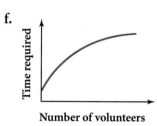

d.
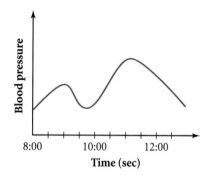

e.

f.

8. Describe the six graphs in problem 7.

9. **APPLICATION** This graph shows Anne's blood pressure level during a morning at school. Give the points or intervals when her blood pressure

a. Reached its highest level.

b. Was rising the fastest.

c. Was decreasing.

10. This graph shows the air temperature in a 24-hour period from midnight to midnight. Write a description of this graph, giving the intervals as the temperature changed.

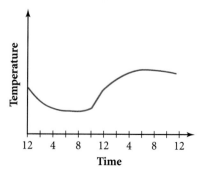

▶ Review

11. For each relationship, identify the independent and dependent variables.

    a. the mass of a spherical lollipop and the number of times it has been licked

    b. the number of scoops in an ice cream cone and the cost of the cone

    c. the distance a rubber band will fly and the amount you stretch it before you release it

    d. the number of coins you flip and the number of heads

12. Use these equations for 12a–c.

    i. $4x + 2y = 16$ **ii.** $4x - 2y = 16$

    a. Solve these equations for y. (Show your work.)

    b. Rewrite the equations using $f(x)$ notation.

    c. Find the value of $f(-1)$ in both equations.

13. Consider the equation $y = -12.4 - 2.5(x + 5.4)$.

    a. Write it in intercept form.

    b. Name the slope and y-intercept of the equation in 13a.

14. Solve the equation $740 = 16.8x + 405$ for x. When finished, check that you are correct.

15. The top 10 grossing movies at the end of the 20th century are shown in the table. Find the five-number summary.

Rank	Total gross (millions of dollars)	Movie
1	601	Titanic
2	461	Star Wars
3	431	The Phantom Menace
4	400	E.T.
5	357	Jurassic Park
6	330	Forrest Gump
7	313	The Lion King
8	307	Return of the Jedi
9	306	Independence Day
10	293	The Sixth Sense

(*http://movieweb.com*)

Defining the Absolute Value Function

Cal and Al both live 3.2 miles from school, but in opposite directions. If you assign the number 0 to the school, you can show that Cal and Al live in opposite directions from it by assigning +3.2 to Al's house and −3.2 to Cal's apartment. For both Cal and Al, the distance from school is 3.2 miles.

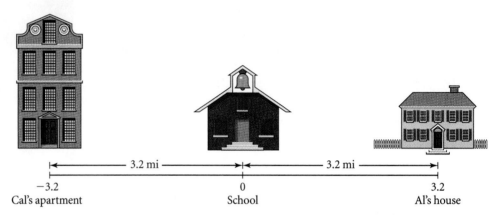

The **absolute value** of a number is its size, or magnitude, regardless of whether the number is positive or negative. In this lesson you'll develop a mathematical understanding of the absolute value function and its graph.

A way to visualize the absolute value of a number is to picture its distance from zero on a number line. A signed number and its opposite both have the same magnitude, or absolute value, because they're both the same distance from zero. For example, 3.2 and −3.2 are both 3.2 units from zero on a number line, so they both have an absolute value of 3.2.

Use the notation $|x|$ when you write absolute value expressions. So you would write $|3.2| = 3.2$ and $|-3.2| = 3.2$.

EXAMPLE A

Evaluate each expression.

a. $|5| + |-5|$ b. $2|-17| + 3$ c. $\dfrac{|-4|}{|4|}$

d. $|-6| - |-6|$ e. $-|8|$ f. $|0|$

▶ **Solution**

Substitute the magnitude of the number for each absolute value.

a. $|5| + |-5| = 5 + 5 = 10$ b. $2|-17| + 3 = 2(17) + 3 = 37$

c. $\dfrac{|-4|}{|4|} = \dfrac{4}{4} = 1$ d. $|-6| - |-6| = 6 - 6 = 0$

e. $-|8| = -8$ f. $|0| = 0$

Absolute value is useful for answering questions about distance, pulse rates, test scores, and other data values that lie on opposite sides of a central point such as a mean. The distance or difference of a data point from the mean of its data set is called its *deviation from the mean*.

Investigation
Deviations from the Mean

In the investigation you will learn how the absolute value function tells how much an item of data or a whole set of data deviates from the mean.

Step 1
Collect at least 10 pulse rates from your class. Record the data in a table and enter the numbers into list L₁ on your calculator.

Step 2
Find the difference between each data point and the mean of the data in list L₁. [▶ 🖳 See **Calculator Note 2B** to review finding the mean of a list. ◀] Record these numbers into a second column of your table and enter them into list L₂. What do these numbers represent?

Step 3
Make a dot plot of the list L₁ data and note the distance from each data point to the mean. Record your results in a third column and enter them into list L₃. How are these entries different from those in list L₂? How are they alike?

Step 4
Next, plot points of the form (L₂, L₃). [▶ 🖳 See **Calculator Note 1E** to review scatter plots. ◀] What numbers are in the domain and range of the graph?

Step 5
Use the trace function on your calculator and use the arrow keys to scan through each data point. Which input numbers are unchanged as output numbers?

Step 6
Which input numbers are changed and how?

Step 7
Does it make sense to connect these points with a continuous graph? Why or why not?

Step 8
How does this graph compare to the graph of Y₁ = abs(x) on your calculator? [▶ 🖳 See **Calculator Note 8C** to access the abs command. ◀]

Step 9
Find the mean of the deviations stored in list L₂. Compare it to the mean of the distances stored in list L₃. Which do you think is a better measure of the spread of the data?

Step 10
In your own words, write the rule for the function you graphed in Step 8. What number is output as y when the input x is positive or equal to zero? What number is output when x is negative? How can you use operations to change these numbers?

Despite deviations from each other in appearance, each impersonator clearly portrays Elvis Presley. This photo was taken at Graceland, the late singer's home in Memphis, Tennessee.

The **absolute value function** is defined by two rules. The first rule says to output the same number when the input value is positive or zero. The second rule says to output the opposite number when the input is negative. You express these rules like this:

$$|x| = \begin{cases} x & \text{if } x \geq 0 \\ -x & \text{if } x < 0 \end{cases}$$

For instance, if x is 3, then $|x|$ is also 3. On the other hand, if x is -3, then multiply by -1 to get 3 again. So there are two solutions to the equation $|x| = 3$.

EXAMPLE B | Solve the equation $12 = |x| + 7$.

▶ **Solution** | There are three ways to solve this equation. The first two are calculator methods that often give only approximate solutions.

Method 1: Looking at a Graph

Set y equal to both sides of the equation. You know from your work solving systems that you can enter $Y_1 = 12$ and $Y_2 = \text{abs}(x) + 7$ on your calculator. The x-coordinates of the two points of intersection, which represent the solutions, are $x = 5$ and $x = -5$. Note that the viewing window must be large enough for you to see both solutions.

$[-8, 8, 1, -1, 14, 1]$

Method 2: Looking at a Table

A table shows these same solutions. You will have to scroll through the table to find both solutions.

 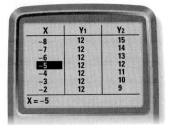

To be guaranteed of finding an exact solution, choose the next method.

Method 3: Solving Symbolically

The process for solving this equation symbolically is a bit different from that for solving most other equations because there is no function for "undoing" the absolute value—and you have to remember that there are two solutions.

$12 = |x| + 7$	The original equation.
$12 - 7 = |x| + 7 - 7$	Subtract 7 from each side of the equation.
$5 = |x|$	Find two numbers whose absolute value is 5.
$x = 5 \text{ or } x = -5$	The two solutions.

Whatever method you use to solve an absolute value equation, you always have to be sure that you are finding all possible solutions. In general, an absolute value equation has two solutions, one solution, or no solution.

If you're not sure how many solutions an equation should have, look at the graph of the situation first and then decide which method you want to use to solve the equation.

EXERCISES

You will need your calculator for problems **1, 3, 5,** and **12.**

▶ Practice Your Skills

1. Find the value of each expression without using a calculator. Check your results with your calculator. [▶ 🖥 See **Calculator Note 8C.** ◀]

   a. $\left|-7\right|$
   b. $\left|0.5\right|$
   c. $\left|-7+2\right|$
   d. $\left|-7\right|+\left|2\right|$

   e. $-\left|5\right|$
   f. $-\left|-5\right|$
   g. $\left|-4\right|\cdot\left|3\right|$
   h. $\dfrac{\left|-6\right|}{\left|2\right|}$

2. What x-values satisfy the equation $\left|x\right|=10$?

3. Evaluate both sides of each statement to determine whether to replace the box with $=$, $<$, or $>$. Use your calculator to check your answers.

   a. $\left|5\right|+\left|7\right|\ \square\ \left|5+7\right|$
   b. $\left|-5\right|\cdot\left|8\right|\ \square\ \left|-40\right|$

   c. $\left|-12-3\right|\ \square\ \left|-12\right|-\left|3\right|$
   d. $\left|-2+11\right|\ \square\ \left|-2\right|+\left|11\right|$

   e. $\dfrac{\left|36\right|}{\left|-9\right|}\ \square\ \left|\dfrac{36}{-9}\right|$
   f. $\left|4\right|^{\left|-2\right|}\ \square\ \left|4^{-2}\right|$

4. Consider the functions $f(x)=3x-5$ and $g(x)=\left|x-3\right|$. Find each value.
   a. $f(5)$
   b. $f(-2.5)$
   c. $g(-5)$
   d. $g(1)$

5. Plot the points determined by the function $y=\left|x\right|$ (or $Y_1=\text{abs}(x)$) using the domain $\{-4,-1.5,0,1.2,3,4.75\}$. Use a friendly window. [▶ 🖥 See **Calculator Note 8D** to learn about friendly windows. ◀] Use your graph, table, or equation to evaluate $\left|4.75\right|$ and $\left|-1.5\right|$.

▶ Reason and Apply

6. Create this graph on graph paper: When $x\geq0$, graph the line $y=x$. When $x<0$, graph the line $y=-x$. What single function has this same graph?

7. Solve this system of equations:
   $$\begin{cases} y=\left|x\right| \\ y=2.85 \end{cases}$$

8. If possible, solve each equation for x.

  a. $|x| = 12$

  b. $10 = |x| + 4$

  c. $10 = 2|x| + 6$

  d. $4 = 2(|x| + 2)$

9. Describe the walk represented by this calculator graph. Each mark on the x-axis represents 1 second and each mark on the y-axis represents 1 meter from a motion sensor.

10. The graph in Example B shows two solutions for x.

  a. Replace $Y_1 = 12$ with a horizontal line that gives exactly one solution for x.

  b. Replace $Y_1 = 12$ with a horizontal line that gives no solution for x.

11. In 11a–d, identify which function, $f(x)$, $g(x)$, or $h(x)$, is used in each (*input, output*) pair.

  $f(x) = 7 + 4x$ $g(x) = |x| + 6$ $h(x) = 18(1 + 0.5)^x$

  a. (5, 11)

  b. (1, 27)

  c. (−2, 8)

  d. (3, 19)

12. Predict the graph of each equation. Sketch your prediction and then check your answer with your calculator.

  a. $y = x + |x|$

  b. $y = \dfrac{x}{|x|}$

13. **APPLICATION** The table shows the weights of fish caught by wildlife biologists in Spider Lake and Doll Lake. In which lake did the fish weights vary more from the mean? Explain how you arrived at your answer.

Weights of Fish

Spider Lake (lb)	Doll Lake (lb)
1.2	0.9
2.1	1.1
0.8	1.6
1.4	1.9
2.7	2.1
1.0	1.4
0.4	1.4
2.4	2.2

Review

14. Solve each system of equations using the method of your choice. For each, tell which method you chose and why.

a. $\begin{cases} -2x + 3y = 12 \\ 4x - 3y = -21 \end{cases}$

b. $\begin{cases} 5x + y = 12 \\ 2x - 3y = 15 \end{cases}$

15. Solve each inequality and graph the solution on a number line.

a. $-2 < 6x + 8$

b. $3(2 - x) + 4 \geq 13$

c. $-0.5 \geq -1.5x + 2(x - 4)$

IMPROVING YOUR REASONING SKILLS

Consider the table of the squares of numbers between 0 and 50 that end in 5.

Number	5	15	25	35	45
Square	25	225	625	1225	2025

Do you notice a pattern that helps you mentally calculate these kinds of square numbers quickly? Can you square 65 in your head? When you think you have discovered the pattern, check your results with a calculator. Then try reversing the process to find the square root of 7225.

Practice this pattern and then race someone using a calculator to see who is quicker at computing a square number ending in 5. Will this pattern work for all numbers ending in 5? Why or why not? Are there numbers that make this pattern too difficult to use?

Squares, Squaring, and Parabolas

Think of a number between 1 and 10. Multiply it by itself. What number did you get? Try it again with the opposite of your number. Did you get the same result? This number is called the **square** of a number. The process of multiplying a number by itself is called **squaring.** The square of a number x is "x squared," and you write it as x to the power of 2, or x^2. When squaring numbers on your calculator, remember the order of operations. Try entering -3^2 and $(-3)^2$ on your calculator. Which result is the square of -3?

Do you think that the rule for squaring is a function? In order to answer this question, you will graph the relationship between numbers and their squares.

The mathematical process for the squaring takes its name from the application of finding a square's area. From the Latin *quadrare*, which means to make square, we also have the word "quadratic" to describe x^2.

Charmion Von Wiegand, *Advancing Magic Squares*, ca. 1958.
The National Museum of Women in the Arts, Washington, D.C.

Investigation
Graphing a Parabola

In this investigation you will explore connections between any number x and its square by graphing the coordinate pairs (x, x^2).

Step 1 | Make a table with column headings like the one shown. Put the numbers -10 through 10 in the first column, and then enter these numbers into list L1 on your calculator.

Number (x)	Square (x^2)

The parabolic shape of the SETI (Search for Extraterrestrial Intelligence) radio telescope at Harvard University in Massachusetts collects radio signals from space.

Step 2	Without a calculator, find the square of each number and place it in the second column. Check your results by squaring list L1 with the x^2 key. [▶ 🔲 See **Calculator Note 8B.** ◀] Store these numbers in list L2.
Step 3	How do the squares of numbers and their opposites compare? What is the relationship between the positive numbers and their squares? Between the negative numbers and their squares?
Step 4	Choose an appropriate window and plot points of the form (L1, L2). [▶ 🔲 See **Calculator Note 1E** to review scatter plots. ◀] Graph Y1 = x^2 on the same set of axes. What relationship does this graph show?
Step 5	Is the graph of $y = x^2$ the graph of a function? If so, describe the domain and range. If not, explain why not.

The graph of $y = x^2$ is called a **parabola.** In later chapters you will learn how to create other parabolas based on variations of this basic equation.

Step 6	The points of the parabola for $y = x^2$ are in what quadrants?
Step 7	What makes the point (0, 0) on your curve unique? Where is this point on the parabola?
Step 8	Draw a vertical line through the point you found in Step 7. How is this line like a mirror?
Step 9	Compare your parabola with the graph of the absolute value function, $y = |x|$. How are they alike and how are they different?
Step 10	Which x- and y-values in your parabola could represent side lengths and areas of squares?

In the investigation you learned that on the graph of $y = x^2$ two different input values can have the same output. For instance, the square of -3 and the square of 3 are both equal to 9. What happens when you try to "undo" the squaring? If you want to find a number whose square is 9, is 3 or -3 the answer? Or are both the answer? You will learn about a function that undoes squaring in the example.

EXAMPLE

Find the side of the square whose area is 6.25 square centimeters.

▶ **Solution**

Let x represent the side of the square in centimeters. To find it, solve the equation $x^2 = 6.25$. There are three ways to do this.

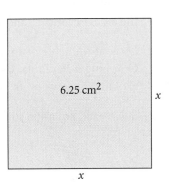

6.25 cm^2

x

x

Method 1: Graph

Graph the line $y = 6.25$ and the parabola $y = x^2$. The graphs intersect at two points. Use the trace key on your calculator to find the x-values of the intersections.

Method 2: Calculator Table

You can also see these x-values in a calculator table. Adjust the table settings and zoom in on the table until you find the row in which both Y_1 and Y_2 equal 6.25. Remember to look for two solutions, one positive and one negative.

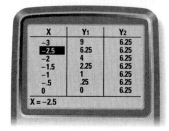 

Method 3: Solve Symbolically

$$x^2 = 6.25$$

The original equation.

$$\sqrt{x^2} = \sqrt{6.25}$$

To solve for x, take the **square root** of both sides.

$$x = 2.5 \text{ or } x = -2.5$$

There are two solutions.

The equation has two solutions, but because the side of a square must be positive, the only realistic solution is 2.5 cm.

Your calculator will not give the negative solution when you press the square root key. The **square root function,** ($f(x) = \sqrt{x}$), undoes squaring, giving only the positive solution.

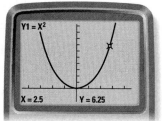

You can learn about the use of parabolas in the real world with the Internet links at www.keymath.com/DA .

EXERCISES

You will need your calculator for problem **5.**

Practice Your Skills

1. Complete the table by filling in the missing values for the side, perimeter, and area of each square.

2. Look at the table of squares in the investigation Graphing a Parabola. Use values from this table to explain why $y = x^2$ is nonlinear.

3. Solve each equation for x.

    a. $|x| = 6$

    b. $x^2 = 36$

    c. $|x| = 3.8$

    d. $x^2 = 14.44$

Side (cm)	Perimeter (cm)	Area (sq cm)
1		
2		
	12	
		16
14		
	60	
		441
	100.8	
		2209

4. Solve each equation, if possible.

    a. $4.7 = |x| - 2.8$

    b. $-41 = x^2 - 28$

    c. $11 = x^2 - 14$

Reason and Apply

5. For what values of x is $|x| \geq x^2$? To check your answer, graph $Y_1 = |x|$ and $Y_2 = x^2$ on the same set of axes.

6. For what values of y does the equation $y = x^2$ have

    a. No real solutions?

    b. Only one solution?

    c. Two solutions?

7. Graph the functions $f(x) = 3x - 5$ and $g(x) = |x - 3|$. What do the two graphs tell you about the equation $3x - 5 = |x - 3|$?

8. Solve each equation symbolically.

    a. $5 = |x| - 3$

    b. $-4 = x^2 - 8$

    c. $4 = 2|x| + 6$

9. Find the sum of each set of numbers in 9a–c.

    a. the first five odd positive integers

    b. the first 15 odd positive integers

    c. the first n odd positive integers

    d. Use the diagram to explain the connection among the sum of odd integers, square numbers, and these square figures.

   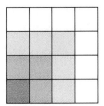

10. Write an equation for the function represented in each table. Check your answers.

    a.

x	-3	-1	0	1	4	6
y	14	10	8	6	0	-4

    b.

x	-3	-1	0	1	4	6
y	9	1	0	1	16	36

    c.

x	-3	-1	0	1	4	6
y	3	1	0	1	4	6

11. This 4-by-4 grid contains squares of different sizes.

    a. How many of each size square are there? Include overlapping squares.

    b. How many total squares would a 3-by-3 grid contain? A 2-by-2 grid? A 1-by-1 grid?

    c. Find a pattern to determine how many squares an n-by-n grid contains. Use your pattern to predict the number of squares in a 5-by-5 grid.

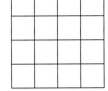

12. Explain why the equation $x^2 = -4$ has no solutions. (Hint: Why is it impossible for the product of a number multiplied by itself to be negative?)

▶ Review

13. The table shows exponential data.

a. What equation in the form $y = ab^x$ can you use to model the data in the table?

b. Use your equation to find the missing values.

x	y
0	
4	126.5625
3	168.75
1	
	1000

14. Use properties of exponents to find an equivalent expression if possible of the form ax^n. Use positive exponents.

a. $24x^6 \cdot 2x^3$

b. $(-15x^4)(-2x^4)$

c. $\dfrac{72x^{11}}{3x^2}$

d. $4x^2(2.5x^4)^3$

e. $\dfrac{15x^5}{-6x^2}$

f. $(-3x^3)(4x^4)^2$

g. $\dfrac{42x^{-6}y^2}{7y^{-4}}$

h. $3(5xy^2)^3$

IMPROVING YOUR **VISUAL THINKING** SKILLS

Square numbers are so named because they result from the geometric application of finding the area of a square. A square of side length 3 has an area equal to 3^2, or 9. You can represent 9, and other perfect square numbers, with diagrams like this:

1

2

3

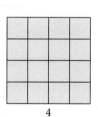
4

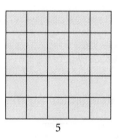
5

What numbers result when you represent them with cubes instead of squares? Use sugar cubes to make these shapes:

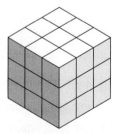

How many sugar cubes does it take to make each figure? What is the relationship between the side length (measured in sugar cubes) and the total number of cubes needed for each figure? If you double the side length of a cubic figure, how many times larger is its resulting volume? If you triple the side length?

In this chapter you used functions to describe real-world relationships. You began by designing and decoding secret messages. You discovered that the easiest way to code is to use a **function**—it codes each input into a single output.

You investigated functions represented by rules, equations, tables, and graphs. You learned to tell whether a relationship is a function by applying the **vertical line test.** On a graph, this means that no vertical line can intersect the graph of a function at more than one point.

You learned how to use function notation $f(x)$ and some new vocabulary—**independent variable, dependent variable, domain,** and **range.** You learned when a function is **increasing** or **decreasing, linear** or **nonlinear,** and the difference between a **discrete** and **continuous** graph. You explored the **absolute value function,** $f(x) = |x|$, and the **squaring function,** $f(x) = x^2$, and their graphs. You learned that these two functions can have zero, one, or two solutions. You learned how to graph a **parabola.** You also used the **square root function** to undo the squaring function and get only the positive square root.

EXERCISES

You will need your calculator for problem **10.**

1. Answer each question for the graph of $f(x)$.

   a. What is the domain of the function?

   b. What is the range of the function?

   c. What is $f(3)$?

   d. For what values of x does $f(x) = 1$?

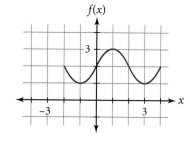

2. Which of these tables of x- and y-values represent functions? Explain your answers.

a.

x	y
0	5
1	7
3	10
7	9
5	7
4	5
2	8

b.

x	y
3	7
4	9
8	4
3	5
9	3
11	9
7	6

c.

x	y
2	8
3	11
5	12
7	3
9	5
8	7
4	11

3. The graph at right shows the relationship between an object's distance from a motion sensor in meters and time in seconds. Sketch a graph to represent the velocity of this object dependent on time *t*.

Elapsed time (sec)

4. In a letter-shift code, ARCHIMEDES codes into ULWBCGYXYM. Use this information to determine the names of famous mathematicians in 4a–c.

   a. XYMWULNYM

   b. BSJUNCU

   c. YOWFCX

   d. Create a grid and state a rule for this code.

5. The graph below shows the velocities of three girls inline skating over a given time interval. Assume that they start at the same place at the same time.

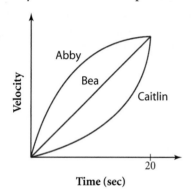

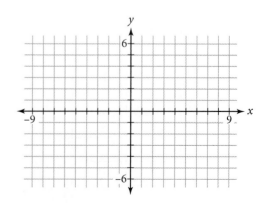

   a. Create a story about these three girls that explains the graph.

   b. Are Caitlin and Bea ever in front of Abby? Explain.

6. APPLICATION A recent catalog price for tennis balls was $4.25 for a can with three balls. The shipping charge per order was $1.00.

   a. Write an equation that you can use to project the costs for ordering different numbers of cans.

   b. Draw a graph showing this relationship.

   c. How does raising the shipping charge by 50¢ affect the graph?

   d. What equation models the cost situation in 6c?

7. Draw graphs that fit these descriptions:

   a. a function that has a domain of $-5 \le x \le 1$, a range of $-4 \le y \le 4$, and $f(-2) = 1$

   b. a relationship that is not a function and that has inputs on the interval $-6 \le x \le 4$ and outputs on the interval $0 \le y \le 5$

8. Cody's code multiplies each letter's position by 2. Complete a table like the one shown. If a number is greater than 26, subtract 26 from it so that it represents a letter of the alphabet. Is the code a function? Is the rule for decoding a function?

A	B	C	D	E
1	2	3	4	5
2	4	6		

9. Consider the function $f(x) = |x|$.
   a. What is $f(-3)$?
   b. What is $f(2)$?
   c. For what x-value(s) is $f(x) = 10$?

10. Use your calculator for 10a–c.
   a. Graph the functions $y = \sqrt{x}$ and $y = x^2$ in a friendly window.
   b. Compare the graphs. How are they similar? How are they different?
   c. Explain why the graph of $y = \sqrt{x}$ has only one "branch."
   d. Sketch the graph of $y^2 = x$. Is this the graph of a function? Explain why or why not.

TAKE ANOTHER LOOK

You learned to solve linear equations by "undoing" the order of operations in them. You learned to code and decode secret messages. Both are examples of reversing the order of a process, or finding an **inverse.**

How do you find the inverse of a function? The equation $y = \frac{9}{5}x + 32$ converts temperatures from x in degrees Celsius to y in degrees Fahrenheit.

If you want to make a graph showing how to convert temperature from degrees Fahrenheit to degrees Celsius, you can swap the two variables in the equation and solve for y.

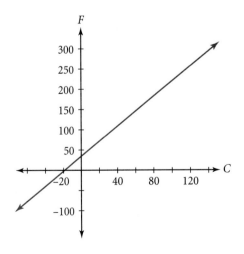

$$y = \frac{9}{5}x + 32 \qquad \text{The original equation.}$$

$$x = \frac{9}{5}y + 32 \qquad \text{Interchange } x \text{ and } y.$$

$$x - 32 = \frac{9}{5}y \qquad \text{Subtract 32 from both sides.}$$

$$5(x - 32) = 9y \qquad \text{Multiply both sides by 5.}$$

$$\frac{5}{9}(x - 32) = y \qquad \text{Divide both sides by 9.}$$

$$y = \frac{5}{9}(x - 32) \qquad \text{Isolate } y \text{ on the left side.}$$

Note that after the switch x represents degrees in Fahrenheit and y represents degrees in Celsius. The domain of the inverse is the range of the original function and vice-versa. Is this a function? Does each input in °F give exactly one output in °C?

How can you tell from the graph of a function whether or not it has an inverse that's a function? Does every linear function in the form $y = a + bx$ have an inverse function? Does the squaring function or the absolute value function have an inverse function? Look for patterns in the graphs of these functions and others. Can you restrict the values on the domain of a function so that the inverse is a function?

Learn more about inverse functions with the Internet links at **www.keymath.com/DA** .

Assessing What You've Learned

ORGANIZE YOUR NOTEBOOK Update your notebook with an example, investigation, or exercise that best demonstrates the concept of a function. Also, add one problem that illustrates the absolute value function and one that shows the squaring function.

WRITE IN YOUR JOURNAL Add to your journal by answering one of these prompts.

▶ You have seen many forms of equations—direct and indirect variation, linear relationships, and exponential modeling. Do you think all equations represent functions? Can you represent all functions as equations?

▶ Is $y = f(x)$ function notation helpful to you or do you find it challenging? When do parentheses ( ) mean multiplication and when do they show the independent variable in function notation?

UPDATE YOUR PORTFOLIO Choose your best explanation of a graph of a real-world situation from this chapter to add to your portfolio. Identify the independent and dependent variables of the situation. Describe all possible domain and range values as shown in the graph. Discuss whether the graph should be continuous or discrete.

GIVE A PRESENTATION Create your own code for making secret messages. Explain the rule for your code with a grid or an equation or both. Is your code a function? Is it simple to code? Is it hard to decode? How does the concept of functions apply to code making and code breaking?

Transformations

The Dome of the Rock is a famous site in Jerusalem. Built in the 7th century, it is well known for its beautiful tile work. Moving a small design left, right, up, or down could create some of the large patterns you see. Flipping or turning a design could create yet other patterns. Moving, flipping, or turning a design is important in creating many art forms. As you will see, changes like these are equally important in mathematics.

OBJECTIVES

In this chapter you will

- move a polygon by changing its vertices' coordinates
- learn to change, or transform, graphs by moving, flipping, shrinking, or stretching
- write a new equation to describe the changed, or transformed, graph
- model real-world data with equations of transformations
- use matrices to transform the vertices of a polygon

Translating Points

In computer animation, many individual points define each figure. You animate a figure by moving these points around the screen, little by little, through a series of frames. When you see the frames one after the other, the entire figure appears to move.

In mathematics, changing or moving a figure is called a **transformation.** So, every frame of an animation is a transformation of the one before it.

If one is lucky, a solitary fantasy can totally transform one million realities.

MAYA ANGELOU

This computer-animated motorcycle was created with software called Maya. The "skeleton" of the motorcycle would look like the face in the photo on page 331. The software allows an animator to move the motorcycle by transforming the points of the underlying skeleton.

Investigation
Figures in Motion

In this investigation you will learn how to move a polygon around the coordinate plane. You will first see what happens when you change the *y*-coordinates of the vertices.

> **Procedure Note**
>
> For this investigation, use a friendly window with a factor of 2.
> [▶ 🔲 See **Calculator Note 8D** to review friendly windows. ◀]

Step 1 | Name the coordinates of the vertices of this triangle.

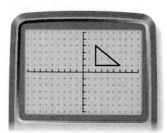

Step 2 | Enter the *x*-coordinates of the vertices into list L₁ and the corresponding *y*-coordinates of the vertices into list L₂. Enter the first coordinate pair again at the end of each list. Graph the triangle by connecting the vertices.
[▶ 🔲 See **Calculator Note 1G** to review connected graphs. ◀]

Step 3	Define list L3 and list L4 as follows

$$L_3 = L_1$$
$$L_4 = L_2 - 3$$

Graph a second triangle using list L3 for the *x*-coordinates of the vertices and list L4 for the *y*-coordinates of the vertices.

Step 4	Name the coordinates of the vertices of the new triangle. Tell how the original triangle has moved. How did the coordinates of the vertices change?
Step 5	Repeat Steps 3 and 4 with these definitions.

a. $L_3 = L_1$ **b.** $L_3 = L_1$
     $L_4 = L_2 + 2$ $L_4 = L_2 - 1$

Step 6	Write definitions for list L3 and list L4 in terms of list L1 and list L2 to create each graph below. Check your definitions by graphing on your calculator.

a.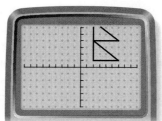
b.
c.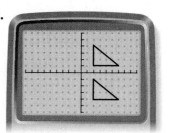

Next, you will include changes to the *x*-coordinates too.

Step 7	Name the coordinates of the vertices of this quadrilateral.

Step 8	Graph the quadrilateral using list L1 for the *x*-coordinates of the vertices and list L2 for the *y*-coordinates of the vertices.
Step 9	Define list L3 and list L4 as follows

$$L_3 = L_1 - 3$$
$$L_4 = L_2$$

Graph a second quadrilateral using list L3 for the *x*-coordinates of the vertices and list L4 for the *y*-coordinates of the vertices.

Step 10	Name the coordinates of the vertices of this new quadrilateral. Describe how the original quadrilateral moved. How did the coordinates of the vertices change?
Step 11	Repeat Steps 9 and 10 with these definitions.

a. $L_3 = L_1 + 2$ **b.** $L_3 = L_1 - 1$
     $L_4 = L_2$ $L_4 = L_2 + 3$

Step 12 | Write definitions for list L₃ and list L₄ in terms of list L₁ and list L₂ to create each graph below. Check your definitions by graphing on your calculator.

a. 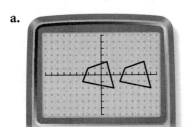   **b.** 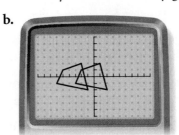   **c.**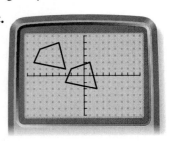

Step 13 | Summarize what you have learned about moving a figure around the coordinate plane.

In the investigation, each new polygon is the result of transforming the original polygon by moving it left, right, up, or down, or combinations of these movements. The figure that results from a transformation is called the **image** of the original figure. Transformations that move a figure horizontally, vertically, or both are called **translations.** You can define the translation of a point simply by adding to or subtracting from its coordinates.

EXAMPLE | Sketch the image of this figure after a translation right 4 units and down 3 units. Define the coordinates of any point in the image using (x, y) as the coordinates of any point in the original figure.

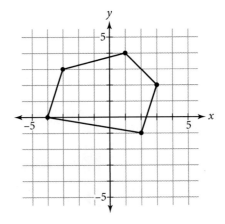

▶ Solution | Translate every point right 4 units and down 3 units. For example, move the vertex at $(1, 4)$ to $(5, 1)$. This is the same as adding 4 to the x-coordinate and subtracting 3 from the y-coordinate. That is, $(1 + 4, 4 - 3)$ gives $(5, 1)$.

A definition for any point in the image is

$$(x + 4, y - 3)$$

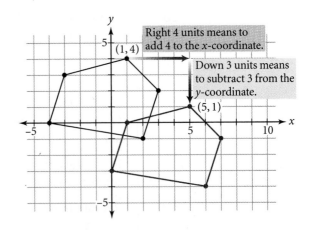

Right 4 units means to add 4 to the x-coordinate.

Down 3 units means to subtract 3 from the y-coordinate.

On your calculator, you can put the coordinates of the vertices of the original pentagon into list L1 and list L2. Then define list L3 as L3 = L1 + 4 and list L4 as L4 = L2 − 3.

Graphing confirms that your definition works.

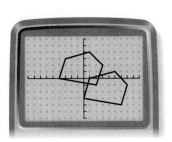

Science
CONNECTION

Many scientists support the theory of plate tectonics. According to this theory, the continents of the world were, at one time, together as a single continent. The German geophysicist and meteorologist Alfred Wegener (1880–1930) called this mass of land Pangaea. Over thousands of years, the individual continents drifted (or translated) to their current locations.

EXERCISES

You will need your calculator for problem **7.**

▶ **Practice Your Skills**

1. Name the coordinates of the vertices of each figure.

a.

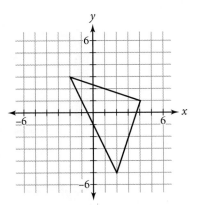

b.

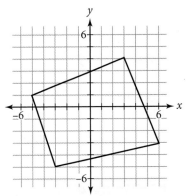

2. The *x*-coordinates of the vertices of a triangle are entered into list L1. The *y*-coordinates are entered into list L2. Describe the transformation for each definition.

   a. L3 = L1 − 5
       L4 = L2

   b. L3 = L1 + 1
       L4 = L2 + 2

3. The red triangle at right is the image of the black triangle after a transformation.

   a. Describe the transformation.

   b. Tell how the *x*-coordinates of the vertices change between the original figure and the image.

   c. Tell how the *y*-coordinates of the vertices change.

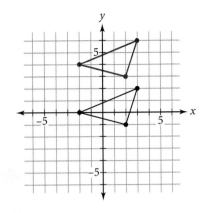

4. Consider the square at right.

   a. Sketch the image of the figure after a translation left 2 units.

   b. Define the coordinates of any point in the image using (x, y) as the coordinates of any point in the original figure.

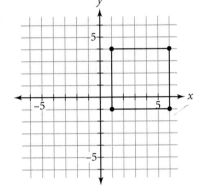

5. The "spider" in the upper left has its *x*-coordinates in list L1 and its *y*-coordinates in list L2.

$[-9.4, 9.4, 1, -6.2, 6.2, 1]$

   a. Describe the transformation to its image in the lower right.

   b. Write definitions for list L3 and list L4 in terms of list L1 and list L2.

   c. How would your answer to 5b change if the "spider" in the lower right were the original figure and the figure in the upper left were the image?

Reason and Apply

6. Consider the triangle on the calculator screen at right.

   a. Describe how to graph this triangle on your calculator.

   b. For each graph below, describe the transformation made to the original triangle.

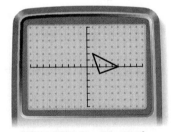

$[-9.4, 9.4, 1, -6.2, 6.2, 1]$

   i.

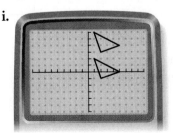

   ii.

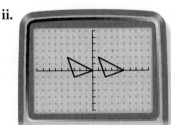

   iii.

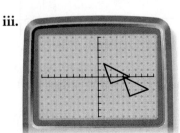

7. The coordinates of the vertices of a triangle are (2, 1), (4, 3), and (3, 0). Sketch the image that results from each definition. Use calculator lists to check your work.

a. $(x, y + 3)$ **b.** $(x - 2, y)$ **c.** $(x + 3, y - 1)$

8. If the triangle at right is the original figure, name the coordinates of the vertices of the image after

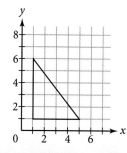

a. A translation up 4 units.

b. A translation left 7 units.

9. Lisa is designing a computer animation program. She has a set of coordinates for the arrow in the lower left. Now she wants the arrow to move to the upper right position.

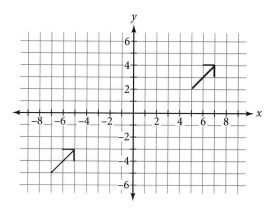

a. Describe the transformation that moves the arrow to the upper right.

b. Define the coordinates of any point in the image using (x, y) as the coordinates of any point in the original figure.

c. Lisa decides that a single move is too sudden. She thinks that moving the arrow little by little, in 20 frames, would look better. How should she define the coordinates of any point in each new image using (x, y) as the coordinates of any point in the figure in the previous frame?

10. Nick is also designing a computer animation program. His program first draws an N by connecting the points (7, 1), (7, 2), (8, 1), and (8, 2). Then, in each subsequent frame, the previous N is erased and an image is drawn whose coordinates are defined by $(x - 0.25, y + 0.05)$. The program uses recursion to do this over and over again.

What will be the coordinates of the N in the

a. 10th new frame?   **b.** 25th new frame?   **c.** 40th new frame?

► Review

11. Use $f(x) = 2 + 3x$ to find

a. $f(5)$ **b.** $f(-4)$ **c.** $f(x + 2)$ **d.** $f(2x - 1)$

12. Solve each equation.

a. $5 = -3 + 2x$ **b.** $-4 = -8 + 3(x - 2)$ **c.** $7 + 2x = 3 + x$

13. Find an equation for each graph.

a.

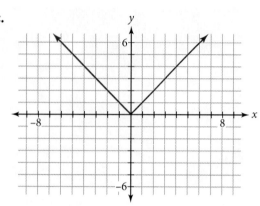

b.

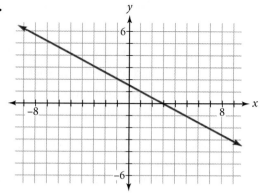

c.

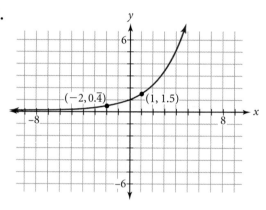

d.

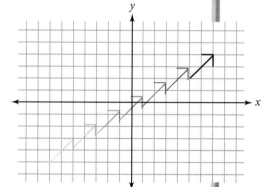

(−2, 0.$\overline{4}$)   (1, 1.5)

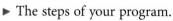

ANIMATING WITH TRANSFORMATIONS

As you've learned in this lesson, you can use transformations to create computer animation. Programs use mathematics to transform the points of a figure little by little. For example, Lisa's arrow in problem 9 appears to move because it makes 20 very small translations.

Now it's your turn to be the computer animator. Use a computer programming language to create an animation of any figure you choose. You can even use your calculator. [▶ 🖳 See **Calculator Note 9A** for a calculator program that moves Lisa's arrow. ◀]

Your project should include

▶ The steps of your program.

▶ An explanation of what each step of the program does.

▶ A description of the transformations used.

▶ A sketch of your original figure and the final image.

As you learn about other transformations in this chapter, you can try including them in your project too. You might also want to research the programming languages and software that professional animators use.

LESSON 9.2

Translating Graphs

There are infinitely many linear and exponential functions. In the previous chapters, you wrote many of them "from scratch" using points, the y-intercept, the slope, the starting value, or the constant multiplier.

Poetry is what gets lost in translation.

ROBERT FROST

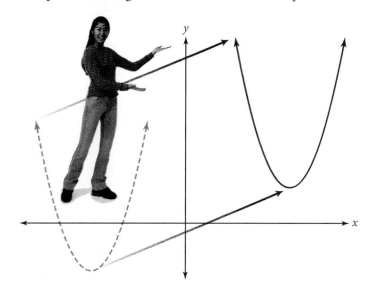

There are also infinitely many absolute value and squaring functions. But rather than starting from scratch, you can transform $y = |x|$ and $y = x^2$ to create many different equations. In this investigation you will use what you know about translating points to translate functions. If you discover any unexpected transformations along the way, make a note so that you can use them later in the chapter.

Investigation
Translations of Functions

First you'll transform the absolute value function by making changes to x.

Step 1	Enter $y = |x|$ into Y1 and graph it on your calculator.
Step 2	If you replace x with $x - 3$ in the function $y = |x|$, you get $y = |x - 3|$. Enter $y = |x - 3|$ into Y2 and graph it.
Step 3	Think of the graph of $y = |x|$ as the original figure and the graph of $y = |x - 3|$ as its image. How have you transformed the graph of $y = |x|$?

The **vertex** of an absolute value graph is the point where the function changes from decreasing to increasing or from increasing to decreasing.

Step 4	Name the coordinates of the vertex of the graph of $y = |x|$. Name the coordinates of the vertex of the graph of $y = |x - 3|$. Do these two points help verify the transformation you found in Step 3?

Procedure Note

For this investigation, use a friendly window with a factor of 2.

Step 5	Find a function for Y2 that will translate the graph of $y = |x|$ left 4 units. What is the function? In the equation $y = |x|$, what did you replace x with to get your new function?
Step 6	Write a function for Y2 to create each graph below. Check your work by graphing both Y1 and Y2.

a.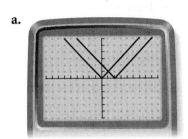

b.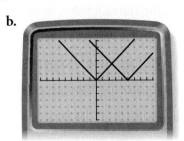

c.

Next, you'll transform the absolute value function by making changes to y.

Step 7	Clear all of the functions in your Y= menu. Enter $y = |x|$ into Y1 and graph it.
Step 8	If you replace y with $y - 3$ in the function $y = |x|$, you get $y - 3 = |x|$. Solve for y and you get $y = |x| + 3$. Enter $y = |x| + 3$ into Y2 and graph it.
Step 9	Think of the graph of $y = |x|$ as the original figure and the graph of $y = |x| + 3$ as its image. How have you transformed the graph of $y = |x|$?
Step 10	Name the coordinates of the vertex of the graph of $y = |x|$. Name the coordinates of the vertex of the graph of $y = |x| + 3$. Do these two points help verify the transformation you found in Step 9?
Step 11	Find a function for Y2 that will translate the graph of $y = |x|$ down 3 units. What is the function? In the function of $y = |x|$, what did you replace y with to get your new function?
Step 12	Write a function for Y2 to create each graph below. Check your work by graphing both Y1 and Y2.

a.

b.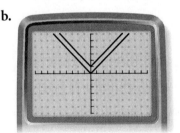

c.

Step 13	Summarize what you have learned about translating the absolute value graph vertically and horizontally.

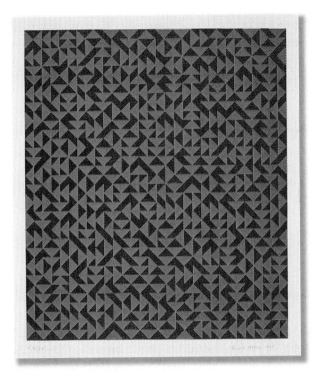

Anni Albers (1899–1994), a German-American artist, used many transformations of a single triangle to create this serigraph. Can you find some translations?

Anni Albers, *Untitled*, ca. 1969. The National Museum of Women in the Arts, Washington, D.C.

The most basic form of a function is often called a **parent function.** By transforming the graph of a parent function, you can create infinitely many new functions, or a **family of functions.** Functions like $y = |x - 3|$ and $y = |x| + 3$ are members of the absolute value family of functions with $y = |x|$ as the parent. Other families of functions include the linear family with $y = x$ as the parent, the squaring family with $y = x^2$ as the parent, and the base-3 exponential family with $y = 3^x$ as the parent.

Learning how to create a family of functions will help you to see relationships between equations and graphs. The translations you learned in the investigation apply to any function.

Science
●━ CONNECTION ●

Earthquakes often translate the earth's crust along a *fault*. You can see faults most easily when buildings and other structures are translated too. These cable car tracks were bent by a fault during the great 1906 earthquake in San Francisco, California. Learn more about earthquakes and faults with the links at www.keymath.com/DA .

Dip-slip fault

Strike-slip fault

EXAMPLE A | The graph of the parent function $y = x^2$ is shown in black. Its image after a transformation is shown in red. Describe the transformation. Then write an equation for the image.

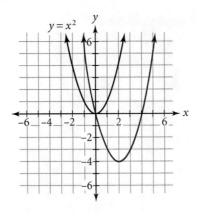

▶ **Solution** | The vertex of a parabola is the point where the squaring function changes from decreasing to increasing or increasing to decreasing. The vertex of the graph of $y = x^2$ is $(0, 0)$. The vertex of the image is $(2, -4)$. So the graph of $y = x^2$ is translated right 2 units and down 4 units to create the red image. You can check this with any other point. For example, the image of the point $(2, 4)$ is $(4, 0)$, which is also a translation right 2 units and down 4 units.

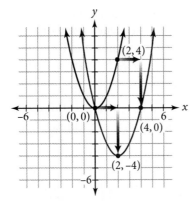

Every point on $y = x^2$ is translated right 2 units and down 4 units.

The equation of the image is
$$y - (-4) = (x - 2)^2$$
or
$$y = (x - 2)^2 - 4$$

Use the translation to write an equation for the red image.

$y = x^2$ Equation of the original parabola.

$y = (x - 2)^2$ Replace x with $x - 2$ to translate the graph right 2 units.

$y - (-4) = (x - 2)^2$ Replace y with $y - (-4)$, or $y + 4$, to translate the graph down 4 units.

$y = (x - 2)^2 - 4$ Solve for y.

The equation of the image is $y = (x - 2)^2 - 4$. You can graph this on your calculator to check your work.

In the next example you'll see how to translate an exponential function. Later you will use these skills to fit a function to a set of data.

EXAMPLE B

The starting number of bacteria in a culture dish is unknown, but the number grows by approximately 30% each hour. After 4 hours there are 94 bacteria present. Write an equation to model this situation. Then find the starting number of bacteria.

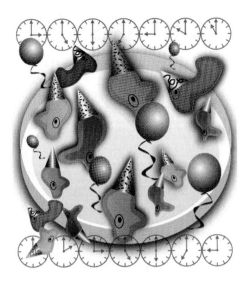

▶ **Solution**

Since the starting number is not known, suppose it was 94 bacteria. Then the function $y = 94(1 + 0.30)^x$ would be a correct model, in which x represents time elapsed in hours and y represents the number of bacteria.

However, there were 94 bacteria after 4 hours, not at 0 hours. So translate the point $(0, 94)$ right 4 units to $(4, 94)$. To translate the whole graph right 4 units, replace x with $x - 4$ in the function. You get

$$y = 94(1 + 0.30)^{(x-4)}$$

The graph shows how the new function translates every point in the graph right 4 units.

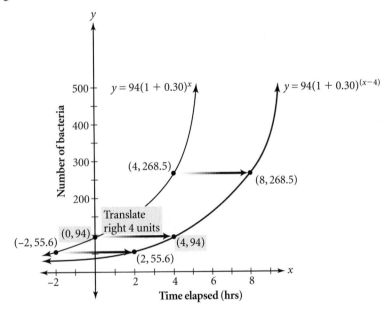

To find the starting value, substitute 0 for x in the new function.

$$94(1 + 0.30)^{(0-4)} = 94(1 + 0.30)^{-4} \approx 33$$

The starting number of bacteria was approximately 33 bacteria.

Using the starting value you found in the example, you could now write the function $y = 33(1 + 0.30)^x$. Can you use properties of exponents to show that $y = 94(1 + 0.30)^{(x-4)}$ is approximately equivalent to $y = 33(1 + 0.30)^x$? Do you think these functions would be considered members of the same family of functions?

▶ **Practice Your Skills**

1. Use $f(x) = 2|x + 4| + 1$ to find
   a. $f(5)$ **b.** $f(-6)$ **c.** $f(-2) + 3$ **d.** $f(x + 2)$

2. List L1 and list L2 contain coordinates for three points on the graph of $f(x)$. List L3 and list L4 contain coordinates for the three points after a transformation of f.

L1 x	L2 y		L3 x	L4 y
-1	3		7	-1
3	5		11	1
2	4		10	0

   a. Write definitions for list L3 and list L4 in terms of list L1 and list L2.
   b. Describe the transformation.

3. Give the coordinates of the vertex for each graph.

   a.
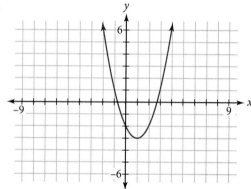

   b.
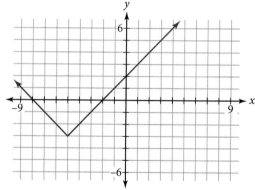

   c.
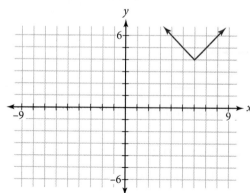

   d.
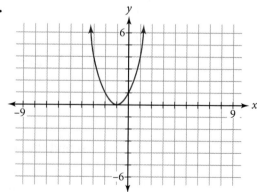

4. Graph each equation and describe the graph as a transformation of $y = |x|$, $y = x^2$, or $y = 3^x$.
   a. $y = |x - 1.5| - 2.5$ **b.** $y = (x + 3)^2$
   c. $y = |x| + 3.5$ **d.** $y = 3^{(x+1)} + 2$

5. Write an equation for each of these transformations.

    a. Translate the graph of $y = x^2$ down 2 units.

    b. Translate the graph of $y = 4^x$ right 5 units.

    c. Translate the graph of $y = |x|$ left 4 units and up 1 unit.

 Reason and Apply

6. Describe each graph in problem 3 as a transformation of $y = |x|$ or $y = x^2$. Then write its equation.

7. This graph shows Beth's distance from her teacher as she turns in her test.

    a. What are the input and output variables?

    b. What are the units of the variables?

    c. What are the domain and range shown in the graph?

    d. Describe the situation.

    e. Write a function that models this situation.

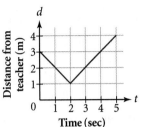

Beth's Walk

8. Graph $Y_1(x) = \text{abs}(x)$ on your calculator. Predict what each graph will look like. Check by comparing the graphs on your calculator.
[▶ 🖳 See **Calculator Note 9B** for specific instructions for your calculator. ◀]

    a. $Y_2(x) = Y_1(x) - 4$
                         b. $Y_2(x) = Y_1(x - 4)$

9. Describe how the graph of $y = x^2$ will be transformed if you replace

    a. x with $(x - 3)$
                              b. x with $(x + 2)$

    c. y with $(y + 2)$
                              d. y with $(y - 3)$

10. Recall that an exponential equation in the form $y = A(1 - r)^x$ models some decreasing patterns. As you increase the value of x, the **long-run value** of y gets closer and closer to zero. Some situations, however, do not decrease all the way to zero. For example, as a cup of hot chocolate cools, the coolest it can get is room temperature. The long-run value will not be 0°C. Consider this table of data.

Time (min)	0	1	2	3	4	5	6
Temperature (°C)	68	52	41	34	30	27	25

    a. Define variables and make a scatter plot of the data. What type of function would fit the data?

    b. Find the ratio of each temperature to the previous temperature. Do these ratios support your answer to 10a?

    Assume the temperature of the room in this situation is 21°C. That means the long-run value of this data will also be 21°C.

    c. Make a new table by subtracting 21 from each temperature. Then make a scatter plot of the changed data. How have the points been transformed? What will be the long-run value?

    d. For your data in 10c, find the ratios of temperatures between successive readings. How do the ratios compare? What is the mean of these ratios?

e. Write an exponential equation in the form $y = A(1 - r)^x$ that models the data in 10c.

f. In 10c you subtracted 21 from each temperature. What transformation takes the original data back to the original data?

g. Your equation in 10e models translated data. Change that equation so that it models the original data. Check the fit by graphing on your calculator.

11. **APPLICATION** Clay works for a company that manufactures pottery. He has designed a new bowl and needs to know how long it takes to cool after being removed from the kiln. He removes a sample bowl from the kiln. After 5 hours of cooling, the temperature of the bowl is 241°C. Clay calls this time 0. After 1 more hour, the temperature of the bowl is 208°C. Room temperature is 20°C.

a. What will be the long-run value for the temperature of the bowl?

b. Transform the temperatures so that the long-run value will be 0°C.

c. Find the rate of cooling per hour using the temperatures in 11b.

d. Define variables and write an exponential function that models the temperatures in 11b.

e. Change your function in 11d to show a translation up 20 units. Why do you need this translation?

f. Change your function in 11e to show a translation right 5 units. Why do you need this translation?

g. Use your function in 11f to find $f(0)$. What is the real-world meaning of this value?

h. The company's safety regulations say that a piece of pottery cannot be handled until it is at most 25°C. How long after the bowl is removed from the kiln can it be handled?

12. **APPLICATION** In 1999, the world population was estimated to be 6.0 billion, with an annual growth rate of 1.3%. (*2000 World Almanac*, p. 878)

a. Define input and output variables for this situation.

b. Without finding an equation, sketch a graph of this situation for 1990 to 2010.

c. What one point on the graph do you know for sure?

d. Write a function that models this situation. Graph your function on your calculator and name an appropriate window.

e. Use your graph to estimate the population to the nearest tenth of a billion in 1990 and 2010. (Assume a constant growth rate during this period.)

13. The graph of a linear equation of the form $y = bx$ passes through $(0, 0)$.

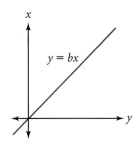

    a. Suppose the graph of $y = bx$ is translated right 4 units and up 8 units. Name a point on the new graph.

    b. Write an equation for the line in 12a after the transformation.

    c. Suppose the graph of $y = bx$ is translated horizontally H units and vertically V units. Name a point on the new graph.

    d. Write an equation for the line in 12c after the transformation.

▶ Review

14. Drew's teacher gives skill-building quizzes at the start of each class.

    a. On Monday, Drew got 77 problems correct out of 85. What is her percent correct?

    b. On Tuesday, Drew got 100% on a quiz that had only 10 problems. Estimate her percent correct for the two-day total.

    c. Calculate her percent correct for the two-day total.

15. Solve each system of equations.

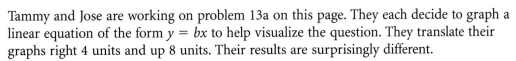

    a. $\begin{cases} y = 5 + 2x \\ y = 8 - 2x \end{cases}$ **b.** $\begin{cases} y = -2 + 3(x - 4) \\ y = 3 + 5(x - 2) \end{cases}$ **c.** $\begin{cases} 2x + 7y = 13 \\ 5x - 14y = 1 \end{cases}$

IMPROVING YOUR **VISUAL THINKING** SKILLS

Tammy and Jose are working on problem 13a on this page. They each decide to graph a linear equation of the form $y = bx$ to help visualize the question. They translate their graphs right 4 units and up 8 units. Their results are surprisingly different.

Jose's Graph

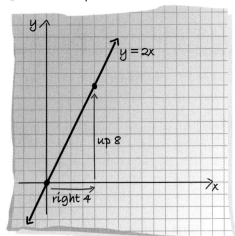

Tammy's Graph

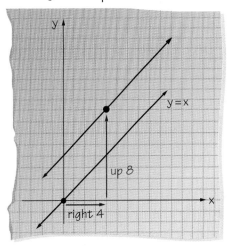

Why did Jose get the same graph after the translation?

If the graph of an equation of the form $y = bx$ is translated horizontally H units and vertically V units, when would you get the same graph after the translation?

Reflecting Points and Graphs

The art of a people is a true mirror to their minds.

JAWAHARLAL NEHRU

Translations move points and graphs around the coordinate plane. Have you noticed that the image of the translation always looks like the original figure? Although the image of a translation moves, it doesn't flip, turn, or change size. To get these changes, you need other types of transformations.

Investigation
Flipping Graphs

In this investigation you will explore the relationships between the graph of an equation and its image when you flip it two different ways.

Step 1 | Name the coordinates of the vertices of this triangle.

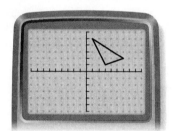

> **Procedure Note**
> For this investigation, use a friendly window with a factor of 2.

Step 2 | Graph the triangle on your calculator. Use list L1 for the *x*-coordinates of the vertices and list L2 for the *y*-coordinates of the vertices.

Step 3 | Define list L3 and list L4 as follows

$$L3 = -L1$$
$$L4 = L2$$

Graph a second triangle using list L3 for the *x*-coordinates of the vertices and list L4 for the *y*-coordinates of the vertices.

Step 4 | Name the coordinates of the vertices of the new triangle. Describe the transformation. How did the coordinates of the vertices change?

Step 5 | Repeat Steps 3 and 4 with these definitions.

a. $L3 = L1$ **b.** $L3 = -L1$
     $L4 = -L2$ $L4 = -L2$

Next, you'll see if what you have learned about flipping points is true for the graphs of functions.

Step 6 | Graph $y = 2^x$ on your calculator.

Step 7 | Replace *x* with $-x$ in the function. Graph this second function. Describe how the second graph is related to the graph of $y = 2^x$.

| Step 8 | Now replace y with $-y$ in the function $y = 2^x$ and solve for y. Graph this third function. Describe how its graph is related to the graph of $y = 2^x$. |
| Step 9 | Repeat Steps 6–8 using these functions. Make a note of anything unusual that you find. |

a. $y = (x - 1)^2$

b. $y = |x|$

c. $y = x$

| Step 10 | Summarize what you have learned about flipping graphs. |

A transformation that flips a figure to create a mirror image is called a **reflection.** A point is **reflected across the x-axis** when you change the sign of its y-coordinate. A point is **reflected across the y-axis** when you change the sign of its x-coordinate. You saw both types of reflections in the investigation. Similar reflections result when you change the sign of x or y in a function.

You can combine reflections with other transformations. Sometimes, different combinations will give the same result.

EXAMPLE A

The graph of a parent function is shown in black. Its image after a transformation is shown in red. Describe the transformation and then write a function for the image.

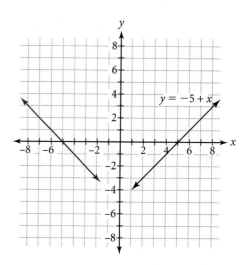

$y = -5 + x$

▶ **Solution**

This is a reflection across the y-axis. The image is produced by replacing each x-value in the original function with $-x$.

$$y = -5 + (-x)$$

$$\text{or}$$

$$y = -5 - x$$

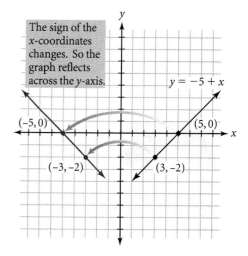

The sign of the x-coordinates changes. So the graph reflects across the y-axis.

$y = -5 + x$

$(-5, 0)$ $(5, 0)$

$(-3, -2)$ $(3, -2)$

EXAMPLE B

The graph of a parent function is shown in black. Its image after a transformation is shown in red. Describe the transformation and then write a function for the image.

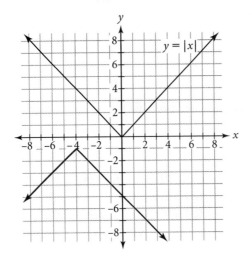

► Solution

There are several ways to think about this transformation. The order of the transformations will create different, yet equivalent, equations.

Here is one possible solution. Reflect the graph of the function across the x-axis, then translate it left 4 units and down 1 unit. To write the equation of the image, change the original function in the same order.

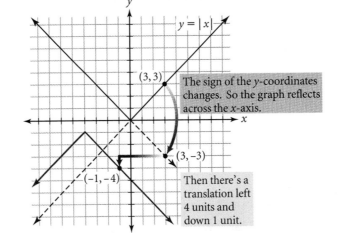

The sign of the y-coordinates changes. So the graph reflects across the x-axis.

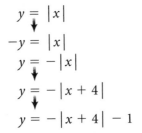

Then there's a translation left 4 units and down 1 unit.

$$y = |x|$$ Original equation.

$$-y = |x|$$ Replace y with $-y$ to reflect across the x-axis.

$$y = -|x|$$ Solve for y.

$$y = -|x + 4|$$ Translate left 4 units.

$$y = -|x + 4| - 1$$ Translate down 1 unit.

A function for the image is $y = -|x + 4| - 1$.

EXAMPLE C

The graph of a parent function is shown in black. Its image after a transformation is shown in red. Describe the transformation and then write a function for the image.

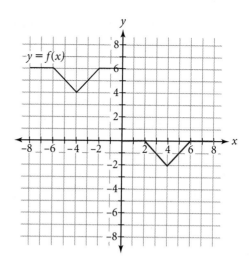

► **Solution**

As in Example B, you can think of this transformation in several ways. One solution is to translate right 8 units and down 6 units, as shown in the graph on the left below. That gives the function $y = f(x - 8) - 6$.

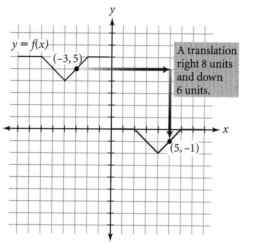

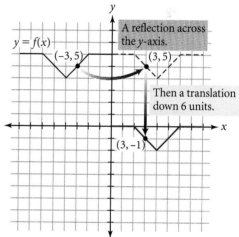

Another solution is to reflect the graph across the y-axis and then translate down 6 units, as shown in the graph on the right above. That gives the function $y = f(-x) - 6$.

In the investigation you probably saw no change when you reflected the graph of $y = |x|$ across the y-axis. In Example C, a reflection across the y-axis has the same result as a horizontal translation. Do you notice anything special about these graphs that could explain these strange results?

EXERCISES

You will need your calculator for problems **4, 5,** and **12.**

Practice Your Skills

1. Use $f(x) = 0.5(x - 3)^2 - 3$ to find

    a. $f(5)$ **b.** $f(-6)$ **c.** $4 \cdot f(2)$

    d. $f(-x)$ **e.** $-f(x)$

2. Describe each graph as a transformation of $y = |x|$ or $y = x^2$. Then write its equation.

a.

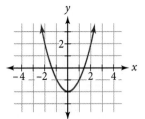

b.

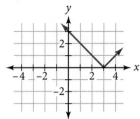

c.

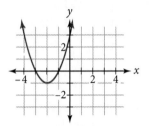

d.

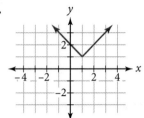

3. Describe each graph below as a transformation of $y = |x + 3|$, shown at right.

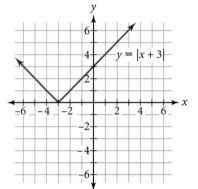

a.

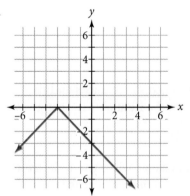

b.

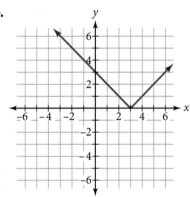

c.

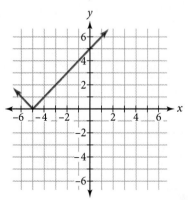

d.

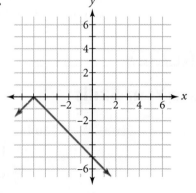

4. Graph $Y_1(x) = 1 + 2.5x$ on your calculator. Predict what each graph will look like. Check by comparing graphs on your calculator. [▶ 🖳 See **Calculator Note 9B** for specific instructions for your calculator. ◀]

    a. $Y_2(x) = Y_1(-x)$

    b. $Y_2(x) = -Y_1(x)$

5. Describe the graph of each function below as a transformation of the graph of the parent function $y = x^2$. Check your answers by graphing on your calculator.

    a. $y = -x^2$

    b. $y = -(x + 3)^2$

    c. $y = -x^2 + 3$

    d. $y = (-x)^2 + 3$

▶ Reason and Apply

6. Consider the triangle at right.

    a. Describe how you can graph this triangle on your calculator.

    b. How could you make these graphs?

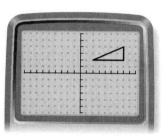

$[-9.4, 9.4, 1, -6.2, 6.2, 1]$

    i.

    ii.

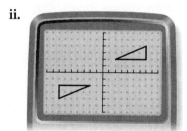

    iii.

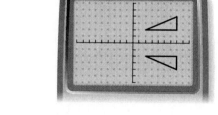

    iv.

7. The points in this table form a star when you connect them in order. Describe the transformation that results when you redefine the points as

    a. $(-x, y)$

    b. $(x - 8, -y)$

    c. $(x + 2, y - 4)$

    d. (y, x) (Hint: Try graphing this.)

x	y
6.0	2.0
2.4	3.2
4.6	0.1
4.6	3.9
2.4	0.8
6.0	2.0

8. Anthony and Cheryl are using a motion sensor for a "walker" investigation.

 a. This graph shows data that Cheryl collected when Anthony walked. Write an equation that models his walk.

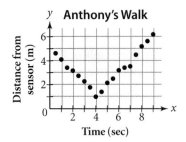

 b. Here is a description of Cheryl's walk.

> Begin at a distance of 0.5 meter from the sensor. Walk away from the sensor at 1 meter per second for 3 seconds. Then walk toward the sensor at the same rate for 3 seconds.

 Write an equation to model her walk.

 c. Give the domain and range for the function that models Cheryl's walk.

9. Bo is designing a computer animation program. She wants the star on the left to move to the position of the star on the right using 11 frames. She also wants the star to flip top to bottom in each frame. Define the coordinates of each image based on the coordinates of the previous figure.

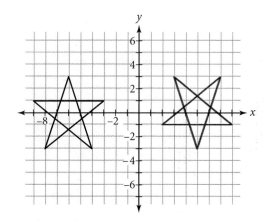

10. For a and b, the graph of a parent function is shown in black. Describe the transformation that creates the red image. Then write a function for the image.

 a.

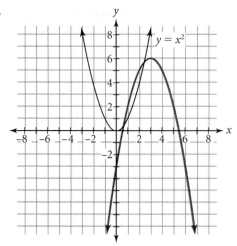

 b.

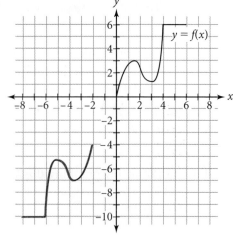

11. A line of reflection does not have to be the x- or y-axis. Consider this example in which $y = |x|$ is reflected across the line $x = 4$.

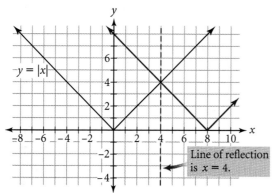

a. Think about each of these transformations as a single reflection. What is the line of reflection?

i.

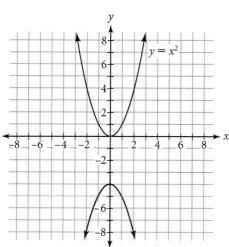

ii.

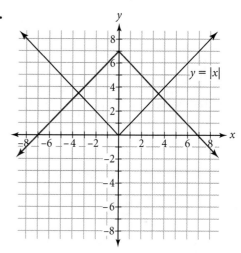

iii.

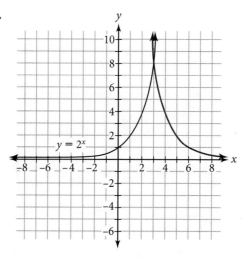

iv.

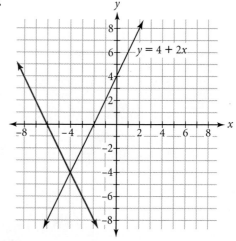

b. Write an equation for the red image in each graph in 11a.

c. The graph of $y = f(x)$ is reflected across the horizontal line $y = b$. What is the equation of the image?

d. The graph of the function $y = f(x)$ is reflected across the vertical line $x = a$. What is the equation of the image?

► Review

12. A chemical reaction consumes 12% of the reactants per minute. A scientist begins with 500 grams of one reactant. So the equation $y = 500(0.88)^x$ gives the amount of reactant remaining, y, after x minutes.

    a. What does the number 0.88 tell you?

    b. What is the long-run value of y? What is the real-world meaning of this value?

    c. What is the long-run value of y for the equation $y = 500(0.88)^x + 100$? What is the real-world meaning of this value?

    d. Graph $y = 500(0.88)^x$ and $y = 500(0.88)^x + 100$. How are these graphs the same? How are they different?

13. Convert 47 tablespoons to quarts. (16 tablespoons = 1 cup; 1 quart = 4 cups)

14. This table shows the temperature of water in a pan set on a stove.

    a. Find the equation of a line that models this data.

    b. How long will it take for the water to boil (100°C)?

Time (min)	0	2	4	6	8	10	12	14	16	18
Temperature (°C)	22	29	36	44	51	58	65	72	80	87

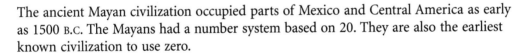

IMPROVING YOUR **REASONING** SKILLS

The ancient Mayan civilization occupied parts of Mexico and Central America as early as 1500 B.C. The Mayans had a number system based on 20. They are also the earliest known civilization to use zero.

Below are the 20 numerals in the Mayan number system. Can you decode the numerals and label them with the numbers 0 to 19? A few are labeled to get you started.

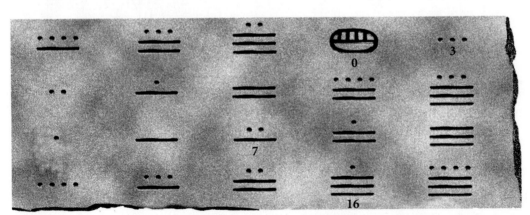

Stretching and Shrinking Graphs

There is no absolute scale of size in the Universe, for it is boundless towards the great and also boundless towards the small.

OLIVER HEAVISIDE

Imagine what happens to the shape of a picture drawn on a rubber sheet as you **stretch** the sheet vertically.

The width remains the same, but the height changes. You can also **shrink** a picture vertically. This makes the picture appear to have been flattened.

You know how to translate and reflect graphs on the coordinate plane. Now let's see how to change their shape.

The German painter Hans Holbein II (1497–1543) used a technique called anamorphosis to hide a stretched skull in his portrait *The Ambassadors* (1533). You can see the skull in the original painting if you look across the page from the lower-left. The painting was originally hung above a doorway so people would notice the skull as they walked through the door. Holbein may have been making a political statement about these two French ambassadors who were members of England's court of King Henry VIII.

 ## Investigation
Changing the Shape of a Graph

In this investigation you will learn how to stretch or shrink a graph vertically.

Procedure Note

For this investigation, use a friendly window with a factor of 2.

Step 1 | Name the coordinates of the vertices of this quadrilateral.

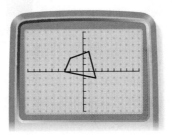

Step 2 | Graph the quadrilateral on your calculator. Use list L1 for the *x*-coordinates of the vertices and list L2 for the *y*-coordinates of the vertices.

Step 3 | Each member of your group should choose one of these values of *a*: 2, 3, 0.5, or −2. Use your value of *a* to define list L3 and list L4 as follows

$$L3 = L1$$
$$L4 = a \cdot L2$$

Graph a second quadrilateral using list L3 for the *x*-coordinates of the vertices and list L4 for the *y*-coordinates of the vertices.

Step 4 | Share your results from Step 3. For each value of *a*, describe the transformation of the quadrilateral in Step 2. What was the result for each vertex?

Step 5 | Organize your results from this first part of the investigation.

| Step 6 | Graph this triangle on your calculator. Use list L_1 for the x-coordinates of the vertices and list L_2 for the y-coordinates of the vertices. | 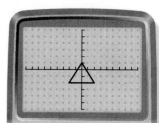 |

| Step 7 | Describe how definitions a and b below transform the triangle. Use list L_3 for the x-coordinates of the vertices of the image and list L_4 for the y-coordinates of the vertices of the image. Check your answers by graphing on your calculator. |

a. $L_3 = L_1$
   $L_4 = -0.5 \cdot L_2$

b. $L_3 = L_1$
   $L_4 = 2 \cdot L_2 - 2$

| Step 8 | Write definitions for list L_3 and list L_4 in terms of list L_1 and list L_2 to create each image below. Check your definitions by graphing on your calculator. |

a.

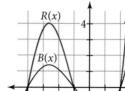

b.

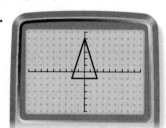

Next, see how you can stretch and shrink the graph of a function.

| Step 9 | Each member of your group should choose an equation from the list below. Enter your equation into Y_1 and graph it on your calculator. |

$Y_1(x) = -1 + 0.5x$ $Y_1(x) = |x| - 2$

$Y_1(x) = -x^2 + 1$ $Y_1(x) = 1.4^x$

| Step 10 | Enter $Y_2(x) = 2 \cdot Y_1(x)$ and graph it. [▶ 🔲 See **Calculator Note 9B** for specific instructions for your calculator. ◀] |

| Step 11 | Look at a table on your calculator and compare the y-values for Y_1 and Y_2. |

| Step 12 | Repeat Steps 10 and 11, but use these equations for Y_2.

a. $Y_2(x) = 0.5 \cdot Y_1(x)$ **b.** $Y_2(x) = 3 \cdot Y_1(x)$ **c.** $Y_2(x) = -2 \cdot Y_1(x)$ |

| Step 13 | Write an equation for $R(x)$ in terms of $B(x)$. Then write an equation for $B(x)$ in terms of $R(x)$. |

a.

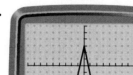

b.

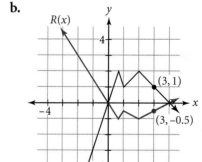

To vertically stretch or shrink a polygon, you multiply the *y*-coordinates of the vertices by a constant factor. To vertically stretch or shrink the graph of a function, you again have to multiply the function by a factor.

EXAMPLE A

Describe how the graph of $y = 0.5|x|$ relates to the graph of $y = |x|$. Then graph both functions.

▶ **Solution**

Tables of values for both functions show that $y = 0.5|x|$ is a vertical shrink. Each *y*-value for $y = 0.5|x|$ is one-half the corresponding *y*-value for $y = |x|$. Multiplying the function by 0.5 has the same effect as multiplying the *y*-coordinate of every point on the graph of $y = |x|$ by 0.5.

x	$y = |x|$	$y = 0.5|x|$
2	2	1
0	0	0
1	1	0.5
5	5	2.5

Graphing the functions together also shows a vertical shrink by a factor of 0.5. Each point on the graph of $y = 0.5|x|$ is one-half the distance from the *x*-axis of the corresponding point on $y = |x|$.

Technology
● ━ **CONNECTION** ● ━

Many computer applications allow you to change the size and shape of clip art. Some applications have commands to change only the horizontal or the vertical scale. If you change only one scale, you distort the picture with a stretch or a shrink. If you change both scales by the same factor, you create a larger or smaller picture that is geometrically similar to the original.

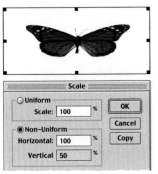

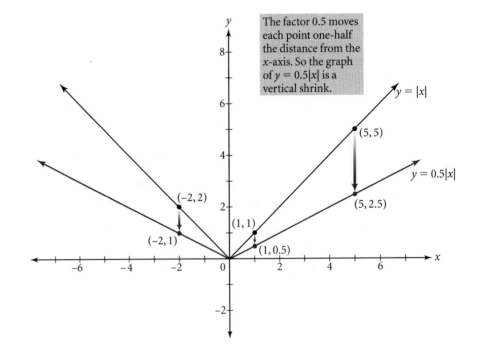

The factor 0.5 moves each point one-half the distance from the *x*-axis. So the graph of $y = 0.5|x|$ is a vertical shrink.

EXAMPLE B

Find an equation for the function shown in this graph.

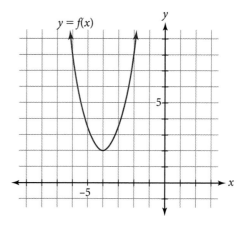

► **Solution**

The graph is a parabola, so the parent function is $y = x^2$. First determine if a vertical stretch or shrink is necessary. An informal way to do this is to think about corresponding points on the graphs of $y = x^2$ and $y = f(x)$.

The parent function, $y = x^2$
When you move 1 unit left of the vertex, you move 1 unit up to find a point on the graph. When you move 2 units right of the vertex, you move 4 units up to find a point on the graph.

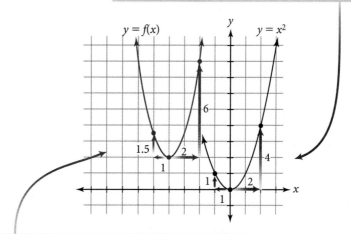

The image function, $y = f(x)$
When you move 1 unit left of the vertex, you move 1.5 units up to find a point on the graph. When you move 2 units right of the vertex, you move 6 units up to find a point on the graph.

For the same x-distances from the vertex on each graph, the corresponding y-distances from the vertex on the image graph $y = f(x)$ are 1.5 times the y-distances on the parent graph $y = x^2$. So the stretch factor is 1.5.

x-distance from vertex	y-distance from vertex of parent function, $y = x^2$	y-distance from vertex of image function, $y = f(x)$	Stretch factor calculation
1	1	1.5	$\frac{1.5}{1} = 1.5$
2	4	6	$\frac{6}{4} = 1.5$

$$y = x^2$$ Equation of the parent function.

$$y = 1.5x^2$$ Multiply the parent function, x^2, by a factor of 1.5 for the vertical stretch.

The vertex of the graph of $y = f(x)$ is at $(-4, 2)$. So you must now change the equation to show a translation left 4 units and up 2 units.

$$y = 1.5(x + 4)^2$$ Replace x with $x - (-4)$, or $x + 4$, to translate the graph left 4 units.

$$y - 2 = 1.5(x + 4)^2$$ Replace y with $y - 2$ to translate the graph up 4 units.

$$y = 1.5(x + 4)^2 + 2$$ Solve for y.

The equation for the function is $y = 1.5(x + 4)^2 + 2$.

How can you check that this equation is correct?

Now that you've learned how to translate, reflect, and vertically stretch or shrink a graph, you can transform a function into many forms. This skill gives you a lot of power in mathematics. You can look at a complicated equation and see it as a variation of a simpler function. This skill also allows you to adjust the fit of mathematical models for many situations.

EXERCISES

You will need your calculator for problems **4, 5, 7, 10, 11,** and **12.**

Practice Your Skills

1. Ted and Ching-I are using a motion sensor for a "walker" investigation. They find that the graph at right models data for Ted's walk. Write an equation for this graph.

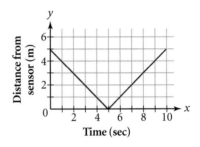

2. Ching-I walks so that her distance from the sensor is always twice Ted's distance from the sensor.

   a. Sketch a graph that models Ching-I's walk.

   b. Write an equation for the graph in 2a.

3. Ted walks so that the data can be modeled by this graph.

   a. Write an equation for this graph.

   b. Describe how Ted walked to create this graph.

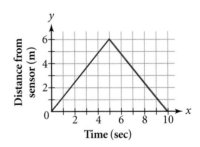

4. Run the ABS program five times. On your own paper, sketch a graph of each randomly generated absolute value function. Find an equation for each graph. [▶🖳 See **Calculator Note 9C** to learn how to use the ABS program. ◀]

5. Run the PARAB program five times. On your own paper, sketch a graph of each randomly generated parabola. Find an equation for each graph. [▶🖳 See **Calculator Note 9D** to learn how to use the PARAB program. ◀]

Reason and Apply

6. This table lists the vertices of a triangle. Name the vertex or vertices that will not be affected by doing a vertical stretch.

x	y
2	0
4	2
0	1

7. Graph each function on your calculator. Then describe how each graph relates to the graph of $y = |x|$ or $y = x^2$. Use the words *translation*, *reflection*, *vertical stretch*, and *vertical shrink*.

   a. $y = 2x^2$
   b. $y = 0.25|x - 2| + 1$
   c. $y = -(x + 4)^2 - 1$
   d. $y = -2|x - 3| + 4$

8. Each row of the table below describes a single transformation of the parent function $y = |x|$. Copy and complete the table.

Change to the equation $y = |x|$	New equation in $y =$ form	Transformation of the graph of $y = |x|$
Replace x with $x - 3$	$y = |x - 3|$	Translation right 3 units
		Translation down 2 units
	$y = -|x|$	
Replace y with $y - 2$		
		Vertical shrink by a factor of $\frac{1}{2}$
		Translation left 4 units
	$y = 1.5|x|$	
		Translation right 1 unit
Multiply the right side by 3		

9. Describe the order of transformations of the graph of $y = x^2$ represented by

   a. $y = -(x + 3)^2$
   b. $y = 0.5(x - 2)^2 + 1$

10. Draw this **J** on graph paper or on your calculator. Then draw the image defined by each of the definitions in 10a–e. Describe how each image relates to the original figure. (If you use graph paper, give yourself a lot of room or make five individual graphs. If you use a calculator, adjust your friendly window so that you can see both figures at the same time.)

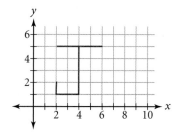

   a. $(x, 3y)$
   b. $(3x, y)$
   c. $(3x, 3y)$
   d. $(0.5x, 0.5y)$
   e. $(-2x, -2y)$
   f. Explain why the transformations in 10c, 10d, and 10e are often called "size transformations."

11. Graph $Y_1(x) = \text{abs}(x)$ on your calculator. Predict what each graph will look like. Check by comparing the graphs on your calculator. [▶ 🖳 See **Calculator Note 9B** for specific instructions for your calculator. ◀]

a. $Y_2(x) = -0.5\,Y_1(x)$

b. $Y_2(x) = 2\,Y_1(x - 4)$

c. $Y_2(x) = -3\,Y_1(x + 2) + 4$

12. In Interlochen, Michigan, it begins to snow in early November. The depth of snow increases over the winter. When winter ends, the snow melts and the depth decreases. This table shows data collected in Interlochen.

Snow in Interlochen

Date	Nov 1	Dec 1	Jan 1	Feb 1	Mar 1	Apr 1
Depth of snow (cm)	25	50	70	60	35	10

a. Plot the data. For the dependent variable, let Nov 1 = 1, Dec 1 = 2, and so on. Find a function that models the data.

b. Use your function to find $f(2.5)$. Explain what this value represents.

c. Find x if $f(x) = 47$. Explain what this x-value represents.

d. According to your model, when was the snow the deepest? How deep was it at that time?

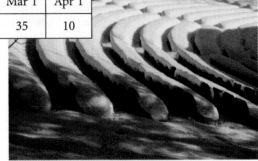

January snow covers the seats of the outdoor theater at Interlochen Center for the Arts.

13. Deshawn is designing a computer animation program. She has a set of coordinates for the tree shown on the right side. She wants to use 13 frames to move the tree from the right to the left. In each frame, she wants the tree to shrink by 80%. How should she define the coordinates of each image using the coordinates from the previous frame?

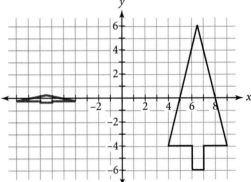

14. Byron says,

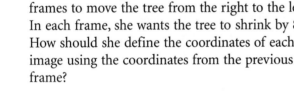

If the graph of a function is stretched vertically, but not translated, the factor a is the same as the y-value when x equals 1.

Does Byron's conjecture work for every function in the forms shown below? Tell why or why not.

a. $y = a \cdot x^2$

b. $y = a \cdot |x|$

c. $y = a \cdot f(x)$

► Review

15. Use the properties of exponents to rewrite each expression without negative exponents.

　a. $(2^3)^{-3}$ 　　　　　　　　　　　　　　**b.** $(5^2)^5$

　c. $(2^4 \cdot 3^2)^3$ 　　　　　　　　　　　　**d.** $(3^2 \cdot x^3)^{-4}$

16. The equation $y = -52 + 1.6x$ approximates the wind chill temperature in degrees Fahrenheit for a wind speed of 40 miles per hour.

　a. Which variable represents the actual temperature? Which variable represents the wind chill temperature?

　b. What x-value gives a y-value of -15? Explain what your answer means in the context of this problem.

IMPROVING YOUR **REASONING** SKILLS

In this lesson you learned how to transform points and functions with a vertical stretch or shrink. In problem 10 in this set of exercises, you also saw how to transform points with a horizontal stretch or shrink. It is also possible to change the equation of a function to show a horizontal stretch or shrink.

Consider the graph of $y = x^2$ and its image after a horizontal stretch by a factor of 2. Write an equation for the image.

Describe the image in terms of a vertical stretch or shrink. Write an equation that shows this transformation. Is this equation equivalent to the one that shows a horizontal stretch?

When you vertically stretch or shrink the graph of $y = f(x)$ by a factor of a, you get a graph of $y = a \cdot f(x)$. If you horizontally stretch or shrink the graph of $y = f(x)$ by a factor of b, you will get the graph of what equation?

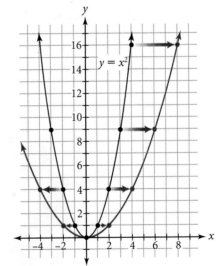

Activity Day

Using Transformations to Model Data

In this lesson you'll do experiments to gather data, and then you'll find a function to model to the data. To fit the model, you'll first need to identify a parent function. Then you'll transform the parent function and fit the image function to the data.

There are three experiments to choose from. Your group should choose one experiment. Do the other experiments if time permits.

Activity
Roll, Walk, or Sum

You will need

- a large marble
- tape
- one sheet of paper
- four books
- a sheet of poster board
- a paper cup
- a meterstick or a yardstick
- a table and chair
- a motion sensor
- a stopwatch or a watch with a second hand

Experiment 1: The Rolling Marble

In this experiment you'll write the equation for the path of a falling marble. Then you'll catch the marble at a point you calculate using your equation.

> **Procedure Note**
>
> Use the books and poster board to build a ramp about 30 cm from the edge of the table. Fold the sheet of paper into fan pleats—the smaller the pleats, the better. This paper, when unfolded, will help you locate where the marble hits the floor.

Step 1	Do a trial run. Roll the marble from the top edge of the ramp. Let it roll down the ramp and across the table and drop to the floor. Spot the place where it hits the floor (approximately). Tape the folded paper to the floor in this area.
Step 2	Now collect data to fix the drop point more precisely. Roll the marble two or three times, and mark the point where it hits the paper. Each roll should be as much like the other rolls as you can make it. So start each marble roll at the same place, and release it the same way each time.
Step 3	Next, find the coordinates of points for a graph. Let x represent horizontal distance, and let y represent vertical distance. Locate the point on the floor directly below the *edge* of the table. Call this point $(0, 0)$. Measure from $(0, 0)$ up to the point at the edge of the table where the marble rolls off. Name the coordinates of this point. Lastly, measure from $(0, 0)$ to each point where the marble hit the floor. Find the average coordinates for these points on the floor.
Step 4	As your marble falls, it will follow the path of a parabola. The point where it leaves the table is the vertex of the parabola. Define variables and write an equation that fits your two points.

Next, you'll test your model by using it to calculate a point on the path of the marble. See if you can catch the marble at that point.

Step 5	Measure the height of the chair seat. Put the chair next to the table and place a small cup on the chair. Use your calculations to adjust the chair so that when you roll the marble, it will land in the cup.
Step 6	You have only one chance to land in the cup. Release the marble as you did in Step 2. Good luck!

Experiment 2: Walking

In this experiment you'll walk past a motion sensor and model the data you collect.

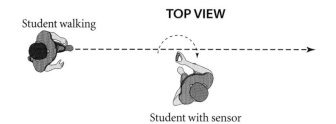

Student walking

TOP VIEW

Student with sensor

Step 1	Walk steadily in the same direction toward the sensor. Pass it and go about 3 meters farther. Record data for the entire walking time.
Step 2	Download the data to each person's calculator. You should expect some erratic data points while the walker is close to the sensor.
Step 3	Fit the data using function transformations. If the vertex is "missing" from the data, estimate its location.

| Step 4 | Write walking instructions for each of these functions. In your instructions say where to start, how fast to go, and when to pass the sensor. |

a. $f(x) = 1.5|x - 1.2|$ **b.** $f(x) = 2.1|x - 0.85|$

| Step 5 | If time permits, try following your instructions to see if the graph fits your data. |

Experiment 3: Calculating

Who is the fastest calculator operator (CO) in your group? The COs had better warm up their fingers!

> **Procedure Note**
>
> You must start with $1 + 2 + 3 \ldots$ each time. It is not fair to use the last result!

| Step 1 | The CO should carefully calculate sums a–g. Record the answers. |

a. $1 + 2 + 3 + \ldots + 8 + 9 + 10$

b. $1 + 2 + 3 + \ldots + 13 + 14 + 15$

c. $1 + 2 + 3 + \ldots + 18 + 19 + 20$

d. $1 + 2 + 3 + \ldots + 23 + 24 + 25$

e. $1 + 2 + 3 + \ldots + 28 + 29 + 30$

f. $1 + 2 + 3 + \ldots + 33 + 34 + 35$

g. $1 + 2 + 3 + \ldots + 38 + 39 + 40$

| Step 2 | Next, the CO calculates the first sum, 1 to 10, again *while being timed*. (Record the time only if the CO gets the correct answer. If not, run the trial again.) |

| Step 3 | Repeat Step 2 for sums b–g, that is, 1 to 15, 1 to 20, . . . , 1 to 40. You should have seven data points in the form (*number of numbers added, time*). |

| Step 4 | Find an equation to model the data. Transform it as needed for a better fit. |

| Step 5 | Use your model to predict the time it would take to sum the numbers from 1 to 47. Test your prediction and record the results. |

| Step 6 | What is the *y*-intercept of your model? Does this value have any real-world meaning? If yes, then what is the meaning? If no, then why not? |

Discuss your results with the class. How were the experiments alike? How were they different? How could you recognize the parent function in the data?

Introduction to Rational Functions

In Chapter 3, you learned about inverse variation. The simplest inverse variation equation is $y = \frac{1}{x}$. Look at the graph of this equation.

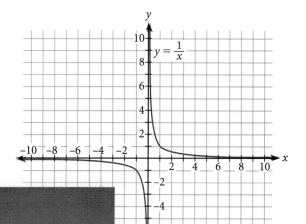

Notice that the graph of $y = \frac{1}{x}$ has two parts. One part is in Quadrant I, and the other is in Quadrant III. In Chapter 3, you wrote inverse variation equations for countable and measurable quantities, such as number of nickels and distance in inches. Since these quantities are always positive, you worked only with the part of the graph in Quadrant I.

Notice that as the x-values get closer and closer to 0, the graph gets closer and closer to the y-axis. As the x-values get farther and farther from 0, the graph gets closer and closer to the x-axis. An **asymptote** is a line that a graph approaches more and more closely. So the graph of $y = \frac{1}{x}$ has two asymptotes: the lines $x = 0$ and $y = 0$. Can you explain why the x- or y-axes are asymptotes for this graph?

Also notice that $y = \frac{1}{x}$ is a function because it passes the vertical line test. You can use the inverse variation function as a parent function to understand many other functions.

Some amusement parks have free-fall rides shaped like a first-quadrant inverse variation graph. This is the Demon Drop at Cedar Point Amusement Park in Ohio.

Investigation
I'm Trying to Be Rational

In the first part of this investigation, you will explore transformations of the parent function $y = \frac{1}{x}$.

Procedure Note

For this investigation, use a friendly window with a factor of 2.

| Step 1 | Graph the parent function $y = \frac{1}{x}$ on your calculator. |

| Step 2 | Use what you have learned about transformations to predict what the graphs of these functions will look like. |

    a. $y = \dfrac{-3}{x}$ **b.** $y = \dfrac{2}{x} + 3$ **c.** $y = \dfrac{1}{x - 2}$ **d.** $y = \dfrac{1}{x + 1} - 2$

Step 3	Graph each equation in Step 2 on your calculator along with $y = \frac{1}{x}$. Compare the graph to your prediction in Step 2. How can you tell where asymptotes occur on your calculator screen?
Step 4	Without graphing, describe what the graphs of these functions will look like. Use the words *linear, nonlinear, increasing,* and *decreasing.* Define the domain and range. Give equations for the asymptotes. **a.** $y = \dfrac{5}{x - 4}$ **b.** $y = \dfrac{-1}{x + 3} - 5$ **c.** $y = \dfrac{a}{(x - h)} + k$
Step 5	Do you think the equations in Step 2 and Step 4 are inverse variations? Explain why or why not.

The function $y = \frac{5}{x - 4}$ is an example of a **rational function** because it shows a ratio between two expressions, 5 and $x - 4$. Not all rational functions are transformations of $y = \frac{1}{x}$, but you will see similarities in their graphs.

Step 6	Graph this equation on your calculator. $$y = \frac{(x + 3)}{(x - 2)(x + 3)}$$
Step 7	The graph should look familiar. What graph does it look like?
Step 8	Trace the graph on your calculator. Describe anything unusual that you find. Can you explain your findings?

Step 9	Graph this equation on your calculator. $$y = \frac{1}{(x + 5)(x - 1)} + 4$$
Step 10	Describe the graph. Explain anything unusual that you find.
Step 11	Rational functions often have asymptotes. Some even have holes, that is, x-values for which there are no y-values. Describe how you can tell if a graph has an asymptote or a hole just by looking at the equation and how you can tell where the asymptote or hole will be.
Step 12	Write a rational function that has asymptotes at $x = 2$, $x = -1$, and $y = -3$, and a hole at $x = 8$.

You can use rational equations to model many real-world applications.

EXAMPLE

A salt solution is made from salt and water. A bottle contains 1 liter of a 20% salt solution. This means that the concentration of salt is 20%, or 0.2, of the whole solution.

a. Show what happens to the concentration of salt as you add water to the bottle in half-liter amounts.

b. Find an equation that models the concentration of salt as you add water.

c. How much water should you add to get a 2.5% salt solution?

Mono Lake is a natural saltwater lake located near Lee Vining, California.

▶ Solution

a. Use a table to show what happens. The bottle originally contains 20% salt, or 0.2 liter. As you add water, the amount of salt stays the same, but the amount of whole solution increases. Each time you add water, recalculate the concentration of salt by finding the ratio of salt to whole solution.

Added amount of water (L)	0	0.5	1.0	1.5	2.0	2.5	3.0	3.5	4.0	4.5	5.0
Amount of salt (L)	0.2	0.2	0.2	0.2	0.2	0.2	0.2	0.2	0.2	0.2	0.2
Whole solution (L)	1.0	1.5	2.0	2.5	3.0	3.5	4.0	4.5	5.0	5.5	6.0
Concentration of salt	0.2	0.133	0.1	0.08	0.067	0.057	0.05	0.044	0.04	0.036	0.033

b. As the amount of whole solution increases, the concentration of salt decreases, but the *amount* of salt stays the same. This is an inverse variation, and the constant of variation is the amount of salt.

$$concentration = \frac{salt}{whole\ solution}$$

The equation you need to write should show a relationship between the amount of water you add, x, and the concentration of salt, y. From the table, you can see that the amount of whole solution starts at 1 liter and increases by the amount of water you add. The equation is

Concentration of salt ⟶ $y = \dfrac{0.2}{1+x}$

Constant amount of salt

Added amount of water
Starting amount of solution ⎱ Whole solution

A graph of the data points and of the equation confirms that this equation is a perfect model.

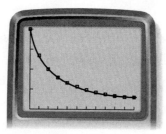

$$[0, 5.5, 0.5, 0, 0.2, 0.05]$$

This equation is not an inverse variation because the product of x and y is not constant. It is, however, a transformation of the parent inverse variation function.

c. Use the equation to find the amount of water that you should add. A 2.5% salt solution has a concentration of salt of 0.025.

$$0.025 = \frac{0.2}{1 + x}$$ Substitute 0.025 for y.

$$0.025 + 0.025x = 0.2$$ Multiply both sides by $(1 + x)$ and distribute.

$$x = 7$$ Solve for x.

You would need to add 7 liters of water to have a 2.5% salt solution.

EXERCISES

You will need your calculator for problems **4, 10, 14,** and **15.**

▶ Practice Your Skills

1. Describe each graph as a transformation of the graph of the parent function $y = |x|$ or $y = x^2$. Then write its equation.

a.

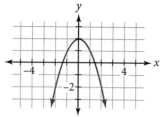

b.

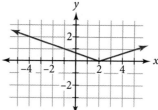

c.

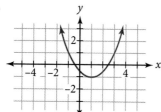

d.

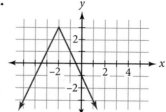

2. Write an equation that generates this table of values.

x	-4	-3	-2	-1	0	1	2	3	4
y	$-\frac{1}{2}$	$-\frac{2}{3}$	-1	-2	Undefined	2	1	$\frac{2}{3}$	$\frac{1}{2}$

3. Write an equation for this graph in the form $y = \frac{a}{x}$.

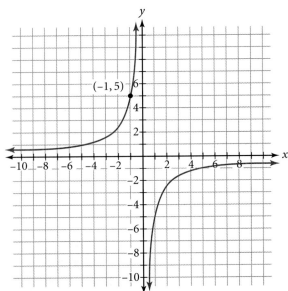

4. Describe each function as a transformation of the graph of the parent function $y = \frac{1}{x}$. Then sketch a graph of each function. Check your answers by graphing on your calculator.

a. $y = \frac{4}{x}$

b. $y = \frac{1}{x - 5} - 2$

c. $y = \frac{0.5}{x} + 3$

d. $y = \frac{-3}{x + 3}$

5. Describe each of the functions in problem 4 as increasing or decreasing. Give equations for the asymptotes, and give the domain and range.

Reason and Apply

6. Write an equation for each graph. Each calculator screen shows a friendly window with a factor of 1.

a.

b.

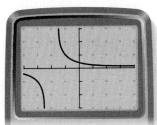

c.

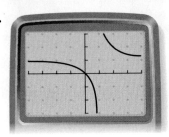

d.

7. Consider the graph of the inverse variation function $f(x) = \frac{1}{x}$. (See page 513.)

    a. Write an equation that reflects the graph across the x-axis. Sketch the image.

    b. Write an equation that reflects the graph across the y-axis. Sketch the image.

    c. Compare your sketches from 7a and 7b. Explain what you find.

8. Write equations for the graphs that meet the descriptions in 8a–d.

    a. A vertical stretch of the graph of $y = \frac{1}{x}$ by a factor of 5. Then a translation right 10 units and down 100 units.

    b. A transformation of the graph of $y = \frac{1}{x}$ that has asymptotes at $x = -3$ and $y = 0$. The graph shows an increasing function.

    c. A rational function that has asymptotes at $x = 2$, $x = 3$, $x = -4$, and $y = 2$.

    d. Looks like $y = \frac{1}{x}$ but has a hole where $x = 1$.

9. APPLICATION A nurse needs to treat a patient's eye with a 1% saline solution (salt solution). She finds only a half-liter bottle of 5% saline solution. Write an equation and use it to calculate how much water she should add to create a 1% solution.

10. Solve this equation symbolically.

$$-95 = \frac{5}{x - 10} - 100$$

Check your solution using a calculator graph or table.

11. APPLICATION A business group wants to rent a meeting hall for its job fair during the week of spring break. The rent is $3500, which will be divided among the businesses that agree to participate. So far, only five businesses have signed up.

The saline solution that is used to clean contact lenses is usually a 1% salt solution.

    a. At this time, what is the cost for each business?

    b. Make a table to show what happens to the cost per business as the additional businesses agree to participate.

    c. Write a function for the cost per business related to the number of additional businesses that agree to participate.

    d. How many additional businesses must agree to participate before the cost per business is less than $150?

12. The intensity, *I*, of a 100-watt light bulb is related to the distance, *d*, from which it is measured. This rational function shows the relationship when intensity is measured in lux (lumens per square meter) and distance is measured in meters.

$$I = \frac{90}{d^2}$$

    a. Find the intensity of the light 4 meters from the bulb.

    b. Find the distance from the bulb if the intensity of the light measures 20 lux.

▶ Review

13. Solve each inequality.

    a. $4 - 2x > 8$

    b. $-8 + 3(x - 2) \geq -20$

    c. $7 + 2x \leq 3 + 3x$

14. Name the coordinates of the vertex of the graph of $y = 2(x - 3)^2 + 1$. Without graphing, name the points on the parabola whose x-coordinates are 1 unit more or less than that of the vertex. Check your answers by graphing.

15. Name the coordinates of the vertex of the graph of $y = -3\,|x + 1| + 2$. Without graphing, name the points on the graph whose x-coordinates are 1 unit more or less than that of the vertex. Check your answers by graphing.

IMPROVING YOUR **VISUAL THINKING** SKILLS

Describe each striped or plaid fabric pattern as a set of transformations. Which patterns are translations? Which are reflections?

Fabric A

Fabric B

Fabric C

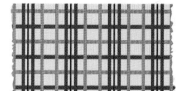

Fabric D

What is the smallest rectangular "unit" that repeats throughout each pattern? Can there be more than one "unit" for a pattern? Suppose a tailor is making a shirt from each fabric pattern. Which shirt should be most expensive? Why?

Transformations with Matrices

You can use a matrix to organize the coordinates of a geometric figure. You can represent this quadrilateral with a 2 × 4 matrix.

$$\begin{bmatrix} 1 & -2 & -3 & 2 \\ 2 & 1 & -1 & -2 \end{bmatrix}$$

Each column contains the x- and y-coordinates of a vertex. The first row contains all the x-coordinates, and the second row contains all the y-coordinates. All four vertices are in consecutive order in the matrix. When you add or multiply this matrix, the coordinates change. So matrices are useful when you transform coordinates.

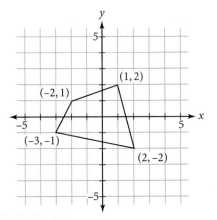

Many textiles, such as this Turkish carpet, use transformations to create interesting patterns.

Investigation
Matrix Transformations

In this investigation you'll use matrix addition and multiplication to create some familiar transformations.

You will need

• graph paper

Step 1 | Create a set of coordinate axes on graph paper. Draw the triangle that is represented by this matrix:

$$[A] = \begin{bmatrix} -4 & 3 & 2 \\ -1 & 4 & 0 \end{bmatrix}$$

Step 2 | Add.

$$[A] + \begin{bmatrix} 5 & 5 & 5 \\ 0 & 0 & 0 \end{bmatrix}$$

Step 3	Draw the image represented by your answer in Step 2. Describe the transformation.
Step 4	How is the transformation related to the matrix that you added?
Step 5	Repeat Steps 1–4, but in Step 2 change what you add to matrix [A] each time. Use a new set of coordinate axes for each transformation.

a. $[A] + \begin{bmatrix} 0 & 0 & 0 \\ -4 & -4 & -4 \end{bmatrix}$ **b.** $[A] + \begin{bmatrix} 5 & 5 & 5 \\ -4 & -4 & -4 \end{bmatrix}$

c. $[A] + \begin{bmatrix} -6 & -6 & -6 \\ 4 & 4 & 4 \end{bmatrix}$

Next, see if you can work backward.

Step 6	Write matrix equations to represent these translations.

a.

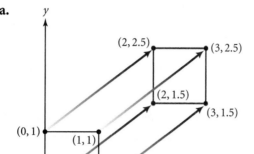

b.

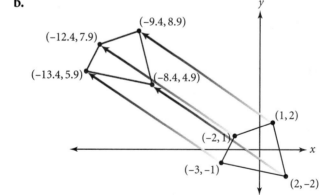

Now, see what effect multiplication has.

Step 7	Draw this quadrilateral on your own graph paper. Write the coordinates of the vertices in a matrix, [B]. Add a fifth column to your matrix to represent any point of the form (x, y).
Step 8	Multiply.

$$\begin{bmatrix} 1 & 0 \\ 0 & -1 \end{bmatrix} \cdot [B]$$

Step 9 | Draw the image represented by your answer in Step 8. Describe the transformation that happened.

Step 10 | How is the last column of the image matrix related to the transformations you made using lists in this chapter?

Step 11 | Repeat Steps 7–10, but in Step 8 change what you multiply by matrix $[B]$ each time. Use a new set of coordinate axes for each transformation.

a. $\begin{bmatrix} -1 & 0 \\ 0 & 1 \end{bmatrix} \cdot [B]$ **b.** $\begin{bmatrix} 1 & 0 \\ 0 & 0.5 \end{bmatrix} \cdot [B]$ **c.** $\begin{bmatrix} 0.5 & 0 \\ 0 & 2 \end{bmatrix} \cdot [B]$

EXERCISES

You will need your calculator for problem **5.**

▶ Practice Your Skills

1. The matrix $\begin{bmatrix} -2 & 1 & -2 \\ 2 & 2 & 6 \end{bmatrix}$ represents a triangle.

   a. Name the coordinates and draw the triangle.

   b. What matrix would you add to translate the triangle down 3 units?

   c. Calculate the matrix for the image if you translate the triangle down 3 units.

2. Refer to these triangles.

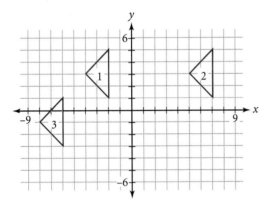

   a. Write a matrix to represent triangle 1.

   b. Write the matrix equation to translate from triangle 1 to triangle 2.

   c. Write the matrix equation to translate from triangle 1 to triangle 3.

3. Add or multiply.

   a. $[4 \quad 7] + [2 \quad 8]$ **b.** $\begin{bmatrix} 4 \\ 7 \end{bmatrix} + \begin{bmatrix} 2 \\ 8 \end{bmatrix}$ **c.** $[4 \quad 7] \cdot \begin{bmatrix} 2 \\ 8 \end{bmatrix}$ **d.** $\begin{bmatrix} 4 \\ 7 \end{bmatrix} \cdot [2 \quad 8]$

4. In the investigation you saw that matrix multiplication can result in a reflection.

   a. What matrix reflects a figure across the y-axis?

   b. What matrix reflects a figure across the x-axis?

Reason and Apply

5. The matrix $\begin{bmatrix} -1 & 2 & 1 & -2 \\ 2 & -1 & -2 & 1 \end{bmatrix}$ represents a quadrilateral.

    a. What kind of quadrilateral is it?

    b. Without using your calculator, tell how to find the image of the point $(2, -1)$ when you multiply $\begin{bmatrix} 1 & 0 \\ 2 & 2 \end{bmatrix} \cdot \begin{bmatrix} -1 & 2 & 1 & -2 \\ 2 & -1 & 2 & 1 \end{bmatrix}$. What are the coordinates of this point's image? In what row and column of the image matrix will you find the new x-coordinate? The new y-coordinate?

    c. Multiply $\begin{bmatrix} 1 & 0 \\ 0 & 2 \end{bmatrix} \cdot \begin{bmatrix} -1 & 2 & 1 & -2 \\ 2 & -1 & 2 & 1 \end{bmatrix}$. Check your work with your calculator.

    [▶ ▢] See **Calculator Note 1P** to review matrix multiplication. ◀

    d. Draw the image represented by your answer to 5c. What kind of polygon is it?

6. Consider this square.

    a. Write a matrix, $[S]$, to represent it.

    b. Describe the transformation when you calculate

        i. $\begin{bmatrix} 1 & 0 \\ 0 & 3 \end{bmatrix} \cdot [S]$ **ii.** $\begin{bmatrix} 1 & 0 \\ 0 & -3 \end{bmatrix} \cdot [S]$

        iii. $\begin{bmatrix} 4 & 0 \\ 0 & 2 \end{bmatrix} \cdot [S]$ **iv.** $[S] + \begin{bmatrix} 4 & 4 & 4 & 4 \\ 2 & 2 & 2 & 2 \end{bmatrix}$

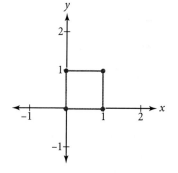

7. Consider this quadrilateral.

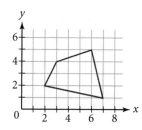

    a. Write a matrix, $[Q]$, to represent it.

    b. Write a matrix multiplication equation that will vertically shrink the quadrilateral by a factor of 0.5.

    c. Write a matrix multiplication equation that will both vertically and horizontally shrink the quadrilateral by a factor of 0.5.

    d. Multiply $\begin{bmatrix} 1 & 0 \\ 0 & -1 \end{bmatrix} \cdot \begin{bmatrix} -1 & 0 \\ 0 & 1 \end{bmatrix}$. Then multiply the result by matrix $[Q]$. Draw the image of the quadrilateral. Describe the transformation that happened. How is the transformation related to the matrices $\begin{bmatrix} 1 & 0 \\ 0 & -1 \end{bmatrix}$ and $\begin{bmatrix} -1 & 0 \\ 0 & 1 \end{bmatrix}$?

    e. Multiply $\begin{bmatrix} 0 & -1 \\ 1 & 0 \end{bmatrix} \cdot [Q]$. Draw the image. Describe the transformation.

8. The points $(-2, 4)$, $(-1, 1)$, $(0, 0)$, $(1, 1)$, and $(2, 4)$ are on a parabola.

   a. What is the equation of the parabola that passes through these points?

   b. Write a matrix, $[P]$, to organize the coordinates of these points.

   c. Add $[P] + \begin{bmatrix} 3 & 3 & 3 & 3 & 3 \\ 2 & 2 & 2 & 2 & 2 \end{bmatrix}$. Write an equation for the parabola that passes through the points organized in the image matrix.

   d. Multiply $\begin{bmatrix} -1 & 0 \\ 0 & 1 \end{bmatrix} \cdot [P]$. Write an equation for the parabola that passes through the points organized in the image matrix.

9. In problem 7e, you saw a transformation called a **rotation.** A rotation turns a figure about a point called the *center.* The center of a rotation can be inside, outside, or on the figure that is rotated.

   a. Draw a polygon of your own design on graph paper. Represent your polygon with a matrix, $[R]$.

   b. Multiply $\begin{bmatrix} 0.5 & -0.866 \\ 0.866 & 0.5 \end{bmatrix} \cdot [R]$.

   Draw the image of your polygon.

   c. The transformation matrix in 9b rotates the polygon. How many degrees is it rotated? What point is the center of the rotation?

   d. Multiply $\begin{bmatrix} 0.5 & -0.866 \\ 0.866 & 0.5 \end{bmatrix} \cdot \begin{bmatrix} 0.5 & -0.866 \\ 0.866 & 0.5 \end{bmatrix}$ and round the entries in the answer matrix to the nearest thousandth. Then multiply the result by matrix $[R]$. Draw the image of the polygon. Describe the transformation.

   e. Describe how you could rotate your polygon 180° (a half-turn).

   f. Describe how you could rotate your polygon 360° so that the image is the same as the original polygon.

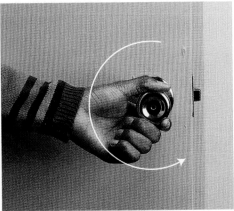

Did you know that you use rotations every day? Opening a door, turning a faucet, tightening a bolt with a wrench—all of these require rotations. The rotational force you use to do these things is called *torque.* Think of other everyday situations that require rotations.

Review

10. Tacoma and Jared are doing a "walker" investigation. Tacoma starts 2 m from the motion sensor. He walks away at a rate of 0.5 m/sec for 6 sec. Then he walks back toward the sensor at a rate of 0.5 m/sec for 3 sec.

   a. Sketch a time-distance graph for Tacoma's walk.

   b. Write an equation that fits the graph.

11. This table shows the approximate population of the ten most populated countries in 1999.

   a. Give the five-number summary.

   b. Make a box plot of the data.

   c. Are there any outliers?

Most Populated Countries, 1999

Country	Population (millions)
China	1,247
India	1,001
United States	273
Indonesia	216
Brazil	172
Russia	146
Pakistan	138
Bangladesh	127
Japan	126
Nigeria	114

(*2000 World Almanac*, pp. 878–879)

TILES

Some floor tiles are simple polygons, like squares. Others have more complex shapes, with curves or unusual angles. But all tiles have one thing in common—they fit together without gaps or overlap.

You can use transformations to create your own tile shape. Start with a polygon that works as a tile. For example, you can start with a rhombus and use transformations to create a complex shape that still works as a tile. In this example, a design drawn on the right side of the rhombus is translated and copied on the left side; a translation is also used for the top and bottom. The result is an interesting shape that still fits together.

The Geometer's Sketchpad was used to create these tiles. Sketchpad has tools to help you create simple polygons and apply transformations. Learn how to use these tools to create your own tiling pattern.

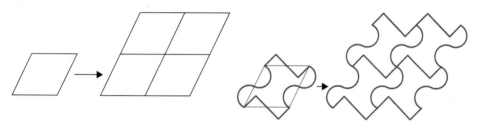

Your project should include

▶ Your tile pattern. Show what a single tile looks like and how several tiles look when they are joined together.

▶ A report of how you created your tile. What polygon did you start with? What transformations did you use?

For an extra challenge, start with a polygon that is not a quadrilateral, like a triangle. Or try other transformations or combinations of transformations. Dot paper, graph paper, a computer drawing program, or The Geometer's Sketchpad software are useful tools for this project.

Learn more about the mathematics of tilings with the Internet links at
www.keymath.com/DA .

CHAPTER
9
REVIEW

In this chapter you moved individual points, polygons, and graphs of functions with **transformations.** You learned to **translate, reflect, stretch,** and **shrink** a **parent function** to create a **family of functions** based on it. For example, if you know what the graph of $y = x^2$ looks like, understanding transformations gives you the power to know what the graph of $y = 3(x + 2)^2 - 4$ looks like.

You transformed the graphs of the parent functions $y = |x|$ and $y = x^2$ to create many different absolute value and squaring functions. You can apply the same transformations to the graphs of other parent functions, like $y = x$ or $y = 2^x$, to create many different linear or exponential functions. You can even fit an equation to data by transforming a simple graph into a graph that fits the data better.

You learned that the inverse variation function, $y = \frac{1}{x}$, is one type of **rational function.** The graphs of most rational functions have **asymptotes;** some even have holes. Understanding transformations helps you know where asymptotes and holes will occur.

Finally, you used matrices to organize the coordinates of points and to do transformations. You can use matrices to do translations, reflections, stretches, and shrinks. You can also use them to do more complex transformations, like **rotations.**

EXERCISES

You will need your calculator for problems **4** and **7.**

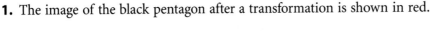

For problems 1 and 2, consider the black pentagon below as the original figure.

1. The image of the black pentagon after a transformation is shown in red.

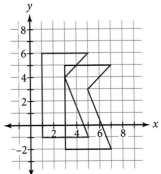

a. Describe the transformation.

b. Define the coordinates of the image using the coordinates of the original figure.

2. Here are three more transformations of the black pentagon from problem 1.

i.

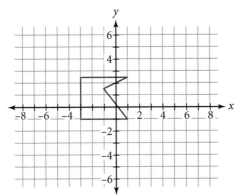

ii.

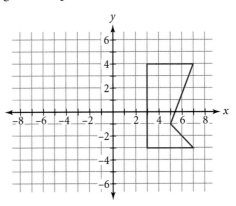

iii.

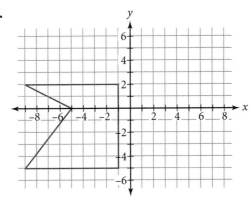

a. Describe the transformations.

b. Patty plots the original pentagon on her calculator. She uses list L1 for the x-coordinates of the vertices and list L2 for the y-coordinates. Tell Patty how to define list L3 and list L4 for each image shown above.

3. You can create this figure on a calculator by connecting four points. Assume the x-coordinates of each point are entered into list L1 and the corresponding y-coordinates are entered into list L2. Explain how to make an image that is

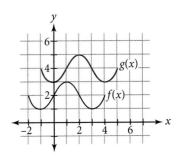

a. A reflection across the x-axis.

b. A reflection across the y-axis.

c. A reflection across the x-axis and a translation right 3 units.

4. Describe each function as a transformation of the graph of the parent function $y = |x|$ or $y = x^2$. Then sketch a graph of each function. Check your answers by graphing on your calculator.

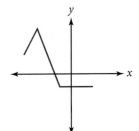

a. $y = 2|x| + 1$ **b.** $y = -|x + 2| + 2$
c. $y = 0.5(-x)^2 - 1$ **d.** $y = -(x - 2)^2 + 1$

5. At right, the graph of $g(x)$ is a transformation of the graph of $f(x)$. Write an equation for $g(x)$ in terms of $f(x)$.

6. Write the equation for each graph.

a.

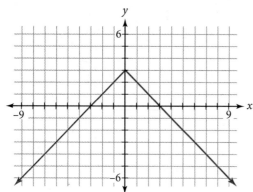

b.

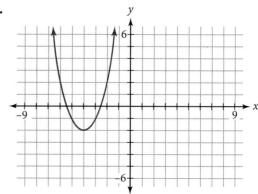

c.

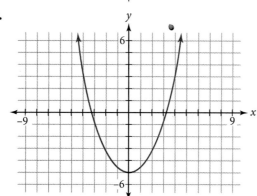

d.

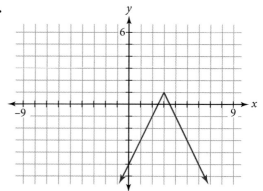

7. Consider the graph of $f(x)$ at right.

  a. Sketch the graph of $-f(x)$.

  b. Enter a linear function into Y1 on your calculator to create a graph like $f(x)$. Enter Y2 $= -$Y1 and graph it too. Describe your results.

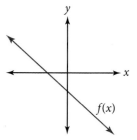

8. Describe each function as a transformation of the graph of the parent function $y = \frac{1}{x}$. Give equations for the asymptotes.

  a. $y = \dfrac{1}{x - 3}$ **b.** $y = \dfrac{3}{x + 2}$ **c.** $y = \dfrac{1}{x - 5} - 2$

9. Describe each graph as a transformation of the graph of the parent function $y = 2^x$ or $y = \frac{1}{x}$. Then write an equation for each graph.

a.

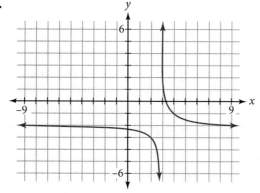

b.

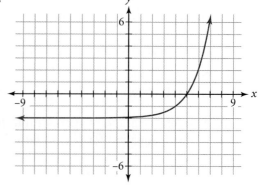

c.

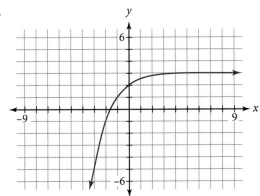

d.

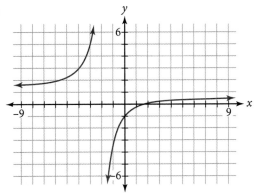

10. Consider this square.

    a. Write a matrix, $[A]$, that represents this square.

    b. Describe the transformation when you calculate

        i. $\begin{bmatrix} 1 & 0 \\ 0 & 1 \end{bmatrix} \cdot [A]$ **ii.** $\begin{bmatrix} -1 & 0 \\ 0 & -1 \end{bmatrix} \cdot [A]$

        iii. $\begin{bmatrix} 1 & 0 \\ 0 & 3 \end{bmatrix} \cdot [A]$ **iv.** $[A] + \begin{bmatrix} 1 & 1 & 1 & 1 \\ 1 & 1 & 1 & 1 \end{bmatrix}$

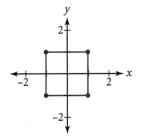

TAKE ANOTHER LOOK

In this chapter you saw reflections across the *x*-axis and across the *y*-axis. You also saw reflections across other vertical and horizontal lines (see problem 11 in Lesson 9.3). Let's examine another very important line of reflection.

Here is the graph of a function in black. The red image was created by a reflection across the dotted line. What is the equation of the line of reflection?

Identify at least three points on the graph of $y = f(x)$. Then name the image of each point after the reflection. How would you define the coordinates of the image based on the coordinates of the original graph?

The image that results from this type of reflection is called an **inverse.** Is the inverse of a function necessarily a function too? Find an example of a function whose inverse is also a function. Find an example of a function whose inverse is not a function.

Learn more about inverse functions with the links at www.keymath.com/DA .

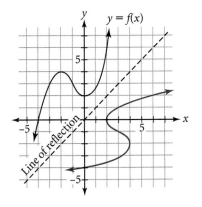

Mirrors are used to create reflections. This mirror helps drivers see around a corner on an Italian street.

Assessing What You've Learned

 ORGANIZE YOUR NOTEBOOK Organize your notes on each type of transformation that you have learned about. Review how each transformation affects individual points and how it changes the equation of a function. Then create a table that summarizes your notes. Use rows for each type of transformation and columns to show effects on points and equations. You can use subrows and subcolumns to further organize the information. For example, you might want to use one row for reflections across the x-axis and another row for reflections across the y-axis. You might want to use one column for changes to $y = f(x)$ and other columns for changes to specific functions like $y = x^2$, $y = |x|$, or $y = \frac{1}{x}$.

 UPDATE YOUR PORTFOLIO Choose one piece of work that illustrates each transformation that you have studied in this chapter. Add these to your portfolio. Describe each work in a cover sheet, giving the objective, the result, and what you might have done differently.

 PERFORMANCE ASSESSMENT Show a classmate, a family member, or your teacher how you can transform a single parent function into a whole family of functions. Explain how you can write a function for a graph by identifying the transformations. In contrast, show how you can sketch a graph just by looking at the equation.

Quadratic Models

Buckingham Fountain in
Chicago's Grant Park contains
1.5 million gallons of water.
When pumped through one of the
fountain's 133 jets, the water forms
the shape of a parabola as it falls back
into the pool. The central spout shoots
135 feet in the air. The relationship
between time and the height of free
falling objects in the air is described
by quadratic equations.

OBJECTIVES

In this chapter you will

- model applications with
  quadratic functions
- compare features of
  parabolas to their
  quadratic equations
- learn strategies for solving
  quadratic equations
- learn how to combine and
  factor polynomials
- make connections between
  some new polynomial
  functions and their graphs

Solving Quadratic Equations

When you throw a ball straight up into the air, its height depends on three major factors—its starting position, the velocity at which it leaves your hand, and the force of **gravity.** The Earth's gravity causes objects to accelerate downward, gathering speed every second. This acceleration due to gravity, called g, is 32 ft/sec². It means that the object's downward speed increases 32 ft/sec *for each second* in flight. If you plot the height of the ball at each instant of time, the graph of the data is a parabola.

EXAMPLE A

A baseball batter pops a fly ball straight up. The ball reaches a height of 68 feet before falling back down. Roughly 4 seconds after it is hit, the ball bounces off home plate. Sketch a graph that models the ball's height in feet during its flight time in seconds. When is the ball 68 feet high? How many times will it be 20 feet high?

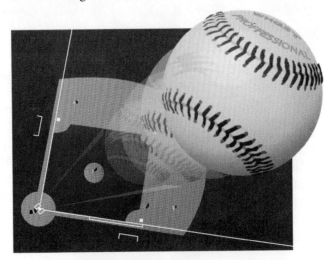

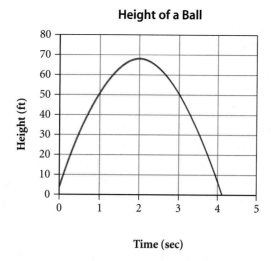

Height of a Ball

The sketch above pictures the ball's height once it is hit and before it lands on the ground. When the bat hits the ball, it is a few feet above the ground. So the y-intercept is just above the origin. The ball's height is 0 when it hits the ground just over 4 seconds later. So the parabola crosses the x-axis near the coordinates $(4, 0)$. The ball is at its maximum height of 68 feet after about 2 seconds, or halfway through its flight time. So the vertex of the parabola is near $(2, 68)$.

The ball reaches a height of 20 feet twice—once on its way up and again on its way down. If you sketch $y = 20$ on the same set of axes, you'll see that this line crosses the parabola at two points.

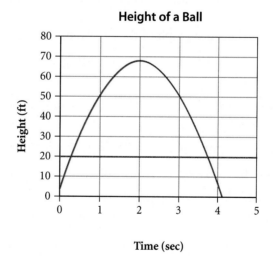

Height of a Ball

The parabola in Example A is a transformation of the equation $y = x^2$. The function $f(x) = x^2$ and transformations of it are called **quadratic functions,** because the highest power of x is x-squared. The Latin word meaning "to square" is *quadrare.* The function that describes the motion of a ball, and many other projectiles, is a quadratic function. You will learn more about this function in the investigation.

Investigation
Rocket Science

A model rocket blasts off from a position 2.5 meters above the ground. Its starting velocity is 49 meters per second. Assume that it travels straight up and that the only force acting on it is the downward pull of gravity. In the metric system, the acceleration due to gravity is 9.8 m/sec². The quadratic function $h(t) = \frac{1}{2}(-9.8)t^2 + 49t + 2.5$ describes the rocket's **projectile motion.**

Step 1	Define the function variables and their units of measurement for this situation.
Step 2	What is the real-world meaning of $h(0) = 2.5$?
Step 3	How is the acceleration due to gravity, or g, represented in the equation? How does the equation show that this force is *downward*?

Known as the father of rocketry, Robert Hutchings Goddard fired the first successful liquid-fueled rocket in 1926.

Next you'll make a graph of the situation.

Step 4	Graph the function $h(t)$. What viewing window shows all the important parts of the parabola?
Step 5	How high does the rocket fly before falling back to Earth? When does it reach this point?
Step 6	How much time passes while the rocket is in flight?
Step 7	Write the equation you must solve to find when $h(t) = 50$.
Step 8	When is the rocket 50 meters above the ground? Use a calculator table to approximate your answers to the nearest tenth of a second.
Step 9	Describe how to answer Step 8 graphically.

In the investigation you approximated solutions to a quadratic equation using tables and graphs. Later in this chapter you will learn to solve quadratic equations in the **general form,** $y = ax^2 + bx + c$, using symbolic manipulation. Until then, quadratic equations must be in a certain form for you to solve them symbolically. You will combine the "undo" and "balance" methods on this form in the next example.

EXAMPLE B | Solve $5(x + 2)^2 - 10 = 47$ symbolically. Check your answers with a graph and a table.

▶ **Solution** | Undo each operation as you would when solving a linear equation. To undo the squaring operation, take the square root of both sides. You will get two possible answers.

$$5(x + 2)^2 - 10 = 47 \qquad \text{The original equation.}$$

$$5(x + 2)^2 - 10 + 10 = 47 + 10 \qquad \text{Add 10.}$$

$$\frac{5(x + 2)^2}{5} = \frac{57}{5} \qquad \text{Divide by 5.}$$

$$(x + 2)^2 = 11.4 \qquad \text{Reduce.}$$

$$\sqrt{(x + 2)^2} = \sqrt{11.4} \qquad \text{Take the square root of both sides.}$$

$$x + 2 = \pm\sqrt{11.4} \qquad \text{The } \pm \text{ symbol shows the two numbers } +\sqrt{11.4} \text{ and } -\sqrt{11.4}, \text{ whose square is 11.4.}$$

$$x = -2 \pm \sqrt{11.4} \qquad \text{Subtract 2 from both sides.}$$

The two solutions are $-2 + \sqrt{11.4}$, or approximately 1.38, and $-2 - \sqrt{11.4}$, or -5.38.

The calculator screens of the graph and the table support each solution.

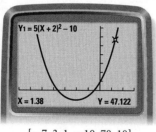
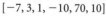

$[-7, 3, 1, -10, 70, 10]$

A symbolic approach allows you to find the exact solutions rather than just approximations from a table or a graph. Exact solutions such as $x = -2 \pm \sqrt{11.4}$ are called **radical expressions** because they contain the square root symbol, $\sqrt{}$, and "radical" comes from the Latin word for "root." As you practice solving quadratic equations symbolically, first think about the order of operations. Then concentrate on what each operation does to the equation and how to undo this order. In some situations only one of the solutions you find has a real-world meaning. Always ask yourself whether the answers you find make sense in real-world situations.

EXERCISES

You will need your calculator for problems **1, 2, 5, 6, 8,** and **9.**

▶ **Practice Your Skills**

1. Use a graph to find the number of solutions for each equation. Explain your answer.

a. $x^2 + 3x - 7 = 11$

b. $-x^2 + x + 4 = 7$

c. $x^2 - 6x + 14 = 5$

d. $-3x^2 - 5x - 2 = -5$

2. For each equation in problem 1, zoom in on a table to approximate the solutions, if they exist, to the nearest hundredth.

3. Use a symbolic method to solve each equation. Show each solution exactly as a radical expression.

    a. $x^2 = 18$ **b.** $x^2 + 3 = 52$ **c.** $(x - 2)^2 = 25$ **d.** $2(x + 1)^2 - 4 = 10$

4. Sketch the graph of a quadratic function with

    a. One x-intercept.

    b. Two x-intercepts.

    c. Zero x-intercepts.

    d. The vertex in the first quadrant and two x-intercepts.

Reason and Apply

5. A baseball is dropped from the top of a very tall building. The ball's height, in meters, t seconds after it has been released is $h(t) = -4.9t^2 + 147$.

    a. Find $h(0)$ and give a real-world meaning for this value.

    b. Solve $h(t) = 20$ symbolically and graphically.

    c. Does your answer to 5b mean the ball is 20 meters above the ground twice? Explain your reasoning.

    d. During what interval of time is the ball less than 20 meters high?

    e. When does the ball hit the ground? Justify your answer with a graph.

6. **APPLICATION** A small rocket is fired into the air from the ground. It reaches its highest point, 108 meters, at 4.70 seconds. It falls back to the ground at 9.40 seconds.

    a. Name three points that the graph goes through.

    b. Name a graphing window that lets you see those three points.

    c. What are the coordinates of the vertex of this parabola?

    d. Find an equation in the form

$$y = a(x - h)^2 + k$$

that fits the three known points. You may need to guess and check to find the value of a.

    e. Find $h(3)$ and give a real-world meaning for this value.

    f. Find the t-values for $h(t) = 47$, and describe the real-world meaning for these values.

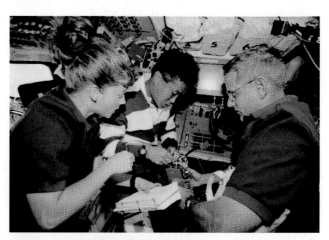

Astronaut trainees Pamela Melroy, Koichi Wakata, and William McArthur are practicing inflight maintenance on a part of a flight deck panel.

7. The path of a ball in flight is given by $p(x) = -0.23(x - 3.4)^2 + 4.2$, where x is the horizontal distance in meters and $p(x)$ is the vertical height in meters. Note that in this case the graph is the path of the ball, not the graph of the ball's height over time.

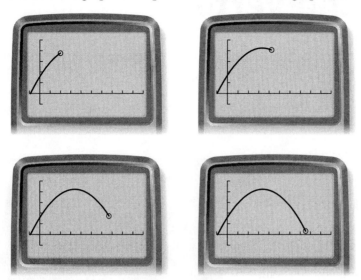

a. Find $p(2)$ and give a real-world meaning for this value.

b. Find the x-values for $p(x) = 2$, and describe their real-world meanings.

c. How high is the ball when it is released?

d. How far will the ball travel horizontally before it hits the ground?

8. Solve the equation $4 = -2(x - 3)^2 + 4$ using

a. A graph. **b.** A table. **c.** Symbolic manipulation.

9. APPLICATION The graph at right shows the parabola for $h(t) = -4.9t^2 + 49t - 97.5$. The variable t represents time in seconds, and $h(t)$ represents the height in meters of a projectile.

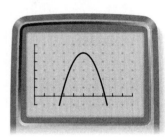

$[0, 10, 1, -5, 30, 5]$

a. What is a real-world meaning for the x-intercepts in the graph?

b. Find the x-intercepts to the nearest 0.01 second.

c. How can you use 9b to find the vertex of this parabola?

d. What is a real-world meaning for the vertex in the graph?

e. What does $h(3.2)$ tell you?

f. When is the projectile 12.5 m high? Explain how to find these solutions on a graph.

Review

10. Show a step-by-step symbolic solution of the inequality $-3x + 4 > 16$.

11. The solid line in the graph passes through $(0, 6)$ and $(6, 1)$. Write an inequality to describe the shaded region.

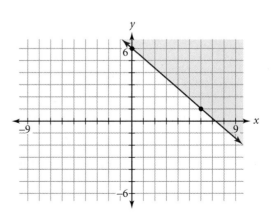

Finding the Roots and the Vertex

In this lesson you will discover that quadratic functions can model relationships other than projectile motions. You will explore relationships between parabolas and their equations. You will practice writing equations, finding x-intercepts, and determining real-world meanings for the x-intercepts and the vertex of a parabola.

Investigation
Making the Most of It

You will need

• graph paper

Suppose you have 24 meters of fencing material and you want to use it to enclose a rectangular space for your vegetable garden. Naturally, you want to have the largest area possible for your vegetables. What dimensions should you use for your garden?

Step 1 | Find the dimensions of at least eight different rectangular areas, each with a perimeter of 24 meters. You must use all of the fencing material for each garden.

Step 2 | Find the area of each garden. Make a table to record the width, length, and area of the possible gardens. It's okay to have widths that are greater than their corresponding lengths.

Width (m)	Length (m)	Area (sq. m)

Step 3 | Enter the data for the possible widths into list L1. Enter the area measures into list L2. Which garden width values would give no area? Add these points to your lists.

Step 4 | Label a set of axes and plot points in the form (x, y), with x representing width in meters and y representing area in square meters. Describe as completely as possible what the graph looks like. Does it make sense to connect the points with a smooth curve?

Step 5 | Where does your graph reach its highest point? Which rectangular garden has the largest area? What are its dimensions?

Next you'll write an equation to describe this relationship.

Step 6 | Describe a relationship between the values for the garden widths and their corresponding lengths. What is the length of the garden that has a width of 2 meters? A width of 4.3 meters? Write an expression for length in terms of width x.

Step 7	Using your expression for the length from Step 6, write an equation for the area of the garden. Enter this equation into Y1 and graph it. Does the graph confirm your answer to Step 5?
Step 8	Locate the points where the graph crosses the *x*-axis. What is the real-world meaning of these points?
Step 9	Do you think the general shape of a garden with a maximum area would change for different perimeters?

In the investigation you found three important points on the graph. The two points on the *x*-axis are called **x-intercepts.** The *x*-values of those points are the solutions of the equation $y = f(x)$ when the function value is equal to zero. These solutions give the **roots** to the equation $f(x) = 0$.

In the investigation the roots are the widths that make the garden area equal to zero. The roots help you to find a third important point—the vertex of the parabola.

In Lesson 10.1, you symbolically solved quadratic equations written in the form $y = a(x - h)^2 + k$. In the next example you will learn to approximate roots of the quadratic equation, $0 = ax^2 + bx + c$.

EXAMPLE A

Use a graph and your calculator's table function to approximate the roots of

$$0 = x^2 + 3x - 5$$

▶ **Solution**

Graph $y = x^2 + 3x - 5$ and find the *x*-intercepts. On the graph you can see that there are two roots— one appears to be a little less than −4, and the other a little greater than 1.

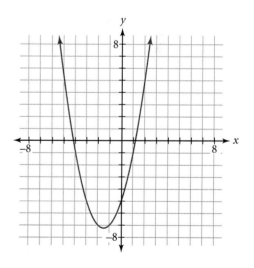

Search in your calculator's table for the positive *x*-value that makes the *y*-value equal to zero. Continue zooming in until you find the positive root, which is about 1.1926. Repeat this process for the negative root, which you'll find to be about −4.1926.

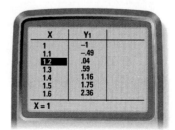 

$$[-5, 5, 1, -5, 5, 1]$$

The line through the vertex that cuts a parabola into two mirror images is called the **line of symmetry.** If you know the roots, you can find the vertex and the line of symmetry.

EXAMPLE B

Find the coordinates (h, k) of the vertex of the parabola $y = x^2 + 3x - 5$. Then write the equation in the form $y = a(x - h)^2 + k$.

▶ Solution

This parabola crosses the x-axis twice and has a vertical line of symmetry. The x-coordinate of the vertex lies on the line of symmetry, halfway between the roots. From Example A you know the two roots are approximately 1.1926 and -4.1926. Averaging the two roots gives -1.5. The graph shows that the line of symmetry goes through this x-value.

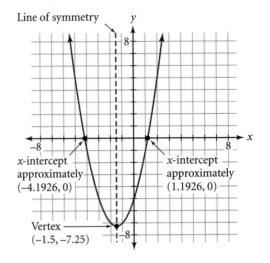

Now use the equation of the parabola, $y = x^2 + 3x - 5$, to find the y-coordinate of the vertex.

$$y = x^2 + 3x - 5 \qquad \text{The equation of the parabola.}$$
$$= (-1.5)^2 + 3(-1.5) - 5 \qquad \text{Substitute } -1.5 \text{ for } x.$$
$$= 2.25 - 4.5 - 5 \qquad \text{Multiply.}$$
$$= -7.25 \qquad \text{Subtract.}$$

So the vertex is $(-1.5, -7.25)$. Sometimes you can find the vertex and see the symmetry in a table of values.

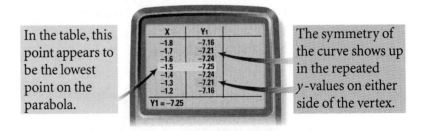

In the table, this point appears to be the lowest point on the parabola.

The symmetry of the curve shows up in the repeated y-values on either side of the vertex.

The graph is a transformation of the parent function, $f(x) = x^2$. The vertex (h, k) is $(-1.5, -7.25)$, so there is a translation left 1.5 units and down 7.25 units. Substitute the values h and k into the equation to get $y = (x + 1.5)^2 - 7.25$. Enter the equation into Y2 and graph it.

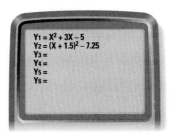
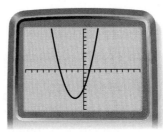

$$[-10, 10, 1, -10, 10, 1]$$

You can see from the graph and the table that the equations $y = x^2 + 3x - 5$ and $y = (x + 1.5)^2 - 7.25$ are equivalent. So the value of a is 1. The equation $y = 1 \cdot (x + 1.5)^2 - 7.25$ is in the **vertex form**, $y = a(x - h)^2 + k$. It tells you that $(-1.5, -7.25)$ is the vertex.

EXERCISES

You will need your calculator for problems **4, 8,** and **11.**

▶ Practice Your Skills

1. What is the x-coordinate of the vertex of the parabola below? The x-intercepts are at 3 and -2. The window shown is $[-4.7, 4.7, 1, -3.1, 3.1, 1]$.

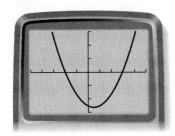

2. The equation for the parabola in problem 1 is $y = 0.4x^2 - 0.4x - 2.4$. Explain how to use the x-coordinate you found in problem 1 to find the y-coordinate of the vertex.

3. Solve $0 = (x + 1.5)^2 - 7.25$ symbolically. Show each step. Compare your solutions with the approximations from Examples A and B.

4. Find the roots of each equation to the nearest thousandth by looking at a graph, zooming in on a table, or both.
   a. $0 = x^2 + 2x - 2$
   b. $0 = -3x^2 - 4x + 3$

5. Solve each equation symbolically and check your answer.
   a. $(x + 3)^2 = 7$
   b. $(x - 2)^2 - 8 = 13$

6. Graph $y = (x + 3)^2$ and $y = 7$. What is the relationship between your solutions to problem 5a and these graphs?

7. The height of a golf ball is given by $h = -16t^2 + 48t$, where t is in seconds and h is in feet.

    a. At what times is the golf ball on the ground?

    b. At what time is the golf ball at its highest point?

    c. How high does the golf ball go?

8. **APPLICATION** Taylor hits a baseball, and its height in the air at time x is given by the equation $y = -16x^2 + 58x + 3$, where x is in seconds and y is in feet. Use the graph and tables to help you answer these questions.

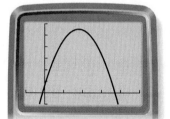
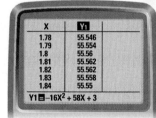
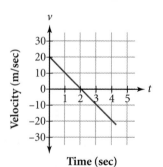

$[-1, 5, 1, -1, 60, 10]$

    a. When does the ball hit the ground?

    b. Use your calculator table to find the answer to three decimal places.

    c. According to the table above, during what time interval is the ball at its highest points? At what time (to the nearest hundredth of a second) is the ball at its highest point, and how high is it?

9. The two graphs at right show aspects of a ball thrown into the air. The first graph shows its height h in meters at any time t in seconds. The second graph shows its velocity v in meters per second at any time t.

    a. What does the first graph tell you about the situation? Use numbers to be as specific as you can.

    b. What does the second graph tell you about the situation? Use numbers to be as specific as you can.

    c. Give a real-world meaning in this context for the negative slope of the lower graph.

    d. What can you say about the ball when the graph of the velocity line intersects the x-axis?

    e. What can you say about the height of the ball when the velocity is 15 meters per second and when it is -15 meters per second?

    f. What are realistic domain and range intervals for the graphs?

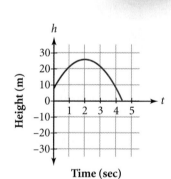

10. Bo and Gale are playing golf. Bo hits his ball, and it is in flight for 3.4 seconds. Gale's ball is in flight for 4.7 seconds.

    a. At what time does each ball reach its highest point?

    b. Can you tell whose ball goes farther or higher? Explain.

11. The table shows the coordinates of a parabola.

   a. On your calculator, plot the points in the table.

   b. Describe the location of the line of symmetry for this graph.

   c. Name the vertex of this graph.

   d. Use your knowledge of transformations to write the equation of this parabola in the vertex form, $y = a(x - h)^2 + k$. Check your answer graphically.

x	y
1.5	−8
2.5	7
3.5	16
4.5	19
5.5	16
6.5	7
7.5	−8

▶ Review

12. Write equations in the form $y = a + bx$ for each of these graphs. One tick mark represents one unit.

   a.

   b.

IMPROVING YOUR **VISUAL THINKING** SKILLS

A parabola is an example of a **conic section.** The Greek geometer Apollonius (255–170 B.C.) defined conic sections by intersecting a double cone with a plane.

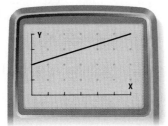

      Plane section   Double cone

The plane is a flat surface that extends into infinity. Likewise, both ends of the double cone widen infinitely in opposite directions. To form a parabola, Apollonius sliced the cone with a plane parallel to the cone's edge.

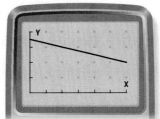

Parabolic section

Other examples of conic sections are circles, ellipses, and hyperbolas. How can you intersect a plane with a cone to form these shapes? Are there any other ways that a plane can intersect a double cone?

    Circle   Ellipse   Hyperbola

Make a drawing that shows how to form each conic section. Can you form any other shapes?

From Vertex to General Form

You have learned two forms of a quadratic equation. The vertex form, $y = a(x - h)^2 + k$, gives you information about transformations of the parent function, $y = x^2$. You used the general form, $y = ax^2 + bx + c$, to model many situations of projectile motion. In this lesson you will learn how to convert an equation from the vertex form to the general form.

The general form, $y = ax^2 + bx + c$, is the sum of three terms. A **term** is an algebraic expression that represents only multiplication and division between variables and constants. A sum of terms with positive integer exponents is called a **polynomial.** Variables cannot appear as exponents in a polynomial.

Here are some examples of polynomials.

$$17x \qquad 4.7x^3 + 3x \qquad x^2 + 3x + 7 \qquad 47x^4 - 6x^3 + 0.28x + 7$$

The expression $17x$ has only one term, so it is called a **monomial.** The second expression has two terms and is called a **binomial.** The third expression is a **trinomial** because it has three terms. If there are more than three terms the expression is generally referred to as a polynomial.

EXAMPLE A

Is each algebraic expression a polynomial? If so, how many terms does it have? If not, give a reason why it is not a polynomial.

a. $3x^2 + 4x^{-1} + 7$ **b.** $2^x - 7.5x + 18$

c. $\dfrac{47}{x} + 28$ **d.** $3x + 1 + 2x$

e. $x^2 - x^{10}$ **f.** $-2x^3 \cdot 3x^2$

▶ Solution

Expression	Is it a polynomial?
a. $3x^2 + 4x^{-1} + 7$	No, because the term $4x^{-1}$ has a negative exponent.
b. $2^x - 7.5x + 18$	No, because 2^x has a variable as the exponent.
c. $\dfrac{47}{x} + 28$	No, because the term $\dfrac{47}{x}$ is equivalent to $47x^{-1}$.
d. $3x + 1 + 2x$	Yes, it is a polynomial. It is equivalent to the binomial $1 + 5x$, which has two terms.
e. $x^2 - x^{10}$	Yes. It has two terms and is a binomial.
f. $-2x^3 \cdot 3x^2$	Yes. It involves only multiplication of constants and variables. It is equivalent to the monomial $-6x^5$.

Terms that differ only in their coefficients, such as $3x$ and $2x$, are *like terms.* When you rewrite $3x + 2x$ as $5x$, you are *combining like terms.* In the investigation you will combine like terms when you convert an equation from the vertex form to the general form.

Investigation
Sneaky Squares

You will need

• graph paper

There are many different, yet equivalent, expressions for a number. For example, 7 is the same as $3 + 4$ and as $10 - 3$. In this investigation you will use these equivalent expressions to model squaring binomials with rectangular diagrams.

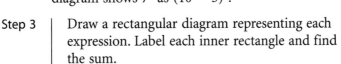

Step 1 | This diagram shows how to express 7^2 as $(3 + 4)^2$. Find the area of each of the inner rectangles. What is the sum of the rectangular areas? What is the area of the overall square? What conclusions can you make?

Step 2 | For each expression below, draw a diagram on your graph paper like the one in Step 1. Label the area of each rectangle and find the total area of the overall square.

a. $(5 + 3)^2$ **b.** $(4 + 2)^2$ **c.** $(10 + 3)^2$ **d.** $(20 + 5)^2$

Even though lengths and areas are not negative, you can use the same kind of rectangular diagram to square an expression involving subtraction. You can use different colors, such as red and blue, to distinguish between the negative and the positive numbers. For example, this diagram shows 7^2 as $(10 - 3)^2$.

Step 3 | Draw a rectangular diagram representing each expression. Label each inner rectangle and find the sum.

a. $(5 - 2)^2$ **b.** $(7 - 3)^2$ **c.** $(20 - 2)^2$ **d.** $(50 - 3)^2$

You can make the same type of rectangular diagram to square an expression involving variables.

Step 4 | Draw a rectangular diagram for each expression. Label each inner rectangle and find the total sum. Combine any like terms you see and express your answer as a trinomial.

a. $(x + 5)^2$ **b.** $(x - 3)^2$ **c.** $(x + 11)^2$ **d.** $(x - 13)^2$

Now use what you have learned to create a rectangular diagram for a trinomial.

Step 5 | Make a rectangular diagram for each trinomial. In so doing, what must you do with the middle term? Label each side of the overall square in your diagram, and write the equivalent expression in the form $(x + h)^2$.

a. $x^2 + 6x + 9$ b. $x^2 - 10x + 25$

c. $x^2 + 8x + 16$ d. $x^2 - 12x + 36$

Step 6 | Use your results from Step 5 to solve each new equation symbolically. Remember, quadratic equations can have two solutions.

a. $x^2 + 6x + 9 = 49$ b. $x^2 - 10x + 25 = 81$

c. $x^2 + 8x + 16 = 121$ d. $x^2 - 12x + 36 = 64$

Numbers like 49 are called **perfect squares** because they are the squares of integers, in this case, 7 or -7. The trinomial $x^2 + 6x + 9$ is the square of $x + 3$. So it is also called a perfect square.

Step 7 | Which of these trinomials are perfect squares?

a. $x^2 + 14x + 49$

b. $x^2 - 18x + 81$

c. $x^2 + 20x + 25$

d. $x^2 - 12x - 36$

Step 8 | Explain how you can recognize a perfect-square trinomial when the coefficient of x^2 is 1. What is the connection between the middle term and the last term?

Step 9 | Square the expression $(x + h)^2$ by making a rectangular diagram. Then describe a shortcut for this process that makes sense to you.

Knowing how to square a binomial is a useful skill. It allows you to convert equations from vertex form to general form.

EXAMPLE B | Rewrite $y = 2(x + 3)^2 - 5$ in the general form, $y = ax^2 + bx + c$.

▶ **Solution** |

				x	3
$y = 2(x + 3)^2 - 5$	The original equation.		x	x^2	$3x$
$y = 2(x^2 + 6x + 9) - 5$	Square the binomial using a rectangular diagram, as shown.				
$y = 2x^2 + 12x + 18 - 5$	Use the distributive property.		3	$3x$	9
$y = 2x^2 + 12x + 13$	Combine like terms.				

You can use a graph or a table on your calculator to verify that the vertex form and the general form of this equation are equivalent. [▶🖳 See **Calculator Note 7B** to review checking different forms of an equation. ◀]

In the example you **expand** $(x + 3)^2$ when you rewrite it as $x^2 + 6x + 9$ in finding the vertex form. The vertex form tells you about translations, reflections, stretches, and shrinks of the graph of the parent function, $y = x^2$. The general form tells you the initial position, velocity, and the force due to gravity in projectile motion applications. Later in the chapter you will learn to convert the general form to the vertex form. Then you'll be able to solve all forms of quadratic equations symbolically.

EXERCISES

You will need your calculator for problems **2, 4,** and **11.**

Practice Your Skills

1. Is each algebraic expression a polynomial? If so, how many terms does it have? If it is not, give a reason why it is not a polynomial.

   a. $x^2 + 3x - 8$

   b. $2x - \dfrac{4}{5}$

   c. $5x^{-1} - 2x^2$

   d. $\dfrac{3}{x^2} - 5x + 2$

   e. $6x$

   f. $\dfrac{x^2}{3^{-2}} + 5x - 8$

   g. $10x^3 + 5x^2$

   h. $3(x - 2)$

2. Expand each expression. On your calculator, enter the original expression into Y_1 and the expanded expression into Y_2. With a graph or a table, check that both forms are equivalent.

   a. $(x + 5)^2$

   b. $(x - 7)^2$

   c. $3(x - 2)^2$

3. Copy each rectangular diagram and fill in the missing values. Then write a squared binomial and an equivalent trinomial that both represent the total area for each diagram.

   a.

	?	2
x	x^2	?
?	?	4

   b.

	x	?
?	?	$12x$
12	?	144

   c.

	x	?
?	x^2	$-7x$
-7	?	?

4. Convert each expression from vertex form to general form. Check your answers by entering the expressions into the $Y=$ screen on your calculator.

   a. $(x + 5)^2 + 4$

   b. $2(x - 7)^2 - 8$

   c. $-3(x + 4)^2 + 1$

   d. $0.5(x - 3)^2 - 4.5$

Reason and Apply

5. You can use the distributive property to write an equivalent expression for the product of two binomials. For example, you can write $(x + 3)(x + 4)$ as $x(x + 4) + 3(x + 4)$ or $x(x + 3) + 4(x + 3)$.

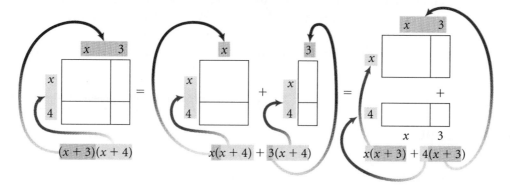

Draw a rectangular diagram to represent each expression. Then write an equation showing the product of the two binomials and the equivalent polynomial in general form.

a. $x(x + 4) + 2(x + 4)$

b. $x(x + 5) + 3(x + 5)$

c. $x(x - 5) + 2(x - 5)$

d. $x(x - 3) - 0(x - 3)$

6. Consider the graph of the parabola $y = x^2 - 4x + 7$.

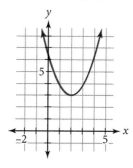

a. What are the coordinates of the vertex?

b. Write the equation in vertex form.

c. Check that the equation you wrote in 6b is correct by expanding it to general form.

7. Heather thinks she has found a shortcut to the rectangular diagram method of squaring a binomial. She says that you can just square everything inside the parentheses. That is, $(x + 8)^2$ would be $x^2 + 64$. Is Heather's method correct?

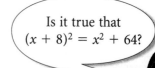

Is it true that $(x + 8)^2 = x^2 + 64$?

8. **APPLICATION** The quadratic equation $y = 0.0056x^2 + 0.14x$ relates a vehicle's stopping distance to its speed. In this equation, y represents the stopping distance in meters and x represents the vehicle's speed in kilometers per hour.

a. Find the stopping distance for a vehicle traveling 100 kph.

b. Write and solve an equation to find the speed of a vehicle that took 50 meters to stop.

9. The function $h(t) = -4.9(t - 0.4)^2 + 2.5$ describes the height of a softball thrown by a pitcher.

   a. How high does the ball go?

   b. What is an equivalent function in general form?

   c. At what height did the pitcher release the ball when t was 0 seconds?

Lori Harrigan of the USA softball team pitches at the 2000 Olympics in Sydney, Australia.

10. Is the expression on the right equivalent to the expression on the left? If not, correct the right side to make it equivalent.

   a. $(x + 7.5)^2 - 3 \stackrel{?}{=} x^2 + 15x + 53.25$

   b. $2(x - 4.7)^2 + 2.8 \stackrel{?}{=} 2x^2 - 9.4x - 41.38$

   c. $-3.5(x + 1.6)^2 - 2.04 \stackrel{?}{=} -3.5x^2 + 11.2x - 11$

   d. $-4.9(x - 5.6)^2 + 8.9 \stackrel{?}{=} -4.9x^2 + 54.88x - 144.764$

11. The Yo-yo Warehouse uses the equation $y = -85x^2 + 552.5x$ to model the relationship between income and price for one of its top-selling yo-yos. In this model, y represents income in dollars and x represents the selling price in dollars of one item.

   a. Graph this relationship on your calculator, and describe a meaningful domain and range for this situation.

   b. Describe a method for finding the vertex of the graph of this relationship. What is the vertex?

   c. What are the real-world meanings of the coordinates of the vertex?

   d. What is the real-world meaning of the two x-intercepts of the graph?

   e. Interpret the meaning of this model if $x = 5$.

12. Use a three-by-three rectangular diagram to square each trinomial.

   a. $(x + y + 3)^2$ **b.** $(2x - y + 5)^2$

13. What is the general form of $(x + 4)^2$? Write a paragraph describing several ways to rewrite this expression in general form.

▶ Review

14. Is the parabola a graph of a function?

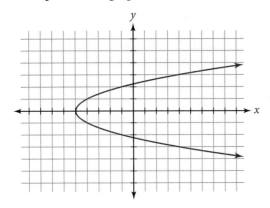

15. Use the graph of $f(x)$ to evaluate each expression. Then think of the numbers 1 through 26 as the letters A through Z to decode a message.

a. $f(18)$

b. $3 \cdot f(3)$

c. $f(4^2)$

d. $(f(3))^2$

e. $f(17)$

f. $f(25)$

g. $f(5) + f(15)$

h. the greater x-value when $f(x) = 1$

i. $f(1) - f(2)$

j. $f(4) \cdot f(5)$

k. $f(5^2 - 2^2)$

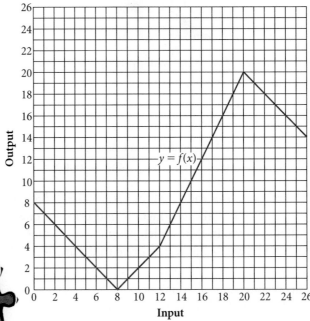

LESSON
10.4

Mathematicians assume the right to choose, within the limits of logical contradiction, what path they please in reaching their results.

HENRY ADAMS

Factored Form

So far you have worked with quadratic equations in vertex form and general form. This lesson will introduce you to another form of quadratic equation, the **factored form:**

$$y = a(x - r_1)(x - r_2)$$

This form helps you identify the roots, r_1 and r_2, of an equation. In the investigation you'll discover connections between the equation in factored form and its graph. You'll also use rectangular diagrams to convert the factored form to the general form and vice versa. Then in the example you'll learn how to use a special property to find the roots of an equation.

Investigation
Getting to the Root of the Matter

You will need

• graph paper

First you'll find the roots of an equation in factored form from its graph.

Step 1	On your calculator, graph the equations $y = x + 3$ and $y = x - 4$ at the same time.
Step 2	What is the x-intercept of each equation you graphed in Step 1?
Step 3	Graph $y = (x + 3)(x - 4)$ on the same set of axes as before. Describe the graph. Where are the x-intercepts of this graph?
Step 4	Expand $y = (x + 3)(x - 4)$ to general form. Graph the equation in general form on the same set of axes. What do you notice about this parabola and its x-intercepts? Is the graph of $y = (x + 3)(x - 4)$ a parabola?

Now you'll learn how to find the roots from the general form.

Step 5 Complete the rectangular diagram whose sum is $x^2 + 5x + 6$. A few parts on the diagram have been labeled to get you started.

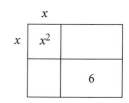

Step 6 Write the multiplication expression of the rectangular diagram in factored form. Use a graph or table to check that this form is equivalent to the original expression.

Step 7 Find the roots of the equation $0 = x^2 + 5x + 6$ from its factored form.

Step 8 Rewrite each equation in factored form by completing a rectangular diagram. Then find the roots of each. Check your work by making a graph.

    a. $0 = x^2 - 7x + 10$ **b.** $0 = x^2 + 6x - 16$

    c. $0 = x^2 + 2x - 48$ **d.** $0 = x^2 - 11x + 28$

Now you have learned three forms of a quadratic equation. You can enter each of these forms into your calculator to check that they are equivalent. Here are three equivalent equations that describe the height in meters, y, of an object in motion for x seconds after being thrown upward. Each equation gives different information about the object.

Vertex form	$y = -4.9(x - 1.7)^2 + 15.876$
General form	$y = -4.9x^2 + 16.66x + 1.715$
Factored form	$y = -4.9(x + 0.1)(x - 3.5)$

Which form is best? The answer depends on what you want to know. The vertex form tells you when the maximum height occurs—in this case, 15.876 meters after 1.7 seconds (the vertex). The general form tells you that the object started at a height of 1.715 meters (the y-intercept). The coefficients of x and x^2 give some information about the starting velocity and acceleration. The factored form tells you the times at which the object's height is zero (the roots).

You have already learned how to convert to and from the general form of a quadratic equation. The example will show you how to get the vertex form from the factored form.

EXAMPLE

Write the equation for this parabola in vertex form, general form, and factored form.

▶ Solution

From the graph you can see that the x-intercepts are 3 and -5. So the factored form contains the binomial expressions $(x - 3)$ and $(x + 5)$.

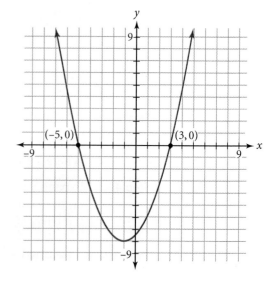

If you graph $y = (x - 3)(x + 5)$ on your calculator, you'll see it has the same x-intercepts as the graph shown here, but a different vertex. The new vertex is at $(-1, -16)$.

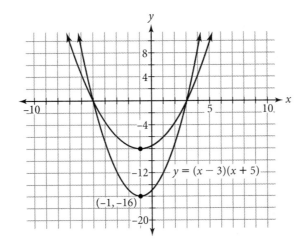

The new vertex needs to be closer to the x-axis, so you need to find the vertical shrink factor a.

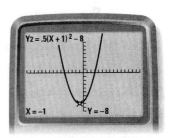

$[-15, 15, 1, -10, 10, 1]$

The original vertex of the graph shown is $(-1, -8)$. So the graph of the function must have a vertical shrink by a factor of $\frac{-8}{-16}$, or 0.5. The factored form is $y = 0.5(x - 3)(x + 5)$. A calculator graph of this equation looks like the desired parabola.

Now you know that the value of a is 0.5 and that the vertex is at $(-1, -8)$. Substitute this information into the vertex form to get $y = 0.5(x + 1)^2 - 8$.

Expand both forms to find the general form.

$y = 0.5(x - 3)(x + 5)$	The original equations.	$y = 0.5(x + 1)^2 - 8$
$y = 0.5\big(x^2 - 3x + 5x - 15\big)$	Expand using rectangular diagrams.	$y = 0.5\big(x^2 + 1x + 1x + 1\big) - 8$

	x	-3
x	x^2	$-3x$
5	$5x$	-15

	x	1
x	x^2	$1x$
1	$1x$	1

$y = 0.5\big(x^2 + 2x - 15\big)$	Combine like terms.	$y = 0.5\big(x^2 + 2x + 1\big) - 8$
$y = 0.5x^2 + x - 7.5$	Distribute and combine.	$y = 0.5x^2 + x - 7.5$

So the three forms of the quadratic equation are

Vertex form	$y = 0.5(x + 1)^2 - 8$
General form	$y = 0.5x^2 + x - 7.5$
Factored form	$y = 0.5(x - 3)(x + 5)$

When finding roots it is helpful to use the factored form. In this example one root is 3 because it makes $(x - 3)$ equal to 0. The other root is -5 because it makes $(x + 5)$ equal to 0. Think of numbers that multiply to zero. If $ab = 0$ or $abc = 0$, the **zero product property** tells you that a, or b, or c must be 0. In an equation like $(x + 3)(x - 5) = 0$, *at least one of the factors must be zero.* The roots of an equation are sometimes called the **zeros** of a function because they make the value of the function equal to zero.

EXERCISES

You will need your calculator for problems **2, 3, 8,** and **11.**

▶ Practice Your Skills

1. Use the zero product property to solve each equation.

   a. $(x + 4)(x + 3.5) = 0$ **b.** $2(x - 2)(x - 6) = 0$

   c. $(x + 3)(x - 7)(x + 8) = 0$ **d.** $x(x - 9)(x + 3) = 0$

2. Graph each equation and then rewrite it in factored form.

   a. $y = x^2 - 4x + 3$ **b.** $y = x^2 + 5x - 24$

   c. $y = x^2 + 12x + 27$ **d.** $y = x^2 - 7x - 30$

3. Name the x-intercepts for the parabola of each quadratic equation. Then check your answers with a graph.

   a. $y = (x - 7)(x + 2)$ **b.** $y = 2(x + 1)(x + 8)$

   c. $y = 3(x - 11)(x + 7)$ **d.** $y = 0.4(x + 5)(x - 9)$

4. Write an equation of a quadratic function that corresponds to each pair of x-intercepts. Assume there is no vertical stretch or shrink.

   a. 2.5 and -1 **b.** -4 and -4

   c. -2 and 2   **d.** r_1 and r_2

5. Consider the equation $y = (x + 1)(x - 3)$.

   a. How many x-intercepts does the graph have?

   b. Find the vertex of this parabola.

   c. Write the equation in vertex form. Describe the transformations of the parent function, $y = x^2$.

▶ Reason and Apply

6. Is the expression on the left equivalent to the expression on the right? If not, change the right side to make it equivalent.

   a. $x^2 + 7x + 12 \overset{?}{=} (x + 3)(x + 4)$

   b. $x^2 - 11x + 30 \overset{?}{=} (x + 6)(x + 5)$

   c. $2x^2 - 5x - 7 \overset{?}{=} (x - 3.5)(x + 1)$

   d. $4x^2 + 8x + 4 \overset{?}{=} (x + 1)^2$

   e. $x^2 - 25 \overset{?}{=} (x + 5)(x - 5)$

   f. $x^2 - 36 \overset{?}{=} (x - 6)^2$

7. The sum and product of the roots of a quadratic equation are related to b and c in $y = x^2 + bx + c$. The first row in the table below will help you to recognize this relationship.

a. Complete the table.

Factored form	Roots	Sum of roots	Product of roots	General form
$y = (x + 3)(x - 4)$	-3 and 4	$-3 + 4 = 1$	$(-3)(4) = -12$	$y = x^2 - 1x - 12$
	5 and -2			
		-5	6	
$y = (x - 5)(x + 5)$		0	-25	

b. Use the values of b and c to find the roots of $0 = x^2 + 2x - 8$.

8. In this problem you will discover whether or not knowing the x-intercepts determines a unique quadratic equation. Work through the steps in 8a–e to find an answer. Graph each equation to check your work.

a. Write an equation for a parabola with x-intercepts at $x = 3$ and $x = 7$.

b. Name the vertex of the parabola in 8a.

c. Modify your equation in 8a so that the graph is reflected across the x-axis. Where are the x-intercepts? Where is the vertex?

d. Modify your equation in 8a to apply a vertical stretch with a factor of 2. Where are the x-intercepts? Where is the vertex?

e. How many quadratic equations do you think there are with x-intercepts at $x = 3$ and $x = 7$? How are they related to one another?

9. Write a quadratic equation for a parabola with x-intercepts at -3 and 9 and vertex at $(3, -9)$. Express your answer in factored form.

10. **APPLICATION** The school ecology club wants to fence in an area along the riverbank to protect some endangered wildflowers that grow there. The club has enough money to buy 200 feet of fencing. It decides to enclose a rectangular space. The fence will form three sides of the rectangle, and the riverbank will form the fourth side.

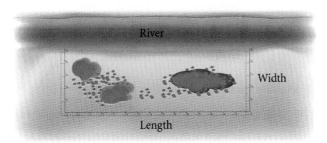

a. If the width of the enclosure is 30 feet, how much fencing material is available for the length? Sketch this situation. What is the area?

b. If the width is w feet, how much fencing material remains for the length, l?

c. Use your answer from 10b to write an equation for the area of the rectangle in factored form. Check your equation with your width and area from 10a.

d. Which two different widths would give an area equal to 0?

e. Which width will give the maximum area? What is that area?

11. Consider the equation $y = x^2 - 9$.

    a. Graph the equation. What are the x-intercepts?

    b. Write the factored form of the equation.

    c. How are the x-intercepts related to the original equation?

    d. Write each equation in factored form. Verify each answer by graphing.

        i. $y = x^2 - 49$
        ii. $y = 16 - x^2$
        iii. $y = x^2 - 47$
        iv. $y = x^2 - 28$

    e. Graph the equation $y = x^2 + 4$. How many x-intercepts can you see?

    f. Explain the difficulty in trying to write the equation in 11e in factored form.

12. Kayleigh says that the roots of $0 = x^2 + 16$ are 4 and -4 because $(4)^2 = 16$ and $(-4)^2 = 16$. Derek tells Kayleigh that there are no roots for this equation. Who is correct and why?

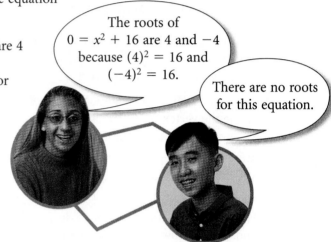

The roots of $0 = x^2 + 16$ are 4 and -4 because $(4)^2 = 16$ and $(-4)^2 = 16$.

There are no roots for this equation.

▶ Review

13. On graph paper, draw a function that has these properties:

    ▶ Domain of $-4 \leq x \leq 4$
    ▶ $f(-4) = 1$
    ▶ Range of $-3 \leq y \leq 3$
    ▶ $f(3) = 3$

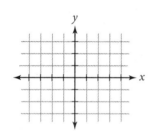

14. On graph paper, draw a graph that is NOT a function with a domain of $-4 \leq x \leq 4$ and a range of $-3 \leq y \leq 3$.

Activity Day

Projectile Motion

You have already learned that quadratic equations model projectile motion. In this lesson you'll do an experiment with projectile motion and find a quadratic function to model the data. If you choose the first experiment, you'll collect data for the *x*-intercepts of a parabola and then find an equation in factored form that matches the graph. If you choose the second experiment, you'll collect parabolic data and then find an equation in vertex form that matches the graph. Read the steps of each experiment and then choose one experiment for your group to do.

Activity
Jump or Roll

You will need

- a motion sensor
- an empty coffee can
- a long table

Each experiment in this activity requires a calculator program. Be sure you have this program in your calculator before you begin. [▶️🖥 See **Calculator Note 10A** for the required programs. ◀] In the first experiment you will collect data for the zeros of a projectile motion function.

Experiment 1: How High?

The object of this experiment is to find how high you jump.

Step 1 Set up the program to collect data. Jump straight up without bending your knees in the air. Be sure to land in front of the sensor again. This way, the sensor records the times your feet left the ground and landed.

> **Procedure Note**
>
> Place the motion sensor on the floor. The jumper stands 2 ft or 0.5 m in front of it. There should be a wall or another object about 4 ft or 1 m from the sensor. When the jumper's feet leave the ground, the motion sensor should register a change in distance at a specific instant in time.

Step 2	The data measured by the motion sensor has the form (*time, distance*), where the distance is that between the motion sensor and the nearest object to it. At first this distance is from the sensor to the jumper's feet. Then during the jump the sensor measures the distance to the wall behind the jumper. After the jumper lands the sensor reads the distance to the jumper's feet again. Look at the graph and use the trace feature to determine the instant the feet left the ground. Do this by finding the sharp change in *y*-values on the graph. Likewise, determine the instant in time when the feet landed back on the ground.
Step 3	If you want to graph the height of your jump over time, what are the variables for the quadratic function in this situation? Substitute the two roots you found in Step 2 for r_1 and r_2 into the equation $h = -192(t - r_1)(t - r_2)$. Use it to calculate the height of your jump in inches. (Or use the equation $h = -490(t - r_1)(t - r_2)$ to find this height in centimeters.) At what time did you reach this height? Explain how you got your answer.
Step 4	Repeat the experiment with each member of your group as a jumper.

Experiment 2: Rolling Along

The object of this experiment is to write a quadratic equation from projectile motion data.

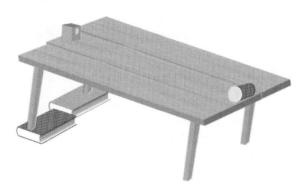

Step 1	Practice rolling the can up the table directly in front of the sensor. The can should roll up the table, stop about 2 feet from the sensor, and then roll back down. Give the can a short push so that it rolls up the table on its own momentum. Then the force of gravity should cause the can to reverse directions as it rolls back down the slanted table.
Step 2	Set up the program to collect the data. When the sensor begins, gently roll the can up the table. Catch it as it falls off the table.
Step 3	The data collected by the sensor will have the form (*time, distance*). If you do the experiment correctly, the graph should show a parabolic pattern.
Step 4	Find the equation for a parabola that fits your data. Which points did you use to find the equation? In which form is it? Sketch a parabola for this equation onto the graph from Step 3.

PARABOLA BY DEFINITION

You have learned that the graph of a quadratic equation is a parabola. One definition of a parabola is the set of all points whose distance from a fixed point, the *focus*, is equal to its distance from a fixed line, the *directrix*. (Use the shortest possible distance for the distance between a point and a line.)

The Geometer's Sketchpad was used to create this parabola. Sketchpad has tools to help you construct points, lines, and the set of points equidistant to both. Learn how to use these tools to create a parabola of your own.

You can use The Geometer's Sketchpad or waxed paper to draw a parabola in various ways based on this definition. Start by drawing a line and a point not on the line. Then locate several points equally distant from the focus and the directrix by using the tools in Sketchpad, or by folding waxed paper. On waxed paper fold the focus to lie on the directrix and crease the paper. Repeat to make many creases. The creases will outline a parabola. If you make a similar set of lines in Sketchpad, you can test what happens to the parabola if the focus moves closer to (or farther from) the directrix.

Your project should include

▶ A drawing of the lines with the parabola's focus, directrix, and vertex labeled.

▶ An explanation of how you constructed the lines.

▶ A discussion of how the distance between the focus and the directrix affects the shape of the parabola.

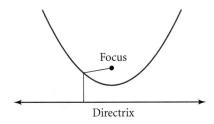

Find more information about parabolas with the helpful Internet links at **www.keymath.com/DA** .

Completing the Square

You can always find approximate solutions to quadratic equations by using tables and graphs. If you can convert the equation to the factored form, $y = a(x - r_1)(x - r_2)$, or the vertex form, $y = a(x - h)^2 + k$, then you can use symbolic methods to find exact solutions. In this lesson you'll learn a symbolic method to find exact solutions to equations in the general form, $y = ax^2 + bx + c$.

Recall that rectangular diagrams help you factor some quadratic expressions.

Perfect square trinomial

$x^2 + 6x + 9$

	x	3
x	x^2	$3x$
3	$3x$	9

Factorable trinomial

$x^2 + x - 6$

	x	3
x	x^2	$3x$
-2	$-2x$	-6

In the first diagram the sum of the rectangular areas, $x^2 + 3x + 3x + 9$, is equal to the area of the overall diagram, $(x + 3)^2$. So -3 is the root. In the second diagram the sum is $x^2 + 3x + (-2x) + (-6)$, which equals $(x - 2)(x + 3)$. Both 2 and -3 are the roots.

How do you find the roots of an equation such as $0 = x^2 + x - 1$? It is not a perfect square trinomial, nor is it factorable with integers. For these equations you can use a method called **completing the square.**

Investigation
Searching for Solutions

You will need

• graph paper

To understand how to complete the square with quadratic equations, you'll first work with rectangular diagrams.

Step 1 | Complete each rectangular diagram so that it is a square. How do you know which number to place in the lower right corner?

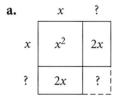

a.

	x	$?$
x	x^2	$2x$
$?$	$2x$	$?$

b.

	x	$?$
x	x^2	$-3x$
$?$	$-3x$	$?$

c.

	x	$?$
x	x^2	$-2.5x$
$?$	$-2.5x$	$?$

d.

	x	$?$
x	x^2	$\frac{b}{2}x$
$?$	$\frac{b}{2}x$	$?$

Step 2 | For each diagram in Step 1, write an equation in the form
$$x^2 + bx + c = (x + h)^2$$
On which side of the equation can you isolate x by undoing the order of operations?

Step 3	Suppose the area of each diagram in parts a–c is 100 square units. For each square, write an equation that you can solve for x by undoing the order of operations.
Step 4	Solve each equation in Step 3 symbolically. You will get two values for x.

All the solutions for x in Step 4 are integers or simple decimals. This means you could have factored the equations with integers. However, the method of completing the square works for other numbers as well. Next you'll consider the solution of an equation that you cannot factor with integers.

Step 5 Consider the equation $x^2 + 6x - 1 = 0$. Describe what's happening in each stage of the solution process.

Stage	**Equation**	**Description**
1	$x^2 + 6x - 1 = 0$	The original equation.
2	$x^2 + 6x = 1$	
3	$x^2 + 6x + 9 = 1 + 9$	

	x	3
x	x^2	$3x$
3	$3x$	9

4	$(x + 3)^2 = 10$	
5	$x + 3 = \pm\sqrt{10}$	
6	$x = -3 \pm\sqrt{10}$	

Step 6 Use your calculator to find decimal approximations for $-3 + \sqrt{10}$ and $-3 - \sqrt{10}$. Then enter the equation $y = x^2 + 6x - 1$ into Y1. Check your answers with a graph and a table.

Step 7 Repeat the solution stages in Step 5 to find the solutions to $x^2 + 8x - 5 = 0$.

The key to solving by completing the square is to express one side of the equation as a perfect-square trinomial. In the investigation the equations are in the form $y = 1x^2 + bx + c$. Note that the coefficient of x^2, called the **leading coefficient,** is 1. However, there are other perfect square trinomials. An example is shown at right.

$4x^2 + 12x + 9$

	$2x$	3
$2x$	$4x^2$	$6x$
3	$6x$	9

$(2x + 3)^2$

In these cases, the leading coefficient is a perfect square number. In Example A you'll learn to complete the square for any quadratic equation in general form.

EXAMPLE A | Solve the equation $3x^2 + 18x - 8 = 22$ by completing the square.

▶ **Solution** | First, transform the equation so that you can write the left side as a perfect-square trinomial in the form $x^2 + 2hx + h^2$.

$3x^2 + 18x - 8 = 22$	The original equation.
$3x^2 + 18x = 30$	Add 8 to both sides of the equation.
$x^2 + 6x = 10$	Divide both sides by 3.

Now you need to decide what number to add to both sides to get a perfect-square trinomial on the left side. Use a rectangular diagram to make a square. When you decide what number to add, you must add it to both sides to balance the equation.

$x^2 + 6x + 9 = 10 + 9$	Add 9 to both sides to complete the square.
$(x + 3)^2 = 19$	Write the perfect-square trinomial as a squared binomial and combine any like terms.
$x + 3 = \pm\sqrt{19}$	Take the square root of both sides.
$x = -3 \pm\sqrt{19}$	Add -3 to both sides.

The two solutions are $-3 + \sqrt{19}$, or approximately 1.36, and $-3 - \sqrt{19}$, or approximately -7.36.

You can also complete the square to convert the general form of a quadratic equation to the vertex form.

EXAMPLE B | Find the vertex form of the equation $y = 2x^2 + 8x + 11$. Then locate the vertex point and any x-intercepts of the parabola.

▶ **Solution** | To convert $y = 2x^2 + 8x + 11$ to the form $y = a(x - h)^2 + k$, complete the square.

$y = 2x^2 + 8x + 11$	The original equation.
$y = 2(x^2 + 4x) + 11$	Factor the 2 from the coefficients.

Now you can complete the square on the expression inside the parentheses. The coefficient of x is 4, so divide that by 2 to get 2. Then add 2^2, or 4, to make a perfect-square trinomial inside the parentheses. You must also subtract 4 inside the parentheses to balance the equation. Note that everything inside the parentheses is multiplied by 2.

$y = 2(x^2 + 4x + 4 - 4) + 11$	Add zero in the form of $4 - 4$.
$y = 2(x^2 + 4x + 4) + 2(-4) + 11$	Rewrite to get a perfect-square trinomial.
$y = 2(x + 2)^2 - 8 + 11$	Express the trinomial as a squared binomial.
$y = 2(x + 2)^2 + 3$	Combine like terms to get the vertex form.

So the vertex is $(-2, 3)$. To find any x-intercept, you can solve the vertex form symbolically.

$$2(x + 2)^2 + 3 = 0 \qquad \text{Substitute 0 for } y \text{ in the original equation.}$$

$$(x + 2)^2 = \frac{-3}{2} \qquad \text{Subtract 3 and then divide both sides by 2.}$$

$$x = -2 \pm \sqrt{\frac{-3}{2}} \qquad \text{Take the square root and then subtract 2 from both sides.}$$

If you try to evaluate $-2 \pm \sqrt{\frac{-3}{2}}$, your calculator may give you an error message about a nonreal answer. These roots are not real numbers because the number under the square root sign is negative. **Real numbers** are all numbers except those that involve even roots of negative numbers. Integers, fractions, and any numbers that can be expressed as decimals are real numbers. Every real number is on the x-axis. So this means that $y = 2(x + 2)^2 + 3$ has no x-intercepts. The graph confirms this result.

Note that the vertex is above the x-axis and the parabola opens upward. So the graph does not cross the x-axis.

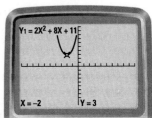

You can now solve any quadratic equation in general form by completing the square. This process leads to a general formula that you will learn in the next lesson.

EXERCISES

You will need your calculator for problems **8**, **11**, and **12**.

Practice Your Skills

1. Solve each quadratic equation written in vertex form.

   a. $2(x + 3)^2 - 4 = 0$ **b.** $-2(x - 5)^2 + 7 = 3$

   c. $3(x + 8)^2 - 7 = 0$ **d.** $-5(x + 6)^2 - 3 = -10$

2. Solve each equation written in factored form.

   a. $(x - 5)(x + 3) = 0$ **b.** $(2x + 6)(x - 7) = 0$

   c. $(3x + 4)(x + 1) = 0$ **d.** $x(x + 6)(x + 9) = 0$

3. Decide what number must be added to each expression to make a perfect-square trinomial. Then rewrite the trinomial as a squared binomial.

   a. $x^2 + 18x$ **b.** $x^2 - 10x$

   c. $x^2 + 3x$ **d.** $x^2 - x$

   e. $x^2 + \frac{2}{3}x$ **f.** $x^2 - 1.4x$

4. Solve each quadratic equation by completing the square. Leave your answer in radical form.

   a. $x^2 - 4x - 8 = 0$ **b.** $x^2 + 2x - 1 = -5$

   c. $x^2 + 10x - 9 = 0$ **d.** $5x^2 + 10x - 7 = 28$

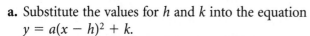

Reason and Apply

5. If you know the vertex and one other point on a parabola, you can find its quadratic equation. The vertex (h, k) of this parabola is $(2, -31.5)$, and the other point is $(5, 0)$.

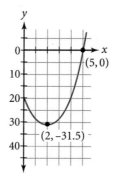

   a. Substitute the values for h and k into the equation $y = a(x - h)^2 + k$.

   b. To find the value of a, substitute 5 for x and 0 for y. Then solve for a.

   c. Use the a-value you found in 5b to write the equation for the graph in vertex form.

   d. Use what you learned in 5a–c to write the equation of the graph whose vertex is $(2, 32)$ and that passes through the point $(5, 14)$.

6. The length of a rectangle is 4 meters more than its width. The area is 12 square meters.

   a. Define variables and write an equation for the area of the rectangle in terms of its width.

   b. Solve your equation in 6a by completing the square.

   c. Which solution makes sense for the length of the rectangle?

7. Consider the equation $y = x^2 + 6x + 10$.

   a. Convert this equation to vertex form by completing the square.

   b. Find the vertex. Graph both equations.

   c. Find the roots of the equation $0 = x^2 + 6x + 10$. What happens and why?

8. **APPLICATION** A professional football team uses computers to describe the projectile motion of a football when punted. After compiling data from several games, the computer models the height of an average punt with the equation

$$h(t) = \frac{-16}{3}(t - 2.2)^2 + 26.9$$

where t is the time in seconds and $h(t)$ is the height in yards. The punter's foot makes contact with the ball when $t = 0$.

   a. When does the punt reach its highest point? How high does the football go?

   b. Find the zeros of $h(t) = \frac{-16}{3}(t - 2.2)^2 + 26.9$. Which solution is the hang time—that is, the time it takes until the ball hits the ground?

   c. How high is the ball when the punter kicks it?

   d. Graph the equation. What are the real-world meanings of the vertex, the y-intercept, and the x-intercepts?

Brad Maynard punts for the N.Y. Giants.

9. **APPLICATION** The Cruisin' Along Company is determining prices for its Caribbean cruise packages. The basic cost is $2500 per person. However, business is slow. To attract corporate clients, the company reduces the cost of each ticket by $5 for each person in the group. The larger the group, the less each person would pay.

Gotta get away?

Caribbean Cruise Package

▶ Fabulous vacation deal
▶ All major ports
▶ All expenses included
▶ Individual/group rates
▶ Special corporate offer

THE **Cruisin' Along Company**

a. Define variables and write an equation for the cost of a single ticket.

b. Write an equation for the total cost the company charges for a group package.

c. Convert the equation in 9b to vertex form.

d. What is the total cost of a cruise for a group of 20 people?

e. The company accountant reports that the cost of running a cruise is $200,000. Solve the equation

$$x(2500 - 5x) = 200,000$$

by completing the square.

f. What limitations on group size should the cruise company use in order to maximize its profits?

10. **APPLICATION** The rate at which a bear population grows in a park is given by the equation $P(b) = 0.001b(100 - b)$. The function value $P(b)$ represents the rate at which the population is growing in bears per year, and b represents the number of bears.

a. Find $P(10)$ and provide a real-world meaning for this value.

b. Solve $P(b) = 0$ and provide real-world meanings for these solutions.

c. For what size bear population would the population grow fastest?

d. What is the maximum number of bears the park can support?

e. What does it mean to say that $P(120) < 0$?

▶ Review

11. Find each product. Check your answers by using calculator tables or graphs.

a. $(x + 1)(2x^2 + 3x + 1)$

b. $(2x - 5)(3x^2 + 2x - 4)$

12. Combine like terms in these polynomials. Check your answers by using calculator tables or graphs.

a. $(x + 1) + (2x^2 + 3x + 1)$

b. $(2x - 5) + (3x^2 + 2x - 4)$

c. $(x + 1) - (2x^2 + 3x + 1)$

d. $(2x - 5) - (3x^2 + 2x - 4)$

LESSON
10.7

The Quadratic Formula

Most people are more comfortable with old problems than with new solutions.

ANONYMOUS

You can solve some quadratic equations symbolically by recognizing their forms:

Vertex form	$-4.9(t - 5)^2 + 75 = 0$
Factored form	$0 = w(200 - 2w)$
Perfect square trinomial	$x^2 + 6x + 9 = 0$

You can also undo the order of operations in other quadratic equations when there is no x-term, as in these:

$$x^2 = 10$$
$$x^2 + 25 = 0$$
$$x^2 - 0.36 = 0$$

If the quadratic expression is in the form $x^2 + bx + c$, you can complete the square by using a rectangular diagram.

	x	$\frac{b}{2}$
x	x^2	$\frac{b}{2}x$
$\frac{b}{2}$	$\frac{b}{2}x$	$\left(\frac{b}{2}\right)^2$

Although it is possible to complete the square for any quadratic equation, it can get messy if your equation is something like $-4.9x^2 + 5x - \frac{16}{3} = 0$.

Let's consider the general case of $ax^2 + bx + c = 0$. The leading coefficient is a. The middle term, bx, is called the **linear term.** The value of c is the **constant term.** Completing the square for the general case gives a formula that solves any quadratic equation. It is called the **quadratic formula.** To use it, all you need to know are the values of a, b, and c.

Investigation
Deriving the Quadratic Formula

You'll solve $2x^2 + 3x - 1 = 0$ and develop the quadratic formula for the general case in the process.

Step 1 | Identify the values of a, b, and c in the general form, $ax^2 + bx + c = 0$, for the equation $2x^2 + 3x - 1 = 0$.

Step 2 | Group all the variable terms on the left side of your equation so that it is in the form

$$ax^2 + bx = -c$$

Step 3	In order to complete the square, the coefficient of x^2 should be 1. So divide your equation by the value of a. Write it in the form
	$$x^2 + \frac{b}{a}x = \frac{-c}{a}$$

Step 4	Use a rectangular diagram to help you complete the square. What number must you add to both sides? Write your new equation in the form
	$$x^2 + \frac{b}{a}x + \left(\frac{b}{2a}\right)^2 = \left(\frac{b}{2a}\right)^2 - \frac{c}{a}$$

Step 5	Rewrite the trinomial on the left side of your equation as a squared binomial. On the right side, find a common denominator. Write the next stage of your equation in the form
	$$\left(x + \frac{b}{2a}\right)^2 = \frac{b^2}{4a^2} - \frac{4ac}{4a^2}$$

Step 6	Take the square root of both sides of your equation, like this:
	$$x + \frac{b}{2a} = \pm\frac{\sqrt{b^2 - 4ac}}{\sqrt{4a^2}}$$

Step 7	Get x by itself on the left side, like this:
	$$x = -\frac{b}{2a} \pm \frac{\sqrt{b^2 - 4ac}}{2a}$$

Step 8	There are two possible solutions given by the equations
	$$x = \frac{-b + \sqrt{b^2 - 4ac}}{2a} \text{ or } x = \frac{-b - \sqrt{b^2 - 4ac}}{2a}$$
	Write your two solutions in radical form.

Step 9	Write your solutions in decimal form. Check them with a graph and a table.

Step 10	Consider the expression $\dfrac{-b \pm \sqrt{b^2 - 4ac}}{2a}$. What restrictions should there be so that the solutions exist and are real numbers?

The quadratic formula gives the same solutions that completing the square or factoring does. You don't need to derive the formula each time. All you need to know are the values for a, b, and c. Then you substitute these values into the formula.

Quadratic Formula

If a quadratic equation is written in the general form, $ax^2 + bx + c = 0$, the roots are given by $x = \dfrac{-b \pm \sqrt{b^2 - 4ac}}{2a}$.

In the next example you'll learn how to use the formula for quadratic equations in general form. You can even use it when the values of a, b, and c are decimals or fractions.

EXAMPLE | Use the quadratic formula to solve $3x^2 + 5x - 7 = 0$.

▶ **Solution** | The equation is already in general form, so identify the values of a, b, and c. For this equation, $a = 3$, $b = 5$, and $c = -7$. Here is one way to use the formula:

$$x = \frac{-b \pm \sqrt{b^2 - 4ac}}{2a}$$

The quadratic formula.

$$= \frac{-(\ ) \pm \sqrt{(\ )^2 - 4(\ )(\ )}}{2(\ )}$$

Replace each letter in the formula with a set of parentheses.

$$= \frac{-(5) \pm \sqrt{(5)^2 - 4(3)(-7)}}{2(3)}$$

Substitute the values of a, b, and c into the appropriate places.

$$= \frac{-5 \pm \sqrt{25 - (-84)}}{6}$$

Do the operations.

$$= \frac{-5 \pm \sqrt{109}}{6}$$

Subtract.

The two exact roots of the equation are $\frac{-5 + \sqrt{109}}{6}$ and $\frac{-5 - \sqrt{109}}{6}$.

You can use your calculator to calculate the approximate values, 0.907 and -2.573, respectively.

To make the formula simpler, think of the expression under the square root sign as one number. This expression $b^2 - 4ac$ is called the **discriminant.** In the example the discriminant is 109. So let $d = b^2 - 4ac$. Then the formula becomes

$$x = \frac{-b \pm \sqrt{d}}{2a}$$

If you store these values into your calculator as shown, then you can use the formula directly on your calculator.

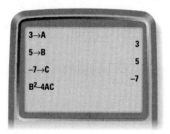
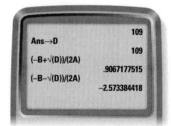

EXERCISES

You will need your calculator for problems **1, 4, 6,** and **12.**

▶ Practice Your Skills

1. Without using a calculator, evaluate each expression in the form $b^2 - 4ac$ for the values given. Then check your answers with a calculator.

a. $a = 3, b = 5, c = 2$

b. $a = 1, b = -3, c = -3$

c. $a = -2, b = -6, c = -3$

d. $a = 9, b = 9, c = 0$

2. Rewrite each quadratic equation in general form if necessary. For each equation, identify the values of a, b, and c.

   a. $2x^2 + 3x - 7 = 0$ **b.** $x^2 + 6x = -11$

   c. $-3x^2 - 4x + 12 = 0$ **d.** $18 - 4.9x^2 + 47x = 0$

   e. $-16x^2 + 28x + 10 = 57$ **f.** $5x^2 - 2x = 7 + 4x$

3. Solve each quadratic equation. Which equation can you solve readily by completing the square? Which equation has no real solutions?

   a. $2x^2 - 3x + 4 = 0$

   b. $-2x^2 + 7x = 3$

   c. $x^2 - 6x - 8 = 0$

   d. $3x^2 + 2x - 1 = 5$

▶ Reason and Apply

4. Graph the equation $y = x^2 + 3x + 5$. Use the graph and the quadratic formula to answer these questions.

   a. How many x-intercepts are on the graph?

   b. Use the quadratic formula to try to find the roots of $0 = x^2 + 3x + 5$. What happens when you take the square root?

   c. Without looking at a graph, how can you use the quadratic formula to tell if a quadratic equation has any real roots?

5. The equation $h = -4.9t^2 + 6.2t + 1.9$ models the height of a soccer ball after Brandi hits it with her head. The t-values represent the time in seconds, and the h-values represent the height in meters. Write an equation that describes each event in 5a–c. Then use the quadratic formula to solve it. Explain the real-world meanings of your solutions.

   a. The ball hits the ground.

   b. The ball is 3 meters above ground.

   c. The ball is 4 meters above ground.

Brandi Chastain of the USA team heads a soccer ball during the 1999 Women's World Cup at Stanford Stadium.

6. Find an equation whose solutions are shown. Evaluate each expression to the nearest 0.1. Make a sketch showing a parabola that has these x-intercepts.

   a. $\dfrac{14 \pm \sqrt{(-14)^2 - 4(1)(49)}}{2(1)}$

   b. $\dfrac{3 \pm \sqrt{(-3)^2 - 4(2)(2)}}{2(2)}$

   c. $\dfrac{3 \pm \sqrt{(-3)^2 - 4(2)(-2)}}{2(2)}$

7. Match each quadratic equation with its graph. Then explain how to find the number of x-intercepts from the discriminant, $b^2 - 4ac$.

a. $y = x^2 + x + 1$ **b.** $y = x^2 + 2x + 1$ **c.** $y = x^2 + 3x + 1$

i. 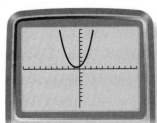   **ii.** 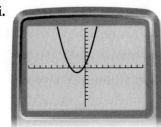   **iii.**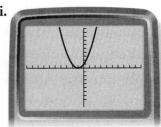

8. The quadratic formula gives two roots for an equation:

$$x = \frac{-b + \sqrt{b^2 - 4ac}}{2a} \quad \text{and} \quad x = \frac{-b - \sqrt{b^2 - 4ac}}{2a}$$

What is the average of these two roots? How does averaging the roots help you find the vertex?

9. The equation $h = -4.9t^2 + 17t + 2.2$ models the height of a stone thrown into the air, where t is in seconds and h is in meters. Use the quadratic formula to find how long the stone is in the air.

10. **APPLICATION** A shopkeeper is redesigning the rectangular sign on her store's rooftop. She wants the largest area possible for the sign. When she considers adding an amount to the width, she subtracts that same amount from the length. Her original sign has a width of 4 m and a length of 7 m.

a. Complete the table.

Increase (x) (m)	Width (m)	Length (m)	Area (sq. m)	Perimeter (m)
0	4	7		
0.5				
1.0				
1.5				
2.0				

b. How do the changes in width and length affect the perimeter?

c. How do the changes in width and length affect the area?

d. Write an equation in factored form for the area A of the rectangle in terms of x, the amount she adds to the width.

e. What are the dimensions of the rectangle with the largest area?

► Review

11. Reduce each rational expression by making a rectangular diagram for the quadratic expression in the numerator. Then cancel one of the binomial factors as shown. For example, to reduce $\frac{x^2 + 3x - 4}{x - 1}$, draw the diagram.

So $\frac{x^2 + 3x - 4}{x - 1} = \frac{(x - 1)(x + 4)}{x - 1} = x + 4$.

a. $\frac{x^2 - 5x + 6}{x - 3}$ **b.** $\frac{x^2 + 7x + 6}{x + 1}$ **c.** $\frac{2x^2 - x - 1}{2x + 1}$

12. On your graph paper, sketch graphs of these equations. Then use your calculator to check your sketches.

a. $y - 2 = (x - 3)^2$ **b.** $y - 2 = -2|x - 3|$

IMPROVING YOUR **REASONING** SKILLS

In Chapter 2, you may have done the project "The Golden Ratio." Now you have the tools to calculate this number. One way to calculate the golden ratio is to add 1 to square it. The symbolic statement of this rule is $x^2 = x + 1$ (or $x = \sqrt{x + 1}$).

You can approximate this value using a recursive routine on your calculator.

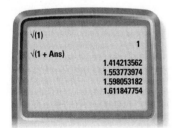

This is the same as calculating $\sqrt{1 + \sqrt{1 + \sqrt{1 + \sqrt{1 + \ldots}}}}$

You can also divide both sides of $x^2 = x + 1$ by x to get $x = 1 + \frac{1}{x}$.

You can use another recursive routine to approximate x this time.

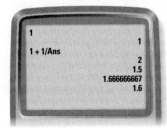

This is the same as calculating $1 + \dfrac{1}{1 + \dfrac{1}{1 + \dfrac{1}{1 + \ldots}}}$

Try different starting values for these recursive routines. Do they always result in the same number? Use one of the methods you learned in this chapter to solve $x^2 = x + 1$ symbolically. What are the answers in radical form? Can you write a recursive routine for the negative solution?

Cubic Functions

In this lesson you'll learn more about what cubic equations have in common with quadratic equations.

The area of the square at right is 16 square meters (m²), so you can cover the square using 16 smaller squares, each 1 m by 1 m. The sides of a square are equal, so you can write its area formula as *area = side²*. The squaring function, $f(x) = x^2$, models area. Each side length, or input, gives exactly one area, or output.

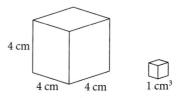

The volume of the cube at right is 64 cubic centimeters (cm³), so you can fill the cube using 64 smaller cubes, each measuring 1 cm by 1 cm by 1 cm. The edges of a cube have equal length, so you can write its volume formula as *volume = (edge length)³*. The cubing function, $f(x) = x^3$, models volume. Each edge length, or input, gives exactly one volume, or output.

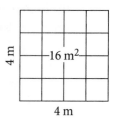

EXAMPLE A

The edge length of a cube with a volume of 64 is 4. So you can write $4^3 = 64$. You call 4 the **cube root** of 64 and the number 64 a **perfect cube** because its cube root is an integer. Then you can express the equation as $4 = \sqrt[3]{64}$. Determine which numbers are perfect cubes.

a. 59319   b. 2197   c. 495   d. 13824

▶ Solution

Find the cube root of each number: [▶ 🖩 See **Calculator Note 10B.** ◀]

a. $\sqrt[3]{59319} = 39$ b. $\sqrt[3]{2197} = 13$

c. $\sqrt[3]{495} \approx 7.910$ d. $\sqrt[3]{13824} = 24$

So 59319, 2197, and 13824 are perfect cubes. The number 495 is not a perfect cube because its cube root is not an integer.

Graphs of cubic functions have different and interesting shapes. In a window $-5 \le x \le 5$ and $-4 \le y \le 4$, the parent function $y = x^3$ looks like the graph shown.

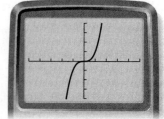

EXAMPLE B

Write an equation for each graph.

a.

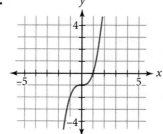

b.

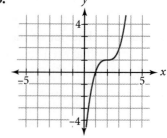

a. Each graph is a transformation of the graph of $y = x^3$. The graph shows a translation of the parent function down 1 unit. So the equation is $y = x^3 - 1$.

b. There is a translation of the graph of $y = x^3$ right 2 units and up 1 unit. So the equation is $y = (x - 2)^3 + 1$.

Investigation
Rooting for Factors

In this investigation you'll discover the connection between the factored form of a cubic equation and its graph.

Step 1 | List the x-intercepts for each of these graphs.

Graph A

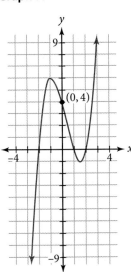

$(0, 4)$

Graph B

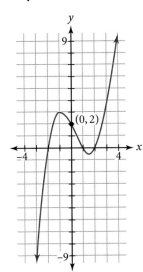

$(0, 2)$

Graph C

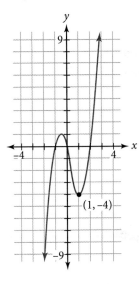

$(1, -4)$

Graph D

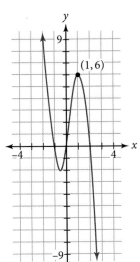

$(1, 6)$

Graph E

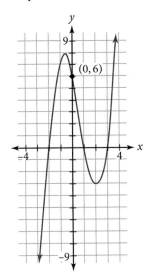

$(0, 6)$

Graph F

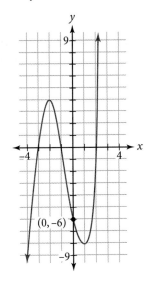

$(0, -6)$

Step 2 | Each equation below matches exactly one graph in Step 1. Use graphs and tables to find the matches.

a. $y = (x + 3)(x + 1)(x - 2)$ **b.** $y = 2x(x + 1)(x - 2)$

c. $y = (x + 2)(x - 1)(x - 2)$ **d.** $y = -3x(x + 1)(x - 2)$

e. $y = 0.5(x + 2)(x - 1)(x - 2)$ **f.** $y = (x + 2)(x - 1)(x - 3)$

Step 3 | Describe how the x-intercepts you found in Step 1 relate to the factored forms of the equations in Step 2.

Now you'll write an equation from a graph.

Step 4 | Use what you discovered in Steps 1–3 to write an equation with the same x-intercepts as the graph shown. Graph your equation; then adjust your equation to match the graph.

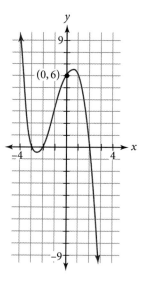

When you can identify the zeros of a function, you can write its equation in factored form. In this lesson you will see cubic equations with integer roots only.

EXAMPLE C | Find an equation for the graph shown.

▶ **Solution** | There are three x-intercepts on the graph. They are at $x = 0, -1,$ and 2. Each intercept helps you find a factor in the equation. These factors are $x, x + 1,$ and $x - 2$. Graph the equation $y = x(x + 1)(x - 2)$ on your calculator. The shape is correct, but you need to reflect it across the x-axis. You also need to vertically stretch the graph. Check the y-value of your graph at $x = 1$. The y-value is -2. You need it to be 4, so multiply by -2. The correct equation is $y = -2x(x + 1)(x - 2)$. Check this equation by graphing it on your calculator.

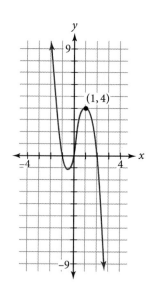

You can also use what you know about roots to convert cubic equations from general form to factored form. Look at the graph of $y = x^3 - 3x + 2$ at right. It has x-intercepts at $x = -2$ and $x = 1$. However, a cubic equation should have three roots. Where is the third one? Notice that the graph just touches the axis at $x = 1$. It doesn't actually pass through the axis. So the root at $x = 1$ is called a *double root*. The factor $x - 1$ is squared. Graph the factored form $y = (x + 2)(x - 1)^2$. It matches. If it didn't, you would need to look at a specific point to find the scale change required to make it match.

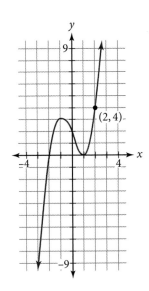

EXERCISES

You will need your calculator for problems **1**, **6**, and **7**.

Practice Your Skills

1. Determine whether each number is a perfect square, a perfect cube, or neither.

   a. 2,209   **b.** 5,832   **c.** 1,224   **d.** 10,201

2. Write and solve an equation to find the value of x in each figure.

  a.   **b.**   **c.**

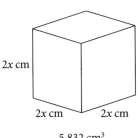
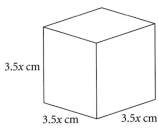
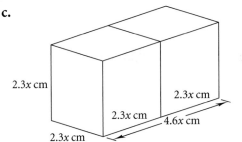

      5,832 cm³   21,952 cm³   3,309 cm³

3. Sometimes you can spot the factors of a polynomial expression without a graph of the equation. The easiest factors to see are those called *common monomial factors*. If you can divide each term by the same expression, then there is a common factor. Factor each expression by removing the largest possible common monomial factor.

   a. $4x^2 + 12x$ **b.** $6x^2 - 4x$ **c.** $14x^4 + 7x^2 - 21x$ **d.** $12x^5 + 6x^3 + 3x^2$

4. Determine whether each table represents a linear function, an exponential function, a cubic function, or a quadratic function.

a.

x	y
2	4
5	25
8	64
11	121
14	196

b.

x	y
2	7
5	11
8	15
11	19
14	23

c.

x	y
2	4
5	32
8	256
11	2,048
14	16,384

d.

x	y
2	8
5	125
8	512
11	1,331
14	2,744

5. Write an equation in factored form for each graph.

a.

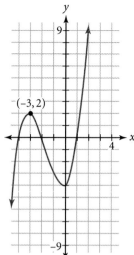

b.

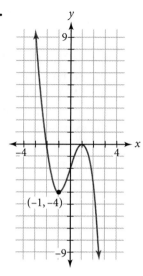

▶ Reason and Apply

6. Some numbers are both perfect squares and perfect cubes.

    a. Find at least three numbers that are both a perfect square and a perfect cube.

    b. Define a rule that you can use to find as many numbers as you like that are both perfect squares and perfect cubes.

7. The length of a box is made so that its length is 6 cm more than its width. Its height is 2 cm less than its width.

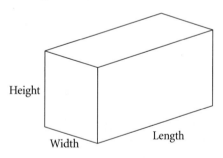

Height

Width   Length

    a. Use the width as the independent variable, and write an equation for the volume of the box.

    b. Suppose you want to ensure that the volume of the box is greater than 47 cm³. Use a graph and a table to describe all possible widths, to the nearest 0.1 cm, of such boxes.

8. Determine whether each statement about the equation $0 = 2x^3 + 4x^2 - 10x$ is true or false.

    a. The equation has three real roots.

    b. One of the roots is at $x = 2$.

    c. There is one positive root.

    d. The graph of $y = 2x^3 + 4x^2 - 10x$ passes through the point $(1, -4)$.

9. To convert from factored form to general form when there are more than two factors, first multiply any pair of factors. Then multiply the result by the other factor. For example, to rewrite the expression $(x + 1)(x + 3)(x + 4)$ in general form, first multiply the first two factors.

$$(x + 1)(x + 3) = x^2 + 1x + 3x + 3 = x^2 + 4x + 3$$

Then multiply the result by the third factor. You might want to use a rectangular diagram to do this.

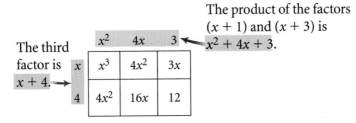

The product of the factors $(x + 1)$ and $(x + 3)$ is $x^2 + 4x + 3$.

The third factor is $x + 4$.

Next, combine like terms to find the sum of the regions.

$$x^3 + 4x^2 + 3x + 4x^2 + 16x + 12 = x^3 + 8x^2 + 19x + 12$$

Convert each expression below from factored form to general form. Use a graph or a table to compare the original factored form to your final general form.

a. $(x + 1)(x + 2)(x + 3)$

b. $(x + 2)(x - 2)(x - 3)$

10. The *girth* of a box is the distance completely around the box in one direction, that is, the length of a string that wraps around the box. Shippers put a maximum limit on the girth of a box rather than trying to limit its length, width, and height. Suppose you must ship a box with a girth of 120 cm in one direction and 160 cm in another direction.

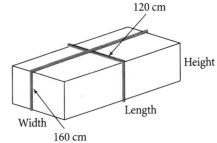

a. If the height of the box is 10 cm, what is the width of the box?

b. If the height of the box is 10 cm, what is the length of the box?

c. What is the volume of the box described in 10a and b?

d. If the height is 15 cm, what are the other two dimensions and what is the volume of the box?

e. If the height is x cm, find an expression for the width of the box.

f. If the height is x cm, find an expression for the length of the box.

g. Using your answers to 10e and f, find an equation for the volume of the box.

h. What are the roots of the equation you found in 10g, and what do they tell you?

i. Find the dimensions of a box with a volume of 48,488 cm^3.

▶ Review

11. Perform the operations, then combine like terms. Check your answers by using tables or graphs.

a. $(8x^3 - 5x) + (3x^3 + 2x^2 + 7x + 12)$

b. $(8x^3 - 5x) - (3x^3 + 2x^2 + 7x + 12)$

c. $(2x^2 - 6x + 11) + (-8x^2 - 7x + 9)$

d. $(2x^2 - 6x + 11) - (-8x^2 - 7x + 9)$

e. $(2x^2 - 6x + 11)(-8x^2 - 7x + 9)$

CHAPTER 10
REVIEW

In this chapter you learned about **quadratic functions.** You learned that they model **projectile motion** and the acceleration due to **gravity.** You discovered important connections between the **roots** and the **x-intercepts** of quadratic equations and graphs. You learned how to use the three different forms of quadratic equations:

General form	$y = ax^2 + bx + c$
Vertex form	$y = a(x - h)^2 + k$
Factored form	$y = a(x - r_1)(x - r_2)$ or $y = ax(x - r_2)$ if $r_1 = 0$

The expression $ax^2 + bx + c$ is a type of **polynomial** because it is the sum of many **terms** or **monomials.** The vertex form gives you information about the **line of symmetry** of the parabola. The factored form shows you the roots of the equation. By the **zero product property,** you know that if the polynomial equals zero, then one of the **binomial** factors, $(x - r_1)$ or $(x - r_2)$, must equal 0. The roots r_1 and r_2 are also called **zeros** of the quadratic function. You learned to expand the vertex and factored forms to the general form by combining like terms.

You first learned to locate solutions to quadratic equations using calculator tables and graphs. You then learned to solve equations symbolically by one of three methods—factor with rectangular diagrams, **complete the square,** or use the **quadratic formula.**

To use the quadratic formula, $x = \dfrac{-b \pm \sqrt{b^2 - 4ac}}{2a}$, you learned to identify the **leading coefficient,** the **linear term,** and the **constant term** for the **trinomial** $ax^2 + bx + c$. You also learned to calculate the **discriminant,** $b^2 - 4ac$, and saw that it gives information about the number of solutions to the equation.

You saw that solutions to quadratic equations often contain **radical expressions.** You learned that the square root of a negative number does not result in a **real number.** You also learned how to find **cube roots, perfect cubes,** and **perfect squares.** In the last lesson you studied cubic functions.

EXERCISES

You will need your calculator for problems **4** and **9.**

1. Tell whether each statement is true or false. If it is false, change the right side to make it true, but keep it in the same form. That is, if the statement is in factored form, write your corrected version in factored form.

a. $x^2 + 5x - 24 \overset{?}{=} (x + 3)(x - 8)$

b. $2(x - 1)^2 + 3 \overset{?}{=} 2x^2 + x + 1$

c. $(x + 3)^2 \overset{?}{=} x^2 + 9$

d. $(x + 2)(2x - 5) \overset{?}{=} 2x^2 - x - 10$

2. The equation of the graph at right is

$$y = -2(x + 5)^2 + 4$$

Describe the transformations on the graph of $y = x^2$ that give this parabola.

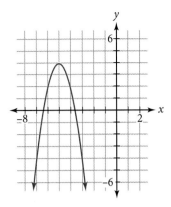

3. Write an equation for each graph below. Choose the form that best fits the information given.

a.

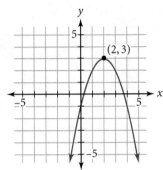

b.

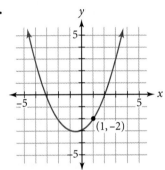

4. Write an equation in the form $y = a(x - h)^2 + k$ for each graph below. Name an appropriate viewing window.

a.

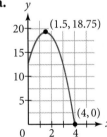

b.

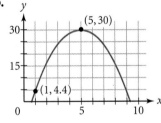

5. Use the zero product property to solve each equation.

a. $(2w + 9)(w - 3) = 0$ b. $(2x + 5)(x - 7) = 0$

6. Write an equation for a parabola that satisfies the given conditions.

a. The vertex is $(1, -4)$, and one of its x-intercepts is 3.

b. The x-intercepts are -1.5 and $\frac{1}{3}$.

7. Solve each equation by completing the square. Show each step. Leave your answers in radical form.

a. $x^2 + 6x - 9 = 13$ b. $3x^2 - 24x + 27 = 0$

8. Solve each equation by using the quadratic formula. Determine whether there are real number solutions. Leave your answer in radical form.

a. $5x^2 - 13x + 18 = 0$ **b.** $-3x^2 + 7x + 9 = 0$

9. The function $f(x) = 0.0015x(150 - x)$ models the rate at which the population of fish grows in a large aquarium. The x-value is the number of fish, and the $f(x)$-value is the rate of increase in the number of fish per week.

  a. Find $f(60)$, and give a real-world meaning for this value.

  b. For what values of x does $f(x) = 0$? What do these values represent?

  c. How many fish are there when the population is growing fastest?

  d. What is the maximum number of fish the aquarium has to support?

  e. Graph this function.

10. A toy rocket blasts off from ground level. After 0.5 second it is 8.8 feet high. It hits the ground after 1.6 seconds. Write an equation in factored form to model the height of the rocket as a function of time.

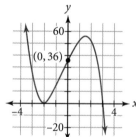

11. Name values of c so that $y = x^2 - 6x + c$ satisfies each condition below. Use the discriminant, $b^2 - 4ac$, or translate the graph of $y = x^2 - 6x$ to help you.

  a. The graph of the equation has no x-intercepts.

  b. The graph of the equation has exactly one x-intercept.

  c. The graph of the equation has two x-intercepts.

12. Use the quadratic formula to find the roots of each equation.

  a. $x^2 + 10x - 6 = 0$ **b.** $3x^2 - 8x + 5 = 0$

13. For each graph, identify the x-intercepts and write an equation in factored form.

  a.

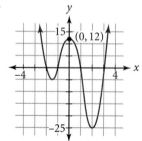

  b.

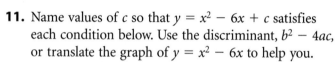

14. Make a rectangular diagram to factor each expression.

  a. $x^2 + 7x + 12$

  b. $x^2 - 14x + 49$

  c. $x^2 + 3x - 28$

  d. $x^2 - 81$

TAKE ANOTHER LOOK

In this chapter you have encountered many equations, such as $x^2 = -4$, that have no real solutions. The solutions to these equations exist in another set of numbers called **imaginary numbers.** To find the solution to $x^2 = -4$, mathematicians write $x = 2i$ or $x = -2i$. The symbol i represents the imaginary unit.

Express i as a square root of a negative number. (Hint: $2i = \sqrt{-4}$, so factor the 4 out of the radical to see what i is.) What happens if you multiply i by itself to find i^2? Use this result to find i^3 and i^4. What happens if you keep multiplying by i?

Use the pattern you discovered to calculate i^{10}, i^{25}, and i^{100}.

Learn more about imaginary numbers with the links at www.keymath.com/DA .

Assessing What You've Learned

ORGANIZE YOUR NOTEBOOK Choose your best graph of a parabola from this chapter. Label the vertex, roots, line of symmetry, and y-intercept of the graph. Show the quadratic equation for the graph in each quadratic form—general, vertex, and factored.

WRITE IN YOUR JOURNAL Add to your journal by answering one of these prompts.

▶ There are many ways to solve quadratic equations—calculator tables and graphs, factoring, completing the square, and the quadratic formula. Which method do you like best? Do you always use the same method, or does the method depend on the problem?

▶ Compare each form of a quadratic equation—general, vertex, and factored. What information does each form tell you? How can you convert an equation from one form to another?

PERFORMANCE ASSESSMENT Show a classmate, a family member, or your teacher that you can solve any quadratic equation. Demonstrate how to find solutions with a calculator (graph or table) and by hand (factoring, completing the square, or using the quadratic formula).

GIVE A PRESENTATION Work with a partner or in a group to create your own problem about projectile motion. It can be about the height of a ball, the path of a rocket, or some other object. If possible, conduct an experiment to collect data. Decide which information will be given and which form of quadratic equation to use. Make up a question about your problem. Solve the problem using one of the methods you learned in this chapter. Present the problem and its solution to the class on a poster.

Introduction to Geometry

These brightly colored wall paintings are a traditional art form of South Africa's Ndebele tribe. Ndebele women paint murals like these to celebrate special occasions such as weddings and harvests. Learning and continuing the ancient art form is an important part of training for young girls. The painters' use of universally recognized geometric shapes help these murals transcend time and cultural boundaries.

OBJECTIVES

In this chapter you will

- learn concepts, definitions, and symbols important in geometry
- use algebra to describe geometric relationships
- discover some properties of parallel and perpendicular lines
- find the coordinates of a line segment's midpoint
- calculate the distance between two points
- learn more about square roots
- explore important relationships between the sides of a right triangle

Parallel and Perpendicular

When you draw geometric figures on scaled coordinate axes, you are doing **analytic geometry.** You describe relationships and properties of the figures using the axes to identify points and write the equations of lines. In this lesson you will discover some interesting connections between algebra and geometry.

Parallel lines are lines in the same plane that never intersect. They are always the same distance apart. You draw arrowheads on each line to show that they are parallel.

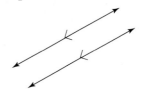

Perpendicular lines are lines that meet at a right angle, that is, at an angle that measures 90°. In fact, four right angles are formed where perpendicular lines intersect. You draw a small box in one of the angles to show that the lines are perpendicular.

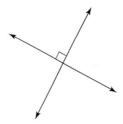

The Russian artist Wassily Kandinsky (1866–1944) used parallel and perpendicular line segments in his 1923 work titled *Circles in a Circle.*

Investigation
Slopes

You will need

• graph paper
• a straightedge

A rectangle has two pairs of parallel line segments and four right angles. When you draw a rectangle on the coordinate plane and notice the slopes of its sides, you will discover how the slopes of parallel and perpendicular lines are related.

Step 1 Draw coordinate axes centered on graph paper. Each member of your group should choose one of the following sets of points. Plot the points and connect them, in order, to form a closed polygon. You should have a rectangle.

a. $A(6, 20)$, $B(13, 11)$, $C(-5, -3)$, $D(-12, 6)$

b. $A(3, -1)$, $B(-3, 7)$, $C(9, 16)$, $D(15, 8)$

c. $A(-11, 21)$, $B(17, 11)$, $C(12, -3)$, $D(-16, 7)$

d. $A(3, -10)$, $B(-5, 22)$, $C(7, 25)$, $D(15, -7)$

The slope of a line segment is the same as the slope of the line containing the segment. You can write the segment between A and D as $\overline{AD}$.

Step 2 Find the slopes of $\overline{AD}$ and $\overline{BC}$.

Step 3 Find the slopes of $\overline{AB}$ and $\overline{DC}$.

Step 4 What conclusion can you make about the slopes of parallel lines based on your answers to Steps 2 and 3?

The ties underneath these railroad tracks in British Columbia are a real-world example of parallel segments.

To find the **reciprocal** of a number, you write the number as a fraction and then invert the numerator and the denominator. For example, the reciprocal of $\frac{2}{3}$ is $\frac{3}{2}$. The product of reciprocals is 1.

Step 5 Express the slope values of $\overline{AB}$ and $\overline{BC}$ as reduced fractions.

Step 6 Express the slope values of $\overline{AD}$ and $\overline{DC}$ as reduced fractions.

Step 7 What conclusion can you make about the slopes of perpendicular lines? What is their product? Check your conclusion by finding the slopes of any other pair of perpendicular sides in your rectangle.

Step 8 On the coordinate plane, draw two new pairs of parallel lines that have the slope relationship you discovered in Step 7. What figure is formed where the two pairs of lines intersect?

These street intersections in New York City are a real-world example of perpendicular lines.

You can draw any polygon on a graph and assign coordinate pairs to its vertices. Then you can use these points to calculate slopes, lengths of sides, perimeters, areas, and even the sizes of angles. You can use this information to draw conclusions about the polygon.

A **right triangle** has one right angle. The sides that form the right angle are called **legs,** and the side opposite the right angle is called the **hypotenuse.**

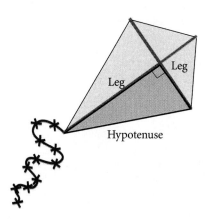

EXAMPLE A

Triangle ABC (written as △ABC) is formed by connecting the points (1, 3), (9, 5), and (10, 1). Is it a right triangle?

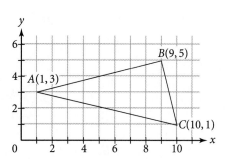

▶ **Solution**

The slope of $\overline{AB}$ is $\frac{1}{4}$, the slope of $\overline{AC}$ is $\frac{-2}{9}$, and the slope of $\overline{BC}$ is -4. The slopes $\frac{1}{4}$ and -4 are negative reciprocals of each other, so the sides with these slopes are perpendicular. That means angle B is a right angle. So these three points define a right triangle. Did you notice that the product of the two slopes, $\frac{1}{4}$ and -4, is -1?

A **trapezoid** is a quadrilateral with one pair of opposite sides that are parallel and one pair of opposite sides that are not parallel. A trapezoid with one of the nonparallel sides perpendicular to both parallel sides is a **right trapezoid.**

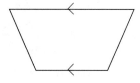

Trapezoid

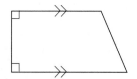

Right trapezoid

EXAMPLE B | Classify as specifically as possible the polygon formed by the points $A(-4, 1)$, $B(-2, 4)$, $C(4, 0)$, and $D(-1, -1)$.

▶ **Solution** | Plot the vertices on a coordinate plane and connect them to form a quadrilateral.

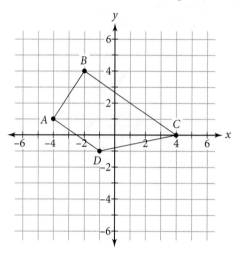

Calculate the slopes of the sides. Notice equal slopes (parallel sides) and negative reciprocal slopes (perpendicular sides).

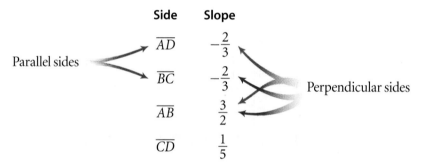

Side	Slope
$\overline{AD}$	$-\dfrac{2}{3}$
$\overline{BC}$	$-\dfrac{2}{3}$
$\overline{AB}$	$\dfrac{3}{2}$
$\overline{CD}$	$\dfrac{1}{5}$

Parallel sides

Perpendicular sides

Quadrilateral $ABCD$ has one set of parallel sides and one side perpendicular to that pair. So $ABCD$ is a right trapezoid.

EXERCISES

▶ Practice Your Skills

1. Find the slope of each line.

  a. $y = 0.8(x - 4) + 7$

  b. $y = 5 - 2x$

  c. $y = -1.25(x - 3) + 1$

  d. $y = -4 + 2x$

  e. $6x - 4y = 11$

  f. $3x + 2y = 12$

  g. $-9x + 6y = -4$

  h. $10x - 15y = 7$

2. Write the equation of a line with a slope of $\frac{3}{4}$ that passes through the point $(-2, 5)$.

3. Determine whether each pair of lines is parallel, perpendicular, or neither.

a. $y = 0.8(x - 4) + 7$
$y = -1.25(x - 3) + 1$

b. $y = 5 - 2x$
$y = -4 + 2x$

c. $6x - 4y = 11$
$-9x + 6y = -4$

d. $3x + 2y = 12$
$10x - 15y = 7$

▶ Reason and Apply

For problems 4–11, plot each set of points on graph paper and connect them in order to form a polygon. Classify each polygon using the most specific term that describes it. Justify your answers by finding the slopes of the sides of the polygons.

4. $(-5, 0), (1, 4), (6, 3), (-3, -3)$

5. $(-3, -2), (3, 1), (5, -3), (-1, -6)$

6. $(-3, 4), (0, 4), (3, 0), (3, -4)$

7. $(-1, 4), (2, 7), (5, -2), (2, -5)$

8. $(-4, -1), (-2, 7), (2, 6), (3, 3)$

9. $(0, 4), (2, 8), (6, -2), (2, -1)$

10. $(-8, -2), (-4, 4), (5, -2), (1, -8)$

11. $(-2, 2), (1, 5), (4, 2), (1, -3)$

12. Name coordinates for the vertices of a quadrilateral that has two right angles and no parallel sides.

▶ Review

13. Find the solution for each system, if there is one.

a. $\begin{cases} y = 0 \\ y = 2 + 3x \end{cases}$

b. $\begin{cases} y = 0.25x - 0.25 \\ y = 0.75 + x \end{cases}$

c. $\begin{cases} 2y = x - 2 \\ 3y = x - 3 \end{cases}$

14. Multiply and combine like terms.

a. $x(x + 2)(2x - 1)$

b. $(0.1x - 2.1)(0.1x + 2.1)$

15. At right, what is the ratio of the total area of shaded triangles to the area of the largest triangle?

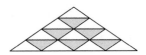

16. Find the value halfway between

a. 3 and 11

b. -4 and 7

c. -12 and -1

d. 2 and 47

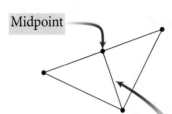
Finding the Midpoint

In analytic geometry you can use the algebraic concept of slope to identify parallel and perpendicular lines. That helps you recognize and draw geometric figures like rectangles and right triangles. Another useful skill is to find a **midpoint,** the middle point, of a line segment. Midpoints are used, for example, to draw these two geometric figures.

Balance is beautiful.

MIYOKO OHNO

Midpoint

Midpoint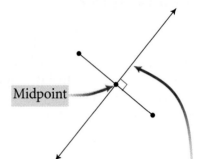

A **median** of a triangle is a segment that connects a vertex to the midpoint of the opposite side.

A **perpendicular bisector** is a line that divides a segment in half and that passes through the segment at a right angle.

Investigation
In the Middle

You will need

- graph paper
- a straightedge

In this investigation you will discover a method for finding the coordinates of the midpoint of a segment. As you work through the steps, think about which algebra concepts help you find a midpoint.

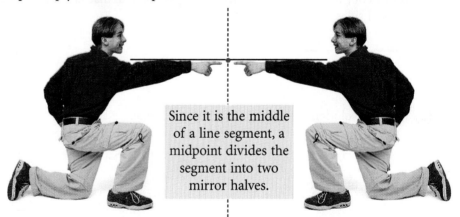

Since it is the middle of a line segment, a midpoint divides the segment into two mirror halves.

Step 1	Plot the points $A(1, 2)$, $B(5, 2)$, and $C(5, 7)$ and connect them.
Step 2	Find the midpoint of $\overline{AB}$. How did you find this point?
Step 3	Find the midpoint of $\overline{BC}$. How did you find this point?
Step 4	Find the midpoint of $\overline{AC}$. How does the midpoint's x-coordinate compare to the x-coordinates of A and C? How does its y-coordinate compare to the y-coordinates of A and C?

Step 5	Consider the points $D(2, 5)$ and $E(7, 11)$. Find the midpoint of $\overline{DE}$.
Step 6	Explain how to find the coordinates of the midpoint of a line segment between any two points.
Step 7	Does your method in Step 6 work even if you don't graph the points? If not, change your method so that it will work.
Step 8	Find the midpoint of the segment between each pair of points.

       a. $F(-7, 42)$ and $G(2, 14)$

       b. $H(2.4, -1.8)$ and $J(-4.4, -2.2)$

There are several ways to find the midpoint of a segment. However, the midpoint is always halfway between the two endpoints, so its x-coordinate will be the average of the x-coordinates of the endpoints. Likewise, its y-coordinate will be the average of the y-coordinates of the endpoints.

In the next example you'll combine your knowledge of midpoints and slopes.

EXAMPLE

A triangle has vertices $A(-4, 3)$, $B(5, 9)$, and $C(0, -3)$.

a. Write the equation of the median from vertex B.

b. Write the equation of the perpendicular bisector of $\overline{AB}$.

▶ **Solution**

First, plot $\triangle ABC$.

a. The median from vertex B connects to the midpoint of $\overline{AC}$. Find the midpoint of $\overline{AC}$, then sketch the median.

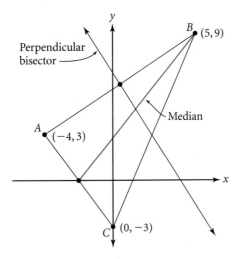

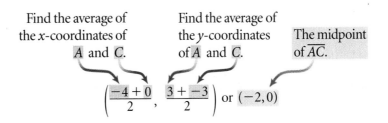

Find the average of the x-coordinates of A and C.

Find the average of the y-coordinates of A and C.

The midpoint of $\overline{AC}$.

$$\left(\frac{-4+0}{2}, \frac{3+-3}{2}\right) \text{ or } (-2, 0)$$

Now use the coordinates of vertex B and the midpoint of $\overline{AC}$ to find the slope of the median.

Find the slope between
vertex B and the midpoint.

$$\text{Slope} = \frac{9 - 0}{5 - (-2)} = \frac{9}{7}$$

Use the coordinates of the midpoint and the slope to write the equation of the median in point-slope form.

$$y = 0 + \frac{9}{7}(x - (-2)) \quad \text{or} \quad y = \frac{9}{7}(x + 2)$$

b. Look back at the sketch on page 589. The perpendicular bisector goes through the midpoint of $\overline{AB}$ and is perpendicular to $\overline{AB}$. First find the midpoint of $\overline{AB}$.

$$\left(\frac{-4 + 5}{2}, \frac{3 + 9}{2}\right) \text{ or } \left(\frac{1}{2}, 6\right)$$

The slope of $\overline{AB}$ is $\dfrac{9 - 3}{5 - (-4)}$ which equals $\dfrac{6}{9}$ or $\dfrac{2}{3}$. The slope of the perpendicular bisector is the negative reciprocal, or $\dfrac{-3}{2}$.

The equation of the perpendicular bisector of $\overline{AB}$ in point-slope form is

$$y = 6 + \frac{-3}{2}\left(x - \frac{1}{2}\right)$$

What you have learned about finding the midpoint of a segment is summarized by this formula.

> ### Midpoint Formula
>
> If the endpoints of a segment have the coordinates $\left(x_1, y_1\right)$ and $\left(x_2, y_2\right)$, the midpoint of the segment has the coordinates
>
> $$\left(\frac{x_1 + x_2}{2}, \frac{y_1 + y_2}{2}\right)$$

EXERCISES

▶ Practice Your Skills

1. Line ℓ has a slope of 1.2. What is the slope of line p that is parallel to line ℓ?

2. Line ℓ has a slope of 1.2. Line m is perpendicular to line ℓ.
   a. What is the slope of line m?
   b. What is the product of the slopes of line ℓ and line m?

3. Find the midpoint of the segment between each pair of points.
   a. $(4, 5)$ and $(-3, -2)$ **b.** $(7, -1)$ and $(5, -8)$

4. The equation of line ℓ has the form $Ax + By = C$. What is the slope of a line

    a. Perpendicular to line ℓ?   **b.** Parallel to line ℓ?

5. Find the midpoint of a segment with the endpoints (a, b) and (c, d).

▶ Reason and Apply

6. The vertices of $\triangle ABC$ are $A(0, 0)$, $B(1, 5)$, and $C(6, 4)$. Is it a right triangle? Explain how you know.

7. There is a situation in which two lines are perpendicular but the product of their slopes is not -1. Describe this situation.

8. The points $A(2, 1)$ and $B(4, 6)$ are the endpoints of a segment.

    a. Find the midpoint of $\overline{AB}$.

    b. Write the equation of the perpendicular bisector of $\overline{AB}$.

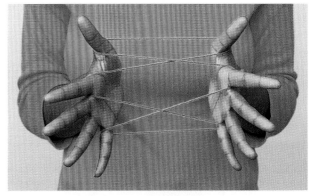

The starting position of the game Cat's Cradle shows triangles, parallel lines, and midpoints. What other geometric shapes do you see?

9. Sketch this quadrilateral on your own paper.

    a. Find the midpoint of each side.

    b. Connect the midpoints in order. What polygon is formed? How do you know?

    c. Draw the diagonals of the polygon formed in 9b. Are the diagonals perpendicular? Explain how you know.

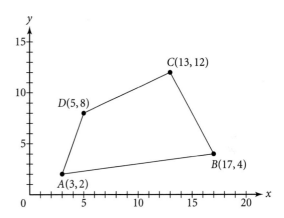

10. On graph paper, draw a triangle with vertices $A(11, 6)$, $B(4, -8)$, and $C(-6, 6)$.

    a. Find the midpoint of each side. Label the midpoint of $\overline{AB}$ point D, the midpoint of $\overline{BC}$ point E, and the midpoint of $\overline{CA}$ point F.

    b. Find the slopes of $\overline{AB}$ and $\overline{FE}$. What is special about these two line segments?

    c. Do your findings in 10b hold true between $\overline{BC}$ and $\overline{DF}$ and between $\overline{CA}$ and $\overline{ED}$? Explain why or why not.

    d. Compare the length of $\overline{ED}$ to the length of $\overline{CA}$.

    e. Without calculating, compare the area of $\triangle ABC$ to the area of $\triangle DEF$.

    f. If you connect the midpoints of the sides of $\triangle DEF$, how does the area of the new triangle compare to the area of $\triangle ABC$?

11. In 11a–c, you are given the midpoint of a segment and one endpoint. Find the other endpoint.

    a. midpoint: $(7, 4)$ endpoint: $(2, 4)$

    b. midpoint: $(9, 7)$ endpoint: $(15, 9)$

    c. midpoint: $(-1, -2)$ endpoint: $(3, -7.5)$

12. Two intersecting lines have the equations $2x - 3y + 12 = 1$ and $x = 2y - 7$.

   a. Find the coordinates of the point of intersection.

   b. Write the equations of two different lines that intersect at this same point.

   c. Write the equation of a parabola that passes through this same point.

13. Draw four congruent rectangles—that is, all the same size and shape.

   a. Shade half the area in each rectangle. Use a different way of dividing the rectangle each time.

   b. Which of your methods in 13a divide the rectangle into congruent polygons?

   c. Ripley divided one of her rectangles like this. Is the area divided in half? Explain.

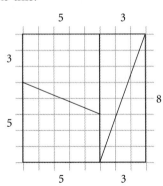

IMPROVING YOUR **GEOMETRY** SKILLS

This puzzle was created by the English mathematician Charles Dodgson (1832–1898). You may know him better as Lewis Carroll, the author of *Alice's Adventures in Wonderland.*

Cut an 8-by-8 square into pieces like this:

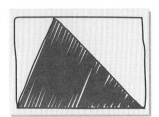

Reassemble them like this.

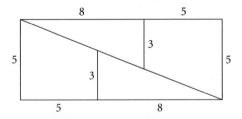

What is the area of the square? What is the area of the rectangle? Why aren't they equal?

Squares, Right Triangles, and Areas

Triangles, squares, rectangles, and other polygons have been important throughout the history of design and construction. Finding the areas of farms, lots, floors, and walls is important for city planners, architects, building contractors, interior designers, and people in building trades and other occupations. An architect designs space for the people who will use a building. A contractor must be able to determine an approximate price per square foot to bid a job.

In this lesson you will practice finding the area of squares and the lengths of their sides on graph paper. You'll use this relationship to find the lengths of the sides of a right triangle.

Framing a house requires many parallels, perpendiculars, and area calculations.

EXAMPLE A | Find the area of each shape on the grid at right.

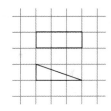

▶ **Solution** | The rectangle has an area of 3 square units. The area of the triangle is half the area of the rectangle, so it has an area of 1.5 square units.

You can often draw or visualize a rectangle or square related to an area to help you find the area.

EXAMPLE B | Find the area of square *ABCD*.

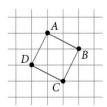

▶ **Solution** | Using the grid lines, draw a square around square *ABCD*. The outer square, *MNOP*, has an area of 9 square units. Each triangle has an area equal to half of 2 square units, or 1 square unit.

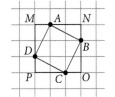

Area of square ABCD = Area of square MNOP − 4(1)

So the area of square *ABCD* is 9 − 4, or 5 square units.

Investigation
What's My Area?

You will need

• graph paper
• a straightedge

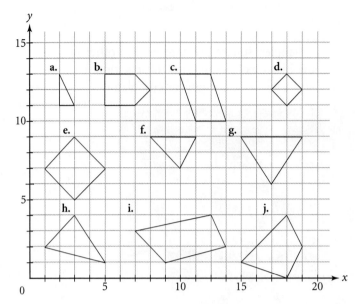

Step 1 | Copy these shapes onto your graph paper. Work with a partner to find the area of each figure.

If you know the side length, s, of a square, then the area of the square is s^2. Likewise, if you know that the area of a square is s^2, then the side length is $\sqrt{s^2}$, or s. So the square labeled d in Step 1, which has an area of 2, has a side length of $\sqrt{2}$ units.

Step 2 | What are the area and side length of the square labeled e in Step 1?

Step 3 | What are the area and side length of each of these squares?

a.   b.

Step 4 | Shown below are the smallest and largest squares with grid points for vertices that can be drawn on a 5-by-5 grid. Draw at least five other different-size squares on a 5-by-5 grid. They may be tilted, but they must be square, and their vertices must be on the grid. Find the area and side length of each square.

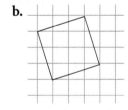

EXAMPLE C | Draw a line segment that is exactly $\sqrt{10}$ units long.

▶ **Solution** | A square with an area of 10 square units has a side length of $\sqrt{10}$ units. Ten is not a perfect square, so you will have to draw this square tilted. Start with the next largest perfect square, that is 16 square units (4-by-4), and subtract the areas of the four triangles to get 10. Here are two ways to draw a square tilted in a 4-by-4 square. Only the square on the left has an area of 10 square units.

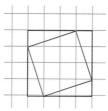 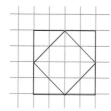

So a line segment with a length of $\sqrt{10}$ units looks like this:

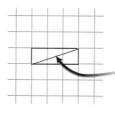

This red segment has a length of $\sqrt{10}$ units.

If a 4-by-4 square had not worked, you could have tried a larger square.

You can draw segments on graph paper with lengths equal to many square root values, but you may have to guess and check!

This fabric quilt, *Spiraling Pythagorean Triples*, shows several tilted squares. It was made by Diana Venters, a mathematician who uses mathematical themes in her quilts. You will learn about the Pythagorean theorem in Lesson 11.4.

See more mathematical quilts with the links at **www.keymath.com/DA** .

EXERCISES

You will need your calculator for problems **2** and **10.**

Practice Your Skills

1. Find an exact solution for each equation. Leave your answers in radical form.
   a. $x^2 = 47$
   b. $(x - 4)^2 = 28$
   c. $(x + 2)^2 - 3 = 11$
   d. $2(x - 1)^2 + 4 = 18$

2. Calculate decimal approximations for your solutions to problem 1. Round your answers to the nearest thousandth. Check each answer by substituting it into the original equation.

3. Find the area of each figure at right.

4. Find the side length of the square in 3f.

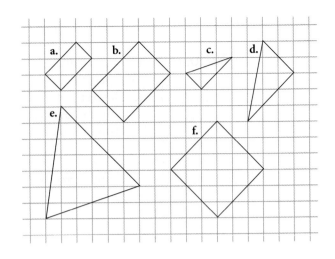

▶ Reason and Apply

5. Find the side lengths of the polygons in 3a, 3b, and 3e.

6. Find the area of each triangle below. You may want to draw each triangle separately on graph paper.

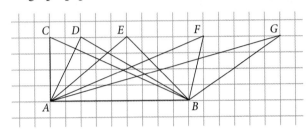

  a. △ABC **b.** △ABD **c.** △ABE **d.** △ABF **e.** △ABG

7. In the figure at right △ABC is a right triangle.

  a. Find the area of each of the squares built on the sides of this triangle.

  b. Find the lengths of $\overline{AB}$, $\overline{BC}$, and $\overline{AC}$.

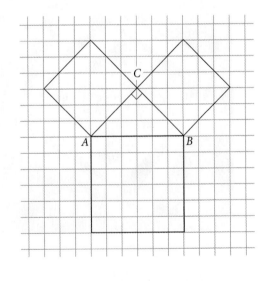

8. A square is drawn on graph paper. One side of the square is the segment with endpoints (2, 5) and (8, 1). Find the other two vertices of the square. There are two possible solutions. Can you find both?

9. Below is a right triangle.

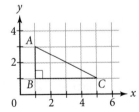

  a. Which side is the hypotenuse of △ABC? Which sides are the legs?

  b. Draw this triangle on graph paper and draw a square on each side, as in problem 7.

  c. Find the area of each square you drew in 9b.

  d. Find the lengths of $\overline{AB}$, $\overline{BC}$, and $\overline{AC}$.

  e. What is the relationship between the areas of the three squares?

▶ Review

10. The population of City A is currently 47,000 and is growing at a rate of 4.5% per year. The population of City B is currently 56,000 and is decreasing at a rate of 1.2% per year.

    a. What will the population of the two cities be in 5 years?

    b. When will the population of City A first exceed 150,000?

    c. If the population decrease in City B began 10 years ago, how large was the population before the decline started?

11. Use all these clues to find the equation of the one function that they describe.

    ▶ The graph of the equation is a parabola that crosses the x-axis twice.

    ▶ If you write the equation in factored form, one of the factors is $x + 7$.

    ▶ The graph of the equation has y-intercept 14.

    ▶ The axis of symmetry of the graph passes through the point $(-4, -2)$.

IMPROVING YOUR **VISUAL THINKING** SKILLS

The Chokwe people of northeastern Angola, Africa, are respected for their mat weaving designs. They weave horizontal white strands with vertical brown strands. In the design below, the first brown strand passes over one white strand and then under four white strands; the next brown strand to the right repeats the weaving pattern, but the design is translated down two units.

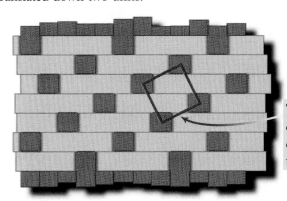

> The exposed brown strands could be connected to create tilted squares throughout the design.

Notice that the Chokwe design creates tilted squares similar to those you saw in this lesson. These tilted squares are repeated throughout the design.

Paulus Gerdes, a Mozambican mathematician, calls this design a "$(1, -2)$-solution" for finding a pattern of tilted squares. (*Geometry from Africa*, 1999, p. 75) That means if you move 1 unit right and 2 units down from any brown square, you hit another brown square.

Is this the only design that could be called a $(1, -2)$-solution? For example, can passing each brown strand over one white strand and then under three white strands create a pattern of tilted squares? How about over one, under two? How about over one, under five? Describe your results.

What under-over design creates a $(1, -3)$-solution?

The Pythagorean Theorem

You've seen that the area of a right triangle is half the area of the rectangle drawn around it. This is true for many other triangles too, but not all triangles.

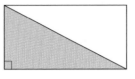

The area of this right triangle is half the area of the rectangle.

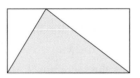

The area of this right triangle is also half the area of the rectangle.

The area of this triangle is actually less than half the area of the rectangle around it.

You might have figured out that triangles with the same base and the same height have the same area. This is true whether or not the triangles all fit inside the same rectangle. The area formula for a triangle is

$$Area = \frac{base \cdot height}{2} \quad \text{or} \quad A = \frac{1}{2}bh$$

You can use the formula to find the area of a triangle without adding grid squares or subtracting areas.

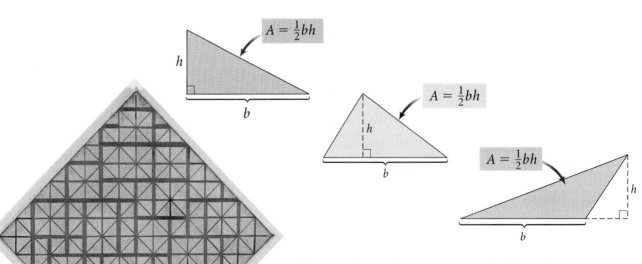

$A = \frac{1}{2}bh$

You'll remember a formula more easily if you discover it yourself. In this lesson you will discover a right triangle formula that planners and builders have used for thousands of years.

The Dutch artist Piet Mondrian (1872–1944) is famous for his use of straight lines, right angles, and geometric shapes. *Composition in Black and Gray* (1919) shows many right triangles and quadrilaterals.

Investigation
The Sides of a Right Triangle

You will need

- graph paper
- a straightedge

This investigation will help you discover a very useful formula that relates the lengths of the sides of a right triangle.

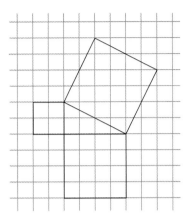

This is only a sample. Your right triangle should be larger or smaller.

Step 1 Draw a right triangle on graph paper with its legs on the grid lines and its vertices at grid intersections.

Step 2 Draw a square on each side of your triangle.

Step 3 Find the area of each square and record it.

Step 4 As a group or as a class, combine your results in a large table like this one. Look for a relationship between the numbers in each row of the table.

	Area of square on leg 1	Area of square on leg 2	Area of square on hypotenuse
Trisha's triangle			
Joe's triangle			

Step 5 Calculate the lengths of the legs and the hypotenuse for each triangle based on the areas you calculated in Step 3.

	Length of leg 1	Length of leg 2	Length of hypotenuse
Trisha's triangle			
Joe's triangle			

Step 6 Use what you discovered about the areas of the squares to write a rule relating the lengths of the legs to the length of the hypotenuse.

What you discovered in the investigation is the famous **Pythagorean theorem.** A theorem is a mathematical formula or statement that has been proven to be true. This theorem is named after Pythagoras, a Greek mathematician who lived around 500 B.C. This relationship was discovered and used by people in cultures before Pythagoras, but the theorem is usually given his name.

The Pythagorean Theorem

The sum of the squares of the lengths of the legs a and b of a right triangle equals the square of the length of the hypotenuse c.

$$a^2 + b^2 = c^2$$

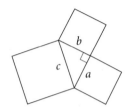

The next example shows how you can use what you learned in the investigation to find the missing length of a side of a right triangle.

EXAMPLE

A baseball diamond is a square with 90 ft between first and second base. What is the distance from home plate to second base?

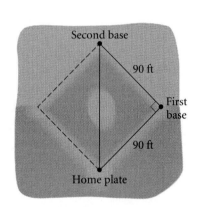

▶ **Solution**

The distance between home plate and second base is the hypotenuse of a right triangle. Call it c. This means that the area of a square on c equals the sum of the areas of the squares on each leg. The legs a and b are equal in this case.

$$c^2 = a^2 + b^2 \qquad \text{The Pythagorean theorem.}$$
$$ = 90^2 + 90^2 \qquad \text{Each leg is 90 ft.}$$
$$ = 8{,}100 + 8{,}100 \qquad \text{Square each leg length.}$$
$$c^2 = 16{,}200 \qquad \text{Add.}$$
$$c = \sqrt{16{,}200} \approx 127.3 \qquad \text{Find the square root.}$$

So the distance from home plate to second base is approximately 127.3 ft.

History
CONNECTION

The earliest known proof of the "Pythagorean" theorem came from China over 2500 years ago. Can you use this version of the Chinese proof to show $a^2 + b^2 = c^2$?

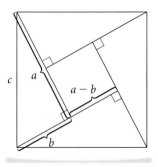

EXERCISES

You will need your calculator for problems **2, 5, 7, 9,** and **11.**

Practice Your Skills

In problems 1–4, *a* and *b* are the legs of a right triangle and *c* is the hypotenuse.

1. Suppose the square on side *c* has an area of 2601 cm² and the square on side *b* has an area of 2025 cm². What is the area of the square on side *a*?

2. Using the areas from problem 1, find each side length, *a, b,* and *c*.

3. Suppose $a = 10$ cm and $c = 20$ cm. Find the exact length of side *b* in radical form.

4. Suppose the right triangle is isosceles (two equal sides).

   a. Which two sides are the same length: the two legs or a leg and the hypotenuse?

   b. If the two equal sides are each 8 cm in length, what is the exact length of the third side in radical form?

Reason and Apply

5. APPLICATION Triangles that are similar to a right triangle with sides 3, 4, and 5 are often used in construction. The roof shown here is 36 ft wide. The two halves of the roof are congruent. Each half is a right triangle with sides proportional to 3, 4, and 5. (The shorter leg is the vertical leg.)

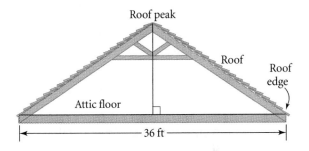

   a. How high above the attic floor should the roof peak be?

   b. How far is the roof peak from the roof edge?

   c. What is the shingled area of the roof if the building is 48 feet long?

6. Cal and Al are trying to solve the problem $\sqrt{x + 4} = 5$. Cal says that $\sqrt{x + 4} = \sqrt{x} + \sqrt{4}$. Al disagrees but can't explain why. Who is right? Explain your reasoning.

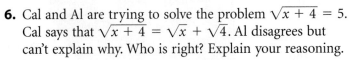

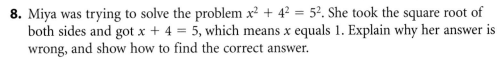

7. You will need a centimeter ruler for this problem.

   a. Measure the length and width of your textbook cover in centimeters.

   b. Use the Pythagorean theorem to calculate the diagonal length using the length and width you measured in 7a.

   c. Measure one of the diagonals of the cover.

   d. How close are the values you found in 7b and c? Should they be approximately the same?

8. Miya was trying to solve the problem $x^2 + 4^2 = 5^2$. She took the square root of both sides and got $x + 4 = 5$, which means *x* equals 1. Explain why her answer is wrong, and show how to find the correct answer.

9. The launching pad for a hot air balloon is 1.2 miles away from where you're standing. If the balloon rises vertically 3000 feet into the air, how far (in feet) will it be from you?

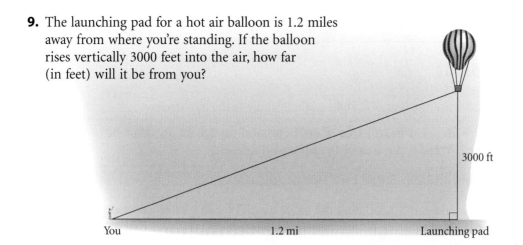

You   1.2 mi   Launching pad

3000 ft

10. Strips of graph paper may help in 10a.

a. Draw or make triangles with the side lengths that are given in i–iv. Then use a protractor or the corner of a sheet of paper to find whether or not each triangle is a right triangle.

   i. 5, 12, 13   **ii.** 7, 24, 25

   iii. 8, 10, 12   **iv.** 6, 8, 10

b. Based on your results from 10a does the Pythagorean theorem work in reverse? In other words, if the relationship $a^2 + b^2 = c^2$ is true, is the triangle necessarily a right triangle?

11. A 27-inch TV has a screen that measures 27 inches on its diagonal. Complete the following steps and use the Pythagorean theorem to find the **dimensions** of the screen with maximum area for a 27-inch TV.

a. Enter the positive integers from 1 to 26 into list L1 on your calculator to represent the possible screen widths.

b. Imagine a screen 1 inch wide. Calculate the length of a 27-inch TV screen with a width of 1 inch, and enter your answer into the first row of list L2.

c. Define list L2 to calculate all the possible screen lengths.

d. What is the area of a 27-inch screen with a width of 2 inches?

e. Define list L3 to calculate all possible screen areas.

f. Plot points in the form (*width, area*) and find an equation that fits these points.

g. What screen dimensions give the largest area for a 27-inch TV?

27 in.

The size of a television is measured on its diagonal.

27-inch TV

Width (L1)	Length (L2)	Area (L3)
1		
2		
3		
⋮		
26		

12. In problem 10, you showed that a triangle with side lengths of 5, 12, and 13 units is a right triangle.

    a. Explain why a triangle with side lengths of 10, 24, and 26 units is similar to this triangle.

    b. Is the triangle in 12a a right triangle? Explain.

13. APPLICATION Nadia Ferrell wants to build an awning over her porch. She wants the slope of the awning to be $\frac{5}{12}$. The porch is 8 feet deep, and the roof line is 14 feet above the porch. She draws this sketch to help her plan.

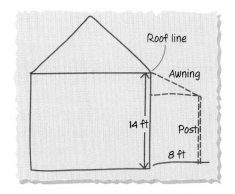

    a. How long will the awning be from the roof line to the porch support posts? Show your work.

    b. How tall will the posts be that hold up the front of the awning? Show your work.

Review

14. Ibrahim Patterson is planning to expand his square deck. He will add 3 feet to the width and 2 feet to the length to get a total area of 210 square feet. Find the dimensions of his original deck. Show your work.

PYTHAGORAS REVISITED

You know that the Pythagorean theorem says that the sum of the areas of the squares on the two legs of a right triangle is equal to the area of the square on the hypotenuse.

But if the shapes aren't squares, but are similar, does the sum of the areas on the legs still equal the area on the hypotenuse? Does a Pythagorean-like relationship still hold for the shape in these drawings?

Design your own similar shapes on the sides of a right triangle. Carefully measure or calculate their areas. Then examine and report on your results. Your project should include

▶ Your right triangle drawing with similar shapes on each side.

▶ Your measurements and calculations.

▶ A written explanation of how you drew the similar shapes, how you calculated the areas, and a conclusion whether or not a Pythagorean-like relationship holds for any shape.

Dot paper, graph paper, a computer drawing program, or The Geometer's Sketchpad software are useful tools for this project.

THE GEOMETER'S SKETCHPAD

In The Geometer's Sketchpad, you can drag parts of these figures and the triangles will remain right triangles and the three shapes will remain similar. Sketchpad also measures areas. Learn how to use its tools to create your own Pythagorean drawing!

Operations with Roots

When you use the Pythagorean theorem to find the length of a right triangle's side, the result is often a radical expression. You can always find an approximate value for a radical expression with a calculator, but sometimes it's better to leave the answer as an exact value, or in radical form. But there is more than one way to write an exact value or a radical expression. In this lesson you'll discover ways to rewrite radical expressions so that you can recognize solutions in a variety of forms.

EXAMPLE A | Draw a segment that is $\sqrt{13} + \sqrt{13}$ units long.

▶ **Solution** | First think of two perfect squares whose sum is 13, such as $4 + 9 = 13$. Then draw a right triangle on graph paper using the square roots of your numbers, 2 and 3, for the leg lengths. By the Pythagorean theorem, the hypotenuse of your triangle is $\sqrt{13}$.

Now draw a second congruent triangle so that the hypotenuses form a single segment. This pair of hypotenuses is $\sqrt{13} + \sqrt{13}$, or $2\sqrt{13}$, units long.

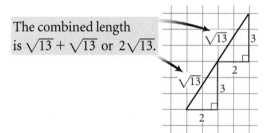

The combined length is $\sqrt{13} + \sqrt{13}$ or $2\sqrt{13}$.

Investigation
Radical Expressions

You will need

• graph paper

How can you tell if two different radical expressions are equivalent? Is it possible to add, subtract, multiply, or divide radical expressions? You'll answer these questions as you work through this investigation.

Step 1 | On graph paper, draw line segments for each length given below. You may need more than one triangle to create some of the lengths.

 a. $\sqrt{18}$ **b.** $\sqrt{40}$ **c.** $\sqrt{20}$

 d. $2\sqrt{5}$ **e.** $3\sqrt{2}$ **f.** $2\sqrt{10}$

 g. $\sqrt{10} + \sqrt{10}$ **h.** $\sqrt{2} + \sqrt{2} + \sqrt{2}$ **i.** $\sqrt{5} + \sqrt{5}$

Step 2	Do any of the segments seem to be the same length?
Step 3	Use your calculator to find a decimal approximation to the nearest ten thousandth for each expression in Step 1. Which expressions are equivalent?

Step 4	Find another way to write each expression below. Choose positive values for the variables, and use decimal approximations to check that your expression is equivalent to the original expression.

a. $\sqrt{x} + \sqrt{x} + \sqrt{x} + \sqrt{x}$ **b.** $\sqrt{x} \cdot \sqrt{y}$ **c.** $\sqrt{x \cdot x \cdot y}$

d. $\left(\sqrt{x}\right)^2$ **e.** $\dfrac{\sqrt{xy}}{\sqrt{y}}$

Step 5	Summarize what you discovered about adding, multiplying, and dividing radical expressions.
Step 6	Use what you've learned to find the area of each quadrilateral below. Give each answer in radical form as well as a decimal approximation to the nearest hundredth.

a.
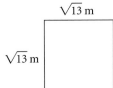
$\sqrt{13}$ m
$\sqrt{13}$ m

b.
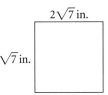
$2\sqrt{7}$ in.
$2\sqrt{7}$ in.

c.
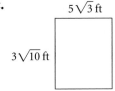
$5\sqrt{3}$ ft
$3\sqrt{10}$ ft

d.
$7\sqrt{8}$ cm
$5 + 6\sqrt{12}$ cm

EXAMPLE B

Rewrite each expression with as few square root symbols as possible and no parentheses.

a. $2\sqrt{3} + 5\sqrt{3}$ **b.** $3\sqrt{5} \cdot 2\sqrt{7}$

c. $\dfrac{\sqrt{15}}{\sqrt{3}}$ **d.** $\sqrt{3}\left(5\sqrt{2} + 3\sqrt{3}\right)$

▶ Solution

a. Add the terms using the distributive property.

$$2\sqrt{3} + 5\sqrt{3} = (2 + 5)\sqrt{3} = 7\sqrt{3}$$

The distributive property allows you to factor out $\sqrt{3}$.

The decimal approximations that your calculator gives support the idea that $2\sqrt{3} + 5\sqrt{3}$ is equivalent to $7\sqrt{3}$.

b. All of the numbers are multiplied. So use the commutative property of multiplication to group coefficients together and radical expressions together.

First the commutative property allows you to swap $\sqrt{5}$ and 2.

$$3\sqrt{5} \cdot 2\sqrt{7} = 3 \cdot 2 \cdot \sqrt{5} \cdot \sqrt{7}$$

Then to multiply two radical expressions, you multiply the numbers under the square root symbols.

$$3 \cdot 2 \cdot \sqrt{5} \cdot \sqrt{7} = 6\sqrt{5 \cdot 7} = 6\sqrt{35}$$

Use your calculator. Do the decimal approximations support the idea that $3\sqrt{5} \cdot 2\sqrt{7}$ is equivalent to $6\sqrt{35}$?

c. You can combine the numbers under one square root symbol and divide.

$$\frac{\sqrt{15}}{\sqrt{3}} = \sqrt{\frac{15}{3}} = \sqrt{5}$$

To divide radical expressions, you can rewrite the numbers under one square root symbol.

d. $\sqrt{3}\left(5\sqrt{2} + 3\sqrt{3}\right) = 5\sqrt{2} \cdot \sqrt{3} + 3\sqrt{3} \cdot \sqrt{3}$ Distribute $\sqrt{3}$.

$= 5\sqrt{6} + 3\sqrt{9}$ Multiply the radical expressions.

$= 5\sqrt{6} + 3 \cdot 3$ $\sqrt{9}$ is equal to 3.

$= 5\sqrt{6} + 9$ Multiply.

The investigation and Example B have illustrated several rules for rewriting radical expressions.

Rules for Rewriting Radical Expressions

For $x \geq 0$ and $y \geq 0$, and any values of a or b, these rules are true:

Addition of Radical Expressions

$$a\sqrt{x} + b\sqrt{x} = (a + b)\sqrt{x}$$

Multiplication of Radical Expressions

$$a\sqrt{x} \cdot b\sqrt{y} = a \cdot b\sqrt{x \cdot y}$$

For $x \geq 0$ and $y > 0$, this rule is true:

Division of Radical Expressions

$$\frac{\sqrt{x}}{\sqrt{y}} = \sqrt{\frac{x}{y}}$$

EXAMPLE C | Show that $6 + \sqrt{20}$ is a solution to the equation $0 = 0.5x^2 - 6x + 8$.

▶ **Solution** | If $6 + \sqrt{20}$ is a solution, then you will get a true statement when you substitute it into the equation and evaluate both sides.

$0 = 0.5x^2 - 6x + 8$	The original equation.
$0 \overset{?}{=} 0.5\left(6 + \sqrt{20}\right)^2 - 6\left(6 + \sqrt{20}\right) + 8$	Substitute $-6 + \sqrt{20}$ for x.
$0 \overset{?}{=} 0.5\left(6 + \sqrt{20}\right)^2 - 36 - 6\sqrt{20} + 8$	Distribute the 6 over $6 + \sqrt{20}$.
$0 \overset{?}{=} 0.5\left(36 + 6\sqrt{20} + 6\sqrt{20} + 20\right) - 36 - 6\sqrt{20} + 8$	Use a rectangular diagram to square the expression $6 + \sqrt{20}$.

	6	$\sqrt{20}$
6	36	$6\sqrt{20}$
$\sqrt{20}$	$6\sqrt{20}$	20

$0 \overset{?}{=} 18 + 3\sqrt{20} + 3\sqrt{20} + 10 - 36 - 6\sqrt{20} + 8$	Distribute the 0.5 over the expression in parentheses.
$0 \overset{?}{=} 18 + 10 - 36 + 8$	Combine the radical expressions.
$0 = 0$	Add and subtract.

The right side of the equation does evaluate to 0, so $6 + \sqrt{20}$ is a solution to the equation $0 = 0.5x^2 - 6x + 8$.

▶ Practice Your Skills

1. Rewrite each expression with as few square root symbols as possible and no parentheses. Use your calculator to support your answers with decimal approximations.

   a. $2\sqrt{3} + \sqrt{3}$

   b. $\sqrt{5} \cdot \sqrt{2} \cdot \sqrt{5}$

   c. $\sqrt{2}\left(\sqrt{2} + \sqrt{3}\right)$

   d. $\sqrt{5} - \sqrt{2} + 3\sqrt{5} + 6\sqrt{2}$

   e. $\sqrt{3}\left(\sqrt{2}\right) + 5\sqrt{6}$

   f. $\sqrt{2}\left(\sqrt{21}\right) + \sqrt{3}\left(\sqrt{14}\right)$

   g. $\dfrac{\sqrt{35}}{\sqrt{7}}$

   h. $\sqrt{5}\left(4\sqrt{5}\right)$

2. Find the exact length of the missing side for each right triangle.

   a.

   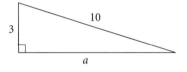

   b.

   c.

   d.

   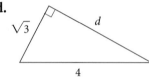

3. Write the equation for each parabola in general form. Use your calculator to check that both forms give the same graph or table.

   a. $y = \left(x + \sqrt{3}\right)\left(x - \sqrt{3}\right)$

   b. $y = \left(x + \sqrt{5}\right)\left(x + \sqrt{5}\right)$

4. Name the *x*-intercepts for each parabola in problem 3. Give both the exact value and a decimal approximation to the nearest thousandth for each *x*-value.

5. Name the vertex for each parabola in problem 3. Give both exact values and decimal approximations to the nearest thousandth for the coordinates of each vertex.

▶ Reason and Apply

6. Write the equation for each parabola in general form. Use your calculator to check that both forms have the same graph or table.

   a. $y = \left(x + 4\sqrt{7}\right)\left(x - 4\sqrt{7}\right)$

   b. $y = 2\left(x - 2\sqrt{6}\right)\left(x + 3\sqrt{6}\right)$

   c. $y = \left(x + 3 + \sqrt{2}\right)\left(x + 3 - \sqrt{2}\right)$

7. Name the *x*-intercepts for each parabola in problem 6. Give both the exact values and a decimal approximation to the nearest thousandth for each *x*-value.

8. Name the vertex for each parabola in problem 6. Give both exact values and decimal approximations to the nearest thousandth for the coordinates of each vertex.

9. A radical expression with a coefficient can be rewritten without a coefficient. Here's an example:

$2\sqrt{5}$ — The original expression.

$\sqrt{4} \cdot \sqrt{5}$ — $\sqrt{4}$ is equivalent to 2.

$\sqrt{20}$ — Multiply.

Use this method to rewrite each radical expression.

a. $4\sqrt{7}$ **b.** $5\sqrt{22}$ **c.** $18\sqrt{3}$ **d.** $30\sqrt{5}$

10. You can rewrite some radical expressions using the fact that they contain perfect-square factors. Here's an example:

$\sqrt{125}$ — The original expression.

$\sqrt{25 \cdot 5}$ — 25 is a perfect-square factor of 125.

$\sqrt{25} \cdot \sqrt{5}$ — Rewrite the expression as two radical expressions.

$5\sqrt{5}$ — Find the square root of 25.

Use this method to rewrite each radical expression.

a. $\sqrt{72}$ **b.** $\sqrt{27}$ **c.** $\sqrt{1800}$ **d.** $\sqrt{147}$

11. Show that $6 - \sqrt{20}$ is a solution to the equation $0 = 0.5x^2 - 6x + 8$.

12. APPLICATION The Great Pyramid of Cheops in Egypt has a square base with a side length of 800 feet. Its triangular faces are almost equilateral. These diagrams show an unfolded and a folded scale model of the pyramid.

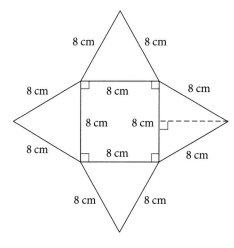

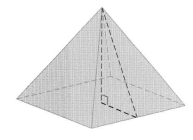

a. Note the model's measurements. Explain how to find the distance from the centers of the sides of the base to the tip of each triangle. (That is, find the length of the segment shown in red.)

b. Find the height of the model. (That is, find the length of the segment shown in blue.)

c. Use your result from 12b to find the approximate height of the Great Pyramid.

The Great Pyramid of Cheops was built using over 2,300,000 blocks of stone weighing 2.5 tons each.

13. The longest pole that fits in a rectangular box goes from one corner to the corner farthest from it. Find the longest pole that fits a 30-cm-by-50-cm-by-20-cm box. Show all your work. Give the answer as an exact value.

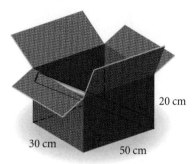

30 cm
50 cm
20 cm

14. Find the exact lengths of sides *a*, *b*, and *c* in the figure at right.

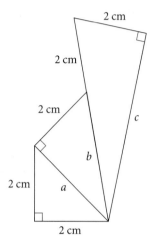

2 cm
2 cm
2 cm
c
2 cm
b
2 cm
a
2 cm

Review

15. Write an equation for each transformation of the graph of $y = x^2$.

   a. a translation up 3 units and right 2 units

   b. a reflection across the *x*-axis and then a translation up 4 units

   c. a vertical stretch by a factor of 3 and then a translation right 1 unit

16. How many *x*-intercepts does the graph of each equation in 15a–c have?

---

project

SHOW ME PROOF

Throughout history, many different civilizations have used the right triangle relationship $a^2 + b^2 = c^2$. The people of Babylonia, Egypt, China, Greece, and India all found this relationship useful and fascinating.

Along the way, there also have been many different proofs of this theorem. Drawing squares on each side of a right triangle is only one of them. Research the history of the Pythagorean theorem and locate a proof you find interesting. Prepare a paper or a presentation of the proof.

Your project should include

▶ A clear and accurate presentation of the proof. Include diagrams and mathematical equations when appropriate.

▶ A written or verbal explanation of why the proof works. (You may need to do some research to fully understand what a proof is.)

▶ The history associated with the proof.

▶ A list of the resources you used.

A Distance Formula

If you hike 2 kilometers east and 1 kilometer north from your campsite, do you know how far you are from camp? In this lesson you will use coordinate geometry and the Pythagorean theorem to find the distance between any two points.

The shortest distance between two points is under construction.

NOELIE ALTITO

EXAMPLE A

If you start at your campsite at the point (0, 0) and walk 2 km east and 1 km north to the point (2, 1), how far are you from your campsite?

▶ **Solution**

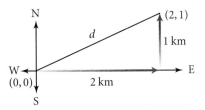

The east-west leg of the right triangle is $2 - 0$ or 2 km in length, and the north-south leg of the triangle is $1 - 0$ or 1 km in length. Let d be your distance from camp, the distance between points (0, 0) and (2, 1). By the Pythagorean theorem, $d^2 = 2^2 + 1^2$, so $d = \sqrt{2^2 + 1^2}$. So your distance from camp is $\sqrt{5}$ km, or approximately 2.24 km.

The four thunderbirds in the center of this Eastern Sioux buckskin pouch (ca. 1820) stand for the four cardinal directions—north, east, south, and west.

Investigation
Amusement Park

You will need

• graph paper

In this investigation you will discover a general formula for the distance between two points.

Step 1 | Copy the map of amusement park attractions and assign coordinates to each attraction on the map.

Step 2 | Find the distance between each pair of attractions in a–e. When appropriate, draw a right triangle. Use what you know about right triangles and the Pythagorean theorem to find the exact distance between each pair of attractions.

    a. Bumper Cars and Sledge Hammer

    b. Ferris Wheel and Hall of Mirrors

    c. Mime Tent and Hall of Mirrors

    d. Refreshment Stand and Ball Toss

    e. Bumper Cars and Mime Tent

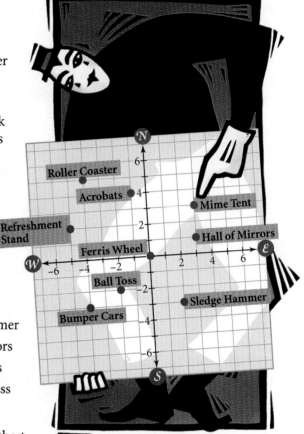

Step 3 | Which pairs of attractions are farthest apart? If each grid unit represents 0.1 mile, how far apart are they?

Step 4 | Chris parked his car at the coordinates $(17, -9)$. If each grid unit represents 0.1 mile, how far is it from the Refreshment Stand to his car? (Try to do this without plotting the location of his car.)

Two new attractions are being considered. The first attraction, designated P_1, will be located at the coordinates (x_1, y_1) as shown at right, and the second building, designated P_2, will be located at the coordinates (x_2, y_2).

Step 5 | Sketch a right triangle with horizontal and vertical legs and hypotenuse $\overline{P_1P_2}$.

Step 6 | Write an expression for the vertical distance between P_1 and P_2.

Step 7 | Write an expression for the horizontal distance between P_1 and P_2.

Step 8 | Write an expression for the distance between these two points. (This formula should work for any two points.)

Step 9 | Verify that your formula works by using it to find the distance between the Bumper Cars and the Mime Tent.

The Pythagorean theorem is an efficient way to solve many problems involving distance.

EXAMPLE B

If you walk 6 kilometers northeast from (2, 1), at a compass reading of 45°, what is your new location?

▶ Solution

First you find the horizontal and vertical change from your starting position. If you walk northeast, you walk just as far to the east as you walk to the north. Your path creates an isosceles right triangle. Use the Pythagorean theorem to find a in the sketch below.

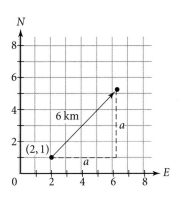

$a^2 + b^2 = c^2$ Pythagorean theorem.

$a^2 + a^2 = 6^2$ Substitute 6 for c, and substitute a for b because both legs are the same length.

$2a^2 = 36$ Add $a^2 + a^2$.

$a = \sqrt{18}$ Divide by 2 and take the square root of both sides.

To get the coordinates of your new location, add $\sqrt{18}$ to each coordinate of your starting location, (2 , 1). Your new location is exactly $\left(2 + \sqrt{18}, 1 + \sqrt{18}\right)$, or about 6.2 kilometers east and 5.2 kilometers north of the campsite.

You can also use similar trianges to find a. The triangle created by your path is similar to the triangle created when one grid square is cut in half. The smaller triangle has legs of 1 unit each and a hypotenuse of $\sqrt{2}$ units. Set up a proportion and solve for a.

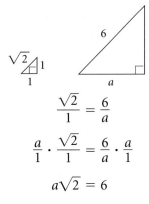

$\dfrac{\sqrt{2}}{1} = \dfrac{6}{a}$ Set up a proportion that compares the hypotenuses to the legs.

$\dfrac{a}{1} \cdot \dfrac{\sqrt{2}}{1} = \dfrac{6}{a} \cdot \dfrac{a}{1}$ Multiply both sides by a.

$a\sqrt{2} = 6$ Reduce.

$$a = \frac{6}{\sqrt{2}}$$ Divide both sides by $\sqrt{2}$.

$$a = \frac{\sqrt{36}}{\sqrt{2}}$$ Rewrite 6 as $\sqrt{36}$.

$$a = \sqrt{\frac{36}{2}}$$ Rewrite $\frac{\sqrt{36}}{\sqrt{2}}$ as $\sqrt{\frac{36}{2}}$.

$$a = \sqrt{18}$$ Rewrite $\sqrt{\frac{36}{2}}$ as $\sqrt{18}$.

Your result is the same as above.

Sometimes equations from distance problems have variables within a square root. Example C shows how to work with these situations.

EXAMPLE C | Solve the equation $\sqrt{15 + x} = x$.

▶ **Solution**

$$\sqrt{15 + x} = x$$ The original equation.

$$\left(\sqrt{15 + x}\right)^2 = x^2$$ Square both sides to undo the square root.

$$15 + x = x^2$$ The result of squaring.

$$0 = x^2 - x - 15$$ Subtract 15 and x from both sides to get a trinomial set equal to 0.

$$x \approx -3.4 \text{ and } x \approx 4.4$$ Use the quadratic formula, a graph, or a table to find two possible solutions.

Check:

$$\sqrt{15 + (-3.4)} \neq -3.4$$ The square root of a number can't be negative. So 3.4 is not a solution.

$$\sqrt{15 + 4.4} = \sqrt{19.4} \approx 4.4$$ This solution checks.

Whenever you solve a square root equation, be careful to check whether or not each solution satisfies the original equation.

In the second part of the investigation you used the Pythagorean theorem to derive the **distance formula.** When you know the coordinates of two points, this formula allows you to find the distance between the points even without plotting them.

> **Distance Formula**
>
> The distance d between $P_1(x_1, y_1)$ and $P_2(x_2, y_2)$ is given by the formula
> $$d = \sqrt{(x_2 - x_1)^2 + (y_2 - y_1)^2}$$

Look back at Example A. Can you show how to use the distance formula to solve this problem without graphing?

The length of a segment is the same as the distance between the endpoints of the segment. So, the distance formula has many applications in analytic geometry.

EXERCISES

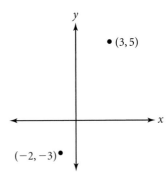

▶ Practice Your Skills

1. Is a triangle with side lengths of 9 cm, 16 cm, and 25 cm a right triangle? Explain.

2. Plot these points on graph paper.

   a. Draw the segment between the two points. Then draw a horizontal segment and a vertical segment to create a right triangle.

   b. Find the lengths of the horizontal and vertical segments.

   c. Find the exact distance between the two points.

3. On his homework, Matt wrote that the distance between two points was

$$\sqrt{(6-1)^2 + (3-7)^2}$$

What two points was Matt working with?

▶ Reason and Apply

4. Quadrilateral *ABCD* is pictured at right.

   a. What is the slope of each side?

   b. What type of quadrilateral is it?

   c. Find the length of each side.

For problems 5 and 6, refer to the Amusement Park investigation on page 612.

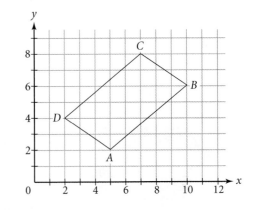

5. Use the distance formula to find the distance between each pair of attractions.

   a. Refreshment Stand and Bumper Cars

   b. Acrobats and Hall of Mirrors

6. Jake's sawdust spreader breaks down halfway between the Roller Coaster and the Sledge Hammer. At what point on the map on page 612 does the breakdown occur?

7. A rectangular box has the dimensions shown in the diagram.

  a. What is the length of the diagonal $\overline{BD}$?

  b. What is the length of the diagonal $\overline{BH}$?

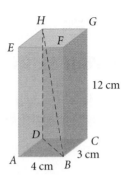

8. Consider the equation
$$\sqrt{20 - x} = x$$

  a. Solve the equation symbolically.

  b. Solve the equation using a graph or a table.

  c. Explain why you get two possible solutions when you solve the equation symbolically and only one solution when you look at a graph or table. Substitute both possible solutions into the original equation, and describe what happens.

▶ Review

9. Solve each equation.

  a. $\dfrac{3}{5} = \dfrac{a}{105}$ **b.** $\dfrac{1}{\sqrt{2}} = \dfrac{b}{7\sqrt{2}}$ **c.** $\dfrac{\sqrt{3}}{2} = \dfrac{c}{\sqrt{12}}$

10. **APPLICATION** Martin Weber is building a wheelchair ramp at the Town Hall. The ramp will meet a door that is 3 feet off the ground. Building codes in his area require an exterior ramp to have a slope of 0.05. How long will the ramp need to be? Give your answer in exact form and as an approximation to the nearest inch.

Ralph Hotchkiss, shown above in his Oakland, California, workshop, is a designer of wheelchairs and an advocate for physical independence. He founded Whirlwind Wheelchair International, an organization that provides wheelchairs to individuals in developing countries. For a mobility designer like Hotchkiss, handrail dimensions, seat angles, and the slope of wheelchair ramps are important issues.

Similar Triangles and Trigonometric Functions

Similar figures have corresponding angles that are equal and corresponding side lengths that are proportional. So the figures have the same shape, but one is an enlargement of the other. You can use ratios and proportions to compare and calculate length measurements for similar figures.

These Japanese cat figurines, called Maneki Neko, are near examples of three-dimensional similar figures. In what ways are they not mathematically similar?

EXAMPLE A

Elene is walking to school. From where she stands, she can see the 5-meter flagpole on top of her school. She holds her centimeter ruler approximately 50 centimeters from her eye. Against the ruler, the flagpole looks 25 centimeters tall. How far is she from school?

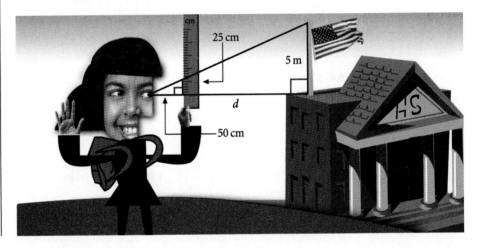

▶ **Solution**

This solution creates two similar triangles. One triangle is formed by Elene's eye and the 0-cm and 25-cm marks on her ruler. The other triangle is formed by Elene's eye and the ends of the flagpole. Elene's eye is a common vertex.

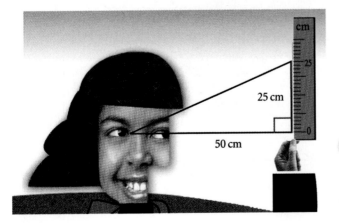

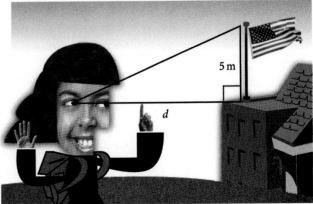

Use the ratios of adjacent sides to write the proportion

$$\frac{d \text{ m}}{5 \text{ m}} = \frac{50 \text{ cm}}{25 \text{ cm}}$$

Here, $d = 10$, so Elene is 10 meters from school.

Look back at the similar right triangles in Example A.

Notice the ratios compare the vertical legs to the horizontal legs. Both ratios, $\frac{25 \text{ cm}}{50 \text{ cm}}$ and $\frac{5 \text{ m}}{10 \text{ m}}$, equal 0.5.

If another triangle similar to those has a horizontal leg of 7 meters, then the vertical leg would be 3.5 meters in length. The angles in all of these triangles are about 26.6°, 63.4°, and 90°. Any other right triangle with a value of 0.5 for the ratio of its vertical leg to its horizontal leg also has angles with these measures.

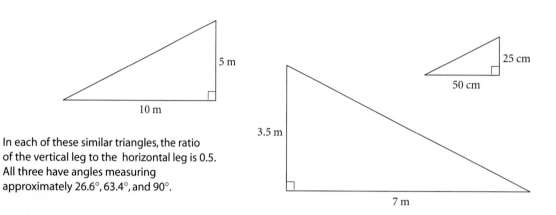

In each of these similar triangles, the ratio of the vertical leg to the horizontal leg is 0.5. All three have angles measuring approximately 26.6°, 63.4°, and 90°.

Likewise, if a right triangle has these angle measures, the ratio of its vertical leg to its horizontal leg is 0.5. There is a connection between the angle measures of a triangle and the ratios of its sides. You'll explore this connection in the investigation.

The angle below measures 36°. You already know that an angle that measures 90° is called a right angle. An angle that measures less than 90° is called an **acute angle.** An angle that measures more than 90° but less than 180° is called an **obtuse angle.**

Read 36° as the measure of this acute angle.

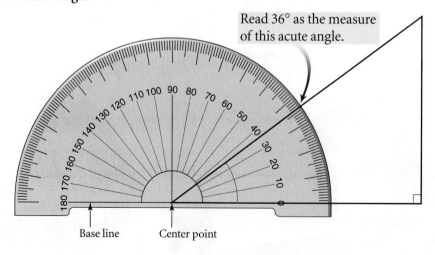

Base line   Center point

Investigation
Ratio, Ratio, Ratio

In this investigation you'll learn about some very important ratios in right triangles.

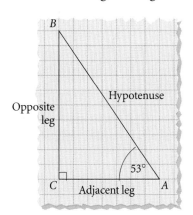

This is only a sample. Your triangle should look different.

Step 1 On graph paper, use a straightedge to draw a right triangle. Use the grid lines of the graph paper to make sure the legs are perpendicular. Make the triangle large enough for you to measure its angles accurately.

Step 2 Label one of the acute angles *A*. Measure it.

If you look at one acute angle in a right triangle, the **adjacent leg** is the leg of the triangle that is part of the measured angle. The **opposite leg** is the leg of the triangle that is not part of the angle you are looking at.

Step 3 Make a table like this one and record the information for each triangle drawn by a member of your group.

	Tony's triangle	Alice's triangle
Measure of angle A		
Length of adjacent leg (a)		
Length of opposite leg (o)		
Length of hypotenuse (h)		

Step 4 Make a new table like this one and calculate the ratios for each triangle drawn by a member of your group.

	Tony's triangle	Alice's triangle
Measure of angle A		
$\frac{o}{h}$		
$\frac{a}{h}$		
$\frac{o}{a}$		

	Tony's triangle	Alice's triangle
Measure of angle A		
sine (A)		
cosine (A)		
tangent (A)		

Step 5 With your calculator in degree mode, find the value of the **sine,** the **cosine,** and the **tangent** of angle *A*. [▶ 🖳 See **Calculator Note 11A** to learn about these functions on your calculator. ◀] Record these values to the nearest 0.01 in a table like this one.

Step 6 Compare your results for Steps 4 and 5. Define each function—sine, cosine, and tangent—in terms of a ratio of the lengths of the adjacent leg, the opposite leg, and the hypotenuse.

Step 7 Draw a larger right triangle with an acute angle *D* equal to your original angle *A*.

Step 8 Measure the side lengths and calculate the sine, the cosine, and the tangent of angle *D*. What do you find?

In the investigation you discovered that some ratios of the sides of a right triangle have special names: sine, cosine, and tangent. Sine, cosine, and tangent are all called **trigonometric functions.** They are fundamental to the branch of mathematics called **trigonometry.** Learning to identify the parts of a right triangle and to evaluate these functions for particular angle measures is an important problem-solving tool.

History
● **CONNECTION** ●

The word *trigonometry* comes from the Greek words for triangle and measurement. Its first use in English was in a 1614 translation of *Trigonometry: Doctrine of Triangles* by the Silesian mathematician Bartholmeo Pitiscus (1561–1613).

Trigonometric Functions

For acute angle *A* in a right angle, the trigonometric functions are

$$\text{sine of angle } A = \frac{\text{length of opposite leg}}{\text{length of hypotenuse}} \quad \text{or} \quad \sin A = \frac{o}{h}$$

$$\text{cosine of angle } A = \frac{\text{length of adjacent leg}}{\text{length of hypotenuse}} \quad \text{or} \quad \cos A = \frac{a}{h}$$

$$\text{tangent of angle } A = \frac{\text{length of opposite leg}}{\text{length of adjacent leg}} \quad \text{or} \quad \tan A = \frac{o}{a}$$

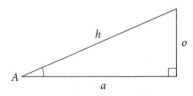

In this lesson you will get practice writing ratios associated with the trigonometric functions. You will also get more practice with ratio, proportion, and similarity. In the next lesson you will apply the ratios to solve problems.

EXAMPLE B

Find these ratios for this triangle.

a. sin A

b. cos A

c. tan B

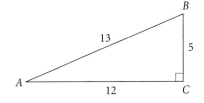

▶ **Solution**

For angle A, the side length values are $a = 12$, $o = 5$, and $h = 13$.

a. $\sin A = \dfrac{o}{h} = \dfrac{5}{13}$

b. $\cos A = \dfrac{a}{h} = \dfrac{12}{5}$

Using angle B changes which leg is opposite and which is adjacent. For angle B, the side length values are $a = 5$, $o = 12$, and $h = 13$.

c. $\tan B = \dfrac{o}{a} = \dfrac{12}{5}$

Note that identifying the opposite and adjacent legs depends on which angle you are using. Be careful to identify the correct sides and angles when using trigonometric functions.

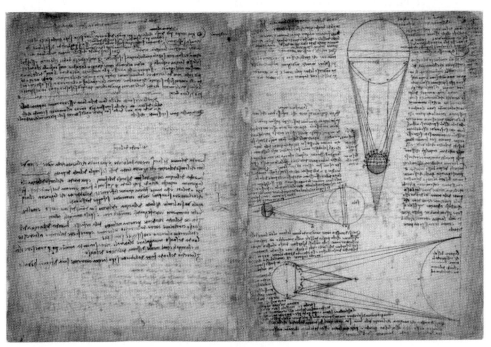

This sketch by Leonardo da Vinci (Italian, 1452–1519) explains why moonlight is less bright than sunlight. A diverse genius, Leonardo was a painter, draftsman, sculptor, architect, and engineer. The triangles used in this sketch show that he was also knowledgeable about geometry and trigonometry.

EXERCISES

You will need your calculator for problems **7, 8, 9, 10,** and **12.**

▶ Practice Your Skills

1. Solve each equation for *x*.

   a. $\dfrac{2}{3} = \dfrac{18}{x}$

   b. $\dfrac{7}{8} = \dfrac{x}{40}$

   c. $\dfrac{1}{4} = \dfrac{\sqrt{10}}{\sqrt{x}}$

   d. $\dfrac{2}{x} = \dfrac{x}{8}$

2. **APPLICATION** One inch on a road map represents 50 miles on the ground. Two cities are 3.6 inches apart on a map. What is the actual distance between the cities?

3. Find these ratios for the triangle at right.

   a. sin *D*

   b. cos *E*

   c. tan *D*

4. The diagram at right shows △*ABC* and △*ADE*.

   a. Are the triangles similar? Why or why not? (Hint: The angles in any triangle sum to 180°.)

   b. Find the ratio of corresponding side lengths of △*ADE* to △*ABC*.

   c. Find the lengths of $\overline{AD}$ and $\overline{AE}$.

   d. Find the areas of △*ADE* and △*ABC*.

   e. What is the ratio of the area of △*ADE* to the area of △*ABC*?

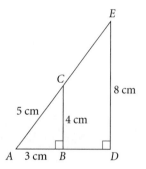

▶ Reason and Apply

5. Find *x* in each pair of similar triangles.

   a.

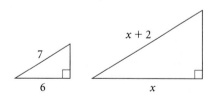

   b.

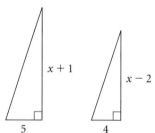

   c.

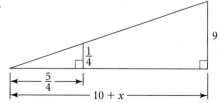

6. An 8-meter ladder is leaning against a building. The bottom of the ladder is 2 meters from the building.

a. How high on the building does the ladder reach?

b. A windowsill is 7 meters high on the building. How far from the building should the bottom of the ladder be placed to meet the windowsill?

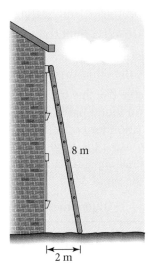

8 m

2 m

7. The wire attached to the top of a telephone pole makes a 65° angle with the level ground. The distance from the base of the pole to where the wire is attached to the ground is *d*. The height of the pole is *h*. The length of the wire is *w*.

a. What trigonometric function of 65° is the same as $\frac{d}{w}$?

b. What trigonometric function of 65° is the same as $\frac{h}{w}$?

c. What trigonometric function of 65° is the same as $\frac{h}{d}$?

d. Use your calculator to approximate the values in 7a–c to the nearest ten thousandth.

e. If the wire is attached to the ground 2.6 meters from the pole, how high is the pole?

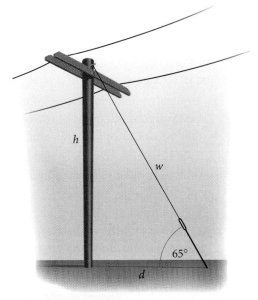

h

w

65°

d

8. Consider this right triangle with a 28° angle.

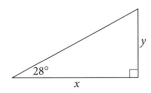

28°

x

y

a. Write an equation that relates *x* and *y*, the legs of the right triangle.

b. Graph your equation on your calculator. Make a sketch of the graph on your paper. Describe the graph.

c. Find *y* if *x* = 100.

d. If *y* = 80, find *x*.

Surveyors use a tool called a theodolite, or transit, to measure angles. The angles are sometimes used with trigonometry to measure distances.

9. You need a protractor and a straightedge for this problem.

   a. Draw an isosceles right triangle. Each acute angle should be 45°.

   b. Label one of the legs of your isosceles right triangle "1 unit." Find the exact lengths of the other two sides.

   c. Make a table like this one on your own paper. First write each ratio using the lengths you found in 9b. Then use your calculator to find a decimal approximation for each exact value to the nearest ten thousandth. Finally, find each ratio using the trigonometric function keys on your calculator. Check that your decimal approximations and the values using the trigonometric function keys are the same.

Trigonometric Functions for a 45° Angle

	Sine	Cosine	Tangent
Exact value of ratio			
Decimal approximation of exact value			
Value by trigonometric function keys			

10. You need a protractor and a straightedge for this problem.

   a. Draw an equilateral triangle. Each angle should be 60°. Draw a segment from one vertex to the midpoint of the opposite side. You should have two triangles with angles measuring 30°, 60°, and 90°.

   b. Label one side of your equilateral triangle "2 units." Find the exact lengths of the other two sides of your 30°-60°-90° triangles.

   c. Make tables like these on your own paper. First write each ratio using the lengths you found in 10b. Then use your calculator to find a decimal approximation for each exact value to the nearest ten thousandth. Finally, find each ratio using the trigonometric function keys on your calculator. Check that your decimal approximations and the values using the trigonometric function keys are the same.

Trigonometric Functions for a 30° Angle

	Sine	Cosine	Tangent
Exact value of ratio			
Decimal approximation of exact value			
Value by trigonometric function keys			

Trigonometric Functions for a 60° Angle

	Sine	Cosine	Tangent
Exact value of ratio			
Decimal approximation of exact value			
Value by trigonometric function keys			

Review

11. Here are four linear equations.

$$y = 2x - 1 \qquad x + 2y = 4$$
$$y = 3 + 2x \qquad x = -2y + 10$$

a. Graph the four lines. What polygon is formed?

b. Find the coordinates of the vertices of the polygon.

c. Find the linear equations for the diagonals of the polygon.

d. Find the coordinates of the point where the diagonals intersect.

12. Find the missing side lengths in this figure.

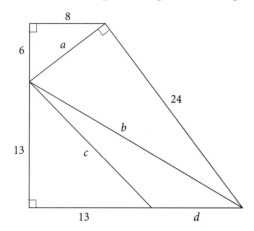

Two Tibetan Buddhist monks create a mandala from colored sand. The delicate geometric design will take days to create and will then be dismantled in a special ceremony.

Learn more about the mathematics and meaning of mandalas with the Internet links at www.keymath.com/DA .

Trigonometry

In Lesson 11.7, you learned about trigonometric ratios in right triangles. The trigonometric functions allow you to find the ratios of side lengths when you know the measure of an acute angle. So if you know the length of one side and the measure of one acute angle, you can solve for the lengths of the other sides.

EXAMPLE A

Consider this triangle.

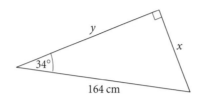

a. Find the length of the side labeled x.
b. Find the length of the side labeled y.

▶ **Solution**

a. The variable x represents the length of the side opposite the 34° angle. The length of the hypotenuse is 164 cm.

$\sin A = \dfrac{o}{h}$	Definition of sine.
$\sin 34° = \dfrac{x}{164}$	Substitute 34° for the measure of the angle and 164 for the hypotenuse.
$164 \sin 34° = x$	Multiply both sides by 164.
$91.7 \approx x$	Multiply and round to the nearest tenth.

The side labeled x is approximately 91.7 cm.

b. The variable y represents the length of the side adjacent to the 34° angle. The length of the hypotenuse remains 164 cm.

$\cos A = \dfrac{a}{h}$	Definition of cosine.
$\cos 34° = \dfrac{y}{164}$	Substitute the measure of the angle and hypotenuse.
$164 \cos 34° = y$	Multiply both sides by 164.
$136.0 \approx y$	Multiply and round to the nearest tenth.

The side labeled y is approximately 136.0 cm.

What if you know the lengths of the sides but want to know the measure of an acute angle? You can use the inverses of the trigonometric functions to find the angle measure when you know the ratio. The inverses of the trigonometric functions are inverse sine, inverse cosine, and inverse tangent. They are written $\sin^{-1}$, $\cos^{-1}$, and $\tan^{-1}$.

EXAMPLE B

Find the measure of angle *A*.

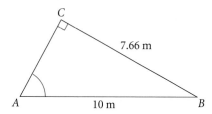

▶ **Solution**

Because you know the length of the side opposite angle *A* and the length of the hypotenuse, you can find the sine ratio.

$$\sin A = \frac{7.66}{10} = 0.766$$

You find the measure of the angle with the inverse sine of 0.766.

[▶ 🖥 See **Calculator Note 11B** to learn about the inverses of the trigonometric functions. ◀]

$$A = \sin^{-1}(0.766) \approx 50°$$

So the measure of angle *A* is approximately 50°. Check your answer using the sine function.

$$\sin 50° \approx 0.766$$

You use an inverse to undo a function. You may have noticed in Example B that sin and $\sin^{-1}$ undo each other the same way that squaring and finding the square root undo each other.

Investigation
Reading Topographic Maps

You will need

• a centimeter ruler
• a protractor

A **topographic map,** or contour map, reveals the shape of the land surface by showing different levels of elevation. The simple map below shows the elevation of a hill. You can use the map to determine the size of this hill and how steep it is. In this investigation you will take an imaginary hike over the summit and calculate the steepness of the hill at different points along the way.

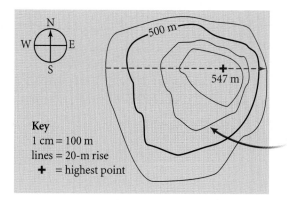

Each of these rings is called a **contour line,** or isometric line. There is a 20-m rise between each contour line.

Step 1

If you want to hike over the summit, will it be easier to hike up from the west or the east? How can you tell?

Step 2	Suppose you want to go from the west side of the hill, over the summit, and down the east side along the dotted-line trail shown on the map. You begin your hike at the edge of the hill, which has an elevation of 480 m above sea level, and travel east. By the contour lines and the peak on the map, your hike will be divided into 8 sections. Find the horizontal and vertical distance traveled for each section of your hike.
Step 3	Draw a slope triangle representing each section of the hike. On graph paper, draw a right triangle with a base representing the horizontal distance and a leg representing the vertical distance. Find the slope of each hypotenuse.
Step 4	Use the Pythagorean theorem to calculate the actual distance you hiked in each section.
Step 5	Find the angle of the climb for each section of the hike. Use an inverse trigonometric function.
Step 6	Use a protractor to measure the angle of the climb in each of your slope triangles in Step 3. How do these answers compare to your answers in Step 5?
Step 7	You have used two methods for finding the angle of the climb: **1.** Drawing triangles and using a protractor to find angle measures. **2.** Using trigonometry to calculate angle measures. Are there times when one method of finding angle measures is more convenient than the other? Explain your thinking.

EXAMPLE C

Consider this triangle.

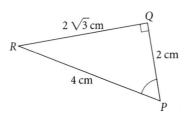

a. Name the lengths of the sides opposite angle P and adjacent to angle P.

b. Use the side lengths to find exact ratios for sin P, cos P, and tan P.

c. Find the measure of angle P using each of the inverse trigonometric functions.

▶ Solution

a. The length of the side opposite angle P is $2\sqrt{3}$ cm, and the length of the side adjacent to angle P is 2 cm.

b. $\sin P = \dfrac{2\sqrt{3}}{4} = \dfrac{\sqrt{3}}{2}$ $\qquad \cos P = \dfrac{2}{4} = \dfrac{1}{2}$ $\qquad \tan P = \dfrac{2\sqrt{3}}{2} = \sqrt{3}$

c. $\sin^{-1}\left(\dfrac{\sqrt{3}}{2}\right) = 60°$ $\qquad \cos^{-1}\left(\dfrac{1}{2}\right) = 60°$ $\qquad \tan^{-1}\left(\sqrt{3}\right) = 60°$

Each of the inverse trigonometric functions gives the measure of angle P as 60°.

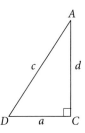

EXERCISES

You will need your calculator for problems **2, 3, 4, 5, 6, 7, 8, 9, 10,** and **11.**

▶ Practice Your Skills

1. Use the triangle at right as a guide. For 1a–f, fill in the correct angle, side, or ratio.

   a. $c^2 - a^2 = \boxed{}^2$ **b.** $\tan \boxed{} = \dfrac{a}{d}$ **c.** $\cos \boxed{} = \dfrac{d}{c}$

   d. $\sin^{-1} \boxed{} = D$ **e.** $\sin D = \cos \boxed{}$ **f.** $\sin \boxed{} = \dfrac{a}{c}$

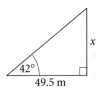

2. You will need a straightedge and a protractor for this problem.

   a. On graph paper, draw a right triangle with legs exactly 3 and 4 units long.

   b. Write trigonometric ratios and use inverse functions to find the angle measures.

   c. Measure the angles to check your answers to 2b.

3. Use a trigonometric ratio to find the length of side x in the triangle at right.

4. Sketch a right triangle to illustrate the ratio

   $$\tan 25° = \frac{6.8}{b}$$

   Then find the length of side b.

▶ Reason and Apply

5. Find the measure of each labeled angle or side length to the nearest tenth of a degree or centimeter.

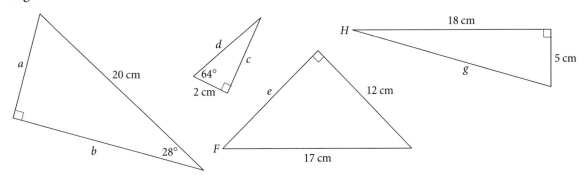

6. The legs of △PQR measure 8 cm and 15 cm.

   a. Find the length of the hypotenuse.

   b. Find the area of the triangle.

   c. Find the measure of angle P.

   d. Find the measure of angle Q.

   e. What is the sum of the measures of angles P, Q, and R?

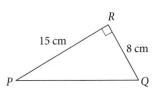

7. APPLICATION The **angle of elevation** is the angle between the horizontal and the line of sight. The angle of elevation of the roof of this building is 31°. Use a trigonometric ratio to find the height of the building.

Line of sight

Building

Angle of elevation

31°

135 m

8. You will need to find an actual stairway to do this problem. Use the diagram below as a guide.

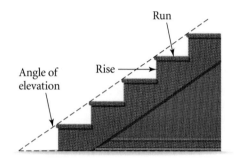

Run

Rise

Angle of elevation

a. Estimate the angle of elevation of the stairs.

b. Measure the rise and run of several steps.

c. Find the slope of the line going up the stairs.

d. Calculate the angle of elevation of the stairs.

e. Find the tangent of the angle of elevation.

9. APPLICATION The grade of a road is a percent calculated from the ratio

$$\frac{vertical\ distance\ traveled}{horizontal\ distance\ traveled}$$

The road in the sketch below has a 5% grade.

A sextant is a tool used to measure the angle of elevation of the sun or a star, thereby allowing one to determine latitude on the earth's surface. Here, Richard Byrd (1888–1957) checks his sextant before a historical 1926 flight over the North Pole.

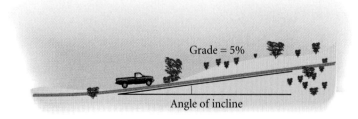

Grade = 5%

Angle of incline

a. Find the angle of incline of the road.

b. A very steep street has a grade of 15%. If you drive 1000 feet on this street, how much has your elevation changed?

10. Find the area of each figure to the nearest 0.1 cm².

a.
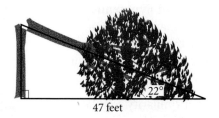
5 cm

20°

b.
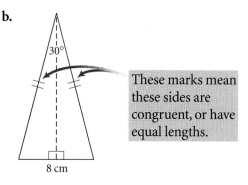
30°

These marks mean these sides are congruent, or have equal lengths.

8 cm

11. A tree is struck by lightning and breaks as shown. The tip of the tree touches the ground 47 feet from the stump and makes a 22° angle with the ground.

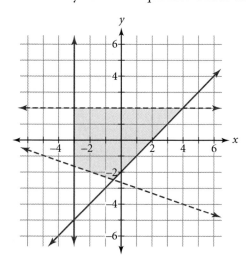

22°

47 feet

a. How high is the part of the trunk that is still standing?

b. How long is the portion of the tree that is bent over?

c. How tall was the tree originally?

▶ Review

12. Write the system of inequalities whose solution is shown here.

13. Give the equation for a rational function with asymptotes at $y = -3$ and $x = 2$ and a hole where $x = -7$.

CHAPTER
11
REVIEW

You began this chapter by exploring relationships between algebra and geometry. You used **analytic geometry** to discover properties related to the slopes of **parallel** and **perpendicular lines.** You also used analytic geometry to find the **midpoint** of a segment.

Then you explored area and side relationships for squares drawn on graph paper. You learned how to draw segments whose lengths are equal to many different square roots. You also found ways to rewrite radical expressions.

You discovered the **Pythagorean theorem,** an important relationship between the lengths of the **legs** and **hypotenuse** of a **right triangle.** This relationship is useful in many professions and has been valuable to many civilizations for thousands of years. You used analytic geometry and the Pythagorean theorem to find a formula for the distance between any two points.

Finally, you reviewed ratio, proportion, and similarity. Similar right triangles introduced you to **trigonometric functions**—sine, cosine, and tangent. For an **acute angle** in a right triangle, these functions are defined by ratios between the **opposite leg, adjacent leg,** and hypotenuse.

EXERCISES

You will need your calculator for problems **6, 7, 8,** and **18.**

1. Rewrite each expression with as few square root symbols as possible.

a. $4\sqrt{5} + 4\sqrt{5}$

b. $10\sqrt{17} - 6\sqrt{17}$

c. $138\sqrt{3} + 21\sqrt{3} - 36\sqrt{3}$

d. $\sqrt{5} \cdot \sqrt{3}$

e. $4\sqrt{5} \cdot 4\sqrt{5}$

f. $(10\sqrt{17})^2$

g. $\sqrt{6} \cdot \sqrt{15}$

h. $4\sqrt{25} \cdot 4\sqrt{5}$

i. $\sqrt{2} + \sqrt{3}$

j. $\sqrt{2} + \sqrt{8}$

k. $\dfrac{\sqrt{18}}{\sqrt{3}}$

l. $\sqrt{3} + \sqrt{27}$

2. Find the area of the tilted square. Use two different strategies to check your answers.

3. Use analytic geometry to show that the sides of the square in problem 2 are perpendicular.

4. Explain how to draw a square with a side length of $\sqrt{29}$ units.

5. APPLICATION Is a triangle with side lengths 5 feet, 12 feet, and 13 feet a right triangle? Explain how you know. Then explain how a construction worker could use a 60-foot piece of rope to make sure that the corners of a building foundation are right angles.

6. Draw this quadrilateral on graph paper.

a. Name the coordinates of the vertices of this quadrilateral.

b. Find the slope of each side.

c. What kind of quadrilateral is this? Explain how you know.

d. Find the coordinates of the midpoint of each side. Mark the midpoints on your drawing. Connect the midpoints in order.

e. Use the distance formula to find the length of each side of the figure formed by connecting the midpoints in 6d.

f. Find the slope of each side of the figure formed in 6d.

g. What kind of figure is formed in 6d? Explain how you know.

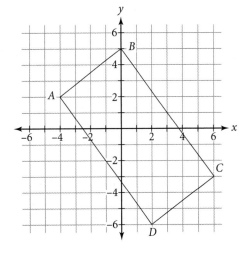

7. Find the approximate lengths of the legs of this right triangle.

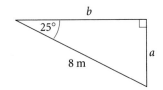

8. You will need a straightedge and a protractor for this problem.

a. Carefully draw a right triangle with sides 5, 12, and 13 units on graph paper.

b. Measure the angle opposite the 5-unit side.

c. Find the measure of the angle opposite the 5-unit side using $\sin^{-1}$, $\cos^{-1}$, and $\tan^{-1}$.

d. Explain how you can find the measure of the angle opposite the 12-unit side. What is the measure of this angle?

9. A rectangular box has the dimensions shown in the diagram at right.

a. What is the length of the diagonal $\overline{AC}$?

b. What is the length of the diagonal $\overline{AG}$?

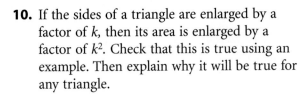

10. If the sides of a triangle are enlarged by a factor of k, then its area is enlarged by a factor of k^2. Check that this is true using an example. Then explain why it will be true for any triangle.

MIXED REVIEW

11. This table gives the normal minimum and maximum January temperatures for 18 U.S. cities.

January Temperatures Across the United States

City	Minimum temp. (°F)	Maximum temp. (°F)	City	Minimum temp. (°F)	Maximum temp. (°F)
Mobile, AL	40	60	Helena, MT	10	30
Little Rock, AK	29	49	Atlantic City, NJ	21	40
Denver, CO	16	43	New York, NY	26	37
Jacksonville, FL	41	64	Cleveland, OH	18	32
Honolulu, HI	66	80	Pittsburgh, PA	19	34
Indianapolis, IN	17	34	Rapid City, SD	11	34
New Orleans, LA	42	61	Houston, TX	40	61
Boston, MA	22	36	Richmond, VA	26	46
Minneapolis, MN	3	21	Lander, WY	8	31

(*2001 World Almanac*, p. 243)

a. Let x represent the normal minimum temperature, and let y represent the normal maximum temperature. Use the Q-point method to find an equation for a line of fit for the data. (Round the slope to the nearest hundredth.)

b. The normal minimum January temperature for Memphis, Tennessee, is 31°F. Use your equation to predict the normal maximum January temperature.

c. The normal maximum January temperature for Charleston, South Carolina, is 58°F. Use your equation to predict the normal minimum January temperature.

12. The HealthyFood Market sells dried fruit by the pound. Jan bought 3 pounds of dried apricots and 1.5 pounds of dried papaya for $13.74. Yoshi bought 2 pounds of dried apricots and 3 pounds of dried papaya for $16.32.

a. Write a system of equations to represent this situation.

b. How much does a pound of dried apricots cost? How much does a pound of dried papaya cost?

13. Tell whether the relationship between x and y is a direct variation, an inverse variation, or neither, and explain how you know. If the relationship is a direct or inverse variation, write its equation.

a.

x	y
0.2	10
0.8	2.5
1	2
4	0.5
5	0.4

b.

x	y
0.3	6
0	3
1	13
3	33
10.0	103

c.

x	y
0.8	0.2
1	0.25
3	0.75
12	3
28.0	7

d.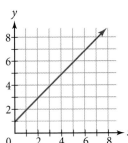

e.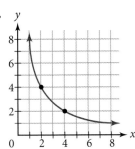

f.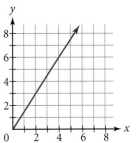

14. Here is a graph of a function f.

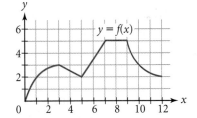

 a. Use words such as *linear, nonlinear, increasing,* and *decreasing* to describe the behavior of the function.

 b. What is the range of this function?

 c. What is $f(3)$?

 d. For what x-values does $f(x) = 2$?

 e. For what x-values does $f(x) = 5$?

15. Use the properties of exponents to rewrite each expression. Your answers should have only positive exponents.

 a. $(3x^2y)^3$
 b. $\dfrac{5^2 p^7 q^3}{5 p^3 q}$
 c. $x^{-4} y^{-2} x^5$
 d. $m^2(n^{-4} + m^{-6})$

16. Here are the running times in minutes of 22 movies in the new-release section of a video store.

120	116	93	108	134	90	112	99	93	104	110
105	97	115	100	82	102	105	104	105	112	179

 a. Find the mean, median, and mode of the data.

 b. Find the five-number summary of the data and create a box plot.

 c. Create a histogram of the data. Use an appropriate bin width.

 d. Make at least three observations about the data based on your results from 16a–c.

17. Write the equation for this parabola in

  a. Factored form.

  b. Vertex form.

  c. General form.

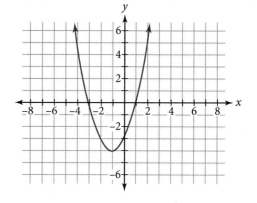

18. Six years ago, Maya's grandfather gave her his baseball card collection. Since then, the value of the collection has increased by 8% each year. The collection is now worth $1900.

  a. How much was the collection worth when Maya first received it?

  b. If the value of the collection continues to grow at the same rate, how much will it be worth 10 years from now?

19. If $f(x) = x^2 + |x| - 4$, find

  a. $f(-5)$ **b.** $f(2)$ **c.** $f(-7) - f(4)$

  d. $f(-7 - 4)$ **e.** $-3 \cdot f(3)$

20. **APPLICATION** The Galaxy of Shoes store is having a 22nd anniversary sale. Everything in the store is reduced by 22%.

  a. Anita buys a pair of steel-toed boots originally priced at $79.99. What is the discounted price of the boots?

  b. The sales tax on Anita's boots is 5%. What total price will Anita pay for the boots?

21. The image of the black rectangle after a transformation is shown in red.

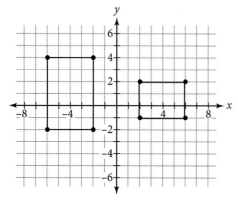

  a. Describe the transformation.

  b. Define the coordinates of any point in the image using (x, y) as the coordinates of any point in the original figure.

22. Solve for *x*.

   a. $0 = (x + 5)(x - 2)$

   b. $0 = x^2 + 8x + 16$

   c. $x^2 - 5x = 2x + 30$

   d. $x^2 = 5$

23. Give the equation for each graph.

   a.

   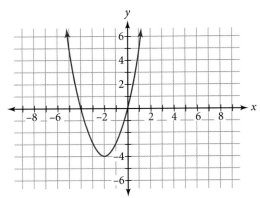

   b.

   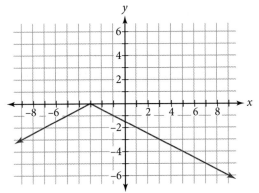

24. Zoe Kovalesky visits companies and teaches the employees how to use their computers and software. She charges a fixed fee to visit a company, plus an amount for each employee in the training class. This table shows the number of employees trained and the total bill for the last five companies she visited.

   a. How much does Zoe charge for each employee in a training class?

   b. What fixed fee does Zoe charge to visit a company?

Computer Training

Employees trained	Total bill ($)
5	400
11	610
17	820
3	330
25	1100

   c. Write a recursive routine to find the total bill for any number of employees.

   d. Write an equation to calculate the total bill, *y*, for any number of employees, *x*.

   e. ACME, Inc., has hired Zoe to train 12 employees. How much will the total bill be?

   f. Last week, Zoe taught a training class at the Widget Company. The total bill was $505. How many employees were in the class?

25. The vertices of $\triangle ABC$ are $A(-6, 1)$, $B(2, 7)$, and $C(10, 1)$. Do 25a–d before you graph the triangle.

   a. Find the length and slope of each side.

   b. What kind of triangle is $\triangle ABC$? Explain how you know.

   c. Find the midpoint of $\overline{AC}$ and call it *D*.

   d. If points *B* and *D* are connected, they form $\overline{BD}$. That creates two triangles. What kind of triangles are $\triangle ABD$ and $\triangle BCD$? Explain how you know.

   e. Draw $\triangle ABC$ on graph paper and draw $\overline{BD}$. Does your drawing support your results from 25a–d?

TAKE ANOTHER LOOK

You have seen that the Pythagorean theorem, $a^2 + b^2 = c^2$, holds true for right triangles. What about triangles that don't have a right angle? Is there a relationship between the side lengths of any triangle?

If a triangle is not a right triangle, you can classify it as one of two other types of triangles, based on its angles. An **acute triangle** has three angles that are all acute. An **obtuse triangle** has one angle that is obtuse.

Use your straightedge and protractor to draw an acute triangle. Label the longest side c, and label the shorter sides a and b. Find the square of the length of each side and compare them. Does the relationship $a^2 + b^2 = c^2$ still hold true? If not, state an equation or inequality that does hold true.

Make a conjecture about the squares of the side lengths for an obtuse triangle. Then draw an obtuse triangle and measure the sides. What relationship do you find this time?

Summarize your results.

You can learn how trigonometry is used in acute and obtuse triangles with the Internet links at www.keymath.com/DA .

Assessing What You've Learned

ORGANIZE YOUR NOTEBOOK Make sure your notebook contains notes and examples of analytic geometry. Your notes should give you quick reference to important concepts like the slope of parallel and perpendicular lines, the Pythagorean theorem, finding a midpoint of a segment, and finding the distance between two points. In your math class next year you may be studying more advanced concepts of geometry, so your notes can help you in the future too.

UPDATE YOUR PORTFOLIO Geometry is the study of points, lines, angles, and shapes, so you have drawn and graphed a lot of figures for this chapter. Choose several pieces of work that illustrate what you have learned, and show how you can use algebra and geometry together. For each piece of work, make a cover sheet that gives the objective, the result, and what you might have done differently.

PERFORMANCE ASSESSMENT Show a classmate, a family member, or your teacher that you understand how analytic geometry combines algebra and geometry. Show both the geometric and algebraic method of finding the midpoint of a segment, or the distance between two points. You may also want to show a geometric proof of the Pythagorean theorem and the algebraic formula that results. Be sure to compare and contrast the geometric method and the algebraic method. Explain which method you prefer and why.

Selected Answers

This section contains answers for the odd-numbered problems in each set of Exercises. When a problem has many possible answers, you are given only one sample solution or a hint on how to begin.

LESSON 0.1

1a. $\frac{1}{8}$; $\frac{1}{16} + \frac{1}{16}$ or $2 \times \frac{1}{16}$

1b. $\frac{3}{64}$; $\frac{1}{64} + \frac{1}{64} + \frac{1}{64}$ or $3 \times \frac{1}{64}$

1c. $\frac{3}{5}$; $\frac{1}{25} + \frac{1}{25} + \frac{1}{25} + \frac{1}{25} + \frac{1}{25} + \frac{1}{25} + \frac{1}{25} + \frac{1}{25} + \frac{1}{25} + \frac{1}{25} + \frac{1}{25} + \frac{1}{25} + \frac{1}{25} + \frac{1}{25} + \frac{1}{25}$ or $15 \times \frac{1}{25}$ or $1 - \frac{10}{25} = \frac{3}{5}$

1d. $\frac{7}{625}$; $\frac{1}{625} + \frac{1}{625} + \frac{1}{625} + \frac{1}{625} + \frac{1}{625} + \frac{1}{625} + \frac{1}{625}$ or $7 \times \frac{1}{625}$

3a.

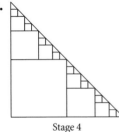

Stage 4

3b. 32   **3c.** 48   **3d.** 56

5. Sample answers:

5a.

5b.

5c.

5d.

7a. Sample answer: Divide each side of the square into thirds and connect those points with lines parallel to the sides. A square is formed in the middle. Erase everything except the center square. To get the next stage, do the same thing in all eight squares formed around the middle square.

7b.

Stage 3

7c. $\frac{1}{9}$; $\frac{17}{81}$, $\frac{217}{729}$ **7d.** $\frac{8}{9}$; $\frac{64}{81}$, $\frac{512}{729}$

9a. $\frac{9}{16}$

9b. Sample answers: $12 \times \left(\frac{9}{16}\right) = 6\frac{3}{4}$; $(12 \div 16) \times 9 = 6.75$

11a. $\frac{1}{4} \times \frac{1}{4} \times 32 = \frac{32}{16} = 2$

Sample answer:

11b. $\frac{3}{4} \times \frac{1}{4} \times \frac{1}{4} \times 32 = \frac{96}{64} = \frac{3}{2} = 1\frac{1}{2}$

Sample answer:

11c. $\frac{1}{2} \times \frac{1}{2} \times \frac{1}{4} \times 32 = \frac{32}{16} = 2$

Sample answer:

13. $1 - \frac{11}{32} = \frac{21}{32}$

1a. 5^4 **1b.** 7^5 **1c.** 3^7 **1d.** 2^3

3a. 3^3 **3b.** 2^5 **3c.** 5^4 or 25^2 **3d.** 7^3

5a. 4 or 2^2; 8 or 2^3 **5b.** 2^5

7a. 64 **7b.** 512 **7c.** 8^2; 8^3 **7d.** 262,144 or 8^6

7e. The exponent is always one less than the stage number.

7f. Yes, because $8^0 = 1$.

9b.

Stage number	Area of one shaded triangle	Total area of the shaded triangles
0	1	1
1	$\frac{1}{4}$	$\frac{3}{4}$
2	$\frac{1}{16}$	$9 \cdot \frac{1}{16}$ or $\frac{3}{4} \cdot \frac{3}{4} = \frac{9}{16}$
3	$\frac{1}{64}$	$\frac{3}{4} \cdot \frac{9}{16} = \frac{27}{64}$

9c. The area of one shaded triangle is $\frac{1}{4}$ the area of one of the previous shaded triangles. The total area of the shaded triangles in each figure is $\frac{3}{4}$ the shaded area in the previous figure.

11. Hint: Think of a situation in which $\frac{3}{4}$ of something is divided into 5 pieces.

1a. $\frac{125}{8}$; 15.63 **1b.** $\frac{25}{9}$; 2.78

1c. $\frac{2401}{81}$; 29.64 **1d.** $\frac{729}{64}$; 11.39

3. $\left(\frac{5}{3}\right)^4 \approx 7.72$; $\left(\frac{5}{3}\right)^5 \approx 12.86$; Stage 5

5a. See below.

5b. Stage 5: $\left(\frac{5}{4}\right)^5 \approx 3.05$; Stage 11: $\left(\frac{5}{4}\right)^{11} \approx 11.64$

7a. See below.

7b. Stage 5

7c. No. Stage 6 has a length of slightly more than 161, and Stage 7 has a length of over 376.

5a. (*Lesson 0.3*)

Stage number	Total length		
	Multiplication form	Exponent form	Decimal form
2	$5 \cdot 5 \cdot \frac{1}{4} \cdot \frac{1}{4} = \frac{25}{16}$	$5^2 \cdot \left(\frac{1}{4}\right)^2 = \left(\frac{5}{4}\right)^2$	1.56
3	$5 \cdot 5 \cdot 5 \cdot \frac{1}{4} \cdot \frac{1}{4} \cdot \frac{1}{4} = \frac{125}{64}$	$5^3 \cdot \left(\frac{1}{4}\right)^3 = \left(\frac{5}{4}\right)^3$	1.95

7a. (*Lesson 0.3*)

Stage number	Total length		
	Expanded form	Exponent form	Decimal form
2	$7 \cdot 7 \cdot \frac{1}{3} \cdot \frac{1}{3} = \frac{49}{9}$	$7^2 \cdot \left(\frac{1}{3}\right)^2 = \left(\frac{7}{3}\right)^2$	5.44
3	$7 \cdot 7 \cdot 7 \cdot \frac{1}{3} \cdot \frac{1}{3} \cdot \frac{1}{3} = \frac{343}{27}$	$7^3 \cdot \left(\frac{1}{3}\right)^3 = \left(\frac{7}{3}\right)^3$	12.70
4	$7 \cdot 7 \cdot 7 \cdot 7 \cdot \frac{1}{3} \cdot \frac{1}{3} \cdot \frac{1}{3} \cdot \frac{1}{3} = \frac{2401}{81}$	$7^4 \cdot \left(\frac{1}{3}\right)^4 = \left(\frac{7}{3}\right)^4$	29.64

Selected Answers

9. 2.8

11a. 4   **11b.** 16   **11c.** 64   **11d.** $4^1, 4^2, 4^3$

11e. The exponent is one less than the stage number. For Stage 1, $4^0 = 1$

LESSON 0.4

1a. 3   **1b.** -3 **1c.** -7 **1d.** -3

3a. -8 **3b.** -25 **3c.** -17 **3d.** -11

5a. In the first recursion, he should get $-0.2 \cdot 2 = -0.4$, not $+0.4$. His arithmetic when evaluating $0.4 - 4$ was correct. In the second recursion, he used the wrong value (-3.6 instead of -4.4) because of his previous error. His arithmetic was also incorrect because $-0.2 \cdot -3.6 = +0.72$, not -0.72. His arithmetic when evaluating $-0.72 - 4$ was correct.

5b. $-4.4, -3.12, -3.376, -3.3248$

5c. $-3.8, -3.24, -3.352$

5d. Yes. The calculations seem to be approaching a value close to -3.3.

7a.

Starting value	2	-1	10
First recursion	-3	3	-19
Second recursion	7	-5	39
Third recursion	-13	11	-77

7b. No, the values get farther and farther apart.

9a. i. 7.5 or $\frac{15}{2}$

9a. ii. -10

9a. iii. 6.25

9b. When the coefficient is 0.2, the attractor value is 1.25 times the constant. In general the attractor value is $\frac{constant\ term}{1 - coefficient\ of\ the\ box}$

9c. Sample answer: $0.2 \cdot \boxed{} + 1.8$.

11. -22.5

LESSON 0.5

1a. 8.0 cm   **1b.** 4.3 cm   **1c.** 7.2 cm

3a–c.
```
A   E       C     D B
●───●───────●─────●─●
```

3d. points D and B; no

5d. The resulting figure should resemble a right-angle Sierpiński triangle.

7a. This game fills the entire square.

7b. This game creates a small Sierpiński triangle at each corner of the triangle.

7c. This game creates four small Sierpiński carpets, one at each corner of the square.

7d. This game creates an overlapping pattern like the Sierpiński triangle.

9. Point should divide segment into an 8-cm and a 4-cm segment.

11. See below.

11. (*Lesson 0.5*)

Stage number	Total length		
	Multiplication form	**Exponent form**	**Decimal form**
1	$6 \cdot \frac{1}{4}$	$6^1 \cdot \left(\frac{1}{4}\right)^1$	1.5
2	$6 \cdot 6 \frac{1}{4} \cdot \frac{1}{4}$	$6^2 \cdot \left(\frac{1}{4}\right)^2 = \left(\frac{6}{4}\right)^2 = \frac{9}{4}$	2.25
3	$6 \cdot 6 \cdot 6 \cdot \left(\frac{1}{4}\right) \cdot \left(\frac{1}{4}\right) \cdot \left(\frac{1}{4}\right)$	$6^3 \cdot \left(\frac{1}{4}\right)^3 = \left(\frac{6}{4}\right)^3 = \frac{27}{8}$	3.38
4	$6 \cdot 6 \cdot 6 \cdot 6 \cdot \left(\frac{1}{4}\right) \cdot \left(\frac{1}{4}\right) \cdot \left(\frac{1}{4}\right) \cdot \left(\frac{1}{4}\right)$	$6^4 \cdot \left(\frac{1}{4}\right)^4 = \left(\frac{6}{4}\right)^4 = \frac{81}{16}$	5.06

CHAPTER 0 REVIEW

1a. 1   **1b.** 3   **1c.** 9   **1d.** 27

1e. 81   **1f.** 6561   **1g.** 531,441   **1h.** 1

1i. $\frac{1}{3}$ **1j.** $\frac{1}{9}$ **1k.** $\frac{1}{27}$

1l. $\frac{1}{81}$ **1m.** $\frac{1}{6561}$ **1n.** $\frac{1}{531,441}$

3a. 72   **3b.** 290   **3c.** -10

3d. 312   **3e.** $2.1\overline{6}$ **3f.** -34

5a. $\frac{1}{16} + \frac{1}{16} + \frac{1}{16} = \frac{3}{16}$ **5b.** $\frac{1}{16} + \frac{1}{16} + \frac{1}{4} = \frac{3}{8}$

5c. $\frac{1}{9} + \frac{1}{9} + \frac{1}{81} + \frac{1}{81} = \frac{20}{81}$

5d. $\frac{1}{4} + \frac{1}{16} + \frac{1}{64} + \frac{1}{64} = \frac{11}{32}$

7a. See below.

7b. $\left(\frac{7}{5}\right)^{20} \approx 836.68$

CHAPTER 1 · CHAPTER **1** **CHAPTER 1 · CHAPTER**

LESSON 1.1

1. maximum: 93 bpm; minimum: 64 bpm; range: 93 bpm $-$ 64 bpm $=$ 29 bpm

3a. Uranus

3b. Mercury and Venus have no satellites.

3c. 9   **3d.** nine times

5a. 80 bpm   **5b.** 29 bpm

5c. She counted her pulse rates for one full minute.

5d. Any whole number could occur, not just multiples of four.

5e. A full minute, sometimes longer, to ensure accuracy.

7. 3 in Classical

9a. 10^4 **9b.** $2^3 \cdot 5^6$ **9c.** $\frac{3^6}{8^3}$

11a. 18;

Doubles of 225	**450**	900	1800	**3600**	7200
Doubles of 1	**2**	4	8	**16**	32

11b. $9\frac{1}{2}$;

Doubles of 6	**6**	12	24	**48**	96
Doubles of 1	**1**	2	4	**8**	16

Half of 6	3
Half of 1	$\frac{1}{2}$

LESSON 1.2

1a. mean and median are 6; mode 5

1b. mean 5.1; median 5; modes 3, 8

1c. mean 10.25; median 9; no mode

1d. mean 17.5; median 20; mode 20

3.

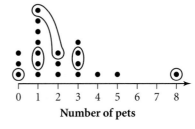

Number of pets

3a. 20   **3b.** 8   **3c.** 1

5a. $18.24. If you multiply the mean by the number of items, you get the sum of the items.

7. 45.5 seconds

9a. Multiply the mean by 10; together they weigh approximately 15,274 pounds.

9b. Five of the fish caught weigh 1449 lbs or less, and five weigh 1449 lbs or more.

9c. 2664 pounds.

7a. (*Chapter 0 Review*)

Stage number	Total length		
	Multiplication form	Exponent form	Decimal form
0	1	1	1
1	$7\left(\frac{1}{5}\right)$	$7^1\left(\frac{1}{5}\right)^1 = \left(\frac{7}{5}\right)^1 = \frac{7}{5}$	1.4
2	$7 \cdot 7 \cdot \left(\frac{1}{5}\right) \cdot \left(\frac{1}{5}\right)$	$7^2\left(\frac{1}{5}\right)^2 = \left(\frac{7}{5}\right)^2 = \frac{49}{25}$	1.96

11a. See below.

11b. mean: 33.95; median: 30; mode: 29

11c. Either the mode or the median is probably best. The mean is distorted by one extremely high value.

13a. 12 cm   **13b.** 8 cm   **13c.** 2.4 cm

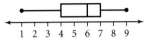

1a. 5, 10, 23, 37, 50   **1b.** 10, 22, 31.5, 37, 50

1c. 14, 22.5, 26, 41, 47   **1d.** 5, 10, 19, 34.5, 47

3. 1, 4, 6, 7, 9

5a. Quartiles are the boundaries dividing a data set into four groups, or quarters, with approximately the same number of values.

5b. The range.

5c. The interquartile range, or IQR.

5d. Outliers are at or near the minimun and maximum values, which are the endpoints of the whiskers.

7a. 151, 426, 644, 1020, 1305

7b. The mean for the Bulls is about 675 points, and the mean for the Raptors is about 715 points; the medians are 416 points and 644 points, respectively. The Chicago mean is much higher than its median because of Michael Jordan's total points. On the average, individual Toronto players scored more points than individual Chicago players.

7c. The median probably best represents the total-points-scored data for the Bulls. Students can justify choosing either the mean or the median for the Raptors. As a team owner, you might think the mean better reflects your team's talents.

7d. The lengths of the boxes are about the same, and the medians seem to divide the boxes into regions that look about the same length for the two teams.

7e. See the graph below. Without Jordan, the range of the data is much smaller, and the length of the box is a little shorter. If Jordan's points scored are eliminated, the Raptors have the higher-scoring players. There is more variation in the number of points scored by individual Raptors than for individual Bulls.

9a. 35 feet

9b. More information is needed. The length is between 11 and 17.5 feet.

9c. 65 mi/hr

9d. The ten longest snakes vary in length from about 8 ft to 35 ft. About half of the snakes range in length from about 11 ft to 25 ft. Running speeds of the ten fastest mammals range from 42 mi/hr to 65 mi/hr. About half of the speeds are between 43 mi/hr and 50 mi/hr. The cheetah appears to run much faster than most other mammals.

9e. No, the units of these data sets are different.

9f. about 47 mi/hr

11. Hint: For the median age to be 14, the middle age must be 14. For the mean age to be 22, the sum of the ages must be 5 · 22, or 110.

LESSON 1.4

1a. matinee: 29; evening: 30

1b. matinee   **1c.** None

1d. The number is less than or equal to 4.

3a. 51

3b. Approximately $\frac{1}{4}$ of the countries had a female life expectancy between approximately 81 and 83 years.

3c. 2

3d. There are no bins to the right of 85 in the histogram. Also, the maximum point in the box plot is located at approximately 83 years.

5a. minimum: 6.0 cm; maximum: 8.5 cm; range: 2.5 cm

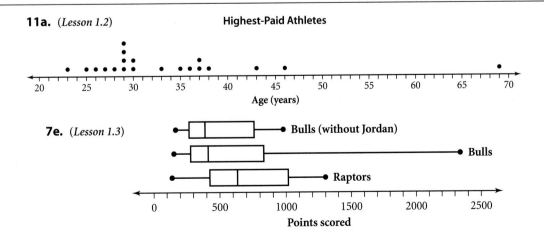

11a. (*Lesson 1.2*)   **Highest-Paid Athletes**

Age (years)

7e. (*Lesson 1.3*)   Bulls (without Jordan)   Bulls   Raptors

Points scored

5b. Ring Finger Length

```
6 | 0 5 5
7 | 0 0 0 5
8 | 5
```

Key
```
6 | 0 means 6.0 cm
```

7a. Hint: When you roll a die, there are 6 possible outcomes, and each outcome has an equal chance of being rolled.

7b. Hint: Estimates are likely to be clumped around the actual value.

7c. Hint: What is the range of ages of people in your school? Approximately how many students are in each grade?

7d. Fairly flat bins on the left getting taller as you go to the right, with the last two bins (8–9 and 9–10) both being 25 units tall.

9a. Hint: There is a definite clustering near the center. Assign a grade for this "average" performance and then work your way out.

9b. The outlier should get an A.

9c. Hint: Try to describe the thinking process you used while answering 9a.

11. See below.

LESSON 1.6

1.

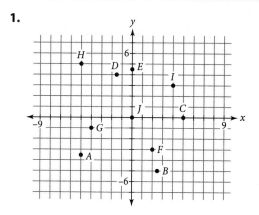

---

3. Your calculator will tell you the correct answers.

5a. $A(-7, -4), B(0, -2), C(3, 6), D(0, 3), E(4, -4),$ $F(-2, -6), G(4, 0), H(7, 1), I(-5, 2),$ and $J(-5, 4)$

5b.

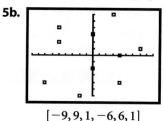

$[-9, 9, 1, -6, 6, 1]$

5c. Points $B, D,$ and G

5d. I: C and H II: I and J III: A and F IV: E

7a. Approximate answers (the second coordinates are in millions): $(1984, 280), (1985, 320), (1986, 340),$ $(1987, 415), (1988, 450), (1989, 445), (1990, 440),$ $(1991, 360), (1992, 370), (1993, 340), (1994, 345),$ $(1995, 273), (1996, 225), (1997, 173), (1998, 159),$ $(1999, 142).$

7b.

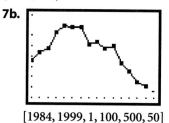

$[1984, 1999, 1, 100, 500, 50]$

7c. Hint: During what years did shipments increase? During what years did shipments decrease?

9a. $(-1.5, 2.6), (-3, 0), (-1.5, -2.6), (1.5, -2.6)$

11a. Hint: With an odd number of values, the middle value must be the median, 15. The minimum and maximum must be 5 and 47, respectively. Now find two values between the minimum and median that give a first quartile of 12; and find two values between the median and maximum that give a third quartile of 30.

11b. Hint: With 10 data values, the median will divide the data set into two groups of 5 data values. The middle value of the lower half must be the first quartile, 12; and the middle value of the upper half must be the third quartile, 30.

11c. Hint: Try adding two values to your data set from 11b, but maintain the same five-number summary.

---

11. (*Lesson 1.4*)

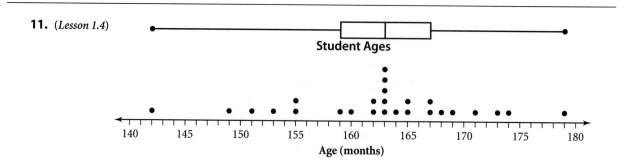

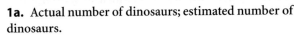

LESSON 1.7

1a. Actual number of dinosaurs; estimated number of dinosaurs.

1b.

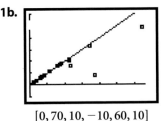

$[0, 70, 10, -10, 60, 10]$

1c. $y = x$

1d. No. In no case is the estimated number of dinosaurs more than the actual number.

1e. Yes. The estimated count of five different species was less than the actual count.

3. Overestimates are *B, C, D, E,* and *G.* Underestimates are *A, F, H,* and *I.*

5a. The more the rubber band is stretched, the farther it flies.

5b. Hint: Find 15 on the *x*-axis and move up into the graph. What *y*-coordinate would give you a point "in line" with the other points?

5c. Hint: Find 400 on the *y*-axis and move right into the graph. What *y*-coordinate would give you a point "in line" with the other points?

7a. It provides the differences between the estimated number of each species and the actual count. This helps identify over- and underestimates.

7b.

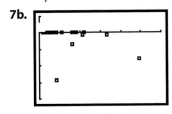

$[0, 60, 10, -40, 10, 10]$

7c. 5; underestimates

7d. $(8, -29)$; the 8 is the estimated number of velociraptors; the number of velociraptors was underestimated by 29.

9a. More states had students with higher verbal scores than mathematics scores.

9b. Those states in which students scored higher in mathematics had larger student populations taking the test. The national average is an average of all students, not an average of all state averages.

11a. Hint: If the median is not in the box, it must be located on one of the ends of the box.

11b. Hint: The difference between Q1 and Q3 will be zero.

11c. Hint: The minimum needs to be far removed from the other data values.

11d. Hint: If there is no right whisker, the maximum will be on the right end of the box.

LESSON 1.8

1. Randall Cunningham threw 19 touchdown passes in 1992.

3. 3×4

5. $\begin{bmatrix} 788 & 489 & 35 & 19 \\ 809 & 492 & 53 & 21 \\ 919 & 590 & 61 & 19 \end{bmatrix}$ This matrix gives the totals from the two years.

7. $\begin{bmatrix} -158 & -115 & -11 & -9 \\ 41 & 26 & 15 & -1 \\ 115 & 54 & 11 & 5 \end{bmatrix}$

This matrix gives the difference between the 1998 totals and the 1992 totals.

9a. $\begin{bmatrix} 8 & -5 & 4.5 \\ -6 & 9.5 & 5 \end{bmatrix}$ **9b.** $\begin{bmatrix} -3 & 4 & -2.5 \\ 2 & -6 & -4 \end{bmatrix}$

9c. $\begin{bmatrix} 15 & -3 & 6 \\ -12 & 10.5 & 3 \end{bmatrix}$ **9d.** $\begin{bmatrix} 4 & -2.5 & 2.25 \\ -3 & 4.75 & 2.5 \end{bmatrix}$

11. Hint: Think about the first matrix representing cost per unit of two items, and the second matrix as units of each item.

13a. Hint: The minimum in the data set must be 0, and the maximum must be 7. Also, the data value 2 must occur more frequently than any other.

13b. Hint: The minimum must be 22.2, and the maximum must be 30.4. No values in the list should be repeated.

CHAPTER 1 REVIEW

1a. Mean: 41.5; divide the sum of the numbers by 14. Median: 40; list the numbers in ascending order and find the mean of the two middle numbers. Mode: 36; find the most frequently occurring number.

1b. 27, 36, 40, 46, 58

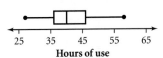

Battery Life

Hours of use

3a.

Mean Annual Wages, 1998

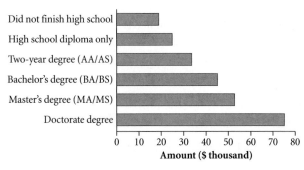

3b. Greatest jump: from a master's degree to a doctorate; smallest difference: from not finishing high school to a high school diploma.

5a. Mean: approximately 154; median: 121; there is no mode.

5b. Bin widths may vary.

Pages Read in Current Book

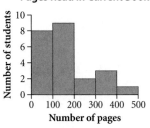

Pages Read in Current Book

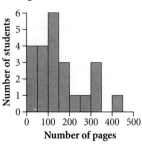

5c. **Pages Read in Current Book**

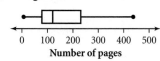

5d. Sample answer: Most of the students questioned had read fewer than 200 pages, with a fairly even distribution between 0 and 200.

7a. $[A] = \begin{bmatrix} 5.00 & 8.00 \\ 3.50 & 4.75 \\ 3.50 & 4.00 \end{bmatrix}, [B] = \begin{bmatrix} 0.50 & 0.75 \\ 0.50 & 0.25 \\ 0.50 & 0.25 \end{bmatrix}$

$[C] = [43 \quad 81 \quad 37]$

7b. $[A] + [B] = \begin{bmatrix} 5.50 & 8.75 \\ 4.00 & 5.00 \\ 4.00 & 4.25 \end{bmatrix}$

7c. $[C] \cdot ([A] + [B]) = [708.5 \quad 938.5]$
matinee: $708.50
evening: $938.50

9a. 2,820,000

9b. See below.

9c. **The Ten Most Populated U.S. Cities, 1998**

0	97
1	08 11 20 22 44 79
2	82
3	60
4	
5	
6	
7	42

Key

2	82 means 2.82 million

9d. **The Ten Most Populated U.S. Cities, 1998**

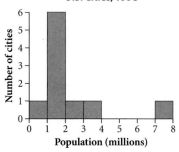

9b. (*Chapter 1 Review*)

The Ten Most Populated U.S. Cities, 1998

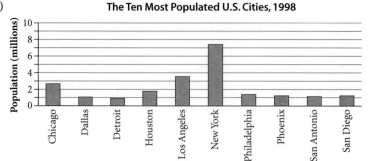

9e. The bar graph helps show how each city compares with the others, since they remain identified by name. The stem plot shows distribution but also shows actual values. The histogram shows distribution and a definite clustering between 1 and 2 million, but it does not show individual cities or populations.

LESSON 2.1

1a. 0.875   **1b.** 0.65   **1c.** 2.6   **1d.** 2.08

3a. $\dfrac{240 \text{ miles}}{1 \text{ hour}}$

3b. $\dfrac{10 \text{ parts capsaicin}}{1,000,000 \text{ parts water}}$ or $\dfrac{1 \text{ part capsaicin}}{100,000 \text{ parts water}}$

3c. $\dfrac{350 \text{ women-owned firms}}{1000 \text{ firms}}$ or $\dfrac{7 \text{ women-owned firms}}{20 \text{ firms}}$

3d. $\dfrac{35,500 \text{ dollars}}{1 \text{ person}}$

5a. $T = 18$ **5b.** $R = 28$ **5c.** $S = 73.5$

5d. $x = 2.1$ **5e.** $M = 6$ **5f.** $n = 21$

5g. $c = 31.2$ **5h.** $W = 9$

7a. 85%

7b. $\dfrac{t}{7.38} = \dfrac{85}{100}, t = \$6.27$

9. $2\frac{2}{3}$ cups of water and $\frac{2}{3}$ cup of oatmeal; $6\frac{2}{3}$ cups water and $1\frac{2}{3}$ cups oatmeal

11.

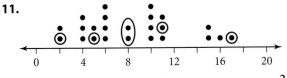

13a. -12 **13b.** -4 **13c.** -8 **13d.** $-\frac{2}{3}$

LESSON 2.2

1a. 32% of what number is 24?

1b. 48% of 450 is what number?

1c. What percent of 117 is 98? or 98 is what percent of 117?

3. 269

5a. Marie should win over half the games.

5b. $\dfrac{28 \text{ games won by Marie}}{28 + 19 \text{ total games}} = \dfrac{M}{12}$; $M = 7.15$ or 7 games

5c. $\dfrac{19}{47} = \dfrac{30}{G}$; $G \approx 74$ games

7. approximately 76 errors

9a. 1 sulfur atom, 2 hydrogen atoms, and 4 oxygen atoms

9b. 200 hydrogen atoms would combine with 100 sulfur atoms and 400 oxygen atoms.

9c. Use all 400 atoms of oxygen, 200 atoms of hydrogen, and 100 atoms of sulfur to make 100 molecules of sulfuric acid.

11. Matt; by the order of operations, you multiply before you subtract.

LESSON 2.3

1a. $x = 49.4$ **1b.** $x = 40$

1c. $x \approx 216$ **1d.** $x = 583.\overline{3}$

3a. $\dfrac{50 \text{ m}}{1 \text{ sec}} \cdot \dfrac{1 \text{ km}}{1000 \text{ m}} \cdot \dfrac{60 \text{ sec}}{1 \text{ min}} \cdot \dfrac{60 \text{ min}}{1 \text{ hr}} = 180 \text{ km/hr}$

3b. $0.025 \text{ day} \cdot \dfrac{24 \text{ hr}}{1 \text{ day}} \cdot \dfrac{60 \text{ min}}{1 \text{ hr}} \cdot \dfrac{60 \text{ sec}}{1 \text{ min}} = 2160 \text{ sec}$

3c. $1200 \text{ oz} \cdot \dfrac{1 \text{ lb}}{16 \text{ oz}} \cdot \dfrac{1 \text{ ton}}{2000 \text{ lb}} = 0.0375 \text{ ton}$

5a. 158.8 cm   **5b.** 244 cm

5c. 4.72 in.   **5d.** 1.28 in.

7a. Measurement in Yards and Feet

Yards	1	2	3	4	5
Feet	3	6	9	12	15

7b. For each additional yard there are 3 more feet.

7c. $\dfrac{f}{y} = \dfrac{3}{1}$

7d. i. 450 feet   **7d.** ii. 128 yards

9a. Fifteen 12-oz cans to make 960 oz.

9b. $\dfrac{12}{64}$ or 0.1875 oz

9c. $\dfrac{\text{number of ounces of concentrate}}{\text{number of ounces of lemonade}} = \dfrac{12}{64}$

9d. $\dfrac{16}{L} = \dfrac{12}{64}$; $L \approx 85$ oz

11. If the profits are divided in proportion to the number of students in the clubs, the Math Club would get $288, leaving $192 for the Chess Club.

13. laughing kookaburra: 46 cm, green kingfisher: 22 cm, belted kingfisher: 33 cm, pygmy kingfisher: 10 cm, ringed kingfisher: 41 cm

LESSON 2.4

1a. F   **1b.** T   **1c.** F   **1d.** T

3. 2001: 277,722,000; 2002: 280,777,000 2003: 283,865,000

5a. approximately 49%   **5b.** approximately 19.2%

5c. approximately 1,398,110,000

7a. $7.76   **7b.** $7.49

7c. Her wage has dropped by $0.01 per hour because the increase was calculated as 3.5% of $7.50, but the decrease was based on $7.76.

9. Hint: Each bin contains data values in a range. For example, the first bin contains two heights between 120 and 129, so you can choose any two heights in this range for your data set.

11a.

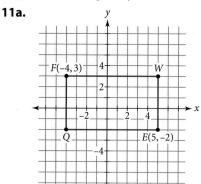

11b. Hint: W and Q do not need to be diagonally opposite to make a rectangle.

LESSON 2.5

1. Type AB = 3750; Type B = 9000; Type A = 30,000; Type O = 32,250

3. No, the total height of all the bars must be 100%.

5a.

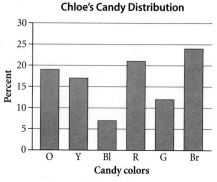

5b.

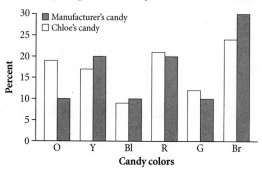

Chloe's bag of candy had the same dominant color as the graph from the manufacturer, and her least color was one of the least manufactured. But the distributions are not very close.

7a. 9th: 27%; 10th: 26%; 11th: 25%; 12th: 22%

7b. 9th: 189; 10th: 172; 11th: 170; 12th: 147; 0.3%; 2 students

7c.

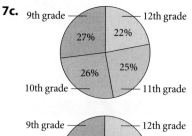

Proportionally, the ninth grade increased 1% and the tenth grade decreased 1%.

9a. 180 pulses per second

9b. $18.\overline{6}$ meters per second

LESSON 2.6

1a. heads or tails

1b. 1, 2, 3, 4, 5, or 6

1c. 2, 3, 4, 5, 6, 7, 8, 9, 10, 11, 12

1d. A, B, C, D, or E

3a. $\frac{27}{100} = 0.27$

3b. P(landing in shaded area) = 0.25 because the shaded area is $\frac{1}{4}$ of the circle.

5f. P(every teacher will give 100 free points) = 0

5g. P(earth will rotate on its axis in 24 hours) = 1

7a. Finding and counting a litter is a trial; an outcome may be having one cub (or two or three or four).

7b. No. If outcomes were equally likely, then the number of litters of each size would have been almost the same, with about nine litters of each size.

7c. $\frac{22}{35} \approx 0.63$

9a. $\frac{1}{6} \approx 17\%$

9b. $\frac{90 + 165}{360} = \frac{255}{360} \approx 71\%$

9c. $\frac{45 + 60}{360} = \frac{105}{360} \approx 29\%$

11. Sample answer: $(-4, 1), (-1, 3), (4, 3), (1, 1)$.

13. 3

1. Probability: $\frac{74}{180} \approx 0.411$; observed probability: $\frac{15}{50} = 0.30$. Sample answer: Perhaps your method of selecting students was not random. For example, your results could be biased because you talked only to students who were participating in after-school activities or only to students in a particular class.

3a. $\frac{15}{250} = 0.06$

3b. $\frac{235}{250} = 0.94$

3c. 0.06

5a. H, H, T, H, H, T

5b. Find the cumulative sum of list L1.

```
2randInt(0,1,100
)-1→L₁
{1 -1 1 1 -1 -1...
cumSum(L₁)
{1 0 1 2 1 0 -1...
```

5c. After many steps, you may be close to 0.

5d. With many trials you might be closer to 0.

7. Hint: See Calculator Note 2A to review random integers.

9. 229

11. Hint: Look back at problem 3 in Lesson 2.6 (page 125).

CHAPTER 2 REVIEW

1a. $n = 8.75$

1b. $w = 84.6$

1c. $k = 5\frac{1}{6}$ or $5.1\overline{6}$

3. Hint: One sample answer is $\frac{5 \text{ hours}}{7 \text{ birdhouses}} = \frac{x \text{ hours}}{30 \text{ birdhouses}}$. Find two more ways to write this proportion.

5a. Hint: Pick any x-coordinate and divide it in half to get the y-coordinate.

5b. All points appear to lie on a line.

7a. 12.5 cm²; $\frac{12.5}{40} = 0.3125$

7b. 32.5 cm²; $\frac{32.5}{45} = 0.7\overline{2}$

9. Sample answer: If the person is talking about the entire state, this cannot occur. All scores cannot be greater than the middle score. The probability is 0.

11. 1365 shih rice; 169 shih millet

LESSON 3.1

1a. $\frac{3 \text{ pounds}}{30 \text{ days}} = 0.1$ pound per day

1b. $\frac{5 \text{ pounds}}{45 \text{ days}} = 0.\overline{1}$ pound per day

1c. Crystal's cat

3a. 0.1 gallon per mile

3b. 22 gallons

3c. 150 miles

5a. $\frac{24901.55 \text{ miles}}{(2 \cdot 365 + 2 \cdot 30.4 + 2) \text{ days}} \approx 31.4$ miles per day

5b. $\frac{31.4 \text{ miles}}{1 \text{ day}} \cdot \frac{(1.5 \cdot 365) \text{ days}}{1} \approx 17191.5$ miles

5c. $\frac{31.4 \text{ miles}}{1 \text{ day}} = \frac{60,000 \text{ miles}}{t}$; $t \approx 1911$ days, or more than 5 years

7a. $2.49 per box, 42¢ per bar, $2.99 per box, 25¢ per ounce

7b. yes

7c. $\frac{1.495 \text{ ounces}}{1 \text{ bar}}$

7d. approximately 25¢

7e. Hint: Compare the price per box, price per bar, price per ounce, and ounces per bar for both brands.

9a. Hint: One sample answer is $\frac{20 \text{ pounds of food}}{1 \text{ week}}$. Find another ratio stated in the problem.

9b. $936

9c. $\frac{20}{85} = \frac{f}{60}$; 14 pounds of food per week

9d. $280.80

11a. $x = \frac{21}{5}$ or 4.2

11b. $x = \frac{22}{9}$ or $2.\overline{4}$

11c. $x = \frac{cd}{e}$

13a.

Jelly Beans in Small Bag

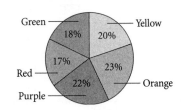

13b.

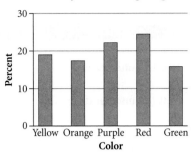

Jelly Beans in Large Bag

13c. In the small bag, orange occurs most frequently and red least frequently. In the large bag, red occurs most often and green least often.

13d. 85 pieces. No, because each color does not occur equally often.

LESSON 3.2

1a. 40   **1b.** 75

3.

Distance (miles)	Distance (kilometers)
2.8	4.5
7.8	12.5
650.0	1040.0
937.5	1500.0

5a. $3.13

5b. 4 yards

5c. $1.25

7a.   Bernard Lavery's Vegetables

Vegetable	Weight (kilograms)	Weight (pounds)
Cabbage	56	123
Summer squash	49	108
Zucchini	29	64
Kohlrabi	28	62
Celery	21	46
Radish	13	28
Cucumber	9	20
Brussels sprout	8	18
Carrot	5	11

(*The Top 10 of Everything 1998*, p. 98)

7b. $y = 2.2x$

7c. 2.95 kilograms

7d. 7920 pounds

7e. 100 lb = $45.4\overline{5}$ kg; 100 kg = 220 lb

9a. Sample answer: {100, 50, 20, 10, 5, 1, 0.50, 0.25, 0.10, 0.05, 0.01}

9b. Sample answer: To convert to Japanese yen, multiply the list by 108.770. {10877, 5438.5, 2175.4, 1087.7, 543.85, 108.77, 54.385, 27.19, 10.88, 5.44, 1.09}

9c. Divide list L_2 by the exchange rate to obtain the original values.

9d. Use dimensional analysis.
$$\frac{2119.150 \text{ liras}}{1 \text{ dollar}} \cdot \frac{1 \text{ dollar}}{2.140 \text{ marks}} = 990.257 \text{ liras per mark.}$$

11a. 81.25 mph

11b. 40 kilometers per hour

11c. 65 mph is 104 kilometers per hour. A speed limit sign might post 100 kilometers per hour.

13a. $\frac{2}{10} = \frac{1}{5}$

13b. No. All the tables are equally likely to have their table number spun; 9 or 10 being spun affects all tables equally.

LESSON 3.3

1a. $y = \frac{3}{4} = 0.75$

1b. $y = 2$

1c. $x = 100$

1d. $x = 108$

3a. Sample answer: $l = \frac{8}{6}s$, where l is a length on the larger polygon an s is a length on the smaller polygon.

3b. $w = 9.\overline{3}$ cm; $x = 2.25$ cm; $y = 6$ cm; $z = 4$ cm

5a. 42 miles

5b. approximately 1.5 inches

5c. a little over 3 inches

5d. 52.5 miles

7. Rhombuses i and iii are similar, and rhombuses iv and v are similar. In each similar pair, corresponding sides are proportional and the angle measures are the same.

9a.

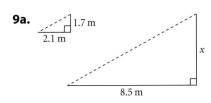

9b. $\frac{1.7}{2.1} = \frac{x}{8.5}$; $x \approx 6.88$ or 6.9 meters high

9c. You could measure the length of the tree's shadow and write a proportion using a person's height and the length of his or her shadow.

11a. Sample answer: 1 cm represents 250 km

11b. Sample answer: Using the scale of 1 cm = 250 km, an equation would be $y = \frac{1}{250}x$, where x is the actual river length in kilometers and y is the scale drawing length in centimeters. Approximate lengths are around 26.7 cm, 25.8 cm, 25.2 cm, 23.9 cm, and 22.2 cm, respectively.

13a. $\frac{30}{100} = \frac{x}{24}$; $x \approx 7$

13b. $\frac{30}{100} = \frac{p}{s}$

13c. $p = \frac{30}{100}s$ or $p = 0.3s$

13d. part = percent · total

15a. $0.31 per day

15b. $60\frac{2}{3}$ pounds of food per cat per year

15c. It costs about $113, or 8.7 fourteen-pound bags. The owner will have to buy 9 bags and spend $116.82.

17. P(mango flavored) = $\frac{4}{48}$ or $\frac{1}{12}$ or 0.083

LESSON 3.4

1a. $y = \frac{15}{x}$

1b. $y = \frac{35}{x}$

1c. $y = \frac{3}{x}$

3. Hint: Pick any value for x, put it into the equation, and evaluate to find y. You could do all 5 points at once using list calculations on your calculator.

5. This is not an inverse variation. The product of the quantities

(*time spent watching TV, time spent doing homework*)

is not a constant. Instead, the sum is a constant. This is a relationship of the form $x + y = k$ or $y = k - x$, not an inverse variation $xy = k$ or $y = \frac{k}{x}$.

7a. 62.$\overline{3}$ N, 93.5 N, and 187 N.

7b. As you move closer to the hinge, it takes more force to open the door. Moving from 15 cm to 10 cm needs an increase of about 31.2 N. Moving 5 cm closer requires an increase of 93.5 N. As you move closer, the force needed increases more rapidly. When you get very close to the hinge, the force needed becomes extremely large.

7c. On the graph, the curve goes up very steeply near the y-axis.

9a. If the balance point is at the center, then the weight of an unknown object will be exactly the same as the weight that balances it on the other side. If the balance point is off-center, you must know the two distances and do some calculation.

9b. $15 \cdot M = 20 \cdot 7$; M $\approx$ 9.3 kg

11a. On this graph, x represents frequency and y represents tube length.

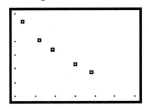

[400, 1000, 100, 30, 90, 10]

11b. Sample answer: $y = \frac{37{,}227.1}{x}$ where 37,227.1 is the mean of the products of the frequencies and tube lengths.

11c. $y = \frac{37{,}227.1}{x}$; $y \approx 42.3$ cm

13. $s = 0.85p$. The sale price is $11.86.

CHAPTER 3 REVIEW

1a. approximately 13 miles per gallon or 0.077 gallon per mile

1b. 65 miles   **1c.** 7.7 gallons

3a. If x represents the weight in kilograms and y represents weight in pounds, one equation is $y = 2.2x$ where 2.2 is the data set's mean ratio of pounds to kilograms.

3b. approximately 13.6 kilograms

3c. 55 pounds

5a. approximately 2.2 inches

5b. 15.75 miles

7a. Because the product of the x- and y-values is approximately constant, it is an inverse relationship.

7b. Sample answer: $y = \dfrac{45.5}{x}$, where 45.5 is the mean of the products.

7c. $y = \dfrac{45.5}{32}, y \approx 1.4$

9a. 4 feet

9b. Yes, Robbie can balance by sitting 2.9 feet from the center.

11. No, they won't fit. 210 centimeters is 6.89 feet.

13. **Algebra Grades**

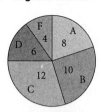

The size and the angle of each piece of the circle graph vary directly with the percent of students who earned each grade.

15a. approximately 1100 thousand visitors

15b. 354, 465, 740, 1272, 3494

15c.

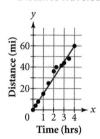

[0, 4000, 100, 1, 12, 1]

15d. Yosemite; the number of visitors exceeds 1272 by more than 1.5 (1272 − 465).

17a. 9 amperes

17b. 6 ohms

CHAPTER 4 · CHAPTER **4** CHAPTER 4 · CHAPTER

LESSON 4.1

1a. 3

1b. −10

1c. $\dfrac{1}{4}$

1d. −6

3a. $2L + 2W$

3b. $2(15 + 4) = 38; 2 \cdot 15 + 2 \cdot 4 = 38$

5a. $84

5b. You can multiply 7 by each day's hours and then add those products to find the week's total. Or you can find the total hours for the week first and multiply this result by $7 \cdot 4 + 7 \cdot 3 + 7 \cdot 5 = 7(4 + 3 + 5)$.

7a. $5.40   **7b.** $3.15

7c. $7.28

9a. First multiply 16 by 4.5, then add 9.

9b. First divide 18 by 3, then add 15.

9c. First square 6, then add −5, then multiply by 4 and subtract 124 from 3.

11a. Sample answer: $(3 + 2)(5) - 7 = 18$

11b. Sample answer: $8 - 5(6 - 7) = 13$

13. $\dfrac{(6 + 3) \cdot 4^2}{8 + 2} - 9 = 5.4$

15a. **Distance Traveled**

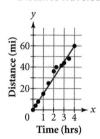

15b. 15 mph

15c. $y = 15x$

15d. Downhill during the intervals of 1 to 2.25 hours away from home and 3.5 to 4 hours (the bike's velocity is above average) and uphill during the interval of 2.25 to 3.5 hours away from home (the bike's speed is slower than average).

17. The 20% discount is $2.59. The 8% tax on 10.36 is $0.83. The tip amount on $12.95 is $1.94, so the total is $13.13.

LESSON 4.2

1a. −12   **1b.** 32

1c. −24   **1d.** 35

1e. 13   **1f.** 3

1g. −19   **1h.** −6

1i. −4

3a. 15   **3b.** 18

3c. 17   **3d.** 5

5a. Juwan forgot the parentheses.

5b. $(2 + 3)4 - 5$ or $4(2 + 3) - 5$.

7. See below.

9a.

Description	Jack's sequence	Nina's sequence
Pick the starting number.	5	3
Multiply by 2.	10	6
Multiply by 3.	30	18
Add 6.	36	24
Divide by 3.	12	8
Subtract your original number.	7	5
Subtract your original number again.	2	2

9b.

Description	Jack's sequence	Nina's sequence
Pick the starting number.	−10	10
Add 2.	−8	12
Multiply by 3.	−24	36
Add 9.	−15	45
Subtract 15.	−30	30
Multiply by 2.	−60	60
Divide by 6 (you should have your original number).	−10	10

11a–c. Hint: Look back at Example A (page 191). The number trick in that example always results in −18.

13. Hint: Look back at Example B (page 192). The number trick in that example always results in the starting number.

15a. value: 25; order of operations: add 7, multiply by 5, divide by 3.

15b. Equation: $\dfrac{5(x + 7)}{3} = -18$ Work backward

Operation on x	Undo operation	
		$x = -17.8$
$+ (7)$	$- (7)$	-10.8
$\cdot (5)$	$\div (5)$	-54
$\div (3)$	$\cdot (3)$	-18

17a. 202.3 ft/minute   **17b.** 102.8 cm/sec

19a. $1\frac{11}{12}$ cups   **19b.** $13.05

LESSON 4.3

1a. 15   **1b.** −16   **1c.** −5

3a. 5 (ENTER) Ans + 3 (ENTER) , (ENTER) , . . .

3b.

Figure #	Perimeter
1	5
2	8
3	11
4	14
5	17

3c. 32   **3d.** Figure 15

5a. Start with 3, then apply the rule Ans + 6; 10th term = 57.

5b. Start with 1.7, then apply the rule Ans − 0.5; 10th term = −2.8.

5c. Start with −3, then apply the rule Ans · −2; 10th term = 1536.

5d. Start with 384, then apply the rule Ans/2 or Ans · 0.5; 10th term = 0.75.

---

7. (*Lesson 4.2*)

Description	Daxun's sequence	Lacy's sequence	Claudia's sequence	Al's sequence
Pick the starting number.	14	−5	−8.6	x
Add 5.	19	0	−3.6	$x + 5$
Multiply by 4.	76	0	−14.4	$4(x + 5)$
Subtract 12.	64	−12	−26.4	$4(x + 5) - 12$
Divide by 4.	16	−3	−6.6	$\dfrac{4(x + 5) - 12}{4}$
Subtract the original number.	2	2	2	$\dfrac{4(x + 5) - 12}{4} - x$

7a. Sample answer: The smallest square has an area of 1. The next larger white square has an area of 4, which is 3 more than the smallest square. The next larger gray square has an area of 9, which is 5 more than the 4-unit white square.

7b. 1 (ENTER) Ans + 2 (ENTER) , (ENTER) , . . .

7c. 17, the value of the ninth term in the sequence

7d. 39

7e. The 48th term is 95. Students might press (ENTER) 48 times or compute $2(48) - 1$.

9a. 4 meters

9b. Press 101 and then Ans − 4. The 19th term represents the height of the 7th floor. The height is 29 meters.

9c. 26 terms

9d. 19 meters

11a. $17 \cdot 7 = 119$ **11b.** 14

11c. Hint: Be sure to try other intervals of 100, such as 300 to 400 and 400 to 500.

11d. Hint: Try to describe a method you can calculate with pencil and paper, and another method using a recursive routine on your calculator.

13a. 1 (ENTER) , Ans · 3 (ENTER) , (ENTER) . . . The 9th term is 6561.

13b. 5 (ENTER) , Ans · (−1) (ENTER) , (ENTER) . . . The 123rd term is 5.

13c. −16.2 (ENTER) , Ans + 1.4 (ENTER) , (ENTER) . . . The 13th term is the first positive term.

13d. −1 (ENTER) , Ans · (−2) (ENTER) , (ENTER) . . . The 8th term is the first to be greater than 100.

15a. $297.25

15b. $7.25 (4 \cdot 8 + 6 \cdot 1.5) = 7.25(41) = 297.25$

LESSON 4.4

1a. negative; −1517 **1b.** positive; 472

1c. positive; $12.\overline{3}$ **1d.** positive; 326

1e. negative; $-3.\overline{3}$ **1f.** negative; −1464

3.

x	y
0	4
1	3
2	2
3	1
4	0

x	y
0	−1.5
1	0
2	1.5
3	3

5a. $-9.\overline{3}$

5b. Equation:

$$\frac{4 - 5(x + 3)}{6} = 12$$

Work backward

$x = -16.6$

Operation on x	Undo operation	
+ (3)	− (3)	−13.6
· (−5)	÷ (−5)	68
+ (4)	− (4)	72
÷ (6)	· (6)	12

7a. {0, 272} (ENTER) {Ans(1) + 1, Ans(2) − 68} (ENTER) , (ENTER) , (ENTER) , (ENTER) , (ENTER)

7b.

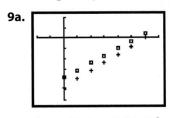

7c. The starting value is the point (0, 272) on the graph.

7d. On the graph, you move right 1 unit and down 68 units to get from one point to the next. In the recursive routine, you add 1 to the first number and subtract 68 from the second number.

7e. This is a linear graph relating a distance to any time between 0 and 5 hours. The line represents the distances at all possible times; points only represent distances at certain times.

7f. The car is within 100 miles of San Antonio after 2.53 hours have elapsed. Explanations will vary. Graphically, it is the time after which the line crosses the horizontal line $y = 100$.

7g. The car takes 4 hours to reach San Antonio. Answers will vary. The answer is the fourth entry in the table. Graphically, it is where the line crosses the x-axis.

9a.

[−10, 35, 5, −60, 20, 10]

9b. The points for each submarine appear to lie on a line. The USS *Dallas* surfaces at a faster rate.

9c. Yes, each line means that any time in this range corresponds to depth below the surface.

9d. The submarine's nose rises slightly above the water when surfacing.

11a. Hint: Since the bicyclist pedals at two different rates, the graph will be made of two distinct sections. How will these sections compare? Where will one section end and the other start?

11b.

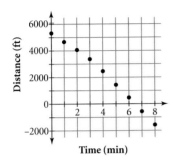

Bicyclist

11c. Sample answer: Where on the graph does the bicyclist pass you? The answer is on the x-axis between 6 and 7 minutes.

13a. Subtract 32 from the temperature in Fahrenheit, multiply the difference by 5, and then divide by 9.

13b. $F = \dfrac{9C}{5} + 32$; $C = \dfrac{5(F - 32)}{9}$

15a.

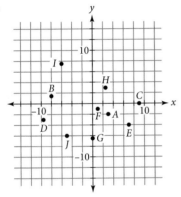

15b. Quad I: H; Quad II: B, I; Quad III: D, J; Quad IV: A, E, F; x-axis: C; y-axis: G.

15c. Hint: One sample answer is, "If both the x-coordinate and the y-coordinate are positive, the point will be in Quadrant I." Modify this statement to describe criteria for the other quadrants and the axes.

LESSON 4.6

1a. ii   **1b.** iv   **1c.** iii   **1d.** i

3a. $d \approx 38.3$ ft   **3b.** $d \approx 25.42$ ft

3c. The walker started 4.7 feet away from the motion sensor.

3d. The walker was walking at a rate of 2.8 feet per second.

5a. $-2\,L_1 + 10$ **5b.** $4L_1 - 4L_2$

7a. Sample answer: Jo has an initial start-up cost of $300 for equipment and expenses. She makes $15 for every lawn she mows, N.

7b. Sample answers: How many lawns will Jo have to mow to break even? [Solve $-300 + 15N = 0$. Jo must mow 20 lawns.] How much profit will Jo earn if she mows 40 lawns? [Substitute 40 for N. $300.]

7c. Subtracting 300 from $15N$ is the same as adding $15N$ to -300.

7d. The input variable is N for number of lawns, and the output variable is P for profit.

9a. $y = 45 + 0.12x$, where x represents dollar amounts customers spend and y represents Manny's daily income in dollars.

9b.

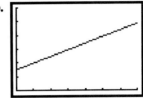

$[0, 840, 120, 0, 180, 30]$

9c. $45 + 0.12 \cdot 312 = \$82.44$

9d. between $500 and $625

11a. $\dfrac{8}{n} = \dfrac{15}{100}$, $n \approx 53.3$ **11b.** $\dfrac{15}{100} = \dfrac{n}{18.95}$, $n = 2.8$

11c. $\dfrac{p}{100} = \dfrac{326}{64}$, $p \approx 509.4$ **11d.** $\dfrac{10}{100} = \dfrac{40}{n}$, $n = 400$

13a. ii   **13b.** iv   **13c.** iii   **13d.** i

15a. The expression equals -4.

Ans $-$ 8	-3
Ans $\cdot$ 4	-12
Ans/3	-4

15b. $y = 14$

LESSON 4.7

1a.

Input x	Output y
20	100
-30	-25
16	90
15	87.5
-12.5	18.75

1b.

L₁ x	L₂ y
0	-5.2
-8	74.8
24	-245.2
-35	344.8
-5.2	46.8

3a. The rate is negative, so the line goes from the upper left to the lower right.

3b. The rate is not negative or positive. It is zero. The line is a horizontal line.

3c. The rate is positive, so the line goes from the lower left to the upper right.

3d. The rate for the speedier walker will be greater than the rate for the person walking more slowly, so the graph for the speedier walker will be steeper than the graph for the slower walker.

5a. i. 3.5; ii. 8; iii. -1.4

5b. i. -6; ii. 1; iii. 23; the y-intercept

5c. i. $y = -6 + 3.5x$; ii. $y = 1 + 8x$; iii. $y = 23 - 1.4x$

7a. $35 + 0.8(25) = 55$ miles

7b. 50 min

9a. 990 square units

9b. Sample answer: $33x = 990$; $x = \dfrac{990}{33}$

9c. 30 units

11a.
-15	-15
Ans + 52	37
Ans/1.6	23.125
$52 + 1.6(23.125) = -15$	Check

11b.
52	52
Ans -7	45
Ans/-3	-15
$7 - 3(-15) = 52$	Check

13a. 70.4 lengths　　**13b.** 2.2 feet per second

13c. 43.7 lengths

13d. 40 lengths for a kilometer, 64 lengths for a mile

LESSON 4.8

1a. $2x = 6$　　　　**1b.** $x + 2 = 5$

1c. $2x - 1 = 3$　　**1d.** $2 = 2x - 3$

3a.

$0.1x + 12 = 2.2$	Original equation.
$0.1x + 12 - 12 = 2.2 - 12$	Subtract 12 from both sides.
$0.1x = -9.8$	Remove the 0 and subtract.
$x = -98$	Divide both sides by 0.1.

3b.

$\dfrac{12 + 3.12x}{3} = -100$	Original equation.
$12 + 3.12x = -300$	Multiply both sides by 3.
$-12 + 12 + 3.12x = -12 - 300$	Subtract 12 from both sides.
$3.12x = -312$	Remove the 0.
$x = -100$	Divide both sides by 3.12.

5a. $x = \dfrac{1}{12}$　　　　**5b.** $x = 36$

7a.

$4 - 1.2x = 12.4$	Original equation.
$4 - 4 - 1.2x = 12.4 - 4$	Subtract 4 from both sides.
$-1.2x = 8.4$	Subtract.
$\dfrac{-1.2x}{-1.2} = \dfrac{8.4}{-1.2}$	Divide both sides by -1.2.
$x = -7$	Reduce.

7b.
Start with 12.4.	12.4
Ans -4	8.4
Ans/-1.2	-7

7c.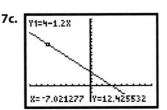

$[-10, 10, 1, -5, 20, 1]$

7d.

X	Y1
-3	7.6
-4	8.8
-5	10
-6	11.2
-7	12.4
-8	13.6
-9	14.8

X=-7

9a. $r = \dfrac{C}{2\pi}$　　　　**9b.** $h = \dfrac{2A}{b}$

9c. $l = \dfrac{P}{2} - w$　　**9d.** $s = \dfrac{P}{4}$

9e. $t = \dfrac{d}{r}$　　　　**9f.** $h = \dfrac{2A}{a + b}$

11a. $-\dfrac{1}{5}$　　　　**11b.** -17

11c. 2.3　　　　**11d.** x

13. $120

15. Your equation is correct when the graph of your line exactly matches the program's line.

CHAPTER 4 REVIEW

1a. $x = -7$　　　　**1b.** $x = -23.4$

3a. iii　　　　　**3b.** i

3c. ii

5a. $y = x$　　　　**5b.** $y = -3 + x$

5c. $y = -4.3 + 2.3x$　　**5d.** $y = 1$

7a. 0 represents no bookcases sold; -850 represents fixed overhead, such as startup costs; Ans(1) represents the previously calculated number of bookcases sold; Ans(1) + 1 represents the current number of bookcases sold, one more than the previous; Ans(2) represents the profit for the previous number of bookcases; Ans(2) + 70 represents the profit for the current number of bookcases—the company makes $70 more profit for each additional bookcase.

7b.

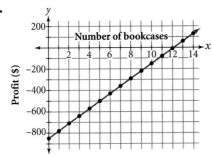

7c. Sample answer: The graph crosses the x-axis at approximately 12.1 and is positive after that. It shows you need to make at least 13 bookcases to make a profit.

7d. -850, the profit if the company makes 0 bookcases, is the y-intercept; 70, the amount of additional profit for each additional bookcase, is the rate of change.

7e. No, partial bookcases cannot be sold.

9a. 4 seconds

9b. Away; the distance is increasing.

9c. approximately 0.5 meters

9d. $\dfrac{2.9 - 0.5}{4} = 0.6$ meter per second

9e. $\dfrac{5.5 \text{ meters}}{0.6 \text{ meters/second}} = 9.1\overline{6}$ seconds. Approximately 9 seconds.

9f. The graph is a straight line.

11a. $L_2 = -5.7 + 2.3 \cdot L_1$

11b. $L_2 = -5 - 8 \cdot L_1$

11c. $L_2 = 12 + 0.5 \cdot L_1$

13a. Hint: You could use lists, tables, graphs, guessing and checking, working backward, or the balancing method.

13b. $2(3.5 - 6) = 2(-2.5) = -5$

15a. $t \approx -1.2°$ **15b.** $t \approx 38.1°$

15c. $t \approx 38.1°$ **15d.** $t \approx 27.4°$

CHAPTER 5 · CHAPTER **5** CHAPTER 5 · CHAPTER

LESSON 5.1

1a. 2   **1b.** $\dfrac{2}{3}$ **1c.** $-\dfrac{4}{3}$

3. Sample answers:

3a. $(1, 7), (-1, 1)$ **3b.** $(3, 3), (1, 13)$

3c. $(12, 3), (4, 9)$ **3d.** $(6, 7.2), (4, 6.8)$

5a. i. The x-values don't change, so the slope is undefined.

5a. ii. The y-values decrease as the x-values increase, so the slope is negative.

5a. iii. The y-values don't change, so the slope is 0.

5a. iv. The y-values increase as the x-values increase, so the slope is positive.

5b. i. Using the points $(4, 0)$ and $(4, 3)$, the slope is $\dfrac{3 - 0}{4 - 4} = \dfrac{3}{0}$. Since you can't divide by 0, the slope is undefined.

5b. ii. Using the points $(1, 3)$ and $(4, -3)$, the slope is $\dfrac{-3 - 3}{4 - 1} = \dfrac{-6}{3} = -2$.

5b. iii. Using the points $(-4, -5)$ and $(-3, -5)$, the slope is $\dfrac{-5 - (-5)}{-3 - (-4)} = \dfrac{-5 + 5}{-3 + 4} = \dfrac{0}{1} = 0$.

5b. iv. Using the points $(0, -2)$ and $(4, 1)$, the slope is $\dfrac{1 - (-2)}{4 - 0} = \dfrac{3}{4}$.

5c. i. $x = 4$; **ii.** $y = 5 - 2x$; **iii.** $y = -5$; **iv.** $y = -2 + \dfrac{3}{4}x$

7a. Use the slope to move backward from $(4, 16.75)$; $(4 - 1, 16.75 - 2.95) = (3, 13.80)$, or \$13.80 for 3 hours; $(3 - 1, 13.80 - 2.95) = (2, 10.85)$, or \$10.85 for 2 hours.

7b. Continuing the process in 7a leads to $(0, 4.95)$, or \$4.95 for 0 hours. This is the flat monthly rate for Hector's Internet service.

7c. $y = 4.95 + 2.95x$, where x is time in hours and y is total fee in dollars.

7d. Substitute 28 for x and solve for y: $y = 4.95 + 2.95(28) = 87.55$. \$87.55 for 28 hours.

9. $y = e - \dfrac{a}{c}x$

11a. i. Line 2 is a better choice.

11a. ii. Either line 3 or line 4 is a reasonable fit.

11b. Hint: Remember that not all data is linear.

13. Hint: One sample answer is $\{3, 3, 6, 16, 22\}$. Try to find another.

15a. 85%   **15b.** 150%   **15c.** 6.5%   **15d.** 107%

LESSON 5.2

1a. No. Although this line goes through four points, too many points are below the line.

1b. No. Although the slope of the line shows the general direction of the data, too many points are below the line.

1c. Yes. About the same number of points are above the line as below the line, and the slope of the line shows the general direction of the data.

1d. No. Although the same number of points are above the line as below the line, the slope of the line doesn't show the direction of the data.

3a. $y = -2 + \dfrac{2}{3}x$

3b. $y = 2 - \frac{2}{3}x$

3c. $y = -2 - \frac{2}{5}x$

3d. $y = 3$

5a. The number of representatives depends on the population.

5b. Let x represent population in millions, and let y represent the number of representatives.

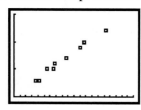

$[0, 10, 0.5, 0, 15, 5]$

5c. Sample answer: $y = 1.6x$. The slope represents the number of representatives per 1 million people. The y-intercept means that a state with no population would have no representatives.

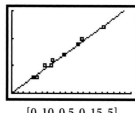

$[0, 10, 0.5, 0, 15, 5]$

5d. The equation $y = 1.6x$ gives $y = 1.6(33) = 52.8$ or 53 representatives. On July 19, 2000, California had 52 representatives.

5e. The equation $y = 1.6x$ gives $8 = 1.6x$; $x = \frac{8}{1.6} = 5$; 5 million. The estimated population of Minnesota in July 1999 was 4.8 million.

5f. The relationship should be a direct variation because it should go through the point $(0, 0)$. A state with no population would have no representatives.

7a. $\frac{y_2 - y_1}{x_2 - x_1} = \frac{4.4 - 3.4}{4.5 - 2}$; the slope is 0.4 meter per second.

7b. The y-intercept is 2.6 meters; students can find this by working backward with the slope or by estimating from a graph.

7c. $y = 2.6 + 0.4x$

9a. Sample answer: $y = -8 + 4x$

9b. Sample answer: $y = -2x$

9c. $y = 6 + x$ **9d.** $y = 10$

11a. neither

11b. inverse variation; $y = \frac{100}{x}$

11c. direct variation; $y = -2.5x$

11d. direct variation; $y = \frac{1}{13}x$

13a. mean: $24.8\overline{6}$; median: 21

13b. mean: 44.5; median: 40

13c. mean: approximately 140.1; median: 145

13d. mean: 85.75; median: 86.5

LESSON 5.3

1a. $4; (5, 3)$ **1b.** $2; (-3.1, 1.9)$

1c. $-3.47; (7, -2)$ **1d.** $-1.38; (2.5, 5)$

3a. 2

3b. $y = -1 + 2(x + 2)$

3c. $y = 13 + 2(x - 5)$

3d. The graphs coincide and the tables are identical.

5. There are no "correct" answers for this game.

7a. AD: $y = 2 + 0.2(x + 1)$ or $y = 3 + 0.2(x - 4)$
BC: $y = -2 + 0.2(x + 3)$ or $y = -1 + 0.2(x - 2)$
AB: $y = 2 + 2(x + 1)$ or $y = -2 + 2(x + 3)$
DC: $y = 3 + 2(x - 4)$ or $y = -1 + 2(x - 2)$

7b. The slopes are the same; the coordinates of the points are different.

7c. $ABCD$ appears to be a parallelogram because each pair of opposite sides is parallel. The equal slopes in 7b mean that the opposite sides are parallel.

9a. The data appears linear.

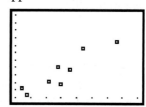

$[10, 50, 5, 250, 800, 50]$

9b. Sample answer: Using the points $(21, 360)$ and $(43, 620)$, an equation is $y = 620 + 11.82(x - 43)$.

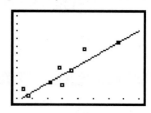

$[10, 50, 5, 250, 800, 50]$

9c. In the graph of $y = 620 + 11.82(x - 43)$ approximately 490 calories.

9d. Compared to the graph of $y = 620 + 11.82(x - 43)$, the point lies above the line. A point above the line means the sandwich has more calories than the model predicts.

9e. Using $y = 620 + 11.82(x - 43)$, four points are above the line, two are on the line, and two are below the line

Selected Answers

9f. Hint: Remember that a good model usually shows the general direction of the data and has about the same number of points above the line as below.

9g. Using $y = 620 + 11.82(x - 43)$, approximately 112 calories; this makes sense because not all calories in food come from fat.

11a. 4.125 liters   **11b.** 180 K

13a. $-1; (5, -1)$

13b. undefined; $(2, 3)$

13c. $-\frac{5}{2}; (0, -3)$

LESSON 5.4

1a. not equivalent; $-3x - 9$

1b. equivalent   **1c.** equivalent

1d. not equivalent; $-2(x + 4)$ or $2(-x - 4)$

3a. $x = 4$; division property

3b. $-x = 92$; addition property
$x = -92$; multiplication property

3c. $x = -7$; subtraction property

3d. $x = 112$; multiplication property

5a. $(-5, 25)$ **5b.** $x = 0$

7a. $y = 5(2 + x)$ **7b.** $y = 5(x + 2)$

7c. The y_1 value is missing, which means it is zero; $y = 0 + 5(x + 2)$.

7d. $(-2, 0)$; this is the x-intercept.

9a. $x = 2$; the point $(2, 0)$ is the x-intercept.

9b. $y = 3$; the point $(0, 3)$ is the y-intercept.

9c.

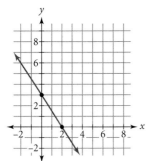

9d. The slope is $-\frac{3}{2}$; $y = 3 - \frac{3}{2}x$.

9e.

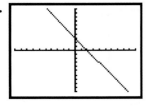

$[-10, 10, 1, -10, 10, 1]$

The two lines are the same; hence the equations are equivalent.

9f. $3x + 2y = 6$ Original equation.
$2y = 6 - 3x$ Subtraction property with $3x$.
$y = 3 - \frac{3}{2}x$ Division property with 2.

11a. $y = 15.20 + 0.85(x - 20)$

11b. $19.45

11c. The equation is used to model the bill only when Dorine is logged on for more than 15 hours. Substituting 15 for x gives the flat rate of $10.95 for all amounts of time less than 15.

11d. 30 hours

13a. Hint: It seems logical that compensation should increase over time because of inflation. Are there any countries for which compensation behaved contrary to this assumption?

13b. Germany is the country with the greatest increase ($25.91), and Mexico is the country with the least increase ($0.04).

13c. Top box is 1975, middle 1985, lowest 1995.

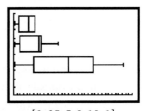

$[0, 35, 5, 0, 10, 1]$

Notice that there is a much larger range in data for 1995 than in 1975. The lowest compensation has not changed much, whereas the top end has moved considerably. The median has moved upward as well.

15. $z = \frac{3.8 + 5.4}{0.2} - 6.2; z = 39.8$

LESSON 5.5

1a. $y = 1 + 2(x - 1)$ or $y = 5 + 2(x - 3)$

1b. $y = 3 + \frac{2}{3}(x - 1)$ or $y = 5 + \frac{2}{3}(x - 4)$

1c. $y = 6 - \frac{4}{3}(x - 1)$ or $y = 2 - \frac{4}{3}(x - 4)$

3. The x-intercept of $y = b(x - x_1)$ is at $x = x_1$.

3a. 3   **3b.** -4 **3c.** 6

5a.

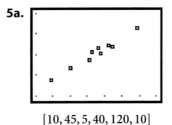

$[10, 45, 5, 40, 120, 10]$

5b. Sample answer: Using the points $(20, 67)$ and $(31.2, 88.6)$, an equation is $y = 67 + 1.9(x - 20)$.

5c. Using $y = 67 + 1.9(x - 20)$:

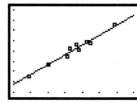

[10, 45, 5, 40, 120, 10]

5d. $y = 32 + 1.8(x - 0)$ or $y = 212 + 1.8(x - 100)$

5e. The sample equation in 5b gives $y = 29 + 1.9x$; the equations in 5d both give $y = 32 + 1.8x$.

5f. The difference could be a result of measurement error or faulty procedures.

7a. $y = -11 + 2x$

7b. $y = 17 + 41x$

7c. $y = 59 - 6x$

7d. $y = 9 - 4x$

9a. -3

9b. $y = 106 - 3(x - 10)$

9c. After 45 full days, there will be only one biscuit left, so it will be empty in the middle of the 46th day.

9d. When the box was new, before Anchor had any biscuits, there were 136 biscuits.

LESSON 5.6

1a. 166, 405, 623, 1052, 1483

1b. 204, 514, 756, 1194, 1991

1c.

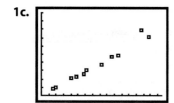

[0, 1650, 100, 0, 2500, 250]

1d. The slope will be positive because as the flying distance increases so does the driving distance.

1e. $(405, 514), (1052, 1194)$

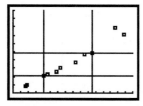

[0, 1650, 100, 0, 2500, 250]

1f.

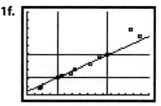

[0, 1650, 100, 0, 2500, 250]

The slope is approximately 1.05;
$y = 1194 + 1.05(x - 1052)$ or $y = 514 + 1.05(x - 405)$.

1g. approximately 1054 miles

1h. approximately 535 or 536 miles

3a. Sample answer: The slope will be positive and you will choose the lower-left and upper-right corners of the rectangle for the Q-points.

3b. Sample answer: The slope will be negative and you will choose the upper-left and lower-right corners of the rectangle.

5. Hint: Look back at the graph on page 291. Based on the scattering of data points, why was the Q-point $(6, 15)$ not one of the data points?

7a. $y = 1.3 + 0.625(x - 4)$ or $y = 6.3 + 0.625(x - 12)$

7b. The elevator is rising at a rate of 0.625 second per floor.

7c. 36.3 seconds after 2:00, or approximately 2:00:36

7d. almost at the 74th floor

9a. Hint: Since the elevators are traveling in opposite directions at equal rates, a good method of estimation is to find the average of the elevators' starting positions.

9b. At 28.8 sec, or at about 2:00:29, the elevators will pass at the 48th floor.

11a. Start with 370, then use the rule Ans $- 54$.

Time (hr)	Distance from Mt. Rushmore (mi)
0	370
1	316
2	262
3	208
4	154
5	100
6	46

11b.

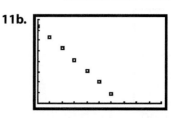

[0, 10, 1, 0, 400, 50]

11c.

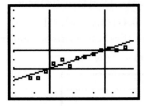

[0, 10, 1, 0, 400, 50]

The line represents the distance remaining at any time during the trip. With the line, you can see how far you are at any time, instead of just at the top of each hour.

11d. -54; the real-world meaning of the slope is that your distance from Mt. Rushmore decreases by 54 miles each hour.

11e. The car will reach the Wall Drug Store in the first half hour of the 5th hour of the trip. You can see this on the graph if you look at the line where it has a y-value of about 80.

11f. The car will reach Mt. Rushmore after almost 7 hours of travel. You can see this on the graph or in the table since after 7 hours, the car would be 8 miles too far if it kept going.

LESSON 5.7

1a. $(6, 6)$

1b. $(5, 9)$

1c. $y = 9 - 3(x - 5)$

1d. $y = 24 - 3x$

1e. $(8, 0)$

3a. $y = \dfrac{18 - 2x}{5}$ or $y = 3.6 - 0.4x$

3b. $y = \dfrac{-12 - 5x}{-2}$ or $y = 6 + 2.5x$

5. Let x represent distance from Los Angeles in miles, and let y represent elapsed time from Seattle in minutes.

5a. $y = 1385.5 - 1.49(x - 411.5)$ or $y = 240 - 1.49(x - 1181.5)$. The slope means the distance from Los Angeles decreases by 1 mile each 1.49 minutes.

5b. approximately 1701 min, or 28 hrs 21 min, by the first equation or approximately 1702 min, or 28 hrs 22 min, by the second equation

5c. approximately 939 miles by the first equation or 940 miles by the second equation

7. 15.4321 grains per gram.

1.
$$-3 = \dfrac{4 - 10}{x_2 - 2}$$
$$-3(x_2 - 2) = -6$$
$$x_2 - 2 = 2$$
$$x_2 = 4$$

3. Line a has slope -1, y-intercept 1, and equation $y = 1 - x$. Line b has slope 2, y-intercept -2, and equation $y = -2 + 2x$.

5a. $(-4.5, -3.5)$

5b. $y = 2x + 5.5$

5c. $y = 2(x + 2.75)$; the x-intercept is -2.75.

5d. The x-coordinate is 5.5; $y = 16.5 + 2(x - 5.5)$.

5e. Hint: You can try graphing, using a calculator table, or putting all equations in intercept form.

7a. $y = 12600 - 1350x$

7b. -1350; the car's value decreases by $1,350 each year.

7c. 12,600; Karl paid $12,600 for the car.

7d. $9\frac{1}{3}$; in $9\frac{1}{3}$ years the car will have no monetary value.

9a. 1952, 1956, 1976, 1990, 2000; 1.67, 1.835, 1.93, 2.02, 2.05

9b. The Q-points are $(1962, 1.835)$ and $(1990, 2.020)$.

9c. $y = 1.835 + 0.007(x - 1962)$ or $y = 2.020 + 0.007(x - 1990)$

9d. Hint: Remember that a good model usually shows the general direction of the data and has about the same number of points above the line as below.

[1940, 2000, 10, 1.5, 2.5, 0.1]

9e. Using $y = 1.835 + 0.007(x - 1962)$, the prediction is 2.19 m.

11a. Written as $y = a + bx$, b is the slope and a is the y-intercept.

11b. If the points are (x_1, y_1) and (x_2, y_2), then the slope of the line is $\dfrac{y_2 - y_1}{x_2 - x_1} = b$. The equation is $y = y_1 + b(x - x_1)$.

LESSON 6.1

1a. Yes, because $47 + 3(-15.6) = 0.2$ and $8 + 0.5(-15.6) = 0.2$.

1b. No, because $23 \neq 12 + (-4)$. The point satisfies only one of the equations.

1c. No, $12.3 \neq 4.5 + 5(2)$. Furthermore, the lines are parallel, so there is no solution to the system.

3. In both cases, the calculator gives exact solutions that satisfy each system.

3a. $(8, 7)$

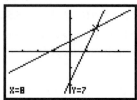

3b. $(1.5, 0.5)$

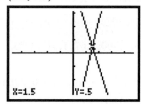

5. The results of substituting the x- and y-values into the original equation tell you that each coordinate satisfies both the standard form and the intercept form of the equation.

5a. $y = 3 - 2x$; $(1, 1)$: $4(1) + 2(1) = 6$

5b. $y = -4 + 0.4x$; $(1, -3.6)$: $2(1) - 5(-3.6) = 20$
A coordinate satisfies both forms of a linear equation.

7a. $P = -5000 + 2.5N$

7b.

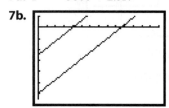

$[0, 7000, 500, -13000, 20000, 2000]$

7c. Sally will always profit more than Gizmo.kom for the same number of web site hits. Because their lines never intersect, there is no solution to the system of equations, and their profits will never be equal.

7d. Sally pays less to start Gadget.kom, but she profits at the same rate as Gizmo.kom. Her profit will always be $7,000 more for the same N-values.

9a. $d = 9 - t$ where d is drill team member's distance from end zone; $d = 3 + 0.5t$ where d is tuba player's distance from end zone

9b. $(4, 5)$. After 4 seconds, the tuba player bumps into the drill team member at the 5-yard mark.

11a. Because lines with different slopes always intersect, the y-intercept a can equal any number and b can be any number except -5.

11b. $a \neq 2$ and $b = -5$, same slope, different y-intercept, lines do not intersect.

11c. $a = 2$ and $b = -5$, same slope and y-intercept, lines overlap.

13a. 85

13b. -8.2

13c. 3

13d. 3.5

13e. 1.5

15a. $\begin{bmatrix} 1 & -11 \\ -6 & 8 \end{bmatrix}$

15b. $\begin{bmatrix} 13 & -1 \\ 7 & 8 \end{bmatrix}$

LESSON 6.2

1. 3. Add $2.5t$ to both sides.
   5. Divide both sides by 4.

3a. $x = -2$

3b. $y = 10$

3c. $d = 3$

3d. $t = 4$

5a. $-x + 8$

5b. $13x - 8$

7a. $N = 7,777\frac{7}{9}, P = 7,444\frac{4}{9}$

7b. The approximate solution, $N \approx 7778$ and $P \approx 7444$, is more meaningful because there cannot be a fractional number of web site hits.

9a. $A + C = 200$

9b. $8A + 4C = 1304$

9c. $A = 126$ and $C = 74$, so the theater sold 126 adult tickets and 74 child tickets.

11a. $\begin{cases} d = 35 + 0.8t \\ d = 1.1t \end{cases}, \left(116\frac{2}{3}, 128\frac{1}{3}\right)$.

The pickup passes the sports car roughly 128 miles from Flint after approximately 117 minutes.

11b. $\begin{cases} d = 220 - 1.2\,t \\ d = 1.1t \end{cases}$,

$\left(\dfrac{2200}{23}, \dfrac{2420}{23}\right) \approx (95.7, 105.2)$. The minivan meets the pickup truck about 105 miles from Flint after approximately 96 minutes.

11c. $\begin{cases} d = 220 - 1.2t \\ d = 35 + 0.8t \end{cases}$,

$(92.5, 109)$. The minivan meets the sports car 109 miles from Flint after 92.5 minutes.

11d. $220 - 1.2t = 2(35 + 0.8t)$; $t \approx 53.6$ min, minivan is about 156 mi, sports car is about 78 mi.

11e. The solutions found using substitution are exact (if not rounded off). Recursive routines sometimes give approximate answers because of their discreteness.

13a. 12.1 ft/sec

13b. 50 seconds

13c. $y = 100 + 12.1x$, where x represents the time in seconds and y represents her height above ground level. To find out how long her ride to the observation deck is, solve the equation $520 = 100 + 12.1x$.

15a. i

15b. iii

15c. ii

LESSON 6.3

1a. $y = \dfrac{10 - 5x}{2}$ or $y = \dfrac{10}{2} - \dfrac{5x}{2}$

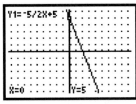

$[-9.4, 9.4, 1, -6.2, 6.2, 1]$

1b. $y = \dfrac{30 - 15x}{6}$ or $y = \dfrac{30}{6} - \dfrac{15x}{6}$

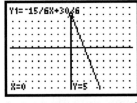

$[-9.4, 9.4, 1, -6.2, 6.2, 1]$

The graph is the same as the graph for 1a. Both equations are equivalent to $y = 5 - \dfrac{5}{2}x$.

3a. $(-2.5, -1)$

3b. $(3, -2)$

5a. Multiply the first equation by -5 and the second equation by 3, or multiply the first equation by 5 and the second equation by -3.

5b. Multiply the first equation by -8 and the second equation by 7, or multiply the first equation by 8 and the second equation by -7.

7a. $(4, 2)$

7b. $(3, -1)$

7c. $(-3, -1)$

9a. $y = 163 - x$ and $y = -33 + x$

9b. $y = 65$

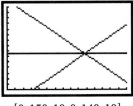

$[0, 150, 10, 0, 140, 10]$

9c. $x = 98$

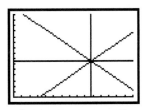

$[0, 150, 10, 0, 140, 10]$

9d. The four lines intersect at the same point, $(98, 65)$. The solution to the system must satisfy all the equations—the original equations in the system and any new equations created by combining pairs of equations.

11a. substitution

11b. Her solution is half right. Anisha didn't find the value $y = -1$ when $x = 4$.

13a. Let c represent gallons burned in the city and h represent gallons burned on the highway.
$\begin{cases} c + h = 11 \\ 17c + 25h = 220 \end{cases}$

13b. $(6.875, 4.125)$; 6.875 gallons in the city, 4.125 gallons on the highway

13c. $\dfrac{17\text{ mi}}{\text{gal}} \cdot 6.875\text{ gal} \approx 117$ city mi,

$\dfrac{25\text{ mi}}{\text{gal}} \cdot 4.125\text{ gal} \approx 103$ hwy mi

13d. Check: $\begin{cases} 6.875 + 4.125 = 11 \\ 17(6.875) + 25(4.125) = 220 \end{cases}$

and $117 + 103 = 220$.

15a. See below.

15b. $T = 95.2 - 0.004E$. The slope is the rate of change in temperature with elevation. The y-intercept (in this case, T-intercept) is the temperature that day at sea level in the same area.

15c. At the summit the temperature was 13.9 degrees Fahrenheit.

17a. P (top) $= \dfrac{400}{769}$ or 0.52 roughly.

P (bottom) $= \dfrac{369}{769}$ or 0.48 roughly.

17b. A trial is assigning a locker. An outcome is whether a locker is in the top row or the bottom row.

LESSON 6.4

1a. $\begin{cases} 2x + 1.5y = 12.75 \\ -3x + 4y = 9 \end{cases}$

1b. $\begin{cases} \dfrac{1}{2}x = \dfrac{1}{2} \\ -x + 2y = 0 \end{cases}$

1c. $\begin{cases} 2x + 3y = 1 \\ 2y = 0 \end{cases}$

3a. $(8.5, 2.8)$

3b. $\left(\dfrac{1}{2}, \dfrac{13}{16}\right)$

3c. $(0, 0)$

5a. $\begin{cases} 3x + y = 7 \\ 2x + y = 21 \end{cases}$

5b. $\begin{bmatrix} 3 & 1 & 7 \\ 2 & 1 & 21 \end{bmatrix}$

7a.

	Adults	Children	Total (kg)
Monday	40	15	10.80
Tuesday	35	22	12.29

7b. Let x represent the average weight of chips an adult eats and y represent the average weight of chips a child eats. The system is $\begin{cases} 40x + 15y = 10.8 \\ 35x + 22y = 12.29 \end{cases}$

7c. $\begin{bmatrix} 40 & 15 & 10.8 \\ 35 & 22 & 12.29 \end{bmatrix}$

7d. $\begin{bmatrix} 1 & 0 & 0.15 \\ 0 & 1 & 0.32 \end{bmatrix}$

7e. Each adult ate an average of about 0.15 kg (150 g) of chips, and each child ate an average of 0.32 kg (320 g) of chips.

9a. Let x represent the number of small trucks and y represent the number of large trucks. The system is $\begin{cases} 5x + 12y = 532 \\ 7x + 4y = 284 \end{cases}$

9b. $\begin{bmatrix} 5 & 12 & 532 \\ 7 & 4 & 284 \end{bmatrix}$

9c. $\begin{bmatrix} 1 & 0 & 20 \\ 0 & 1 & 36 \end{bmatrix}$

9d. Zoe should order 20 small trucks and 36 large trucks.

11a. $\begin{cases} m + t + w = 286 \\ m - t = 7 \\ t - w = 24 \end{cases}$

11b. $\begin{bmatrix} 1 & 1 & 1 & 286 \\ 1 & -1 & 0 & 7 \\ 0 & 1 & -1 & 24 \end{bmatrix}$

The rows represent each equation. The columns represent the coefficients of each variable and the constants.

11c. $\begin{bmatrix} 1 & 0 & 0 & 108 \\ 0 & 1 & 0 & 101 \\ 0 & 0 & 1 & 77 \end{bmatrix}$

11d. They cycled 108 km on Monday, 101 km on Tuesday, and 77 km on Wednesday.

13a. $4, \text{Ans} - 0.5$

13b. $-3, \text{Ans} + 2$

13c. $1/2, \text{Ans} - 1$

13d. $0, \text{Ans} + 1$

15. $\begin{bmatrix} 1 & 3 \\ -2 & 1 \\ 3 & 23 \end{bmatrix} \rightarrow \begin{matrix} 3 & -3 \\ -6 & -1 \\ 9 & -23 \end{matrix} \rightarrow \begin{matrix} 0 \\ -7 \\ -14 \end{matrix}$

$-7y = -14, y = 2$ and $x = 7$

15a. (*Lesson 6.3*)

Marsha's Climb

	Elevation (ft)	Temperature (°F)
Start	4,300	78
Rest station	7,800	64
Highest point	11,900	47.6

1a. Multiply by 4; $12 < 28$

1b. Multiply by -3; $-15 \geq -36$

1c. Add -10; $-14 \geq x - 10$

1d. Subtract 8; $b - 5 > 7$

1e. Divide by 3; $8d < 10\frac{2}{3}$

1f. Divide by -3; $-8x \geq -10\frac{2}{3}$

3a. $x \leq -1$

3b. $x > 0$

3c. $x \geq -2$

3d. $-2 < x < 1$

3e. $0 < x \leq 2$

5a. $y = \dfrac{5.2 - 3x}{4} = 1.3 - 0.75x$

5b. $y = \dfrac{2x}{3} + 5$

7a. $x \leq 3$

7b. $x < -2$

7c. $x \geq -3$

7d. $0 \geq x$ or $x \leq 0$

9a. Add 3 to both sides; $4 < 5$.

9b. Divide both sides by 2 (or multiply by 0.5); $3 > 1$.

9c. Multiply both sides by -3; $3 > -3$.

9d. Multiply both sides by 2; $0 < 6$.

11a. The variable x drops out of the inequality, leaving $-3 > 3$, which is never true. So the original inequality is not true for any number x. You can't draw a graph to represent this situation on a number line.

11b. The variable x drops out of the inequality, leaving $-6.6 \geq -15$, which is always true. So the original inequality is true for any number x. The graph would be a line with arrows on both ends.

13a. Multiply 12 by 3.2 to get 38.4. Subtract 38.4 from 72 to get 33.6.

13b. Square 5 to get 25. Subtract 25 from 3 to get -22. Multiply -22 by 1.5 to get -33. Add -33 to 2 to get -31.

13c. Divide 21 by 7 to get 3 and divide 6 by 2 to get 3. Subtract 3 from 3 to get 0.

15a. $-2x - 16$

15b. $3 - 4y$

15c. $-z + 5$

1a. iii　　**1b.** ii　　**1c.** i　　**1d.** iv

3a. 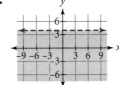　　**3b.**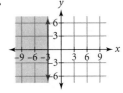

3c. 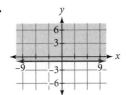　　**3d.**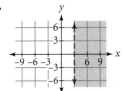

5a–c.

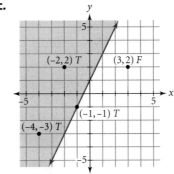

7a. $y \leq 1 - 2x$　　**7b.** $y < -2 + \frac{2}{3}x$

7c. $y > 1 - 0.5x$　　**7d.** $y \geq -2 + \frac{1}{3}x$

9a.

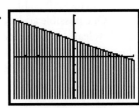

9b.

9c.

9d.

11.

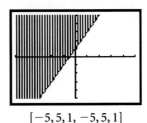

$[-5, 5, 1, -5, 5, 1]$

$[-5, 5, 1, -5, 5, 1]$

13a. about 27 mph

13b. Since $d = r \cdot t$, and the distance was the same for both Ellie and her grandmother, you can set these products equal to each other. If you let $r =$ Ellie's grandmother's speed, then $2.5(65) = 6r$.

1a. iii   **1b.** i   **1c.** ii

3a. $y \geq -x + 2; y \geq x - 2$

3b.

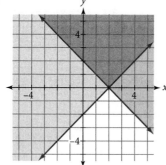

5. $\begin{cases} y > 2 - x \\ y < 2 \\ x < 3 \end{cases}$ **7a.** $\begin{cases} A \leq C \\ A + C \leq 75 \\ A \geq 0 \\ C \geq 0 \end{cases}$

7b.

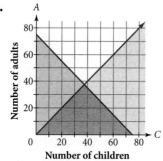

All the points in the dark-shaded triangular region satisfy the two inequalities. The point (50, 10) represents the situation in which 50 children escort 10 adults.

7c. Sample answer: It is possible to have all children and no adults at the restaurant. One possible additional constraint is that there must be at least one adult per five children, or $A \geq \frac{1}{5}C$. The solution for this set of constraints is the triangular region bounded by $A \leq C$, $A + C \leq 75$, and $A \geq 0.2C$.

9. $x \geq 3$ and $y \geq -2 + \frac{1}{2}x$

11. The region is a pentagon.

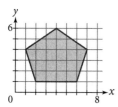

13a. $713.15   **13b.** $957.80

15a. 4   **15b.** 10

15c. $\dfrac{10\left(\dfrac{3x+12}{5} - 1.4\right) - 10}{6}$, which simplifies to x

1. line a: $y = 1 - x$; line b: $y = 3 + \frac{5}{2}x$; intersection: $\left(-\frac{4}{7}, \frac{11}{7}\right)$

3. $(3.75, 4.625)$

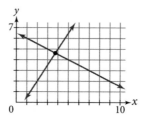

5a. … the slopes are the same, but the intercepts are different (the lines are parallel).

5b. … the slopes are the same and the intercepts are the same (the lines overlap).

5c. … the slopes are different (the lines intersect in a single point).

7. $x \leq -1$ ⟵———●——→
                              -2 0

9a. 10 sq. m/min

9b. 7 sq. m/min

9c. No. He will cut 156 sq. m, and the lawn measures 396 sq. m.

9d. $10h + 7l = 396$

9e. $\frac{1}{30}$ liters per min

9f. $\frac{3}{200}$ liters per min

9g. Yes. He will use $\frac{34}{75}$ liter, and the tank holds 1.2 liters.

9h. $\frac{h}{30} + \frac{3l}{200} = 1.2$

9i. $l = 14.4$ min, $h = 29.52$ min

9j. If Harold cuts for 29.52 minutes at the higher speed and 14.4 minutes at the lower speed, he will finish Mr. Fleming's lawn and use one full tank of gas.

CHAPTER 7 • CHAPTER 7 CHAPTER 7 • CHAPTER

1a. starting value: 16; multiplier: 1.25; 7th term: 61.035

1b. starting value: 27; multiplier: $\frac{2}{3}$ or $0.\overline{6}$; 7th term: $2.\overline{370}$ or $\frac{64}{27}$

3. Start with 72, then apply the rule Ans · (1 + 0.40); first five terms are 72, 100.8, 141.12, 197.568, and 276.5952.

5. Start with 32, then apply the rule Ans · 0.75; Stage 2 has a shaded area of 18 square units.

7a. Start with 115, then apply the rule Ans · (1 − 0.03).

7b. 12 minutes

9a. Start with 6.9, then apply the rule
Ans · (1 + 0.142).

9b. See below.

9c.

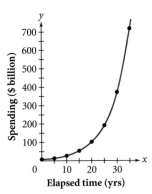

9d. Sample answer: The graph implies a smooth, ever-increasing amount of Medicare spending, which is probably not realistic.

11a. See below.

11b. See below.

11c. The graph of the first plan is linear. The graph of the second is not; its slope increases between consecutive points.

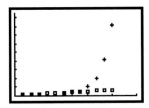

[0, 15, 3, 0, 4750, 500]

11d. Possible answer: Grace's charity will receive more total money with option 2. It will also give the charity more income later in the year when the budget may be tighter.

13a. i   **13b.** iii   **13c.** ii   **13d.** iv

LESSON 7.2

1a. 7^8 **1b.** $3^4 \cdot 5^5$ **1c.** $(1 + 0.12)^4$

3a. ii   **3b.** iii   **3c.** iv   **3d.** i

5a. $y = 1.2 \cdot 2^x$ **5b.** $y = 500 \cdot 0.2^x$

5c. $y = 125 \cdot 0.4^x$

7. Your calculator will tell you when you have the correct answer.

9a. $9,200

9b. Start with 11,500, then apply the rule
Ans · (1 − 0.2).

9c.

Time elapsed (yrs)	0	1	2	3	4
Value ($)	11,500	9,200	7,360	5,888	4,710.40

9d. $y = 11,500(1 - 0.2)^x$

9e.

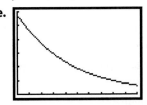

[0, 10, 1, 0, 12000, 2000]

9b. (*Lesson 7.1*)

Medicare Spending

Year	1970	1975	1980	1985	1990	1995	2000	2005
Elapsed time (yrs) x	0	5	10	15	20	25	30	35
Spending ($ billion) y	6.9	13.4	26.0	50.6	98.2	190.8	370.5	719.7

11a. (*Lesson 7.1*)

	Jan	Feb	Mar	Apr	May	June	July	Aug	Sep	Oct	Nov	Dec
Option 1	$50	$25	$25	$25	$25	$25	$25	$25	$25	$25	$25	$25
Option 2	$1	$2	$4	$8	$16	$32	$64	$128	$256	$512	$1,024	$2,048

11b. (*Lesson 7.1*)

	Jan	Feb	Mar	Apr	May	June	July	Aug	Sep	Oct	Nov	Dec
Option 1	$50	$75	$100	$125	$150	$175	$200	$225	$250	$275	$300	$325
Option 2	$1	$3	$7	$15	$31	$63	$127	$255	$511	$1,023	$2,047	$4,095

11. Your calculator will tell you when you have the correct answer.

13a. $y = 5000(1 + 0.05)^x$

13b.
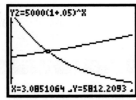
[0, 10, 1, 0, 12000, 2000]

The intersection point represents the time and the value of both cars when their value will be the same. By tracing the graph shown, students should see that both cars will be worth approximately \$5,806 after a little less than 3 years 1 month.

15. Hint: Since you multiply by $(1 - 0.035)$, your situation needs to describe something that decreases over time. The equation $y = 100(1 - 0.035)^x$ will help you solve the problem you create in 15b.

LESSON 7.3

1a. $5x^4$ **1b.** $15x^{10}$ **1c.** $8x^{10}$ **1d.** $-2x^4 - 2x^6$

3a. 3^{40} **3b.** 7^{12} **3c.** x^{12} **3d.** y^{40}

5. Student 2 was correct. $2x^2 = 2 \cdot x \cdot x$; substitute $x = 3$: $2 \cdot 3 \cdot 3 = 18$

7. a, d, and g: 27,521.40084; b and f: 11,711.2344; c and h: 1060.32; e: 129,350.5839

9. Enclose the -5 in parentheses.

11. Sample answers:

11a. $x^3 \cdot x^5 = x^8$ **11b.** $(x^3)^5 = x^{15}$

11c. $(3x)^5 = 3^5 x^5 = 243x^5$

11d. Exponents are added when you multiply two exponential expressions with the same base. Exponents are multiplied when an exponential expression is raised to a power. An exponent is distributed when a product is raised to a power.

13a. $x^4 + x^5$ **13b.** $-2x^4 - 2x^6$

13c. $17x^{7.5} + 8.5x^{8.4}$

15a. $4.5x - 47$

15b. $4.5x - 47 > 0$; $x > 10.\overline{4}$; he must shovel 11 sidewalks to pay for his equipment.

15c. $4.5x - 47 > 100$; $x > 32.\overline{6}$; he must shovel 33 sidewalks to pay for his expenses and buy a lawn mower.

17a. $(5.3625, 0.70625)$

17b. approximately $(3.095, 0.762)$

LESSON 7.4

1a. 3.4×10^{10} **1b.** -2.1×10^6 **1c.** 1.006×10^4

3a. $12x^6$ **3b.** $7y^{16}$ **3c.** $2b^6 + b^5$ **3d.** $10x^4 - 6x^2$

5. $3.5 \times 10^7 = 3.5 \cdot 10 \cdot 10 \cdot 10 \cdot 10 \cdot 10 \cdot 10 \cdot 10$
     $= 35,000,000$;

$3.5^7 = 3.5 \cdot 3.5 \cdot 3.5 \cdot 3.5 \cdot 3.5 \cdot 3.5 \cdot 3.5$
     $= 6433.9296875$

7a. 3.3411×10^{25} **7b.** approximately 3.6×10^{47}

9a. Yes, because they are both equal to 51,800,000,000.

9b. Al's answer   **9c.** Possible answer: 518×10^8

9d. Rewrite the digits before the 10 in scientific notation, then use the multiplication property of exponents to add the exponents on the 10's. In this case, $4.325 \times 10^2 \times 10^3 = 4.325 \times 10^5$.

11a. $2,000,000,000$; 2×10^9

11b. 7.3×10^{11} calls per year

13a. 9.46×10^{12} km; 1.0×10^5 light-years

13b. 9.46×10^{17} km

13c. $\dfrac{(9.46 \times 10^{17})}{(1.27 \times 10^4)} = 7.45 \times 10^{13}$

15.
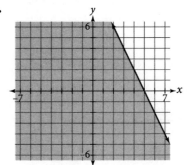

LESSON 7.5

1. $x^3 y$

3. Sample answer: $\frac{3^6}{3^2}$ means there are 6 factors of 3 in the numerator and 2 factors of 3 in the denominator. So there are 2 factors of 1, or $\frac{3}{3}$, in the entire expression, leaving 4 factors of 3 in the numerator, or 3^4.

5a. $648x^{11}$ **5b.** $25x^2$ **5c.** $-4x^2$ **5d.** $-108x^5 y^{14}$

7a. about 132 people per square mile

7b. about 867 people per square mile

7c. The population of Japan was about 6.6 times denser than that of Mexico.

9. four days earlier

Method 1: Use a recursive routine
864 (ENTER), Ans/3 (ENTER), (ENTER), (ENTER), (ENTER)

Method 2: Use an equation: $y = 864\left(\frac{1}{3}\right)^x$;
look at the table to find x when y is less than 20.

11a. approximately 61 years

11b. approximately 1.3 years

11c. approximately 25.4 years

13. approximately 2.272×10^5 flowers

15a. (Answers recorded to tenths.) Mercury: 4.1 cm; Venus: 10 cm; Earth: 10.5 cm; Mars: 5.6 cm; Jupiter: 117.3 cm; Saturn: 94.7 cm; Uranus: 69.3 cm; Neptune: 41.3 cm; Pluto: 2 cm; Sun: 1152 cm

15b. Hint: Convert the largest model diameter to meters. Will Halley be able to make this model? Consider a different size for Pluto that gives a reasonable diameter for the largest model.

LESSON 7.6

1a. $\dfrac{1}{2^3}$ **1b.** $\dfrac{1}{5^2}$ **1c.** $\dfrac{1.35}{10^4}$

3a. -5 **3b.** -4 **3c.** -4

5. Sample answer: Negative exponents mean to use a reciprocal with the exponent positive.
$6^{-3} = \dfrac{1}{6^3} = \dfrac{1}{216}, \; -6^3 = -216$

7a. $3500(1 + 0.04)^{-4}$; approximately \$2,992

7b. $250(1 + 0.04)^{-3}$; approximately \$222

7c. $25(1 + 0.04)^{-5}$; approximately \$21

7d. $126{,}800(1 + 0.04)^{-30}$; approximately \$39,095

9a. true; $(2^3)^2 = 2^3 \cdot 2^3 = 2 \cdot 2 \cdot 2 \cdot 2 \cdot 2 \cdot 2 = 2^6$

9b. false; $(3^3)^4 = 3^3 \cdot 3^3 \cdot 3^3 \cdot 3^3$
$= 3 \cdot 3 \cdot 3 \cdot 3 \cdot 3 \cdot 3 \cdot 3 \cdot 3 \cdot 3 \cdot 3 \cdot 3 \cdot 3 = 3^{12}$

9c. false; $(10^{-2})^4 = \left(\dfrac{1}{10^2}\right)^4$
$= \left(\dfrac{1}{10 \cdot 10}\right)\left(\dfrac{1}{10 \cdot 10}\right)\left(\dfrac{1}{10 \cdot 10}\right)\left(\dfrac{1}{10 \cdot 10}\right) = \dfrac{1}{10^8}$
$= 10^{-8}$

9d. true; $(5^{-3})^{-4} = \left(\dfrac{1}{5^3}\right)^{-4} = \dfrac{1}{\left(\dfrac{1}{5^3}\right)^4}$

$= \dfrac{1}{\left(\dfrac{1}{5 \cdot 5 \cdot 5}\right)\left(\dfrac{1}{5 \cdot 5 \cdot 5}\right)\left(\dfrac{1}{5 \cdot 5 \cdot 5}\right)\left(\dfrac{1}{5 \cdot 5 \cdot 5}\right)} = \dfrac{1}{\dfrac{1}{5^{12}}}$

$= \dfrac{1}{5^{-12}} = 5^{12}$

11. $36(1 + 0.5)^{(4-6)} = 36(1 + 0.5)^{-2}$

13. 1×10^{-1} or $\dfrac{1}{10}$ of a chocolate bar per person

15a. i. 4×10^5 **15a.** ii. 3.1×10^{10}

15a. iii. 1.21×10^5 **15a.** iv. 2×10^{-4}

15b. Sample answer: Divide the coefficients of the powers of 10, and then divide the powers of 10 (subtract the exponents).

15c. 0.6×10^5 or 6×10^4 in scientific notation

LESSON 7.7

1a. $1 + 0.15$; rate of increase: 15%

1b. $1 + 0.08$; rate of increase: 8%

1c. $1 - 0.24$; rate of decrease: 24%

1d. $1 - 0.002$; rate of decrease: 0.2%

1e. $1 + 1.5$; rate of increase: 150%

3. $B = 250(1 + 0.0425)^t$

5a. The ratios are 0.957, 0.956, 0.965, 0.964, 0.963, 0.961, 0.959, 0.958, 0.971, and 0.955.

5b. $0.9609 \approx 0.96$

5c. $1 - 0.04$ **5d.** $y = 47(1 - 0.04)^x$

5e.

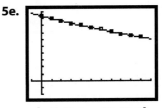

$[-1, 11, 1, -10, 50, 5]$

Adjustments to A or r are not necessary—the fit is good as is.

5f. 55 minutes

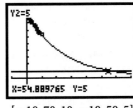

$[-10, 70, 10, -10, 50, 5]$

7a. Sample answer: Let x represent years since 1990, and let y represent median price in dollars. An equation is $y = 85{,}000(1 + 0.06)^x$, where 0.06 is derived from the mean ratio of about 1.06.

7b. Hint: Add 5 to the current year and then subtract 1990 to find the number of years since 1990. Substitute this value for the appropriate variable in your equation.

7c. Hint: Multiply your answer from 7b by 0.10.

7d. Hint: 5 years is the same as 60 months. So divide your answer from 7c by 60.

9. 1000 nanoseconds per microsecond

11a. $y = 2(1 + 0.5)^x$

11b. $y = 2(1 + 0.5)^{0.5} \approx 2.45$; approximately 2.45 liters

11c. approximately 115 liters

11d. After about 30.4, or 30 minutes 24 seconds

13. $x = 3$ cm; $y = 7.2$ cm; $z = 9$ cm

1a. 3^4 **1b.** 3^3 **1c.** 3^2

1d. 3^{-1} **1e.** 3^{-2} **1f.** 3^0

3. Hint: Since you multiply by $(1 - 0.15)$, your situation needs to describe something that decreases over time.

5a. $y = 200(1 + 0.4)^x$ **5b.** $y = 850(1 - 0.15)^x$

x	y
0	200
1	280
2	392
3	548.8
4	768.32
5	1075.648
6	1505.9072

x	y
−2	1176.4706
−1	1000.0000
0	850
1	722.5
2	614.125
3	522.00625
4	443.7053

7. approximately 1.17×10^0 years

9a. False; 3 to the power of 3 is not 9; $27x^6$

9b. False; you can't use the multiplication property of exponents if the bases are different; $3^2 \cdot 2^3$ or 72

9c. False; the exponent -2 applies only to the x; $\dfrac{2}{x^2}$

9d. False; the power property of exponents says to multiply exponents; $\dfrac{x^6}{y^9}$

11a. Let t be the number of T-shirts, and let s be the number of sweatshirts.
$t + s = 12$
$6t + 10s = 88$

11b. 8 T-shirts and 4 sweatshirts

13a. Start with 21, then apply the rule Ans − 4; 10th term = −15.

13b. Start with −5, then apply the rule Ans · (−3); 10th term = 98,415.

13c. Start with 2, then apply the rule Ans + 7; 10th term = 65.

15a. $y = 1.6x$, where x is a measurement in miles and y is a measurement in kilometers

15b. 400 km   **15c.** 3.2 km

15d. approximately 168 m

17a. −4   **17b.** 13   **17c.** −12

19. The following sample answers use the Q-point method and a decimal approximation of the slope.

19a. $y = 96 - 5.7(x - 5)$

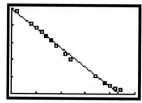

$[0, 25, 5, 0, 125, 25]$

19b. $y = 124.5 - 5.7x$ **19c.** approximately 16 days

19d. The y-intercept would become 200; $y = 200 - 5.7x$.

19e. 86 grams

1a. SBOHF   **1b.** EPNBJO

1c. UBCMF   **1d.** HSBQI

3a. SECRET CODES

3b. The coding scheme is a letter-shift of $+1$.

5a. {1:00, 2:00, 3:00, 4:00, 5:00, 6:00, 7:00, 8:00, 9:00, 10:00, 11:00, 12:00}

5b. {0100, 0200, 0300, 0400, 0500, 0600, 0700, 0800, 0900, 1000, 1100, 1200, 1300, 1400, 1500, 1600, 1700, 1800, 1900, 2000, 2100, 2200, 2300, 2400}

5c. It is not a function because each standard time designation has two military time designations. If students distinguish A.M. from P.M. times, then it is a function.

7a. L1 = {6, 21, 14, 3, 20, 9, 15, 14, 19}
L2 = {15, 30, 23, 12, 29, 18, 24, 23, 28}

7b. You have to subtract 26 from the numbers greater than 26. The new list L2 is {15, 4, 23, 12, 3, 18, 24, 23, 2}.

7c. ODWLCRXWB

7d.

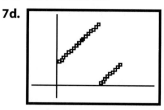

$[-10, 37, 0, -5, 31, 0]$

7e. Avoid any multiple of 26, such as 0, ±26, ±52, and so on.

9. ALGEBRA

11a. Input: {0, 1, −1, 2, −2}; output: {0, 1, 2}; the relationship is a function.

11b. Input: {1, 4, 9}; output: {1, −1, 2, −2, 3, −3}; the relationship is not a function.

13a. Subtract the input letter's position from 27 to get the output letter's position.

13b. Yes, each input matches no more than one output.

13c. Sample answer: LISA codes into ORHZ.

15a. $28b^6$ **15b.** $11a^3$ **15c.** 1.75

15d. $1372d^3$ **15e.** Not possible   **15f.** $4x^5$

1a.

Input x	Output y
−4	1
−1	3.4
1.5	5.4
6.4	9.32
9	11.4

1b.

Domain x	Range y
−4	4.4
−1	2
2.4	−0.72
11	−7.6
14	−10

3.

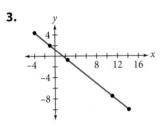

5. Start at the 2-m mark and stand still for 2 sec; walk toward the 4-m mark at 2 m/sec for 1 sec. Stand still for another second; walk toward the 8-m mark at 4 m/sec for 1 sec; then stand still for 3 sec. Yes, the graph represents a function.

7a. No; Los Angeles, for example, has more than one area code (213, 310, . . .).

7b. Yes; each person has only one birth date.

7c. No; the same last name will correspond to many different first names.

7d. Yes; each state has only one capital.

9a. Not a function; the input 3 has two different output values, 10 and 8.

9b. A function; each x-value corresponds to only one y-value.

9c. A function; each x-value corresponds to only one y-value.

11. Hint: Your graph will not pass the vertical line test.

13a. domain: {−5, −4, −3, −2, −1, 0, 1, 2, 3, 4, 5}; range: {0, 1, 2, 3, 4, 5}

13b. domain: $0 \le x \le 360$; range: $-1 \le y \le 1$

13c. domain: all numbers x; range: all numbers y

15a. When $x = 23.125, y = -15$. Answers will vary. Zoom in on the table by changing the start values and the table increments (ΔTbl).

15b. The lines intersect at the solution point.

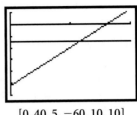

$[0, 40, 5, -60, 10, 10]$

1a. The level of medication drops quickly at first and then decreases more slowly over time. After 10 hours, about 30 milligrams are still in the bloodstream.

1b. The independent variable is time. The dependent variable is the amount of medication in milligrams in the bloodstream.

1c. The domain is the time in hours from 0 through 10. The range is the amount of medication in milligrams from about 30 through 500.

1d. Sample answer: Assuming $(1, 400)$ is a point on the curve, the equation is $y = 500(1 - 0.20)^x$.

3a. The reading on the scale depends on the weight of the dog, so the dog's weight is the independent variable and the reading on the scale is the dependent variable.

3b. The amount of time you spend in the plane depends on the distance you fly, so the distance between the cities is the independent variable and the amount of time in the plane is the dependent variable.

3c. The wax sticks to the candle wick each time you dip it, so the number of dips is the independent variable and the diameter of the candle is the dependent variable.

5. $x \to y$; independent $\to$ dependent; input $\to$ output; time $\to$ distance (or distance $\to$ time)

7. Student graphs should consist of three line segments, each less steep than the one before it.

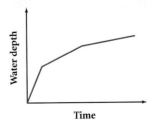

9a. If the *x*-axis is labeled in days, a reasonable domain is from 0 to 120, with 0 representing May 1. If the *x*-axis is labeled in months or another unit of time, the domain will differ accordingly. The range should go from 0 to about 5 inches, or 13 centimeters, depending on how high grass grows before it's cut.

9b. The domain should go from 0 to a large number such as 500 or more, depending on the number of students in the school. The range is a set of integer values from 1 to a maximum that depends on the number of students in the school.

9c. The domain should go from 0 to about 5 seconds. The range should go from 0 to about 300 feet, or 100 meters.

9d. The number of dice is the domain; it is any non-negative integer. The probability is the range, and it is any number between 0 and 1 for observed outcomes. The theoretical probability for a large number of dice is close to 1.

9e. The angle measure is the domain, and it goes from 0° to 180°. The area is the range, and it could be as small as 0 and as large as about 60 sq. in. or 400 sq. cm, depending on the dimensions of the coat hanger.

11a. Hint: The graph should be made up of at least three horizontal segments.

11b. Hint: How will your speed change when the school bus makes a stop? How will your speed change when the school bus starts driving again?

11c. Hint: Your graph might look like an upside-down U.

13a. decreases.   **13b.** increases.

13c. $h = \dfrac{1000}{(\pi r^2)}$

13d.

y-axis: *h*, Height (cm), marked at 18
x-axis: *r*, Radius (cm), marked at 18, origin at 0

15a. -2.5 **15b.** -4 **15c.** $\dfrac{5}{3}$

LESSON 8.4

1a. $3(3) + 2 = 11, Y_2(3) = 11$

1b. $3(-4) + 2 = -10, Y_1(3) = -10$

1c. $(5)^2 - 1 = 24, Y_2(5) = 24$

1d. $(-3)^2 - 1 = 8, Y_2(-3) = 8$

3a. $f(4) = 0$ **3b.** $f(6) = 4$ **3c.** $f(2) = 2, f(5) = 2$

3d. $f(0.5) = 1, f(3) = 1, f(4.5) = 1$

3e. Three   **3f.** $x > 5$

3g. domain: $-1 \le x \le 7$; range: $0 \le y \le 6$

5a. $f(x) = 7x + 5$ **5b.** $f(x) = 5 + 7(x - 1)$

7a. amount of medication in milligrams

7b. time in hours   **7c.** $0 \le x \le 10$; all real numbers x

7d. $53 < y \le 500, y > 0$

7e. 500   **7f.** about 4 hours

9a. Hint: Be sure to try positive and negative numbers, fractions, and zero.

9b. Hint: Be sure to try positive and negative numbers, fractions, and zero.

9c. Hint: You will need to enter numbers into a few lists to use this command.

9d. Hint: There are two different possible answers depending on how you use the random command.

11a. $f(x)$: independent variable x is time in seconds; dependent variable y is height in meters.
$g(x)$: independent variable x is time in seconds; dependent variable y is velocity in meters per second.

11b. $f(x)$: domain $0 \le x < 3.2$; range $0 \le y \le 50$.
$g(x)$: domain $0 \le x < 3.2$; range $-31 \le y \le 0$.

11c. Sample answer: For the graph of $f(x)$, the ball is dropped from an initial height of 50 meters. It hits the ground after about 3.2 seconds. At the moment the ball is dropped, its velocity is 0 m/sec. For the graph of $g(x)$, the velocity starts at 0 m/sec and changes at a constant rate, becoming more and more negative.

11d. In the 1st second, the ball falls about 5 m, from 50 m at $x = 0$ to about 45 m at $x = 1$.

11e. In the 2nd second, the ball falls about 15 m, from about 45 m at $x = 1$ to about 30 m at $x = 2$.

11f. For the graph of $f(x)$, the ball hits the ground after about 3.2 seconds. For the graph of $g(x)$, at $x \approx 3.2$ the velocity is about -31 m/sec, or $g(x) \approx -31$.

13. No, the input -2 has two different output values, 3 and 0, and the input 3 has two different outputs, 1 and -2.

15a. $\dfrac{1}{2^9}$ **15b.** 5^{10}

15c. $2^{12}3^6$ **15d.** $\dfrac{1}{3^8 5^{12}}$

17a. $x = -4$ **17b.** $x = -0.75$ **17c.** $x = 4$

LESSON 8.5

1. Sample answer:

1a.

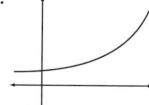

1b.

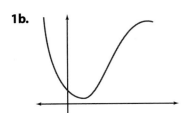

1c.

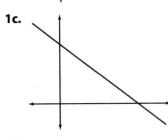

1d.

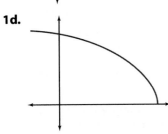

3a. Linear, increasing, constant rate of change

3b. Nonlinear, increasing, fast rate of change, then slowing down

3c. Nonlinear, decreasing, fast rate of change, then slowing down

3d. Linear, neither increasing nor decreasing, zero rate of change

5. Graph B

7. Hint: Think about some actual data values. For example, if it takes 1 volunteer 10 minutes to clean up 1 square yard of campus, how long should it take 2 volunteers? How about 3 volunteers or 4 volunteers? What kind of relationship do you see in the data?

9a. about 11:00 A.M.

9b. between 10:10 and 10:40 A.M.

9c. between 9:00 and 9:45 A.M. and then again after 11:00 A.M.

11a. licks, independent; mass, dependent

11b. scoops, independent; cost, dependent

11c. amount of stretch, independent; flight distance, dependent

11d. number of coins, independent; number of heads, dependent

13a. $y = -25.9 - 2.5x$

13b. The slope is -2.5 and the y-intercept is -25.9.

15. The five-number summary is 293, 307, 343.5, 431, 601.

1a. 7   **1b.** 0.5

1c. 5   **1d.** 9

1e. -5 **1f.** -5

1g. 12   **1h.** 3

3a. $12 = 12$ **3b.** $40 = 40$

3c. $15 > 9$ **3d.** $9 < 13$

3e. $4 = 4$ **3f.** $16 > \dfrac{1}{16}$

5.

x	-4	-1.5	0	1.2	3	4.75
y	4	1.5	0	1.2	3	4.75

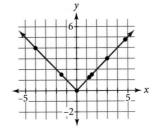

7. The solutions are $(2.85, 2.85)$ and $(-2.85, 2.85)$.

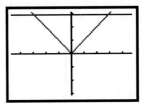

$[-4.7, 4.7, 1, -3.2, 3.2, 1]$

9. The walker starts 5 meters away from the motion sensor and walks toward the motion sensor at a rate of 1 meter per second for 4 seconds and then walks away from the motion sensor at the same rate.

11a. $g(5) = |5| + 6 = 11$

11b. $h(1) = 18(1.5) = 27$

11c. $g(-2) = |-2| + 6 = 8; h(-2) = 18(1.5)^{-2} = 8$

11d. $f(3) = 7 + 4 \cdot 3 = 19$

13. Find the mean of the absolute values of the deviations for each lake. Spider Lake is 0.675 lbs, and Doll Lake is 0.375 lbs. The fish weights varied more in Spider Lake.

15a. $-1\dfrac{2}{3} < x$ or $x > -1\dfrac{2}{3}$

15b. $x \le -1$

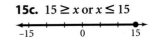

15c. $15 \ge x$ or $x \le 15$

LESSON 8.7

1.

Side (cm)	Perimeter (cm)	Area (sq cm)
1	4	1
2	8	4
3	12	9
4	16	16
14	56	196
15	60	225
21	84	441
25.2	100.8	635.04
47	188	2209

3a. $x = \pm 6$

3b. $x = \pm 6$

3c. $x = \pm 3.8$

3d. $x = \pm 3.8$

5. $-1 \leq x \leq 1$. Good graphing window answers will vary. One possibility is a TI-83 friendly window like $[-2.35, 2.35, 1, 0, 3.1, 1]$.

7.

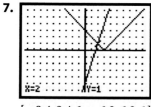

$[-9.4, 9.4, 1, -6.2, 6.2, 1]$

Sample answer: They tell us that there is exactly one solution and that the x-value that makes the equation true is $x = 2$.

9a. $1 + 3 + 5 + 7 + 9 = 25$, or 5^2.

9b. $1 + 3 + 5 + 7 + 9 + \ldots + 29 = 225$, or 15^2.

9c. The sum of the first n positive odd integers is n^2.

9d. Each larger square is created by adding a border of small squares on two sides. The number of little squares added on each time is the next odd number. The resulting figure is the next square number.

11a. 16: 1-by-1, 9: 2-by-2, 4: 3-by-3, and 1: 4-by-4

11b. A 3-by-3 grid has 14 squares: 9: 1-by-1, 4: 2-by-2, and 1: 3-by-3. A 2-by-2 grid has 5 squares: 4: 1-by-1 and 1: 2-by-2. A 1-by-1 grid has 1 square.

11c. Sample answer: One possible response: There are n^2 1-by-1 squares, $(n-1)^2$ 2-by-2 squares, $(n-2)^2$ 3-by-3 squares, and so on. There would be 55 squares in a 5-by-5 grid.

13a. $y = 400(0.75)^x$

13b.

x	y
0	400
4	126.5625
3	168.75
1	300
≈ -3.19	1000

CHAPTER 8 REVIEW

1a. $-2 \leq x \leq 4$

1b. $1 \leq f(x) \leq 3$

1c. 1

1d. at $x = -1$ or 3

3. The graph is a horizontal line segment at 0.5 meters per second.

5a. Hint: Keep in mind that the graph shows time and velocity. It does not show relative position.

5b. No. Because Abby starts out moving faster than both Bea and Caitlin, even when she slows down to their speed, she stays ahead.

7a. Hint: Your graph will pass the vertical line test.

7b. Hint: Your graph will not pass the vertical line test.

9a. $f(-3) = |-3| = 3$

9b. $f(2) = |2| = 2$

9c. $x = 10$ or -10

CHAPTER 9 · CHAPTER **9** CHAPTER 9 · CHAPTER

LESSON 9.1

1a. $(-2, 3), (4, 1), (2, -5)$

1b. $(-5, 1), (3, 4), (6, -3), (-3, -5)$

3a. a translation up 4 units

3b. The x-coordinates are unchanged.

3c. The y-coordinates are increased by 4.

5a. a translation right 10 units and down 8 units

5b. $L_3 = L_1 + 10, L_4 = L_2 - 8$

5c. The signs would change: $L_3 = L_1 - 10, L_4 = L_2 + 8$.

7a.

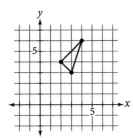

7b.

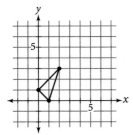

7c.

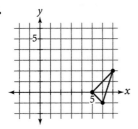

9a. a translation right 12 units and up 7 units

9b. $(x + 12, y + 7)$

9c. $\left(x + \dfrac{12}{20}, y + \dfrac{7}{20}\right) = (x + 0.6, y + 0.35)$

11a. 17   **11b.** -10

11c. $8 + 3x$ **11d.** $-1 + 6x$

13a. $y = -2 + x$ **13b.** $y = 1 - 0.5(x - 1)$

13c. $y = |x|$ **13d.** $y = 1.5^x$

LESSON 9.2

1a. 19   **1b.** 5

1c. 8   **1d.** $2|x + 6| + 1$

3a. $(1, -3)$ **3b.** $(-5, -3)$

3c. $(6, 4)$ **3d.** $(-1, 0)$

5a. $y = x^2 - 2$ **5b.** $y = 4^{(x-5)}$

5c. $y = |x + 4| + 1$

7a. The input variable is t, time, the output variable is d, distance.

7b. Time is in seconds and distance is in meters.

7c. domain: $0 \leq t \leq 5$; range: $1 \leq d \leq 4$

7d. Sample answer: Beth starts 3 m from her teacher and walks toward the teacher at 1 m/sec for 2 sec. When she turns in the test, she is 1 m from the teacher. Beth then turns and walks away from the teacher at 1 m/sec for 3 sec.

7e. $d = |t - 2| + 1$

9a. a translation right 3 units

9b. a translation left 2 units

9c. a translation down 2 units

9d. a translation up 3 units

11a. 20°C   **11b.** 221°C; 188°C

11c. approximately 15% per hour

11d. Let x represent time in hours, and let y represent temperature in degrees Celsius; $y = 221(1 - 0.15)^x$.

11e. $y = 221(1 - 0.15)^x + 20$; the vertical translation is needed to make the long-run value 20°C.

11f. $y = 221(1 - 0.15)^{(x-5)} + 20$; the horizontal translation is needed because the first temperature was measured 5 hours after the bowl was removed from the kiln.

11g. The temperature of the bowl immediately after it was removed from the kiln was approximately 518°C.

11h. To make sure the temperature is 25°C or less, round up to 29 hours.

13a. $(4, 8)$ **13b.** $y = b(x - 4) + 8$

13c. (H, V) **13d.** $y = b(x - H) + V$

15a. $(0.75, 6.5)$ **15b.** $(-3.5, -24.5)$

15c. $(3, 1)$

LESSON 9.3

1a. -1 **1b.** 37.5   **1c.** -10

1d. $0.5(-x - 3)^2 - 3$ **1e.** $-0.5(x - 3)^2 + 3$

3a. a reflection across the x-axis

3b. a translation right 6 units or a reflection across the y-axis

3c. a translation left 2 units

3d. a translation left 2 units and a reflection across the x-axis

5a. a reflection across the x-axis

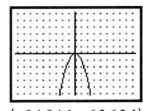

$[-9.4, 9.4, 1, -6.2, 6.2, 1]$

5b. a translation left 3 units and a reflection across the x-axis

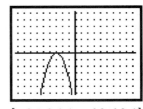

$[-9.4, 9.4, 1, -6.2, 6.2, 1]$

5c. a reflection across the *x*-axis followed by a translation up 3 units

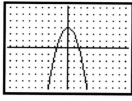

$[-9.4, 9.4, 1, -6.2, 6.2, 1]$

5d. a reflection across the *y*-axis and a translation up 3 units

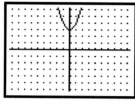

$[-9.4, 9.4, 1, -6.2, 6.2, 1]$

7a. a reflection across the *y*-axis

7b. a translation left 8 units and a reflection across the *x*-axis

7c. a translation right 2 units and down 4 units

7d. a reflection across the line $y = x$

9. $(x + 1, -y)$

11a. i. $y = -2$ **11a. ii.** $y = 3.5$

11a. iii. $x = 3$ **11a. iv.** $x = -4$ or $y = -4$

11b. i. $y = -x^2 - 4$ **11b. ii.** $y = -|x| + 7$

11b. iii. $y = 2^{-(x-6)}$

11b. iv. $y = 2(-(x + 8)) + 4;$
$\quad\quad y = (4 + 2(-x)) - 16;$
$\quad\quad y = -(4 + 2x) - 8;$ or
$\quad\quad y = -(4 + 2(x + 4))$

11c. $y = -f(x) + 2b$

11d. $y = f(-x + 2a)$

13. $47\,\text{T} \cdot \dfrac{1\ \text{cup}}{16\ \text{T}} \cdot \dfrac{1\ \text{quart}}{4\ \text{cups}} = \dfrac{47}{64}\ \text{quart} \approx 0.734\ \text{quart}$

LESSON 9.4

1. $y = |x - 5|$

3a. $y = -1.2|x - 5| + 6$

3b. Begin at the motion sensor and walk away at 1.2 m/sec for 5 sec. At 6 m away, turn and walk back at the same speed.

5. Your equation will be correct when the graphs coincide and when a table of values for Y_1 and Y_0 are identical.

7a. a vertical stretch of $y = x^2$ by a factor of 2

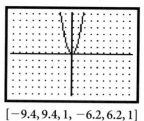

$[-9.4, 9.4, 1, -6.2, 6.2, 1]$

7b. a vertical shrink of $y = |x|$ by a factor of 0.25; then a translation right 2 units and up 1 unit

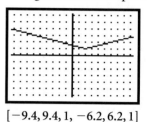

$[-9.4, 9.4, 1, -6.2, 6.2, 1]$

7c. a reflection of $y = x^2$ across the *x*-axis; then a translation left 4 units and down 1 unit

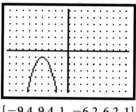

$[-9.4, 9.4, 1, -6.2, 6.2, 1]$

7d. a vertical stretch of $y = |x|$ by a factor of 2 and a reflection across the *x*-axis; then a translation right 3 units and up 4 units

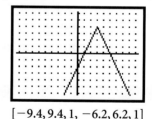

$[-9.4, 9.4, 1, -6.2, 6.2, 1]$

9a. a reflection across the *x*-axis and a translation left 3 units

9b. a vertical shrink by a factor of 0.5; then a translation right 2 units and up 1 unit

11a. a vertical shrink by a factor of 0.5 and a reflection across the *x*-axis

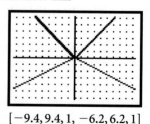

$[-9.4, 9.4, 1, -6.2, 6.2, 1]$

11b. a vertical stretch by a factor of 2 and a translation right 4 units

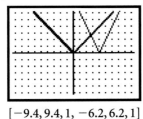

$[-9.4, 9.4, 1, -6.2, 6.2, 1]$

11c. a vertical stretch by a factor of 3 and a reflection across the x-axis; then a translation left 2 units and up 4 units

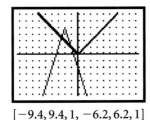

$[-9.4, 9.4, 1, -6.2, 6.2, 1]$

13. $(x - 1, 0.8y)$

15a. $\dfrac{1}{2^9}$

15b. 5^{10}

15c. $2^{12} \cdot 3^6$

15d. $\dfrac{1}{3^8 \cdot x^{12}}$

LESSON 9.6

1a. a reflection of the graph of $y = x^2$ across the x-axis; then a translation up 2 units (or translation then reflection); $y = -x^2 + 2$

1b. a vertical shrink of the graph of $y = |x|$ by a factor of $\frac{1}{3}$ and a translation right 2 units; $y = \frac{1}{3}|x - 2|$

1c. a vertical shrink of the graph of $y = x^2$ by a factor of 0.5; then a translation right 1 unit and down 1 unit; $y = 0.5(x - 1)^2 - 1$

1d. a vertical stretch of the graph of $y = |x|$ by a factor of 2 and a reflection across the x-axis; then a translation left 2 units and up 3 units; $y = -2|x + 2| + 3$

3. $y = -\dfrac{5}{x}$

5a. decreasing; asymptotes at $x = 0$ and $y = 0$; domain: $x \neq 0$; range: $y \neq 0$

5b. decreasing; asymptotes at $x = 5$ and $y = -2$; domain: $x \neq 5$; range: $y \neq -2$

5c. decreasing; asymptotes at $x = 0$ and $y = 3$; domain: $x \neq 0$; range: $y \neq 3$

5d. increasing; asymptotes at $x = -3$ and $y = 0$; domain: $x \neq -3$; range: $y \neq 0$

7a. $y = f(-x)$ or $y = \dfrac{1}{-x}$

7b. $y = -f(x)$ or $y = -\dfrac{1}{x}$ (See sketch above.)

7c. Either reflection produces the same graph because $\dfrac{1}{-x} = -\dfrac{1}{x}$.

9. Let x represent the amount of water to add and y represent the concentration of salt. The amount of salt is $0.05(0.5)$.

$y = \dfrac{0.025}{0.5 + x}$; $0.01 = \dfrac{0.025}{(0.5 + x)}$; $x = 2$; 2 liters.

11a. $700

11b.

Additional businesses	0	5	10	15	20
Cost per business ($)	700.00	350.00	233.33	175.00	140.00

11c. $y = \dfrac{3500}{5 + x}$, where x is the number of additional businesses that have signed up and y is the cost per business.

11d. $150 = \dfrac{3500}{5 + x}$; $x = 18.\overline{3}$; 19 additional businesses.

13a. $x < -2$

13b. $x \geq -2$

13c. $x \geq 4$

15. vertex: $(-1, 2)$; point to the left: $(-2, -1)$; point to the right: $(0, -1)$

LESSON 9.7

1a. $(-2, 2), (1, 2), (-2, 6)$

1b. $\begin{bmatrix} 0 & 0 & 0 \\ -3 & -3 & -3 \end{bmatrix}$

1c. $\begin{bmatrix} -2 & 1 & -2 \\ -1 & -1 & 3 \end{bmatrix}$

3a. $[6 \quad 15]$

3b. $\begin{bmatrix} 6 \\ 15 \end{bmatrix}$

3c. $[64]$

3d. $\begin{bmatrix} 8 & 32 \\ 14 & 56 \end{bmatrix}$

5a. a rectangle

5b. For the x-coordinate, multiply row 1 of the transformation matrix by column 2 of the quadrilateral matrix; $\begin{bmatrix} 1 & 0 \end{bmatrix} \cdot \begin{bmatrix} 2 \\ -1 \end{bmatrix} = 2$; this goes in row 1, column 2, of the image matrix. For the y-coordinate, multiply row 2 of the transformation matrix by column 2 of the quadrilateral matrix; $\begin{bmatrix} 0 & 2 \end{bmatrix} \cdot \begin{bmatrix} 2 \\ -1 \end{bmatrix} = -2$; this goes in row 2, column 2, of the image matrix.

5c. $\begin{bmatrix} -1 & 2 & 1 & -2 \\ 4 & -2 & -4 & 2 \end{bmatrix}$

5d. a parallelogram

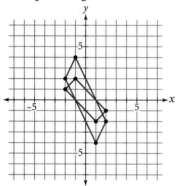

7a. $[Q] = \begin{bmatrix} 2 & 3 & 6 & 7 \\ 2 & 4 & 5 & 1 \end{bmatrix}$

7b. $\begin{bmatrix} 1 & 0 \\ 0 & 0.5 \end{bmatrix} \cdot [Q] = \begin{bmatrix} 2 & 3 & 6 & 7 \\ 1 & 2 & 2.5 & 0.5 \end{bmatrix}$

7c. $\begin{bmatrix} 0.5 & 0 \\ 0 & 0.5 \end{bmatrix} \cdot [Q] = \begin{bmatrix} 1 & 1.5 & 3 & 3.5 \\ 1 & 2 & 2.5 & 0.5 \end{bmatrix}$

7d. $\begin{bmatrix} -1 & 0 \\ 0 & -1 \end{bmatrix}$; $\begin{bmatrix} -1 & 0 \\ 0 & -1 \end{bmatrix} \cdot [Q]$

$= \begin{bmatrix} -2 & -3 & -6 & -7 \\ -2 & -4 & -5 & -1 \end{bmatrix}$; a reflection across both the x- and y-axes; alone, $\begin{bmatrix} 1 & 0 \\ 0 & -1 \end{bmatrix}$ results in a reflection across the x-axis and $\begin{bmatrix} -1 & 0 \\ 0 & 1 \end{bmatrix}$ results in a reflection across the y-axis.

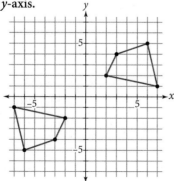

7e. $\begin{bmatrix} -2 & -4 & -5 & -1 \\ 2 & 3 & 6 & 7 \end{bmatrix}$; a quarter-turn counterclockwise

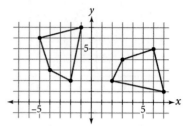

9a. Sample answer: $\begin{bmatrix} 0 & 3 & 4 \\ 0 & 2 & 0 \end{bmatrix}$

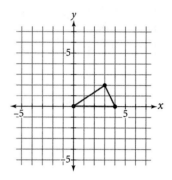

9b.
$\begin{bmatrix} 0.5 & -0.866 \\ 0.866 & -0.5 \end{bmatrix} \cdot \begin{bmatrix} 0 & 3 & 4 \\ 0 & 2 & 0 \end{bmatrix} = \begin{bmatrix} 0 & -0.232 & 2 \\ 0 & 3.598 & 3.464 \end{bmatrix}$

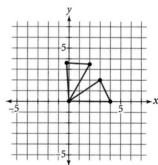

9c. a rotation 60° about the origin, or the point (0, 0)

9d. $\begin{bmatrix} -0.500 & -0.866 \\ 0.866 & -0.5 \end{bmatrix}$;

$\begin{bmatrix} -0.500 & -0.866 \\ 0.866 & -0.5 \end{bmatrix} \cdot \begin{bmatrix} 0 & 3 & 4 \\ 0 & 2 & 0 \end{bmatrix} = \begin{bmatrix} 0 & -3.232 & -2 \\ 0 & 1.598 & 3.464 \end{bmatrix}$;

a rotation 120° about the origin

(See the graph at the top of page 679.)

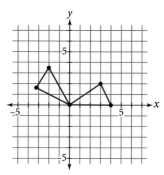

9e. Multiply $\begin{bmatrix} 0.5 & -0.866 \\ 0.866 & -0.5 \end{bmatrix}$ by itself three times

($3 \cdot 60° = 180°$). Then multiply the result by matrix $[R]$.

9f. Multiply $\begin{bmatrix} 0.5 & -0.866 \\ 0.866 & -0.5 \end{bmatrix}$ by itself six times

($6 \cdot 60° = 360°$). Then multiply the result by matrix

$[R]$ or just multiply $[R]$ by $\begin{bmatrix} 1 & 0 \\ 0 & 1 \end{bmatrix}$.

11a. in millions: 114, 127, 159, 273, 1247

11b. **Most Populated Countries, 1999**

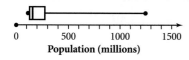

11c. China and India

CHAPTER 9 REVIEW

1a. a translation left 2 units and up 1 unit

1b. $(x - 2, y + 1)$

3. For these sample answers, list L_3 and list L_4 are used for the x- and y-coordinates, respectively, of each image.

3a. $L_3 = L_1, L_4 = -L_2$

3b. $L_3 = -L_1, L_4 = L_2$

3c. $L_3 = L_1 + 3, L_4 = -L_2$

5. $g(x) = f(x - 1) + 2$

7a. The graph should have the same x-intercept as $f(x)$. The y-intercept should be the opposite of that for $f(x)$.

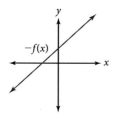

7b. Hint: $Y_2 = -Y_1$ is the same as $y = -f(x)$

9a. a translation of $y = \dfrac{1}{x}$ right 3 units and down 2 units; $y = \dfrac{1}{x - 3} - 2$

9b. a translation of $y = 2^x$ right 4 units and down 2 units; $y = 2^{(x-4)} - 2$

9c. a reflection of $y = 2^x$ across the x-axis and across the y-axis, followed by a translation up 3 units (or a reflection across the x-axis, followed by a translation up 3 units, followed by a reflection across the y-axis); $y = -2^{(-x)} + 3$

9d. a vertical stretch of $y = \dfrac{1}{x}$ by a factor of 4 and a reflection across the x-axis, followed by a translation up 1 unit and left 2 units; $y = -\dfrac{4}{x + 2} + 1$

CHAPTER 10 · CHAPTER **10** CHAPTER 10 · CHAPTER

LESSON 10.1

1. Enter one side of the equation into Y_1 and the other into Y_2 on the calculator. Then find the number of intersection points.

1a. two solutions

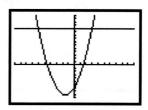

$[-10, 10, 1, -10, 15, 1]$

1b. no solutions

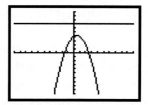

$[-10, 10, 1, -10, 10, 1]$

1c. one solution

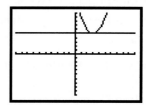

$[-10, 10, 1, -10, 10, 1]$

1d. two solutions

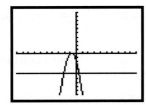

$[-10, 10, 1, -10, 10, 1]$

3a. $x = \pm\sqrt{18}$

3b. $x^2 = 49, x = \pm7$

3c. $x - 2 = \pm5, x = 7$ or $x = -3$

3d. $2(x + 1)^2 = 14, (x + 1)^2 = 7, x + 1 = \pm\sqrt{7},$
$x = -1 \pm\sqrt{7}$

5a. $h(0) = -4.9(0)^2 + 147,$ or $147.$ It is the starting height of the ball, when $t = 0$ seconds.

5b. Solve the equation $20 = -4.9t^2 + 147$ symbolically to get $\pm\sqrt{\dfrac{20 - 147}{-4.9}}$ or approximately $\pm5.09.$ The graph shows two solution points: $(5.09, 20)$ and $(-5.09, 20).$

5c. No, only the solution $(5.09, 20)$ makes sense because the time t must be positive.

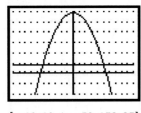

$[-10, 10, 1, -50, 150, 25]$

5d. $t > 5.09$ seconds

5e. The ball hits the ground when $t \approx 5.48$ seconds because the positive x-intercept is near the point $(5.48, 0).$

7a. After the ball has gone 2 m horizontally, it will be approximately 3.75 m above the ground.

7b. The ball will be 2 m above the ground once it has gone 0.31 m horizontally and again once it has gone 6.49 m horizontally. Exact answers are $3.4 \pm\sqrt{\dfrac{220}{23}}.$

7c. When $x = 0,$ the height is 1.5412 m.

7d. When $p(x) = 0,$ the horizontal distance is $3.4 + \sqrt{\dfrac{420}{23}},$ or 7.67 m.

9a. The x-intercepts indicate when the projectile is at ground level.

9b. 2.74 seconds and 7.26 seconds

9c. Find the average of 2.74 and 7.26, which is 5. $h(5) = 25.$

9d. The projectile is 25 meters above the ground, its maximum height, after 5 seconds.

9e. $h(3.2) = 9.124$ meters. This is the height at 3.2 seconds.

9f. Sample answer: The horizontal line $y = 12.5$ intersects the parabola twice—when $x \approx 3.4$ seconds and when $x \approx 6.6$ seconds.

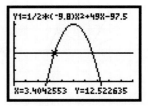

$[0, 10, 1, -5, 30, 5]$

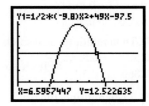

$[0, 10, 1, -5, 30, 5]$

LESSON 10.2

1. The vertex has an x-value of $\dfrac{3 + (-2)}{2},$ or 0.5.

3.
$$0 = (x + 1.5)^2 - 7.25$$
$$7.25 = (x + 1.5)^2$$
$$\pm\sqrt{7.25} = x + 1.5$$
$$-1.5 \pm\sqrt{7.25} = x$$
$$x \approx 1.92582404 \text{ or }$$
$$x \approx -4.192582404$$

5a. $(x + 3) = \pm\sqrt{7}$
$$x = -3 \pm\sqrt{7}$$
$$x \approx -5.646, -0.354$$

5b. $(x - 2)^2 = 21$
$$x = 2 \pm\sqrt{21}$$
$$x \approx -2.583, 6.583$$

7a. The ball is on the ground when $h = 0.$ This happens when $t = 0$ seconds and $t = 3$ seconds.

7b. The ball is at its highest point when $t = 1.5$ seconds, halfway through its flight.

7c. $h = 36$ when $t = 1.5,$ so the ball goes 36 feet high.

9a. Sample answer: The ball is thrown from an initial height of 5 meters. It reaches a maximum height of about 25 meters in 2 seconds and hits the ground at about 4.3 seconds.

9b. The ball starts at a velocity of 20 meters per second and slows down at a constant rate. At 2 seconds it is not moving. Then it starts falling and is moving down at 22 meters per second when it hits the ground.

9c. As the ball moves up to its maximum height, it is slowing down, moving at 0 meters per second at its peak. It then starts falling and speeding up until it hits the ground.

9d. This is when the ball is at its maximum height and not moving. Its velocity is zero.

9e. These are the two times when the ball is about 13 meters high.

9f. domain: $0 \le t \le 4.3$; range: $0 \le h(t) \le 26$; $-22 < v(t) \le 20$

11a.

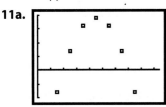

$[0, 9.4, 1, -10, 20, 5]$

11b. It is the vertical line $x = 4.5$.

11c. $(4.5, 19)$

11d. $y = -3(x - 4.5)^2 + 19$

LESSON 10.3

1a. yes; three terms (trinomial)

1b. yes; two terms (binomial)

1c. no; The first term has a negative exponent.

1d. no; The first term is equivalent to $3x^{-2}$, which has a negative exponent.

1e. yes; one term (monomial)

1f. yes; three terms (trinomial)

1g. yes; two terms (binomial)

1h. Not a polynomial as written but it is equivalent to $3x - 6$, a binomial.

3a.

	x	2
x	x^2	$2x$
2	$2x$	4

$(x + 2)^2 = x^2 + 4x + 4$

3b.

	x	12
x	x^2	$12x$
12	$12x$	144

$(x + 12)^2 = x^2 + 24x + 144$

3c.

	x	-7
x	x^2	$-7x$
-7	$-7x$	49

$(x - 7)^2 = x^2 - 14x + 49$

5a.

	x			2
x	x^2	+	x	$2x$
4	$4x$		4	8

$(x + 2)(x + 4) = x^2 + 6x + 8$

5b.

	x			3
x	x^2	+	x	$3x$
5	$5x$		5	15

$(x + 3)(x + 5) = x^2 + 8x + 15$

5c.

	x			2
x	x^2	+	x	$2x$
-5	$-5x$		-5	-10

$(x - 5)(x + 2) = x^2 - 3x - 10$

5d.

	x			0
x	x^2	+	x	$0x$
-3	$-3x$		-3	0

$(x - 0)(x - 3) = x^2 - 3x$

7. No, by squaring the values inside the parentheses, Heather is accounting for only two of the four rectangles in a squaring diagram. She needs to add the two rectangles that sum to the middle term.

9a. The vertex is at $(0.4, 2.5)$, so the ball reaches a maximum height of 2.5 meters.

9b. $h(t) = -4.9t^2 + 3.92t + 1.716$

9c. The pitcher released the ball at a height of 1.716 meters.

11a. meaningful domain: $0 \le x \le 6.5$; meaningful range: $0 \le y \le 897.81$

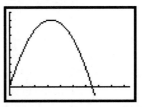

$[0, 9.4, 1, -100, 1000, 100]$

11b. Sample answer: Average the two x-intercepts, then substitute this value into the equation to find the y-coordinate of the vertex. The vertex is $(3.25, 897.8125)$.

11c. The x-coordinate is the price that produces maximum income. The y-coordinate is the maximum income for this relationship.

11d. The prices produce no income.

11e. If the warehouse charges \$5 per item, income will be \$637.50.

13. $x^2 + 8x + 16$

15. The message is POLYNOMIALS.

15a. 16; P   **15b.** 15; O   **15c.** 12; L

15d. 25; Y   **15e.** 14; N   **15f.** 15; O

15g. 13; M   **15h.** 9; I   **15i.** 1; A

15j. 12; L   **15k.** 19; S

1a. $x + 4 = 0$ or $x + 3.5 = 0$, so $x = -4$ or $x = -3.5$

1b. $x - 2 = 0$ or $x - 6 = 0$, so $x = 2$ or $x = 6$

1c. $x + 3 = 0$ or $x - 7 = 0$ or $x + 8 = 0$, so $x = -3$ or $x = 7$ or $x = -8$

1d. $x = 0$ or $x - 9 = 0$ or $x + 3 = 0$, so $x = 0$ or $x = 9$ or $x = -3$

3a. 7 and -2

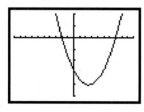

$[-10, 10, 1, -25, 10, 5]$

3b. -1 and -8

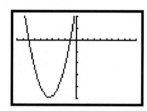

$[-10, 10, 1, -25, 10, 5]$

3c. 11 and -7

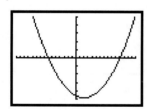

$[-15, 15, 1, -250, 250, 50]$

3d. -5 and 9

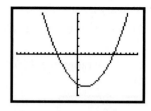

$[-15, 15, 1, -25, 25, 5]$

5a. two intercepts; $x = -1$ and $x = 3$

5b. $(1, -4)$

5c. $y = (x - 1)^2 - 4$; There is a translation left 1 unit and down 4 units.

7a. See below.

7b. The sum of the roots needs to be -2, and the product needs to be -8; 2 and -4 satisfy this requirement.

9. $y = 0.25(x + 3)(x - 9)$

11a. two x-intercepts: $x = 3$ and $x = -3$

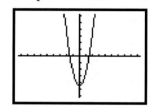

$[-18.8, 18.8, 2, -12.4, 12.4, 2]$

11b. $y = (x - 3)(x + 3)$

11c. The x-intercepts are the positive and negative square roots of the amount the parabola is translated, or $\pm\sqrt{h}$.

11d. i. $y = (x + 7)(x - 7)$

11d. ii. $y = (4 + x)(4 - x)$

11d. iii. $y = (x + \sqrt{47})(x - \sqrt{47})$

11d. iv. $y = (x + \sqrt{28})(x - \sqrt{28})$

7a. *(Lesson 10.4)*

Factored form	Roots	Sum of roots	Product of roots	General form
$y = (x + 3)(x - 4)$	-3 and 4	$-3 + 4 = 1$	$(-3)(4) = -12$	$y = x^2 - 1x - 12$
$y = (x - 5)(x + 2)$	5 and -2	$5 - 2 = 3$	$5(-2) = -10$	$y = x^2 - 3x - 10$
$y = (x + 2)(x + 3)$	-2 and -3	-5	6	$y = x^2 + 5x + 6$
$y = (x - 5)(x + 5)$	5 and -5	0	-25	$y = x^2 - 25$

11e. There are no *x*-intercepts. The graph is above the *x*-axis.

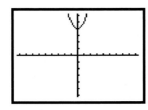

$$[-9.4, 9.4, 1, -6.2, 6.2, 1]$$

11f. Because there are no *x*-intercepts, there is no factored form of the equation.

13. Sample answer:

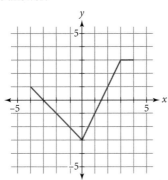

LESSON 10.6

1a. $x = -3 \pm \sqrt{2}$ **1b.** $x = 5 \pm \sqrt{2}$

1c. $x = -8 \pm \sqrt{\frac{7}{3}}$ **1d.** $x = -6 \pm \sqrt{\frac{7}{5}}$

3a. $\left(\frac{18}{2}\right)^2; x^2 + 18x + 81 = (x + 9)^2$

3b. $\left(\frac{-10}{2}\right)^2; x^2 + 10x + 25 = (x - 5)^2$

3c. $\left(\frac{3}{2}\right)^2; x^2 + 3x + \frac{9}{4} = \left(x + \frac{3}{2}\right)^2$

3d. $\left(\frac{-1}{2}\right)^2; x^2 - x + \frac{1}{4} = \left(x - \frac{1}{2}\right)^2$

3e. $\left(\frac{1}{2} \cdot \frac{2}{3}\right)^2; x^2 + \frac{2}{3}x + \frac{1}{9} = \left(x + \frac{1}{3}\right)^2$

3f. $\left(\frac{-1.4}{2}\right)^2; x^2 - 1.4x + 0.49 = (x - 0.7)^2$

5a. $y = a(x - 2)^2 - 31.5$

5b. Solve the equation $0 = a(5 - 2)^2 - 31.5; a = 3.5$.

5c. $y = 3.5(x - 2)^2 - 31.5$

5d. $y = -2(x - 2)^2 + 32$

7a. $y = x^2 + 6x + \left(\frac{6}{2}\right)^2 - 3^2 + 10;$
$y = x^2 + 6x + 9 - 9 + 10; y = (x + 3)^2 + 1$

7b. vertex $(-3, 1);$

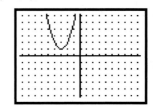

$$[-9.4, 9.4, 1, -6.2, 6.2, 1]$$

7c.
$$x^2 + 6x + 10 = 0$$
$$x^2 + 6x = -10$$
$$x^2 + 6x + 9 = -10 + 9$$
$$(x + 3)^2 = -1$$
$$x + 3 = \pm\sqrt{-1}$$
$$x = -3 \pm\sqrt{-1}$$

Sample answer: There are no real roots because the graph doesn't cross the *x*-axis.

9a. $p = 2500 - 5x$, where *p* represents the cost in dollars of a single ticket, and *x* represents the number of tickets sold. Let *C* represent the total cost of the group package.

9b. $C = xp = x(2500 - 5x)$

9c. $C = -5(x - 250)^2 + 312,500$

9d. $C(20) = 20(2500 - 100) = 48,000; \$48,000$

9e.
$$2500x - 5x^2 = 200,000$$
$$500x - x = 40,000$$
$$x^2 - 500x = -40,000$$
$$x^2 - 500x + 250^2 = 250^2 - 40,000$$
$$(x - 250)^2 = 22,500$$
$$x - 250 = \pm 150$$
$$x = 100 \text{ or } x = 400$$

9f. Sample answer: The company will lose money if more than 400 people or less than 100 people per group go on the cruise. The vertex is $(250, 312500)$, so the company should limit groups to 250 people to maximize earnings at $312,500.

11a. $2x^3 + 5x^2 + 4x + 1$

11b. $6x^3 - 11x^2 - 18x + 20$

LESSON 10.7

1a. $25 - 24 = 1$

1b. $9 - (-12) = 9 + 12 = 21$

1c. $36 - 24 = 12$

1d. $81 - 0 = 81$

3a. $x = \dfrac{3 \pm \sqrt{-23}}{4}$. There are no real solutions.

3b. Use the quadratic equation or complete the square.
$x = \dfrac{1}{2}$ and $x = 3$

3c. Complete the square.
$$x^2 - 6x - 8 = 0$$
$$x^2 - 6x = 8$$
$$x^2 - 6x + 9 = 17$$
$$(x - 3)^2 = 17$$
$$x - 3 = \pm\sqrt{17}$$
$$x = 3 \pm \sqrt{17}$$

3d. Use the quadratic formula.
$$x = \dfrac{-2 \pm \sqrt{76}}{6} = \dfrac{-1 \pm \sqrt{19}}{3}$$

5a. $-4.9t^2 + 6.2t + 1.9 = 0$; $t \approx -0.255$ sec or $t \approx 1.52$ sec. The ball hits the ground after 1.52 seconds.

5b. $-4.9t^2 + 6.2t + 1.9 = 3$; $t \approx 0.21$ sec or $t \approx 1.05$ sec. The ball is 3 meters high when $t \approx 0.21$ seconds and when $t \approx 1.05$ seconds.

5c. $-4.9t^2 + 6.2t + 1.9 = 4$;

$t \approx \dfrac{-6.2 \pm \sqrt{-2.72}}{-9.8}$. This equation has no real solutions, so the ball is never 4 meters high.

7. When $b^2 - 4ac$ is negative, there are no real roots. When it's zero, there is one real root. When it's positive, there are two roots.

7a. i; $1^2 - 4(1)(1) = -3$, no x-intercept

7b. iii; $2^2 - 4(1)(1) = 0$, one x-intercept

7c. ii; $3^2 - 4(1)(1) = 5$, two x-intercepts

9. You need to know the t-value when $h = 0$.
Solve $0 = -4.9t^2 + 17t + 2.2$ with $a = -4.9$, $b = 17, c = 2.2$ to get $t \approx -0.125$ and $t \approx 3.59$. The positive solution of 3.59 seconds makes sense in this situation.

11a. $x - 2$

11b. $x + 6$

11c. $x - 1$

1a. perfect square; $47^2 = 2{,}209$

1b. perfect cube: $18^3 = 5{,}832$

1c. neither

1d. perfect square; $101^2 = 10{,}201$

3a. $4x(x + 3)$

3b. $2x(3x - 2)$

3c. $7x(2x^3 + x - 3)$

3d. $3x^2(4x^3 + 2x + 1)$

5a. $y = 0.5(x + 4)(x + 2)(x - 1)$

5b. $y = -(x + 2)(x - 1)^2$

7a. $V = w(w + 6)(w - 2)$

7b. Three solutions to the equation $47 = w(w + 6)(w - 2)$ are shown on the graph. However, only one solution is a positive value. A table gives the answer $w \approx 3.4$. Widths greater than about 3.4 cm give volumes greater than 47 cm³.

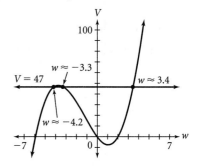

9a. $x^3 + 6x^2 + 11x + 6$

9b. $x^3 - 3x^2 - 4x + 12$

11a. $11x^3 + 2x^2 + 2x + 12$

11b. $5x^3 - 2x^2 - 12x - 12$

11c. $-6x^2 - 13x + 20$

11d. $10x^2 + x + 2$

11e. $-16x^4 + 34x^3 - 28x^2 - 131x + 99$

1a. false; $(x - 3)(x + 8)$

1b. false; $2x^2 - 4x + 5$

1c. false; $x^2 + 6x + 9$

1d. true

3a. $y = -(x - 2)^2 + 3$; vertex form

3b. $y = 0.5(x - 2)(x + 3)$; factored form

5a. $2w + 9 = 0$ or $w - 3 = 0$; $w = -4.5$ or $w = 3$

5b. $2x + 5 = 0$ or $x - 7 = 0$; $x = -2.5$ or $x = 7$

7a. $x^2 + 6x - 9 = 13$
$$x^2 + 6x = 22$$
$$x^2 + 6x + 9 = 22 + 9$$
$$(x + 3)^2 = 31$$
$$x + 3 = \pm\sqrt{31}$$
$$x = -3 \pm\sqrt{31}$$

7b. $3x^2 - 24x + 27 = 0$
$$3x^2 - 24x = -27$$
$$x^2 - 8x = -9$$
$$x^2 - 8x + 16 = -9 + 16$$
$$(x - 4)^2 = 7$$
$$x - 4 = \pm\sqrt{7}$$
$$x = 4 \pm\sqrt{7}$$

9a. $f(60) = 8.1$. When there are 60 fish in the tank, the population is growing at a rate of about 8 fish per week.

9b. $f(x) = 0$ for $x = 0$ and $x = 150$. When there are no fish, the population does not grow. When there are 150 fish, the number of fish hatched is equal to the number of fish that die so the total population does not change.

9c. When there are 75 fish, the population is growing fastest.

9d. The population no longer grows once there are 150 fish, so this is the maximum number of fish the tank has to support.

9e.

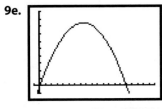

$[-10, 200, 10, -1, 10, 1]$

11a. No x-intercepts means taking the square root of a negative number. So $(-6)^2 - 4(1)(c) < 0$; $-4c\ -36$; $c > 9$. Or translate the graph of $y = x^2 - 6x$ vertically to see that for values of $c > 9$, the parabola does not cross the x-axis.

11b. One x-intercept implies a double root, so $x^2 - 6x + c$ must be a perfect-square trinomial. Make a rectangular diagram to find $\left(\frac{-6}{2}\right)^2 = 9$, so $x^2 - 6x + 9$ is a perfect-square trinomial and $c = 9$. The graph touches the x-axis once. You can solve $b^2 - 4ac = 36 - 4c = 0$ to get $c = 9$.

11c. For $c < 9, b^2 - 4ac > 0$, so the discriminant gives two real roots. Also, the parabola $y = x^2 - 6x + c$ crosses the x-axis twice for values of c that are less than 9.

13a. $x = -2, -1, 1,$ and 3;
$y = 2(x - 3)(x + 2)(x + 1)(x - 1)$

13b. $x = -2$ (double root) and 3;
$y = -3(x + 2)^2(x - 3)$

LESSON 11.1

1a. 0.8 **1b.** -2 **1c.** -1.25 **1d.** 2

1e. $\frac{3}{2}$ **1f.** $-\frac{3}{2}$ **1g.** $\frac{3}{2}$ **1h.** $\frac{2}{3}$

3a. perpendicular **3b.** neither

3c. parallel **3d.** perpendicular

5. rectangle; slopes: $\frac{1}{2}, -2, \frac{1}{2}, -2$

7. parallelogram; slopes: $1, -3, 1, -3$

9. trapezoid; slopes: $2, -\frac{5}{2}, -\frac{1}{4}, -\frac{5}{2}$

11. quadrilateral; slopes: $1, -1, \frac{5}{3}, -\frac{5}{3}$

13a. $\left(-\frac{2}{3}, 0\right)$ **13b.** $(-1.\overline{3}, -0.58\overline{3})$

13c. $(0, -1)$ **15.** $\frac{6}{16}$ or $\frac{3}{8}$

LESSON 11.2

1. 1.2

3a. $(0.5, 1.5)$

3b. $(6, -4.5)$

5. $\left(\frac{a + c}{2}, \frac{b + d}{2}\right)$

7. One line is horizontal and the other is vertical. A horizontal line has a slope of 0 and a vertical line has an undefined slope, so the product is also undefined.

9a. midpoint of $\overline{AB}$: $(10, 3)$
midpoint of $\overline{BC}$: $(15, 8)$
midpoint of $\overline{CD}$: $(9, 10)$
midpoint of $\overline{DA}$: $(4, 5)$

9b. parallelogram: The opposite sides are parallel because the slopes are $1, \frac{-1}{3}, 1,$ and $\frac{-1}{3}$.

9c. No; the slopes of the diagonals are $\frac{3}{11}$ and -7.

11a. $(12, 4)$ **11b.** $(3, 5)$ **11c.** $(-5, 3.5)$

13a. Sample answers:

13b. Any method that uses only one line segment will form congruent polygons. Some other methods may also produce congruent polygons.

13c. Yes. If you imagine a vertical segment through the upper vertex of the triangle, you can see that the triangles on either side of the vertical line are congruent.

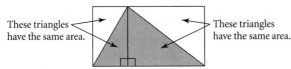

These triangles have the same area.   These triangles have the same area.

So the area of the rectangle is divided in half.

LESSON 11.3

1a. $\pm\sqrt{47}$ **1b.** $4 \pm \sqrt{28}$

1c. $-2 \pm \sqrt{14}$ **1d.** $1 \pm \sqrt{7}$

3a. 4 square units   **3b.** 12 square units

3c. 2 square units   **3d.** 6 square units

3e. 20 square units   **3f.** 18 square units

5. polygon 3a: $\sqrt{8}$ units and $\sqrt{2}$ units

polygon 3b: $\sqrt{8}$ units and $\sqrt{18}$ units

polygon 3e: $\sqrt{50}$ units, $\sqrt{50}$ units, and $\sqrt{40}$ units

7a. 36 square units, 18 square units, 18 square units

7b. Length of $\overline{AB}$: 6 units; length of $\overline{BC}$: $\sqrt{18}$ units; length of $\overline{AC}$: $\sqrt{18}$ units.

9a. $\overline{AC}$ is the hypotenuse; $\overline{AB}$ and BC are the legs.

9b.

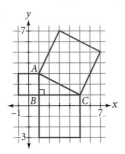

9c. 20 square units, 16 square units, 4 square units

9d. Length of $\overline{AB}$: 2 units; length of $\overline{BC}$: 4 units; length of $\overline{AC}$: $\sqrt{20}$ units

9e. The areas of the two smaller squares add up to the area of the larger square.

11. $y = 2(x + 7)(x + 1)$

LESSON 11.4

1. 576 cm^2

3. $c = \sqrt{300}$ cm

5a. 13.5 ft   **5b.** 22.5 ft   **5c.** 2160 ft^2

7a. approximately 21.6 cm-by-27.6 cm

7b. approximately 35.0 cm

7c. Answers will vary but should be close to 35 cm.

7d. The two results should be approximately the same.

9. approximately 7010 ft

11a. $L_1 = \{1, 2, 3, \ldots, 26\}$

11b. approximately 26.98 in.

11c. $L_2 = \sqrt{(27^2 - L_1^2)}$

11d. approximately 53.85 in^2

11e. $L_3 = L_1 \cdot L_2$, or $L_3 = L_1 \cdot \sqrt{(27^2 - L_1^2)}$

11f. A model that works is $y = x\sqrt{27^2 - x^2}$, where x is the width and y is the area.

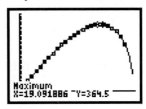

$[0, 27, 5, -50, 450, 50]$

11g. Trace the graph or use a table to find that a 19-by-19-in. square gives maximum area.

13a. $\frac{13}{12} = \frac{a}{8}$, $a = 8.\overline{6}$; 8 ft 8 in. long

13b. $\frac{5}{12} = \frac{14 - b}{8}$, $b = 10\frac{2}{3}$; 10 ft 8 in. tall.

LESSON 11.5

1a. $3\sqrt{3}$ **1b.** $5\sqrt{2}$

1c. $2 + \sqrt{6}$ **1d.** $4\sqrt{5} + 5\sqrt{2}$

1e. $6\sqrt{6}$ **1f.** $2\sqrt{42}$

1g. $\sqrt{5}$ **1h.** 20

3a. $y = x^2 - 3$

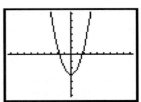

$[-9.4, 9.4, 1, -6.2, 6.2, 1]$

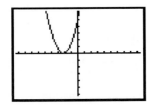

3b. $y = x^2 + 2x\sqrt{5} + 5$

$[-9.4, 9.4, 1, -6.2, 6.2, 1]$

5a. $(0, -3)$

5b. $\left(-\sqrt{5}, 0\right) \approx (-2.236, 0)$

7a. $\pm 4\sqrt{7} \approx 10.583$

7b. $2\sqrt{6} \approx 4.899$ and $-3\sqrt{6} \approx -7.348$

7c. $-3 - \sqrt{2} \approx -4.414$ and $-3 + \sqrt{2} \approx -1.586$

9a. $\sqrt{112}$ **9b.** $\sqrt{550}$ **9c.** $\sqrt{972}$ **9d.** $\sqrt{4500}$

11. See below.

13. $10\sqrt{38}$ cm

15a. $y = (x - 2)^2 + 3$

15b. $y = -x^2 + 4$

15c. $y = 3(x - 1)^2$

1. No, because $9^2 + 16^2 \neq 25^2$.

3. Sample answer: $(6, 3)$ and $(1, 7)$.

5a. $\sqrt{26}$ units or approximately 0.5 mile

5b. 5 units or 0.5 mile

7a. 5 cm

7b. 13 cm

9a. $a = 63$

9b. $b = 7$

9c. $c = 3$

1a. $x = 27$

1b. $x = 35$

1c. $x = 160$

1d. $x = \pm 4$

3a. $\sin D = \dfrac{7}{25}$

3b. $\cos E = \dfrac{7}{25}$

3c. $\tan D = \dfrac{7}{24}$

5a. $x = 12$

5b. $x = 14$

5c. $x = 35$

7a. cosine

7b. sine

7c. tangent

7d. $\cos 65° \approx 0.4226$; $\sin 65° \approx 0.9063$; $\tan 65° \approx 2.1445$

7e. approximately 5.6 m

---

11. (*Lesson 11.5*)

$0 = 0.5x^2 - 6x + 8$ The original equation.

$0 \overset{?}{=} 0.5\left(6 - \sqrt{20}\right)^2 - 6\left(6 - \sqrt{20}\right) + 8$ Substitute $6 - \sqrt{20}$ for x.

$0 \overset{?}{=} 0.5\left(6 - \sqrt{20}\right)^2 - 36 + 6\sqrt{20} + 8$ Distribute the -6 over $6 - \sqrt{20}$.

$0 \overset{?}{=} 0.5(36 - 6\sqrt{20} - 6\sqrt{20} + 20) - 36 + 6\sqrt{20} + 8$ Square the expression.

$0 \overset{?}{=} 18 - 3\sqrt{20} - 3\sqrt{20} + 10 - 36 + 6\sqrt{20} + 8$ Distribute the 0.5 over the expression in parentheses.

$0 \overset{?}{=} 18 + 10 - 36 + 8$ Combine the radical expressions.

$0 = 0$ Add and subtract.

11a. rectangle

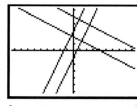

$[-9.4, 9.4, 1, -6.2, 6.2, 1]$

11b. $(1.2, 1.4), (2.4, 3.8), (0.8, 4.6), (-0.4, 2.2)$

11c. $y = 11 - 8x, y = \frac{17}{7} + \frac{4}{7}x$

11d. $(1, 3)$

LESSON 11.8

1a. d **1b.** A **1c.** A

1d. $\frac{d}{c}$ **1e.** A **1f.** A

3. $x \approx 44.6$ m

5. $a \approx 9.4$ cm $b \approx 17.7$ cm
$c \approx 4.1$ cm $d \approx 4.6$ cm
$e = \sqrt{145} \approx 12.0$ cm $F \approx 44.9°$
$g = \sqrt{349} \approx 18.7$ cm $H \approx 15.5°$

7. approximately 81.1 m

9a. approximately 2.86°

9b. about 148 ft

11a. approximately 19 ft

11b. approximately 51 ft

11c. approximately 70 feet

13. Sample answer: $y = -3 + \dfrac{x + 7}{(x - 2)(x + 7)}$

CHAPTER 11 REVIEW

1a. $8\sqrt{5}$ **1b.** $4\sqrt{17}$ **1c.** $123\sqrt{3}$
1d. $\sqrt{15}$ **1e.** 80 **1f.** 1700
1g. $3\sqrt{10}$ **1h.** $80\sqrt{5}$ **1i.** $\sqrt{2} + \sqrt{3}$
1j. $3\sqrt{2}$ **1k.** $\sqrt{6}$ **1l.** $4\sqrt{3}$

3. The slopes of the sides are $\frac{1}{2}$, -2, $\frac{1}{2}$ and -2.
The product of the slopes of each pair of adjacent sides is -1, so the sides are perpendicular.

5. Hint: Find a similar right triangle that has a perimeter of 60 feet.

7. $a \approx 3.38$ m
$b \approx 7.75$ m

9a. $\sqrt{116}$ cm or approximately 10.77 cm

9b. $\sqrt{141}$ cm or approximately 11.87 cm

11a. $y = 60 + 1.08(x - 40)$ or $y = 34 + 1.08(x - 16)$

11b. approximately 50°

11c. approximately 38°

13a. Inverse variation. Sample explanation: The product of x and y is constant; $xy = 2$ or $y = \frac{2}{x}$.

13b. Neither. Sample explanation: The product is not constant, so it is not an inverse variation. The y-value for $x = 0$ is not 0, so it is not a direct variation.

13c. Direct variation. Sample explanation: The ratio of y to x is constant; $y = 0.25x$.

13d. Neither. Sample explanation: The graph is not a curve, so the relationship is not an inverse variation. The line does not pass through the origin, so it is not a direct variation.

13e. Inverse variation. Sample explanation: The product of the x- and y-coordinates for any point on the curve is 8; $xy = 8$ or $y = \frac{8}{x}$.

13f. Direct variation. Sample explanation: The graph is a straight line through the origin; $y = 1.5x$.

15a. $27x6y^3$ **15b.** $5p^4q^2$

15c. $\dfrac{x}{y^2}$ **15d.** $\dfrac{m^2}{n^4} + \dfrac{1}{m^4}$

17a. $y = (x + 3)(x - 1)$

17b. $y = (x + 1)^2 - 4$

17c. $y = x^2 + 2x - 3$

19a. 26   **19b.** 2   **19c.** 36
19d. 128   **19e.** -24

21. Hint: There is more than one way to describe this transformation. Your answer to 21b will rely on your answer to 21a.

23a. $y = (x + 2)^2 - 4$

23b. $y = -0.5|x + 3|$

9c. (*Lesson 11.7*)

Trigonometric Functions for a 45° Angle

	Sine	Cosine	Tangent
Exact value of ratio	$\frac{1}{\sqrt{2}}$	$\frac{1}{\sqrt{2}}$	$\frac{1}{1}$
Decimal approximation of exact value	0.7071	0.7071	1.0000
Value by trigonometric function keys	0.7071	0.7071	1.0000

25a.

segment	length	slope
$\overline{AB}$	10	$\frac{3}{4}$
$\overline{BC}$	10	$-\frac{3}{4}$
$\overline{AC}$	16	0

25b. isosceles triangle; two sides have equal length

25c. $D(2, 1)$

25d. Right triangles. Sample explanation: $\overline{BD}$ has an undefined slope, so it is vertical; $\overline{AC}$ has a slope of 0, so it is horizontal.

25e. A drawing does confirm 25a–d.

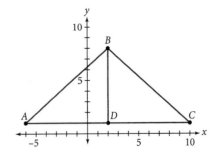

Glossary

The number in parentheses at the end of each definition gives the page where each word or phrase is first used in the text. Some words and phrases are introduced more than once, either because they have different application in different chapters or because they first appeared within features such as Project or Take Another Look; in these cases, there may be multiple page numbers listed.

A

absolute value A number's distance from 0 on the number line. The absolute value of a number gives its size, or magnitude, whether the number is positive or negative. The absolute value of a number x is shown as $|x|$. For example, $|-9| = 9$ and $|4| = 4$. (460)

absolute value function The function $f(x) = |x|$, which gives the absolute value of a number. The absolute value function is defined by two rules: If $x \geq 0$, then $f(x) = x$. If $x < 0$, then $f(x) = -x$. (462)

acute angle An angle that measures less than 90°. (618)

acute triangle A triangle with three acute angles. (638)

addition expression An expression whose only operation is addition. There are also subtraction expressions, multiplication expressions, and division expressions. See **algebraic expression.** (3)

addition property of equality If $a = b$, then $a + c = b + c$ for any number c. (278)

adjacent leg If you consider an acute angle of a right triangle, the adjacent leg is the leg that is part of the angle. (619)

algebraic expression A symbolic representation of mathematical operations that can involve both numbers and variables. (185)

analytic geometry The study of geometry using coordinate axes and algebra. (583)

angle of elevation The angle between a horizontal line and the line of sight. (630)

appreciation An increase in monetary value over time. (380)

associative property of addition For any values of a, b, and c, $a + (b + c) = (a + b) + c$. (278)

associative property of multiplication For any values of a, b, and c, $a(bc) = (ab)c$. (278)

asymptote A line that a graph approaches more and more closely, but never actually reaches. (513)

attractor A number that the results get closer and closer to when an expression is evaluated recursively. (24)

average The number obtained by dividing the sum of the values in a data set by the number of values. Formally called the mean. (44)

axis One of two perpendicular number lines used to locate points in the coordinate plane. The horizontal axis is often called the x-axis, and the vertical axis is often called the y-axis. The plural of axis is axes. (67)

B

balancing method A method of solving an equation that involves performing the same operation on both sides until the variable is isolated on one side. (233)

bar graph A data display in which bars are used to show measures or counts for various categories. (38)

base A number or an expression that is raised to a power. For example, $x + 2$ is the base in the expression $(x + 2)^3$, and 5 is the base in the expression 5^y. (376)

bimodal Used to describe a data set that has two modes. (44)

binary number A number written in base 2. Binary numbers consist only of the digits 0 and 1. Computers store information in binary form. (431)

binomial A polynomial with exactly two terms. Examples of binomials include $-3x + x^4$, $x - 12$, and $x^3 - x^{12}$. (544)

bins Intervals on the horizontal axis of a histogram that data values are grouped into. Boundary values fall into the bin to the right. (57)

box plot A one-variable data display that shows the five-number summary of a data set. A box plot is drawn over a horizontal number line. The ends of the box indicate the first and third quartiles. A vertical segment inside the box indicates the median. Horizontal segments, called whiskers, extend from the left end of the box to the minimum value and from the right end of the box to the maximum value. (51)

box-and-whisker plot See **box plot.**

category A group of data with the same attribute. For example, data about people's eye color could be grouped into three categories: blue, brown, and green. (39)

center (of rotation) The point that a figure turns about during a rotation. (524)

chaotic Systematic and nonrandom, yet producing results that look random. Small changes to the input value of a chaotic process can result in large changes to the output value. (29)

coefficient A number that is multiplied by a variable. For example, in a linear equation in intercept form $y = a + bx$, b is the coefficient of x. (217)

combining like terms Adding terms that have exactly the same variables raised to the same exponents. For example, in the expression $4x + 2x^2 - x + 5 + 7x^2$ you can combine the like terms $4x$ and $-x$ and the like terms $2x^2$ and $7x^2$ to get $3x + 9x^2 + 5$. (544)

common denominator A common multiple of the denominators of two or more fractions. For example, 30 is a common denominator of $\frac{7}{10}$ and $\frac{4}{15}$. (6)

common monomial factor A monomial that is a factor of every term in an expression. For example, $3x$ is a common monomial factor of $12x^3 - 6x^2 + 9x$. (575)

commutative property of addition For any values of a and b, $a + b = b + a$. (278)

commutative property of multiplication For any values of a and b, $ab = ba$. (278)

completing the square Adding a constant term to an expression of the form $x^2 + bx$ to form a perfect-square trinomial. For example, to complete the square in the expression $x^2 + 12x$, add 36. This gives $x^2 + 12x + 36$, which is equivalent to $(x + 6)^2$. To solve a quadratic equation by completing the square, write it in the form $x^2 + bx = c$, complete the square on the left side (adding the same number to the right side), rewrite the left side as a binomial squared, and then take the square root of both sides. (560)

compound inequality A combination of two inequalities. For example, $-5 < x \le 1$ is a compound inequality that combines the inequalities $x > -5$ and $x \le 1$. (342)

congruent Having the same shape and size. Two angles are congruent if they have the same measure. Two segments are congruent if they have the same length. Two figures are congruent if you can move one to fit exactly on top of the other. (3, 156)

conic section Any curve that can be formed by the intersection of a plane and a double cone. Parabolas, circles, ellipses, and hyperbolas are all examples of conic sections. (543)

conjecture A statement that might be true but that has not been proven. Conjectures are usually based on data patterns or on experience. (65)

constant A value that does not change. (147)

constant multiplier In a sequence that grows or decreases exponentially, the number each term is multiplied by to get the next term. The value of $1 + r$ in the exponential equation $y = A(1 + r)^x$. (368)

constant of variation The constant ratio in a direct variation or the constant product in an inverse variation. The value of k in the direct variation equation $y = kx$ or in the inverse variation equation $y = \frac{k}{x}$. (148, 166)

constant term A term that includes no variable. In the expression $ax^2 + bx + c$, the constant term is c. (566)

constraint A limitation on the values of the variables in a situation. A system of inequalities can model the constraints in many real-world situations. (356)

continuous function A function that has no breaks in the domain or range. The graph of a continuous function is a line or curve with no holes or gaps. (455)

contour line See **isometric line.**

contour map A map that uses isometric lines to show elevations above sea level, revealing the character of the terrain. Also called a topographic map. (248, 627)

conversion factor A ratio used to convert measurements from one unit to another. (106)

coordinate plane A plane with a pair of scaled, perpendicular axes allowing you to locate points with ordered pairs and to represent lines and curves by equations. (67)

coordinates An ordered pair of numbers in the form (x, y) that describes the location of a point on the coordinate plane. The x-coordinate describes the point's horizontal distance and direction from the origin, and the y-coordinate describes its vertical distance and direction from the origin. (67)

cosine If A is an acute angle in a right triangle, $cosine\ of\ angle\ A = \frac{length\ of\ adjacent\ leg}{length\ of\ hypotenuse}$, or $\cos A = \frac{a}{h}$. (620)

cryptography The study of coding and decoding messages. (424)

cube (of a number) A number raised to the third power. The cube of a number x is "x cubed" and is written x^3. For example, the cube of 4 is 4^3, which is equal to 64. (572)

cube root The cube root of a number a is the number b such that $a = b^3$. The cube root of a is denoted $\sqrt[3]{a}$. For example, $\sqrt[3]{64} = 4$ and $\sqrt[3]{-125} = -5$. (572)

cubing function The function $f(x) = x^3$, which gives the cube of a number. (572)

data A collection of information, numbers, or pairs of numbers, usually measurements for a real-world situation. (38)

data analysis The process of calculating statistics and making graphs to summarize a data set. (65)

decreasing A term used to describe the behavior of a function. A function is decreasing on an interval of its domain if the y-values decrease as the x-values increase. Visually, the graph of the function goes down as you read from left to right for that part of the domain. (455)

decreasing function A function that is always decreasing. (453)

dependent variable A variable whose values depend on the values of another variable (called the independent variable). In a graph of the relationship between two variables, the values on the vertical axis usually represent values of the dependent variable. (441)

depreciation A decrease in monetary value over time. (379)

deviation from the mean A data value minus the mean of its data set. The deviations of the data values from the mean give an idea of the spread of the data values. (460)

dimensional analysis A strategy for converting measurements from one unit to another by multiplying by a string of conversion factors. (106)

dimensions (of a matrix) The number of rows and columns in a matrix. If a matrix has 2 rows and 4 columns, its dimensions are 2×4. (80)

dimensions (of a rectangle) The width and length of a rectangle. If a rectangle is 2 units wide and 4 units long, its dimensions are 2-by-4. (602)

direct variation A relationship in which the ratio of two variables is constant. That is, a relationship in which two variables are directly proportional. A direct variation has an equation in the form $y = kx$, where x and y are the variables and k is a number called the constant of variation. (148)

directly proportional Used to describe two variables whose values have a constant ratio. (148)

directrix See **parabola.**

discrete function A function whose domain and range are made up of distinct values rather than intervals of real numbers. The graph of a discrete function is made up of distinct points. (455)

Glossary

discriminant The expression under the square root symbol in the quadratic formula. If a quadratic equation is written in the form $ax^2 + bx + c = 0$, then the discriminant is $b^2 - 4ac$. If the discriminant is greater than 0, the quadratic equation has two solutions. If the discriminant equals 0, the equation has one real solution. If the discriminant is less than 0, the equation has no real solutions. (568)

distance formula The distance, d, between points (x_1, y_1) and (x_2, y_2) is given by the formula $d = \sqrt{(x_2 - x_1)^2 + (y_2 - y_1)^2}$. (614)

distributive property For any values of a, b, and c, $a(b + c) = a(b) + a(c)$. (184, 278)

division property of equality If $a = b$, then $\frac{a}{c} = \frac{b}{c}$ for any nonzero number c. (278)

division property of exponents For any nonzero value of b and any values of m and n, $\frac{b^m}{b^n} = b^{m-n}$. (395)

domain The set of input values for a function. (426)

dot plot A one-variable data display in which each data value is represented by a dot above that value on a horizontal number line. (39)

double root A value r is a double root of an equation $f(x) = 0$ if $(x - r)^2$ is a factor of $f(x)$. The graph of $y = f(x)$ will touch, but not cross, the x-axis at $x = r$. For example, 3 is a double root of the equation $0 = (x - 3)^2$. The graph of $y = (x - 3)^2$ touches the x-axis at $x = 3$. (575)

E

elimination method A method for solving a system of equations that involves adding or subtracting the equations to eliminate a variable. In some cases, both sides of one or both equations must be multiplied by a number before the equations are added or subtracted. For example, to solve $\begin{cases} 3x - 2y = 5 \\ -6x + y = 11 \end{cases}$, you could multiply the first equation by 2 and then add the equations to eliminate x. (324)

engineering notation A notation in which a number is written as a number greater than or equal to 1 but less than 1000, multiplied by 10 to a power that is a multiple of 3. For example, in engineering notation, the number 10,800,000 is written 10.8×10^6. (405)

equally likely Used to describe outcomes that have the same probability of occurring. For example, when you toss a coin, heads and tails are equally likely. (124)

equation A statement that says the value of one number or algebraic expression is equal to the value of another number or algebraic expression. (193)

equilateral triangle A triangle with three sides of the same length. (29, 283)

evaluate (an expression) To find the value of an expression. If an expression contains variables, values must be substituted for the variables before the expression can be evaluated. For example, if $3x^2 - 4$ is evaluated for $x = 2$, the result is $3(2)^2 - 4$, or 8. (22)

even temperament A method of tuning an instrument based on an equal tuning ratio between adjacent notes (that is, an exponential equation). (409)

expand (an algebraic expression) To rewrite an expression by multiplying factors and combining like terms. For example, to expand $(x + 8)(x - 2)$, rewrite it as $x^2 + 6x - 16$. (547)

expanded form (of a repeated multiplication expression) The form of a repeated multiplication expression in which every occurrence of each factor is shown. For example, the expanded form of the expression $3^2 \cdot 5^4$ is $3 \cdot 3 \cdot 5 \cdot 5 \cdot 5 \cdot 5$. (376)

experimental frequency The number of times a particular outcome occurred during the trials of an experiment. (123)

exponent A number or variable written as a small superscript of a number or variable, called the base, that indicates how many times the base is being used as a factor. For example, in the expression y^4, the exponent 4 means four factors of y, so $y^4 = y \cdot y \cdot y \cdot y$. (10)

exponential equation An equation in which a variable appears in the exponent. (376)

exponential form The form of an expression in which repeated multiplication is written using exponents. For example, the exponential form of $3 \cdot 3 \cdot 5 \cdot 5 \cdot 5 \cdot 5$ is $3^2 \cdot 5^4$. (376)

exponential growth A growth pattern in which amounts increase by a constant percent. Exponential growth can be modeled by the equation $y = A(1 + r)^x$, where A is the starting value, r is the rate of growth written as a decimal or fraction, x is the number of time periods elapsed, and y is the final value. (377)

 F

factor One of the numbers, variables, or expressions multiplied to obtain a product. (10)

factored form An expression written as a product of expressions, rather than as a sum or difference. For example, $3(x + 2)$ and $y(4 - w)$ are in factored form. See **factoring.** (369)

factored form (of a quadratic equation) The form $y = a(x - r_1)(x - r_2)$, where $a \neq 0$. The values r_1 and r_2 are the zeros of the quadratic function. (551, 578)

factoring The process of rewriting an expression as a product of factors. For example, to factor $7x - 28$, rewrite it as $7(x - 4)$. To factor $x^2 + x - 2$, rewrite it as $(x - 1)(x + 2)$. (280)

family of functions A group of functions with the same parent function. For example, $y = |x - 5|$ and $y = -2|x| + 3$ are both members of the family of functions with parent function $y = |x|$. (485)

fault A break in a rock formation caused by the movement of the earth's crust, in which the rocks on opposite sides of the break move in different directions. (485)

feasible region In a linear programming problem, the set of points that satisfy all the constraints. If the constraints are given as a system of inequalities, the feasible region is the solution to the system. (364)

first quartile (Q1) The median of the values below the median of a data set. (50)

first-quadrant graph A coordinate graph in which all the points are in the first quadrant. (69)

five-number summary The minimum, first quartile, median, third quartile, and maximum of a data set. The five-number summary helps show how the data values are spread. (50)

focus See **parabola.**

fractal The result of infinitely many applications of a recursive procedure to a geometric figure. The resulting figure has self-similarity. From the Latin word *fractus,* meaning broken or irregular. (2, 4, 15)

frequency The number of times a value appears in a data set. (57)

function A rule or relationship in which there is exactly one output value for each input value. (426)

function notation A notation in which a function is named with a letter and the input is shown in parentheses after the function name. For example, $f(x) = x^2 + 1$ represents the function $y = x^2 + 1$. The letter f is the name of the function, and $f(x)$ (read "f of x") stands for the output for the input x. The output of this function for $x = 2$ is written $f(2)$, so $f(2) = 5$. (446)

G

general equation An equation that represents a whole family of equations. For example, the general equation $y = kx$ represents the family of equations that includes $y = 4x$ and $y = -3.4x$. (179)

general form (of a quadratic equation) The form $y = ax^2 + bx + c$, in which $a \neq 0$. (534, 578)

girth The distance around an object in one direction. The girth of a box is the length of string that wraps around the box. (577)

glyph A symbol that presents information nonverbally. (73)

golden ratio The ratio $\frac{1 + \sqrt{5}}{2}$, often considered an aesthetically "ideal" ratio. Examples of the golden ratio can be found in the environment, in art, and in architecture. (99)

gradient The inclination of a roadway. Also called the grade of the road. (232, 630)

graph sketch A rough graph in which the axes are labeled with variable names but not with specific scale values. (213)

gravity The force of attraction between two objects. Gravity causes objects to accelerate toward Earth at a rate of 32 ft/sec² or 9.8 m/sec². (532)

Glossary

H

half-life The time needed for an amount of a substance to decrease by one-half. (414)

half-plane The points on a plane that fall on one side of a boundary line. The solution of a linear inequality in two variables is a half-plane. (348)

hexagon A polygon with exactly six sides. (72)

histogram A one-variable data display that uses bins to show the distribution of values in a data set. Each bin corresponds to an interval of data values; the height of a bin indicates the number, or frequency, of values in that interval. (57)

horizontal axis The horizontal number line on a coordinate graph or data display. Also called the x-axis. (39, 67)

hypotenuse The side of a right triangle opposite the right angle. (585)

I

image The figure or graph of a function that is the result of a transformation of an original figure or graph of a function. (478)

imaginary number A number in the set of numbers that includes square roots of negative numbers. In the set of imaginary numbers, $\sqrt{-1}$ is represented by the letter i. For example, the solution to $x^2 = -4$ is $\sqrt{-4}$ or $2i$. (581)

increasing Used to describe the behavior of a function. A function is increasing on an interval of its domain if the y-values increase as the x-values increase. Visually, the graph of the function goes up as you read from left to right for that part of the domain. (455)

increasing function A function that is always increasing. (453)

independent variable A variable whose values affect the values of another variable (called the dependent variable). In a graph of the relationship between two variables, values on the horizontal axis usually represent values of the independent variable. (441)

inequality A statement that one quantity is less than or greater than another. For example, $x + 7 \geq -3$ and $6 + 2 < 11$ are inequalities. (339)

intercept form The form $y = a + bx$ of a linear equation. The value of a is the y-intercept, and the value of b, the coefficient of x, is the slope of the line. (217)

interest A percent of the balance added to an account at regular time intervals. (369)

interquartile range (IQR) The difference between the third quartile and the first quartile of a data set. (52)

interval The set of numbers between two given numbers, or the distance between two numbers on a number line or axis. (39)

inverse Reversed in order or effect. In an inverse mathematical relationship, as one quantity increases, the other decreases. (163)

inverse (of a function) The relationship that reverses the inputs and outputs of a function. For example, the inverse of the function $y = x + 2$ is $y = x - 2$. (473, 529)

inverse variation A relationship in which the product of two variables is constant. That is, a relationship in which two variables are inversely proportional. An inverse variation has an equation in the form $xy = k$, or $y = \frac{k}{x}$, in which x and y are the variables and k is a number called the constant of variation. (166)

inversely proportional Used to describe two variables whose values have a constant product. (166)

invert To switch the positions of two objects. For example, to invert the fraction $\frac{3}{4}$, switch the numerator and the denominator to get $\frac{4}{3}$. When you invert a fraction, the result is the reciprocal of the fraction. (94)

irrational number A number that cannot be expressed as the ratio of two integers. In decimal form, an irrational number has an infinite number of digits and doesn't show a repeating pattern. Examples of irrational numbers include π and $\sqrt{2}$. (96)

isometric line A line on a contour map that shows elevation above sea level. All the points on an isometric line have the same elevation. Also called a contour line. (248, 627)

Glossary

isosceles triangle A triangle with two sides of the same length. (283)

key A guide for interpreting the values in a data display. For example, a stem plot has a key that shows how to read the stem and leaf values. (59)

Koch curve A fractal generated recursively by beginning with a line segment and, at each stage, constructing an equilateral triangle on the middle third of each line segment and removing the edge of that triangle on the line segment. (14)

leading coefficient In a polynomial, the coefficient of the term with the highest power of the variable. For example, in the polynomial $3x^2 - 7x + 4$, the leading coefficient is 3. (561)

leg One of the perpendicular sides of a right triangle. (585)

light-year The distance light travels in one year. About 9460 billion kilometers. (392)

line of fit A line used to model a set of data. A line of fit shows the general direction of the data and has about the same number of data points above and below it. (261)

line of symmetry A line that divides a figure into mirror-image halves. In a parabola that opens up or down, the line of symmetry is the vertical line through the vertex. (540)

linear In the shape of a line or represented by a line. In mathematics, a linear equation or expression has variables raised only to the power of 1. For example, $y = 3x + 1$ is a linear equation. (207)

linear function A function characterized by a constant rate of change—that is, as the x-values change by a constant amount, the y-values also change by a constant amount. The graph of a linear function is a straight line. (452)

linear programming A process that applies the concepts of constraints, points of intersection, and algebraic expressions to solve application problems. (364)

linear relationship A relationship that you can represent with a straight-line graph. A linear relationship is characterized by a constant rate of change—that is, as the value of one variable changes by a constant amount, the value of the other variable also changes by a constant amount. (207)

linear term A term in which a constant is multiplied by a variable to the first power. In the expression $ax^2 + bx + c$, the linear term is bx. (566)

long-run value The value that the y-values approach as the x-values increase. (489)

lowest terms The form of a fraction in which the numerator and denominator have no common factors except 1. (6)

matrix A rectangular array of numbers or expressions, enclosed in brackets. (80)

maximum The greatest value in a data set or the greatest value of a function. (39, 454)

mean The number obtained by dividing the sum of the values in a data set by the number of values. Often called the average. (44)

measure of center A single number used to summarize a one-variable data set. The mean, median, and mode are measures of center. (44)

median (of a data set) If a data set contains an odd number of values, the median is the middle value when the values are listed in order. If a data set contains an even number of values, the median is the mean of the two middle values when the values are listed in order. (44)

median (of a triangle) A segment from the vertex of a triangle to the midpoint of the opposite side. (588)

midpoint The point on a line segment halfway between the endpoints. If a segment is drawn on a coordinate grid, you can use the midpoint formula to find the coordinates of its midpoint. (2, 588)

midpoint formula If the endpoints of a segment are (x_1, y_1) and (x_2, y_2), then the midpoint of the segment is $\left(\frac{x_1 + x_2}{2}, \frac{y_1 + y_2}{2}\right)$. (590)

Glossary

minimum The least value in a data set or the least value of a function. (39, 454)

mode The value or values that occur most often in a data set. A data set may have more than one mode or no mode. (44)

monomial A polynomial with only one term. Examples of monomials include $-3x$, x^4, and $7x^2$. (544)

multiplication property of equality If $a = b$, then $ac = bc$ for any number c. (278)

multiplication property of exponents For any values of b, m, and n, $b^m \cdot b^n = b^{m+n}$. (384)

N

nonlinear Not in the shape of a line or not able to be represented by a line. In mathematics, a nonlinear equation or expression has variables raised to powers other than 1. For example, $x^2 + 5x$ is a nonlinear expression. (452)

nonlinear function A function characterized by a nonconstant rate of change—that is, as the x-values change by a constant amount, the y-values change by varying amounts. (452)

O

observed probability A probability based on experience or collected data. Also called relative frequency. (123)

obtuse angle An angle that measures more than 90°. (618)

obtuse triangle A triangle with an obtuse angle. (283, 638)

one-variable data Data that measures only one trait or quantity. A one-variable data set consists of single values, not pairs of data values. (67)

opposite leg If you consider an acute angle of a right triangle, the opposite leg is the leg that is *not* part of the angle. (619)

order of magnitude A way of expressing the size of an extremely large or extremely small number by giving the power of 10 associated with the number. For example, the number 6.01×10^{26} is on the order of 10^{26} and the number 2.43×10^{-11} is on the order of 10^{-11}. (421)

order of operations The agreed-upon order in which operations are carried out when evaluating an expression: (1) evaluate all expressions within parentheses or other grouping symbols, (2) evaluate all powers, (3) multiply and divide from left to right, and (4) add and subtract from left to right. (5, 182)

ordered pair A pair of numbers named in an order that matters. For example, (3, 5) is different from (5, 3). The coordinates of a point are given as an ordered pair in which the first number is the x-coordinate and the second number is the y-coordinate. (67)

origin The point on the coordinate plane where the x- and y-axes intersect. The origin has coordinates (0, 0). (67)

outcome A possible result of one trial of an experiment. (123)

outlier A value that is far outside the range of most of the other values in a data set. As a general rule, a data value is considered an outlier if the distance from the value to the first quartile or third quartile (whichever is nearest) is more than 1.5 times the interquartile range. (46)

P

parabola The graph of a function in the family of functions with parent function $y = x^2$. The set of all points whose distance from a fixed point, the focus, is equal to the distance from a fixed line, the directrix. (467, 559)

parallel lines Lines in the same plane that never intersect. They are always the same distance apart. (583)

parallelogram A quadrilateral with opposite sides that are parallel. (159)

parent function The most basic form of a function. A parent function can be transformed to create a family of functions. For example, $y = x^2$ is a parent function that can be transformed to create a family of functions that includes $y = x^2 + 2$ and $y = 3(x - 4)^2$. (485)

pentagon A polygon with exactly five sides. (32)

perfect cube A number that is equal to the cube of an integer. For example, -125 is a perfect cube because $-125 = (-5)^3$. (572)

perfect square A number that is equal to the square of an integer, or a polynomial that is equal to the square of another polynomial. For example, 64 is a perfect square because it is equal to 8^2, and $x^2 - 10x + 25$ is a perfect-square trinomial because it is equal to $(x - 5)^2$. (546)

perpendicular bisector A line that passes through the midpoint of a segment and is perpendicular to the segment. (588)

perpendicular lines Lines that meet at a right angle. (583)

pictograph A data display with symbols showing the number of data items in each category. Each symbol in a pictograph stands for a specific number of data items. (38)

point-slope form The form $y = y_1 + b(x - x_1)$ of a linear equation, in which (x_1, y_1) is a point on the line and b is the slope. (271)

polygon A closed figure made up of segments that do not cross each other. (19)

polynomial A sum of terms that have positive integer exponents. For example, $-4x^2 + x$ and $x^3 - 6x^2 + 9$ are polynomials. (544)

population density The number of people per square mile. (397)

power properties of exponents For any values a, b, m, and n, $(b^m)^n = b^{mn}$ and $(ab)^n = a^n b^n$. (385)

predict To make an educated guess, usually based on a pattern. (9)

probability A number between 0 and 1 that gives the chance that an outcome will happen. An outcome with a probability of 0 is impossible. An outcome with a probability of 1 is certain to happen. (122)

product The result of multiplication. (83)

projectile motion The motion of a thrown, kicked, fired, or launched object—such as a ball—that has no means of propelling itself. (534)

proportion An equation stating that two ratios are equal. For example, $\frac{34}{72} = \frac{x}{18}$ is a proportion. (94)

Pythagorean theorem The sum of the squares of the lengths of the legs a and b of a right triangle equals the square of the length of the hypotenuse c—that is, $a^2 + b^2 = c^2$. (600)

Q-points On a scatter plot, the vertices of the rectangle formed by drawing vertical lines through the first and third quartiles of the x-values and horizontal lines through the first and third quartiles of the y-values. If the points show an increasing linear trend, then the line through the lower-left and upper-right Q-points is a line of fit. If the points show a decreasing linear trend, then the line through the upper-left and lower-right Q-points is a line of fit. (289)

quadrant One of the four regions that a coordinate plane is divided into by the two axes. The quadrants are numbered I, II, III, and IV, starting in the upper right and moving counterclockwise. (67)

quadratic formula If a quadratic equation is written in the form $ax^2 + bx + c = 0$, then the solutions of the equation are given by $x = \frac{-b \pm \sqrt{b^2 - 4ac}}{2a}$. (566, 567)

quadratic function Any function in the family with parent function $f(x) = x^2$. Examples of quadratic functions are $f(x) = 1.5x^2 + 2$, $f(x) = (x - 4)^2$, and $f(x) = 5x^2 - 3x + 12$. (533)

quadrilateral A polygon with exactly four sides. (273)

radical expression An expression containing a square root symbol, $\sqrt{}$. Examples of radical expressions are $\sqrt{x + 4}$ and $3 \pm \sqrt{19}$. (535)

radioactive decay The process by which an unstable chemical element loses mass or energy, transforming it into a different element or isotope. (406)

raised to the power A term used to connect the base and the exponent in an exponential expression. For example, in the expression 7^4, the base 7 is raised to the power of 4. (385)

random Not ordered, unpredictable. (28, 129)

range (of a data set) The difference between the maximum and minimum values in a data set. (40)

range (of a function) The set of output values for a function. (426)

rate A ratio, often with 1 in the denominator. (140)

rate of change The difference between two output values divided by the difference between the corresponding input values. For a linear relationship, the rate of change is constant. (226)

ratio A comparison of two quantities, often written in fraction form. (93)

rational function A function, such as $f(x) = \frac{3}{x+2}$ or $f(x) = \frac{x-1}{(x+3)(x-1)}$, that is expressed as the ratio of two polynomial expressions. (514)

rational number A number that can be written as a ratio of two integers. (96)

real number Any number that can be represented on the number line. The real numbers include integers, rational numbers, and irrational numbers. The real numbers do *not* include imaginary numbers. (563)

reciprocal The multiplicative inverse. The reciprocal of a given number is the number you multiply it by to get 1. To find the reciprocal of a number, you can write the number as a fraction and then invert the fraction. For example, the reciprocal of $\frac{3}{4}$ is $\frac{4}{3}$. (584)

rectangle A quadrilateral with four right angles. In a rectangle, opposite sides are parallel. (31)

recursive Describes a procedure that is applied over and over again, starting with a number or geometric figure, to produce a sequence of numbers or figures. Each stage of a recursive procedure builds on the previous stage. The resulting sequence is said to be generated recursively, and the procedure is called recursion. (2)

recursive routine A starting value and a recursive rule for generating a recursive sequence. (199)

recursive rule The instructions for producing each stage of a recursive sequence from the previous stage. (3)

recursive sequence An ordered list of numbers defined by a starting value and a recursive rule. You generate a recursive sequence by applying the rule to the starting value, then applying the rule to the resulting value, and so on. (199)

reflection A transformation that flips a figure or graph over a line, creating a mirror image. (493)

reflection across the x-axis A transformation that flips a figure or graph across the x-axis. Reflecting a point across the x-axis changes the sign of its y-coordinate. (493)

reflection across the y-axis A transformation that flips a figure or graph across the y-axis. Reflecting a point across the y-axis changes the sign of its x-coordinate. (493)

relative frequency The ratio of the number of times a particular outcome occurred to the total number of trials. Also called observed probability. (123)

relative frequency graph A data display (usually a bar graph or a circle graph) that compares the number in each category to the total for all the categories. Relative frequency graphs show fractions or percents, rather than actual values. (116)

repeating decimal A decimal number with a digit or group of digits after the decimal point that repeats infinitely. (93)

rhombus A quadrilateral with opposite sides parallel and all sides the same length. (159)

right angle An angle that measures 90°. (31, 583)

right trapezoid A trapezoid with two right angles. In a right trapezoid, one of the nonparallel sides is perpendicular to both parallel sides. (585)

right triangle A triangle with a right angle. (31, 252, 283, 585)

roots The solutions of an equation. When the equation is written in the form $f(x) = 0$, the roots are the x-intercepts of the graph of $y = f(x)$. For example, the roots of $(x - 2)(x + 1) = 0$ are 2 and -1. These roots are the x-intercepts of the graph of $y = (x - 2)(x + 1)$. (539)

rotation A transformation that turns a figure about a point called the center of rotation. (524)

row operations Operations performed on the rows of a matrix in order to transform it into a matrix with a diagonal of 1's with 0's above and below, creating a solution matrix. These are allowable row operations: multiply (or divide) all the numbers in a row by a nonzero number, add (or subtract) all the numbers in a row to (or from) corresponding numbers in another row, add (or subtract) a multiple of the numbers in one row to (or from) the corresponding numbers in another row. (332)

 S

sample A part of a population selected to represent the entire population. Sampling is the process of selecting and studying a sample from a population in order to make conjectures about the whole population. (100)

scale (of an axis or a number line) The values that correspond to the intervals of a coordinate axis or number line. (39)

scale factor A rate that relates the measurements in a scaled figure to the measurements in the original figure. (154)

scatter plot A two-variable data display in which values on a horizontal axis represent values of one variable and values on a vertical axis represent values of the other variable. The coordinates of each point represent a pair of data values. (67)

scientific notation A notation in which a number is written as a number greater than or equal to 1 but less than 10, multiplied by an integer power of 10. For example, in scientific notation, the number 32,000 is written 3.2×10^4. (388)

segment Two points on a line (endpoints) and all the points between them on the line. Also called a line segment. (3)

self-similar Describes a figure in which part of the figure is similar to—that is, has the same shape as—the whole figure. (6)

shrink A transformation that decreases the height or width of a figure. A vertical shrink decreases the height but leaves the width unchanged. A horizontal shrink decreases the width but leaves the height unchanged. A vertical shrink by a factor of a multiplies the y-coordinate of each point on a figure or graph by a. A horizontal shrink by a factor of b multiplies the x-coordinate of each point on a figure by b. (501)

Sierpiński triangle A fractal created by Waclaw Sierpiński by starting with a filled-in equilateral triangle and recursively removing every triangle whose vertices are midpoints of triangles remaining from the previous stage. You can create a Sierpiński-like fractal design by starting with an equilateral triangle and recursively connecting the midpoints of the sides of each upward-pointing triangle. (3)

similar figures Figures that have the same shape. Similar polygons have proportional sides and congruent angles. (156)

simulate To model an experiment with another experiment, called a *simulation,* so that the outcomes of the simulation have the same probabilities as the corresponding outcomes of the original experiment. For example, you can simulate tossing a coin by randomly generating a string of 0's and 1's on your calculator. (100)

sine If A is an acute angle in a right triangle, $sine\ of\ angle\ A = \frac{length\ of\ opposite\ leg}{length\ of\ hypotenuse}$, or $\sin A = \frac{o}{h}$. (620)

slope The steepness of a line or the rate of change of a linear relationship. If (x_1, y_1) and (x_2, y_2) are two points on a line, then the slope of the line is $\frac{y_2 - y_1}{x_2 - x_1}$. The slope is the value of b when the equation of the line is written in intercept form $y = a + bx$, and it is the value of m when the equation of the line is written in slope-intercept form $y = mx + b$. (251, 254, 265)

slope triangle A right triangle formed by drawing arrows to show the vertical and horizontal change from one point to another point on a line. (252)

slope-intercept form The form $y = mx + b$ of a linear equation. The value of m is the slope and the value of b is the y-intercept. (265)

solution The value(s) of the variable(s) that make an equation or inequality true. (193)

spread A property of one-variable data that indicates how the data values are distributed from least to greatest, and where gaps or clusters occur. Statistics such as the range, the interquartile range, and the five-number summary can help describe the spread of data. (39)

square A quadrilateral in which all four angles are right angles and all four sides have the same length. (7)

square (of a number) The product of a number and itself. The square of a number x is "x squared" and is written x^2. For example, the square of 6 is 6^2, which is equal to 36. (466)

square root The square root of a number a is a number b so that $a = b^2$. Every positive number has two square roots. For example, the square roots of 36 are -6 and 6 because $6^2 = 36$ and $(-6)^2 = 36$. The square root symbol, $\sqrt{}$, means the positive square root of a number. So, $\sqrt{36} = 6$. (468)

square root function The function that undoes squaring, giving only the positive square root (that is, the positive number that, when multiplied by itself, gives the input). The square root function is written $f(x) = \sqrt{x}$. For example, $\sqrt{144} = 12$. (468)

squaring The process of multiplying a number by itself. See **square** of a number. (466)

squaring function The function $f(x) = x^2$, which gives the square of a number. (471)

standard form The form $ax + by = c$ of a linear equation, in which a and b are not both 0. (277)

statistics Numbers, such as the mean, median, and range, used to summarize or represent a data set. Statistics also refers to the science of collecting, organizing, and interpreting information. (40)

stem plot A one-variable data display used to show the distribution of a fairly small set of data values. Generally, the left digit(s) of the data values, called the stems, are listed in a column on the left side of the plot. The remaining digits, called the leaves, are listed in order to the right of the corresponding stem. A key is usually included. (59)

stem-and-leaf plot See **stem plot.**

strange attractor A figure that the stages generated by a random recursive procedure get closer and closer to. (29)

stretch A transformation that increases the height or width of a figure. A vertical stretch increases the height but leaves the width unchanged. A horizontal stretch increases the width but leaves the height unchanged. A vertical stretch by a factor of a multiplies the y-coordinate of each point on a figure or graph by a. A horizontal stretch by a factor of b multiplies the x-coordinate of each point on a figure by b. (501)

substitution method A method for solving a system of equations that involves solving one of the equations for one variable and substituting the resulting expression into the other equation. For example, to find the solution of $\begin{cases} y + 2 = 3x \\ y - 1 = x + 3 \end{cases}$, you can solve the first equation for y to get $y = 3x - 2$ and then substitute $3x - 2$ for y in the second equation. (316)

subtraction property of equality If $a = b$, then $a - c = b - c$ for any number c. (278)

symbolic manipulation Applying mathematical properties to rewrite an equation or expression in equivalent form. (320)

symmetric Having a sense of balance, or symmetry. Symmetric is most often used to describe figures with mirror symmetry, or line symmetry—that is, figures that you can fold in half so that one half matches exactly with the other half. (52)

system of equations A set of two or more equations with the same variables. (308)

system of inequalities A set of two or more inequalities with the same variables. (354)

 T

tangent If A is an acute angle in a right triangle, $tangent\ of\ angle\ A = \frac{length\ of\ opposite\ leg}{length\ of\ adjacent\ leg}$, or $\tan A = \frac{o}{h}$. (620)

term An algebraic expression that represents only multiplication and division between variables and constants. For example, in the polynomial $x^3 - 6x^2 + 9$, the terms are x^3, $-6x^2$, and 9. (544)

terminating decimal A decimal number with only a finite number of nonzero digits after the decimal point. (93)

theoretical probability A probability calculated by analyzing a situation, rather than by performing an experiment. If the outcomes are equally likely, then the theoretical probability of a particular group of outcomes is the ratio of the number of outcomes in that group to the total number of possible outcomes. For example, when you roll a die, one of the six possible outcomes is a 2, so the theoretical probability of rolling a 2 is $\frac{1}{6}$. (123)

third quartile (Q3) The median of the values above the median of a data set. (50)

topographic map See **contour map.**

transformation A change in the size or position of a figure or graph. Translations, reflections, stretches, shrinks, and rotations are types of transformations. (476)

translation A transformation that slides a figure or graph to a new position. (478)

trapezoid A quadrilateral with one pair of opposite sides that are parallel and one pair of opposite sides that are not parallel. (585)

trial One round of an experiment. (123)

trigonometric functions The sine, cosine, and tangent functions, which express relationships among the measures of the acute angles in a right triangle and the ratios of the side lengths. (620)

trigonometry The study of the relationships among sides and angles of right triangles. (620)

trinomial A polynomial with exactly three terms. Examples of trinomials include $x + 2x^3 + 4$, $x^2 - 6x + 9$, and $3x^3 + 2x^2 + x$. (544)

two-variable data set A collection of data that measures two traits or quantities. A two-variable data set consists of pairs of values. (67)

undoing method A method of solving an equation that involves working backward to reverse each operation until the variable is isolated on one side of the equation. (194)

value of an expression The numerical result of evaluating an expression. (22)

variable A trait or quantity whose value can change, or vary. In algebra, letters often represent variables. (67, 94)

vertex (of an absolute value graph) The point where the graph changes direction from increasing to decreasing or from decreasing to increasing. (483)

vertex (of a parabola) The point where the graph changes direction from increasing to decreasing or

from decreasing to increasing. (486)

vertex (of a polygon) A "corner" of a polygon. An endpoint of one of the polygon's sides. The plural of vertex is vertices. (28)

vertex form (of a quadratic equation) The form $y = a(x - h)^2 + k$, where $a \neq 0$. The point (h, k) is the vertex of the parabola. (541, 578)

vertical axis The vertical number line on a coordinate graph or data display. Also called the y-axis. (39)

vertical line test A method for determining whether a graph on the xy-coordinate plane represents a function. If all possible vertical lines cross the graph only once or not at all, the graph represents a function. If even one vertical line crosses the graph in more than one point, the graph does not represent a function. (433)

x-intercept The x-coordinate of a point where a graph meets the x-axis. For example, the graph of $y = (x + 2)(x - 4)$ has two x-intercepts, -2 and 4. (539)

y-intercept The y-coordinate of the point where a graph crosses the y-axis. The value of y when x is 0. The y-intercept of a line is the value of a when the equation for the line is written in intercept form $y = a + bx$, and it is the value of b when the equation for the line is written in slope-intercept form $y = mx + b$. (217)

Z

zero product property If the product of two or more factors equals zero, then at least one of the factors equals zero. For example, if $x(x + 2)(x - 3) = 0$, then $x = 0$ or $x + 2 = 0$ or $x - 3 = 0$. (554)

zeros (of a function) The values of the independent variable (the x-values) that make the corresponding values of the function (the $f(x)$-values) equal to zero. For example, the zeros of the function $f(x) = (x - 1)(x + 7)$ are 1 and -7 because $f(1) = 0$ and $f(-7) = 0$. See **roots.** (554)

Glossary

Index

Index

Rooting for Factors, 573–574
Row-by-Column Matrix Multiplication, 82–83
A Scientific Quandary, 388–389
Searching for Solutions, 560–561
Seesaw Nickels, 163–164
Ship Canals, 146–147
The Sides of a Right Triangle, 599
Slopes, 583–586
Sneaky Squares, 545–546
A Strange Attraction, 22–25
Testing for Functions, 433
TFDSFU DPEFT, 424–426
Toe the Line, 340–341
Translations of Functions, 483–484
A "Typical" Envelope, 357
What's My Area?, 594
Where Will They Meet?, 310–311
Working Out with Equations, 216–217
IQR (interquartile range), 52
irrational numbers, 96
isometric lines, 248, 627–628
isosceles triangle, 283
Italy, 391, 621

Japan, 48, 269, 307, 380, 392, 404, 617
Jeanne-Claude, artist, 150
Jordan, Michael, 50, 51
Joule, James, 413
joules, 413

K

Kahlo, Frida, 440
Kandinsky, Wassily, 583
Kelvin, Lord, 288
Kennedy, Florynce, 80
key, 59
kilograms, conversion of, 151
kilometers, conversion of, 146–147
King, Billie Jean, 110
Koch curve, 14–15, 19, 374–375
Koch, Niels Fabian Helge von, 14
Kovalevskaya, Sofia, 520

labor unions, 80
Laplace, Pierre Simon de, 122
Lavery, Bernard, 151
leading coefficient, 561, 566
legs of right trangle, 585
Leonardo da Vinci, 621
librarianship, 103, 117–118
life expectancy, 61, 284–285
light-year, 392

like terms, combining, 544
line(s)
    perpendicular bisector, 588
    slope of. *See* slope
    See also line segment(s)
linear equations
    balancing method of solving, 233–237, 278
    coefficients, 217, 225
    direct variations as, 218
    graphs of, 217, 219–220
    input and output variables, 220
    intercept form. *See* intercept form
    line of fit, 261–265
    point-slope form. *See* point-slope form
    and rate of change, 226–229
    slope in, 255
    slope-intercept form, 265
    standard form. *See* standard form
    systems of. *See* systems of equations
    translation of, 491
    writing, 216–217
    y-intercept, 217
linear equations, solving
    balancing method, 233–237, 278
    calculator methods, 233, 237
    systems of. *See* systems of equations, solving
    by undoing operations, 193–194, 233, 237
linear functions, 452
linear inequalities. *See* inequalities
linear plots, 206–209
linear programming, 364
linear relationships, 207
linear term, 566
line segment(s)
    and fractal length, 14–15, 19, 374–375
    length of, square roots and, 594–595, 604–605
    midpoint of, 2, 588–590
    perpendicular bisector of, 588
lines of fit, 261–265
    See also modeling data
line of symmetry, 540
Lin, Maya, 157
long-run value, 489
lowest terms, 6

M

McArthur, William, 536
mandala, 625
Mandelbrot, Benoit, 4, 5, 14, 15
Man Ray, 417
manufacturing, 134, 166, 282, 364, 372, 398, 404, 490

maps, 159, 160, 162, 248, 612, 627–628
Markham, Beryl, 270
matrices, 80
    addition of, 81–82
    Chinese column equation matrices, 338
    dimension of, 80, 82
    forming, 80–81
    multiplication of, 82–84
    row operations in, 332
    solving systems of equations with, 331–334, 338
    transformations with, 520–522
maximum, 39, 454
Mayans, 252, 500
mean, 44–46, 460
measurements, importance of, 29–30
measures of center, 44–46
    See also mean; median; mode
median (measure of center), 44–46
    data quartiles and, 50–51
median (of triangle), 588
    slope of, 589–590
medicine, 39, 41, 371, 397, 442, 518
Melroy, Pamela, 536
Mencken, H. L., 560
meteorology, 73, 122, 153, 205, 225–229, 230, 247, 273, 330, 346, 439, 508, 509, 634
metric system, 105, 413
    See also conversion of units
Mexico, 252, 500
midpoint, 2, 588–590
miles, conversion of, 146–147
minimum, 39, 454
Mitchell, Maria, 389, 393, 532
mode, 44–46
modeling data, 242
    and constant multipliers, 408
    with cubic functions, 572
    with exponents, 406–409, 414–415
    with fractals, 1
    with functions, 440–442, 454
    intercept form and, 216–220, 225, 242–243, 296
    line of fit, 261–265
    methods compared, 296–298
    point-slope form and, 272, 284–285, 296–297
    Q-points method, 288–291, 297
    with quadratic functions, 534, 538, 557–558
    with rational functions, 514–515
    reporting on, 302
    with squaring function, 572
    transformations and, 506, 510–512
mole, 106, 391
Mondrian, Piet, 598

money systems, 152
monomials, 544
Moore, Gordon, 413
Moore's Law, 413
multiplication
  Arab and Persian method, 198
  associative property of, 278
  of binomials, 548
  commutative property of, 278
  formula of ancient India, 162
  of matrices, 82–84
  order of operations for, 5
  property of equality, 278
  property of exponents, 384
  of radical expressions, 607
  of reciprocals, 584
  as recursive routine, 367–370
  symbols for, 10
multiplication property of
  exponents, 384
music, 38, 42, 138, 167, 170, 299,
  408–409, 411, 423
Myanmar, 181

scientific notation for. See
    scientific notation
  See also negative numbers
number tricks, 190–193

Index

median of, 588, 589–590
obtuse, 283, 638
right. *See* right triangle(s)
slope and, 283
vertex of, 28
See also Pythagorean theorem
trigonometric functions, 619–621,
   624
   inverse, 626–628
trigonometry, 620
trinomials, 544, 560
tuning forks, 170
tuning instruments, 408–409
Turing, Alan, 427
two-variable data, 67–69

undoing method, 190-194
unit conversion. *See* conversion of
   units

value, absolute. *See* absolute value
value of the expression, 22
variables, 67
   cause and effect and, 442
   dependent, 441, 442
   elimination of, 324–327
   independent, 441, 442, 446
   input, 220
   one-variable data, 67, 87
   output, 220
   in proportions, 94–96
   two-variable data, 67–69

variation
   constant of, 166
   direct. *See* direct variation
   inverse. *See* inverse variation
Venters, Diana, 595
vertex (of parabola), 486, 539, 540
vertex (of polygon), 28
vertex form of quadratic equations,
   541
   conversion of factored form to,
      552–553
   conversion of general form to,
      562
   conversion to general form,
      544–547
   information given by, 547, 552
   transformations and, 541, 544,
      547
vertical axis (*y*-axis), 39, 67
   dependent variable shown on,
      441
   reflection across, 493
vertical change over horizontal
   change. *See* slope
vertical lines
   as failing vertical line test, 435
   slope of, as undefined, 255
vertical line test, 433–435
Volterra, Vito, 446
volume, 572

Wakata, Koichi, 536
Washington Monument, 419
weed fractals, 12

Wegener, Alfred, 479
Weigand, Charmion Von, 466
Whitehead, Alfred North, 50
wildlife. *See* population, wildlife
Willis, Suzanne, 389
Woods, Tiger, 99
work, 55, 80, 88, 132, 282, 458

x-axis. *See* horizontal axis
x-intercepts
   of cubic equations, 573–574
   of linear equations, 243
   of quadratic equations, 539

y-axis. *See* vertical axis
Yelesina, Yelena, 304
y-intercept, 217
Young, Steve, 85

zero
   coefficients as, 326
   as exponent, 399–401
   Mayans and, 500
   product property of, 554
zeros. *See* roots
zero product property, 554
Zhu Shijie, 215

Photo Credits

Cover

Background image: Pat O'Hara/DRK Photo; boat image: Marc Epstein/DRK Photo; all other images: Ken Karp Photography

Front Matter

x (*m*): Bettmann/Corbis; xi (*t*): Ken Karp Photography; xi (*b*): Ken Karp Photography; xi: Quilt by Diana Venters/*Mathematical Quilts* by Diana Venters and Elaine Krajenke Ellison; v: Ken Karp Photography; vi: Ken Karp Photography; vii: Ken Karp Photography; viii (*b*): Ken Karp Photography; viii (*mb*): Tom Bean/Corbis; viii (*mt*): Tom Bean/Corbis; xi: Ken Karp Photography

Chapter 0

1: Copyright 2000 Lifesmith Classic Fractals, Palmdale, CA, USA. All rights reserved. http://www.lifesmith.com; 4 (*r*): Ken Karp Photography; 5 (*l*): Roger Ressmeyer/Corbis; 5 (*r*): Ken Karp Photography; 6: Cheryl Fenton; 12: Copyright 2000 Lifesmith Classic Fractals, Palmdale, CA, USA. All rights reserved. http://www.lifesmith.com; 15 (*l*): Yann-Arthus-Bertrand/Corbis; 15 (*r*): Ken Karp Photography; 22: Ken Karp Photography; 29 (*l*): AFP/Corbis; 29 (*r*): Gary Braasch/Corbis; 33 (*t*): Copyright 2000 Lifesmith Classic Fractals, Palmdale, CA, USA. All rights reserved. http://www.lifesmith.com; 33 (*m*): Ken Karp Photography

Chapter 1

51: Cheryl Fenton; 37: The Stock Market; 38: Ken Karp Photography; 39 (*t*): Corbis; 39 (*b*): Ken Karp Photography; 42: Ted Horowitz/The Stock Market; 44: Cheryl Fenton; 45: FPG; 47: Cedar Point photo by Dan Feight; 48 (*l*): *The Hollow of the Deep-Sea Wave Off Kanagawa* by Katsushika Hokusai/Minneapolis Institute of Art Acc. No. 74.1.230; 48 (*r*): Ken Karp Photography; 49: Catherine Noren/Stock Boston; 50: Jonathan Daniel/Allsport; 51 (*b*): Jim Amos/Photo Researchers; 55: Roy Pinney/FPG; 56: Ken Karp Photography; 58: Ken Karp Photography; 61 (*b*): Alison Wright/Corbis; 61 (*mr*): Patricio Robles Gil/Bruce Coleman, Inc.; 61 (*mbl*): Betty Press/Woodfin Camp & Associates; 61 (*ml*): Sharon Smith/Bruce Coleman, Inc.; 61 (*mbr*): Catherine Karnow/Woodfin Camp & Associates; 65: Ken Karp Photography; 66: Ken Karp Photography; 69: Joseph Sohm ChromoSohm, Inc./Corbis; 70: Gregg Mancuso/Stock Boston; 72: John Collier/FPG; 74: Michael Yamashita/Woodfin Camp & Associates; 80: Library of Congress; 81: Ken Karp Photography; 84: Christian Michaels/FPG; 85: Joseph Sohm/ChromoSohm, Inc./Corbis; 86: Roger Ball/The Stock Market; 87 (*t*): Ken Karp Photography; 88: Doug Pensinger/Allsport

Chapter 2

92: Morton Beebe/Corbis; 97: Ken Karp Photography; 99: Corbis; 100: Michael Heron/Woodfin Camp & Associates; 101 (*l*): Steve & Dave Maslowski/Photo Researchers, Inc.; 101 (*m*): Mark Stouffer/Animals Animals; 101 (*r*): Gary Meszaros/Photo Researchers, Inc.; 103 (*t*): Ken Karp Photography; 103 (*b*): John Henley/The Stock Market; 104: Library of Congress; 105: Robert Fried/Stock Boston; 106 (*r*): Cheryl Fenton; 106 (*l*): Robert Holmes/Corbis; 107: Corbis; 108: Cheryl Fenton; 109 (*l*): M. Harvery/DRK Photo; 109 (*r*): Peter & Beverly Pickford/DRK Photo; 109 (*m*): Maslowski/Photo Researchers; 109 (*ml*): Anthony Mercieca/Photo Researchers, Inc.; 109 (*mt*): Kevin Schafer/Corbis; 110: Ken Karp Photography; 112: U.S. Postal Service; 113: Ken Karp Photography; 114: Ken Karp Photography; 115: Cheryl Fenton; 117: Kelly-Mooney Photography/Corbis; 119: Larry Mulvehill/Photo Researchers; 121 (*t*): David L. Brown/Tom Stack & Associates; 121 (*b*): Cary Wolinsky/Stock Boston; 123: Cheryl Fenton; 125: Cheryl Fenton; 126: Kennan Ward/The Stock Market; 129: S. Dalton/Animals Animals; 133: Tom Lazar/Animals Animals; 134 (*t*): Steve & Dave Maslowski/Photo Researchers, Inc.; 135: Jim Harrison/Stock Boston

Chapter 3

138 (*l*): Rob Hann/Retna; 139 (*r*): R. W. Jones/Corbis; 139 (*l*): Steve Chenn/Corbis; 138 (*br*): Steve Double/Retna; 138 (*tr*): Rob Hann/Retna; 141: Cheryl Fenton; 142: Rick Hansen Institute; 144: Ken Karp Photography; 145: Burstein Collection/Corbis; 147: Will and Deni McIntyre/Photo Researchers, Inc.; 150 (*t*): Manfred Vollmer/Corbis; 150 (*b*): Archivo Iconografico, SA/Corbis; 152: John Carter/Photo Researchers, Inc.; 153: Ken Karp Photography; 155 (*l*): Stephen Simpson/FPG; 155 (*r*): Telegraph Colour Library/FPG; 157 (*l*): Bettmann/Corbis; 157 (*r*): James Blank/Bruce Coleman, Inc.; 157 (*m*): Library of Congress; 158: Cheryl Fenton; 160: Robert Caputo/Stock Boston; 161: Andrew Holbrooke/The Stock Market; 162: Ken Karp Photography; 167: Ted Horowitz/The Stock Market; 170: Richard Megna/Fundamental Photographs; 172: Ken Karp Photography; 174: AFP/Corbis; 175: Ken Karp Photography; 176: Judith Canty/Stock Boston; 178 (*r*): Tom Bean/DRK Photo; 178 (*l*): Marc Muench/David Muench Photography

Chapter 4

181 (*t*): Alison Wright/Corbis; 181 (*m*): Christie's Images; 185: Cheryl Fenton; 186: Danny Lehman/Corbis; 189: Ken Karp Photography; 198: Cheryl Fenton; 199: Rafael Macia/Photo Researchers, Inc.; 200: Cheryl Fenton; 200 (*m*): Cheryl Fenton; 203: Photofest; 204: Jeffry Myers/Stock Boston; 205: David Falconer/Bruce Coleman, Inc.; 211: D. Burnett/Woodfin Camp & Associates; 213: Ken Karp Photography; 217 (*t*): Duomo/Corbis; 217 (*b*): Ann McCarthy/The Stock Market; 218 (*b*): Anderson/The Image Works; 220: Hertz Rent-A-Car; 223: Claude Charlier/The Stock Market; 224: George D. Lepp/Photo Researchers, Inc.; 225: Bettmann/Corbis; 227: John Eastcott/YVA Momatiuk/Woodfin Camp & Associates; 228: Corbis; 232: Douglas Peebles/Corbis; 233: James P. Blair/Corbis; 242: Ken Karp Photography; 243: Cheryl Fenton; 247: Ken Karp Photography; 248: Bernard Soutrit/Woodfin Camp & Associates

Chapter 5

250: Smithsonian American Art Museum, Washington, D.C./Art Resource, NY; 251: Collection of Gretchen and John Berggruen, San Francisco; 252: Robert Frerck/Woodfin Camp & Associates; 255 (*t*): Ken Karp Photography; 255 (*m*): Ken Karp Photography; 255 (*mr*): Ken Karp Photography; 257 (*t*): Ken Karp Photography; 257 (*b*): Philip Gould/Corbis; 258: James Marshall/The Stock Market; 260: The Museum of Modern Art, New York. gift of Philip Johnson. Photograph © 2000 The Museum of Modern Art, New York; 261: S. Turner/Animals, Animals; 263 (*t*): Peter Menzel/Stock Boston; 263 (*b*): *Hamburger* (1983) by David Gilhooly, Collection of Harry W. and Mary Margaret Anderson, Photo by M. Lee Fatheree; 267: AFP/Corbis; 268: Ken Karp Photography; 274: Ken Karp Photography; 275: John DeWaele/Stock Boston; 276: Bettmann/Corbis © 2001 Andy Warhol Foundation for the Visual Arts/ARS, New York; 279: Interlochen Center for the Arts; 282 (*t*): David Weintraub/Stock Boson; 282 (*b*): Roger Ball/The Stock Market; 284: Steven Rubin/The Image Works; 285: Morton Beebe, S. F./Corbis; 286: Michael Sedam/Corbis; 288: Ken Karp Photography; 290: Bob Daemmrich/Stock Boston; 295: Tom Bean/DRK Photo; 299: Amtrak; 301: Ken Karp Photography; 303 (*b*): Robert Frerck/Woodfin Camp & Associates; 303 (*b*): Bob Daemmrich/Stock Boston; 304: Mike Powell/Allsport

Chapter 6

307: George Holton/Photo Researchers, Inc.; 308 (*l*): Tom Bean/Corbis; 308 (*r*): Tom Bean/Corbis; 320 (*l*): Tom Bean/Corbis; 320 (*r*): Tom Bean/Corbis; 310 (*l*): Ken Karp Photography; 310 (*r*): Ken Karp Photography; 311: George Chan/Photo Researchers, Inc.; 314: Gerard Smith/Photo Researchers, Inc.; 315 (*t*): Corbis; 315 (*b*): Ken Karp Photography; 316: Philip James;

Photo Credits